D1761508

Coulomb Excitation

PERSPECTIVES IN PHYSICS

A Series of Reprint Collections

Statistical Theories of Spectra: Fluctuations, Charles E. Porter (Ed.). 1965

Quantum Theory of Angular Momentum, L. C. Biedenharn and H. van Dam (Eds.). 1965

Magnetohydrodynamic Stability and Thermonuclear Containment, A. Jeffrey and T. Taniuti (Eds.). 1966

Mathematical Physics in One Dimension: Exactly Soluble Models of Interacting Particles, E. Lieb and D. Mattis (Eds.). 1966

Coulomb Excitation, K. Alder and A. Winther (Eds.). 1966

Coulomb Excitation

A Collection of Reprints
With an Introductory Review by

K. Alder
Institute for Theoretical Physics
University of Basel, Switzerland

A. Winther
Niels Bohr Institute
University of Copenhagen, Denmark

1966 AP

ACADEMIC PRESS New York and London

ACADEMIC PRESS INC.
111 Fifth Avenue, New York, New York 10003

United Kingdom Edition published by
ACADEMIC PRESS INC. (LONDON) LTD.
Berkeley Square House, London W.1

Library of Congress Catalog Card Number: 66-27739

PRINTED IN THE UNITED STATES OF AMERICA

PREFACE

In recent years a number of comprehensive review articles on the topic of Coulomb excitation have appeared. Nevertheless, it seems justified that the available literature should also include a reprint collection. This especially in view of the increasing interest in the field which has been stimulated by the construction of accelerators for heavy ions. Many students and research workers may want to become familiar with the original literature which is to be found scattered in various journals and of which reprints are no longer available.

Since the discovery of Coulomb excitation ten years ago more than 200 papers on this subject have appeared. Although most of these papers deal with experiments, the main emphasis in this reprint collection will be on the theory. We have included a few experimental papers which we found especially important for the development of the field.

The guiding principle in the selection of theoretical papers has been a compromise between history and usefulness. A number of papers still satisfying these conditions had to be omitted for reasons of space, and rather than omit them completely we have chosen to include a reproduction of the abstracts.

As far as possible we have corrected misprints and errors in the original versions of the reprints.

KURT ALDER
AAGE WINTHER

May, 1966

Acknowledgments

We thank the following publishers for permission to reproduce material.

American Institute of Physics and the Physical Review
Office of Naval Research
The Royal Danish Academy

CONTENTS

Preface v

Coulomb Excitation
An Introduction
K. Alder and A. Winther 1

Collective Motions in Nuclei
B. Mottelson
Rept. Intern. Phys. Conf., Copenhagen, June 1952 11

Excitation of Nuclei by the Coulomb Field of Charged Particles
K. A. Ter-Martirosyan
Zh. Eksperim. i Teor. Fiz. **22**, 284 (1952) 15

Proposed Coulomb Excitation of Nuclei
B. Mottelson
Office of Naval Research, European Scientific Notes No. 7–9, May 1, 1953 31

Excitation of Nuclear Rotational States by the Electric Field of Impinging Particles
T. Huus and Č. Zupančič
Kgl. Danske Videnskab. Selskab, Mat.-Fys. Medd. **28**, No. 1 (1953) 33

Excitation of Heavy Nuclei by the Electric Field of Low-Energy Protons
C. L. McClelland and C. Goodman
Phys. Rev. **91**, 760 (1953) 51

Theory of Coulomb Excitation
K. Alder and A. Winther
Phys. Rev. **96**, 237 (1954) 53

Coulomb Excitation Process in the Lighter Odd-Mass Nuclei
G. M. Temmer and N. P. Heydenburg
Phys. Rev. **96**, 426 (1954) 55

Angular Distribution of Gamma Rays from Coulomb Excitation
F. K. McGowan and P. H. Stelson
Phys. Rev. **99,** 127 (1955) 63

Soluble Problem in the Theory of Coulomb Excitation
L. C. Biedenharn and C. M. Class
Phys. Rev. **98,** 691 (1955) 73

Study of Nuclear Structure by Electromagnetic Excitation with Accelerated Ions
K. Alder, A. Bohr, T. Huus, B. Mottelson, and A. Winther
Rev. Mod. Phys. **28,** 432 (1956) 77

Reorientation Effect in Coulomb Excitation
G. Breit, R. L. Gluckstern, and J. E. Russell
Phys. Rev. **103,** 727 (1956) 189

Experimental Observation of Double Coulomb Excitation
J. O. Newton and F. S. Stephens
Phys. Rev. Letters **1,** 63 (1958) 201

Multiple Coulomb Excitation in Th^{232} and U^{238}
F. S. Stephens, Jr., R. M. Diamond, and I. Perlman
Phys. Rev. Letters **3,** 435 (1959) 205

On the Theory of Multiple Coulomb Excitation with Heavy Ions
K. Alder and A. Winther
Kgl. Danske Videnskab. Selskab, Mat.-Fys. Medd. **32,** No. 8 (1960) 209

A Note on the Angular Distribution of Inelastically Scattered Particles in Coulomb Excitation
T. A. Griffy and L. C. Biedenharn
Nucl. Phys. **36,** 452 (1962) 281

Multiple Coulomb Excitation of Rotational Levels in Even-Even Nuclei
J. de Boer, G. Goldring, and H. Winkler
Phys. Rev. **134,** B1032 (1964) 287

A Computer Program for Multiple Coulomb Excitation
A. Winther and J. de Boer
California Institute of Technology, Technical Report, November 18, 1965 303

Coulomb Excitation

An Introduction | by K. ALDER AND A. WINTHER

In spite of the danger of repeating information already contained in the following reprints, we shall here attempt to present a qualitative discussion of the theory of Coulomb excitation, which is essentially meant to serve as a guide to the reproduced papers.

We understand the term "Coulomb excitation" to mean the excitation of atomic nuclei through the electromagnetic field of impinging charged particles which do not penetrate into the region of nuclear forces. Thus we shall not discuss electromagnetic excitations caused by high-energy electrons[1] and the contribution of Coulomb excitation to direct interaction.[2] In this connection it should also be mentioned that in collisions between atoms and ions one is confronted with a problem that is closely related to Coulomb excitation of nuclei.[3]

In the Coulomb excitation process (without penetration) the nuclear properties enter only through the matrix elements of electric and magnetic multipole moments, which are identical with those describing the spontaneous electromagnetic transitions between the same nuclear states. The nuclear properties which can be extracted

[1] For a Born approximation treatment see e.g. K. Alder *et al.*, *Rev. Mod. Phys.* **28,** 432 (1956), included in this volume, page 77. For more exact calculations see e.g. T. A. Griffy, D. S. Onley, J. T. Reynolds, and L. C. Biedenharn, *Phys. Rev.* **128,** 833 (1962); K. Alder and T. Schucan, *Nucl. Phys.* **42,** 498 (1963).

[2] See e.g. R. H. Bassel, G. R. Satchler, R. M. Drisko, and E. Rost, *Phys. Rev.* **128,** 2693 (1962), and references contained therein.

[3] See e.g. M. J. Seaton *in* "Atomic and Molecular Processes" (D. R. Bates, ed.), p. 374. Academic Press, New York, 1962.

from a Coulomb excitation experiment are thus, besides spin, parity, and energy of the nuclear states, exactly the electromagnetic matrix elements.[4] In terms of these parameters (in principle of all states of the nucleus in question) equations can be set up, which describe cross sections, angular distributions, etc., for any Coulomb excitation process with this nucleus.

In the most general case these equations are prohibitively complicated to solve. For actual cases of interest one finds, however, that various approximation methods give quite accurate answers. In the following we discuss these methods in more detail.

The simplest approximation method which one would think of is the perturbation treatment in which the interaction between target nucleus and projectile is considered to be weak. For projectiles with low charge, e.g. protons, this condition is fulfilled as will be discussed later.

While the first computations of Coulomb excitation cross sections were made with the additional assumption of the plane wave Born approximation,[5] the important development in the theory was the realization that in heavy particle Coulomb excitation one is justified in using a classical description of the dynamics of the projectile.[6]

In order to decide whether a classical description or a Born approximation is justified one should compare the wavelength $\lambdabar$ of the projectile with a dimension characteristic for the classical orbits, e.g., the distance of closest approach $2a$ in a head-on collision. If $\lambdabar \ll 2a$ one can form a wave packet which moves along a hyperbolic orbit exactly like a classical particle. If on the other hand $\lambdabar \gg 2a$, a Born approximation applies. It is convenient to introduce the ratio between a and $\lambdabar$ as a parameter which measures the strength of the Coulomb interaction, i.e.,

$$\eta = \frac{a}{\lambdabar} = \frac{Z_1 Z_2 e^2}{\hbar v}. \tag{1}$$

Here Z_1 and Z_2 are the charge numbers of projectile and target nucleus, respectively, while v is the velocity of the projectile at large distances.

The fact that only the Coulomb field can ensure that the projectile does not penetrate into the nucleus shows that the lower limit of the parameter η is larger than 1. In practice it is seen from Eq. (1) that the actual limit is approximately 2.

In the limit $\eta \gg 1$ where a classical description of the projectile orbit holds, the nuclear excitation is caused by the time-dependent electromagnetic field acting on the nucleus while the projectile moves along the classical hyperbolic orbit. The

[4] Since projectiles in Coulomb excitation experiments have a small velocity ($v/c \ll 1$), the magnetic interaction can mostly be neglected and the Coulomb excitation depends essentially only on the electric multipole matrix elements.

[5] C. J. Mullin and E. Guth, *Phys. Rev.* **82**, 141 (1951); R. Huby and H. C. Newns, *Proc. Phys. Soc.* **A64**, 619 (1951).

[6] K. A. Ter-Martirosyan, *Zh. Eksperim. i Teor. Fiz.* **22,** 284 (1952), included in this volume, page 15.

differential excitation cross section of a nuclear state is then given by the product of the excitation probability P and the classical Rutherford cross section, for a definite deflection angle ϑ of the projectile, i.e.,

$$\frac{d\sigma}{d\Omega} = \left(\frac{d\sigma}{d\Omega}\right)_{\text{Ruth}} \cdot P = \frac{1}{4}\, a^2 \sin^{-4}\frac{\vartheta}{2} \cdot P \tag{2}$$

The excitation probability is the square of the excitation amplitude b_{if} from the initial state i (ground state) to the final state f, in practice suitably averaged and summed over magnetic substates, i.e.,

$$P = \Sigma |b_{if}|^2 \tag{3}$$

In first order time-dependent perturbation theory the excitation amplitude is readily found to be

$$b_{if} = \frac{1}{i\hbar} \int_{-\infty}^{+\infty} \langle f | H_{\text{int}}(t) | i \rangle \exp\left(i\,\frac{E_f - E_i}{\hbar}\,t\right) dt \tag{4}$$

Here E_i and E_f denote the energies of the initial and final nuclear state, respectively, and $H_{\text{int}}(t)$ is the time-dependent electromagnetic interaction between projectile and target nucleus. In Eq. (4) the matrix element between the initial and final states of H_{int} occurs, and since these states have definite spins and parities it is convenient to expand the interaction in a multipole series. In practice these series converge quite rapidly and, since the magnetic interaction can be neglected, essentially only the lowest electric term contributes, and the matrix element becomes proportional to the electric multipole matrix element occurring in spontaneous gamma emission.

The integrals occurring in Eq. (4) can be expressed only in terms of elementary functions if the energy loss $E_f - E_i = \Delta E$ can be neglected. The integrals are then functions of the deflection angle ϑ, behaving as ϑ^λ for small values of ϑ where λ is the multipole order. The limit $\Delta E = 0$ corresponds to the situation in which the nuclear period is long compared to the collision time, i.e., where the process can be considered as a sudden impact.

For nonvanishing energy loss the main dependence of the integrals is given by the ratio $\xi(\vartheta)$ between the nuclear frequency $\Delta E/\hbar$ and the reciprocal collision time. For an arbitrary deflection angle the order of magnitude of the collision time is given by $a/[v \sin(\vartheta/2)]$, i.e.,

$$\xi(\vartheta) = \frac{\Delta E}{\hbar}\,\frac{a}{v \sin(\vartheta/2)} \tag{5}$$

This parameter measures the extent to which the process is adiabatic. Thus for $\xi(\vartheta) = 0$ the sudden approximation applies, while for $\xi(\vartheta) \gg 1$ the process becomes adiabatic. In this limit the integrals (4) vanish mainly as $\exp[-\xi(\vartheta)]$. For finite energy losses the process thus always becomes adiabatic for small deflection angles and the Coulomb excitation cross section vanishes exponentially in the forward direction.

For a tabulation of the integrals (4) it is convenient to use the independent parameters $\xi = \xi(\pi)$ and ϑ. The numerical evaluation[7] of the integrals has been carried out for the lowest multipole orders ($\lambda \leq 4$) and for $\xi \leq 4$ as a function of ϑ.

For the first Coulomb excitation experiments it was important that among the low lying nuclear states accessible by Coulomb excitation (i.e., with small ξ) many are connected to the ground state with large electric quadrupole matrix elements. This was predicted by the collective nuclear model,[8] and the first experimental verification[9] was directly stimulated by this theory.[10]

Further Coulomb excitation experiments strongly supported the collective model.[11,12] At the same time the semiclassical theory of Coulomb excitation was refined. In the preceding discussion of the classical theory it was tacitly assumed that the relative energy loss of the projectile was so small that the classical orbit was well defined. If the difference between initial and final velocities is so large that a significantly different result is obtained by inserting v_i or v_f, e.g. in ξ, it is suggestive that one should insert an average of the initial and final velocities for v.

To obtain a more systematic improvement of the semiclassical theory one has to study quantum mechanical effects, taking into account the finite value of the parameter η. In the quantum theory of Coulomb excitation one must describe the projectile by a wave function which is the solution of the Schrödinger equation for a particle in a Coulomb field, and which behaves asymptotically as a plane wave (Coulomb wave). If the interaction can still be considered as a perturbation one can easily write down an expression for the excitation cross section (distorted wave Born approximation)

$$\frac{d\sigma}{d\Omega} = \frac{m^2}{4\pi^2\hbar^4}\frac{v_f}{v_i}\left|\langle f, \mathbf{k}_f|H_{\text{int}}|i, \mathbf{k}_i\rangle\right|^2, \tag{6}$$

where m is the reduced mass and $\mathbf{k}_i$ and $\mathbf{k}_f$ the wave numbers of the projectile in the initial and final states, respectively. In contrast to the matrix element of H_{int}

[7] K. Alder and A. Winther, *Kgl. Danske Videnskab. Selskab. Mat.-Fys. Medd.* **31,** No. 1 (1956); K. Alder, A. Bohr, T. Huus, B. Mottelson, and A. Winther, *Rev. Mod. Phys.* **28,** 432 (1956), included in this volume, page 77.

[8] A. Bohr and B. Mottelson, *Kgl. Danske Videnskab. Selskab, Mat.-Fys. Medd.* **27,** No. 16 (1953).

[9] T. Huus and Č. Zupančič, *Kgl. Danske Videnskab. Selskab, Mat.-Fys. Medd.* **28,** No. 1 (1953), included in this volume, page 33; C. L. McClelland and C. Goodman, *Phys. Rev.* **91,** 760 (1953), included in this volume, page 51.

[10] B. Mottelson, *Rept. Intern. Phys. Conf., Copenhagen, June 1952* p. 17, included in this volume, page 11; Office of Naval Research Report on Colloquium by B. Mottelson February 28, 1953, European Scientific Notes, p. 93, American Embassy, London, 1953, included in this volume, page 31.

[11] G. M. Temmer and N. P. Heydenburg, *Phys. Rev.* **96,** 426 (1954), included in this volume, page 55.

[12] For a complete list of the early experiments see K. Alder *et al., Rev. Mod. Phys.* **28,** 432 (1956), included in this volume, page 77.

which occurs in Eq. (4) we have here in addition taken the matrix element between the Coulomb waves describing the projectile.

A connection between the quantum mechanical expression (6) and the semiclassical formula, Eqs. (2)–(4), can be achieved by a WKB treatment of the projectile wave function for large orbital angular momenta. The resulting expression becomes identical to Eqs. (2)–(4) for large values of η only when the relative energy loss of the projectile is small. For finite energy losses one obtains an expression which differs most importantly from Eq. (4) by the substitution of the parameter ξ by

$$\xi = \eta_f - \eta_i \tag{7}$$

where η_f and η_i denote the values of η which are obtained by inserting final and initial velocities, respectively, in Eq. (1). Furthermore, in the expression for the distance of closest approach one should substitute $v_i v_f$ for v^2 and multiply Eq. (2) by the factor v_f/v_i.[13] This symmetrization and the substitution (7) are strongly suggested, as mentioned above, by the structure of Eq. (6), noting that the matrix element in Eq. (6) is symmetric in the initial and final states. It is possible to find a symmetrization which is somewhat more accurate although important only for relatively small values of η.[14]

Experiments[11] showed quite accurate agreement with the symmetrized expressions for the Coulomb excitation cross sections. It was found,[15] however, that the angular distribution of gamma quanta emitted after the excitation process showed marked deviations from the theoretical expressions.[16] These observations stimulated a complete quantum mechanical perturbation calculation of the Coulomb excitation. The exact evaluation of expression (6) is most conveniently performed by expanding the initial and final Coulomb waves into partial waves. The resulting radial matrix elements can be expressed in terms of elementary functions only for $\xi = 0$.[17] For $\xi \neq 0$ the calculation leads to complicated hypergeometric functions, the numerical evaluation of which can be carried out practically only on fast electronic computers.[18] The radial matrix elements obtained show a remarkable agreement with those obtained with the WKB radial wave functions.[19]

[13] K. Alder and A. Winther, *Phys. Rev.* **96,** 237 (1954), included in this volume, page 53.

[14] L. C. Biedenharn and P. J. Brussaard, "Coulomb Excitation." Oxford Univ. Press (Clarendon), London and New York, 1965.

[15] F. K. McGowan and P. H. Stelson, *Phys. Rev.* **99,** 127 (1955), included in this volume, page 63.

[16] It was shown that a part of this discrepancy was due to an error of sign. See G. Breit, M. Ebel, and F. D. Benedict, *Phys. Rev.* **100,** 429 (1955).

[17] L. C. Biedenharn and C. M. Class, *Phys. Rev.* **98,** 691 (1955), included in this volume, page 73.

[18] L. C. Biedenharn, J. L. McHale, and R. M. Thaler, *ibid.* **100,** 376 (1955); K. Alder and A. Winther, *Kgl. Danske Videnskab. Selskab, Mat.-Fys. Medd.* **29,** No. 18–19 (1955); L. C. Biedenharn, M. Goldstein, J. L. McHale, and R. M. Thaler, *Phys. Rev.* **101,** 662 (1956); **102,** 1567 (1956); K. Alder *et al., Rev. Mod. Phys.* **28,** 432 (1956), included in this volume, page 77; T. A. Griffy, and L. C. Biedenharn, *Nucl. Phys.* **36,** 452 (1962), included in this volume, page 281.

[19] For a discussion of the high accuracy of the WKB treatment see G. Breit, *in* "Encyclopedia of Physics," Vol. 41/1. Springer Verlag, Berlin, 1959.

All experiments performed up to the present time fulfilling the conditions of nonpenetration and perturbation theory agree with the most accurate computations.[20] Thus one has been able to determine the electric transition moments for the lowest states in almost all stable isotopes.[21] A necessary condition for the validity of the perturbation treatment is that the excitation probability [see Eqs. (3) and (4)] which one computes in the perturbation theory be a small number for all excited states. In actual cases this is true for projectiles of low charge such as protons. For the excitation, especially of collective nuclear states with particles of higher charge, the condition may be strongly violated. In cases in which the perturbation theory breaks down, one finds, however, that the semiclassical approximation is always well justified ($\eta \gg 1$); therefore, in the following we shall limit the discussion to the classical treatment.

Under circumstances where the condition for the first order perturbation theory is not too heavily violated it may be sufficient to take into account the deviations from the first order theory by carrying the perturbation treatment to second order. This gives rise to two types of modifications. First, one obtains a correction to the excitation amplitudes of the states which are already strongly excited in first order (reorientation effects).[22] Second, by double excitation, the excitation of states which were otherwise inaccessible or accessible only by excitations of high multipolarity becomes possible. The second order semiclassical treatment of Coulomb excitation leads to double integrals (see ref. 22) which recently have been extensively tabulated.[23] The double excitation process has been observed in a number of cases[24] and is now used as a tool in nuclear spectroscopy.[25,26] In cases where the condition for the first-order theory is heavily violated the perturbation expansion becomes completely impractical and multiple excitation occurs. A given nuclear state can then be populated in a number of different ways by virtual excitations through many intermediate states.

For the discussion of multiple Coulomb excitation it is convenient to introduce parameters which describe the rate of transition between the nuclear states during the collision. A parameter of this type is the first order transition probability (3). This number which in general may become larger than 1 is defined for any pair of nuclear states a and b and depends on the nuclear matrix element as well as on

[20] K. Alder *et al.*, *Rev. Mod. Phys.* **28,** 432 (1956), included in this volume, page 77; E. M. Bernstein and E. Z. Skurnik, *Phys. Rev.* **121,** 841 (1961).

[21] Typical investigations of this kind are e.g. P. H. Stelson and F. K. McGowan, *Phys. Rev.* **110,** 489 (1958); D. A. Bromley, J. A. Kühner, and E. Almquist, *ibid.* **115,** 586 (1959).

[22] G. Breit, R. L. Gluckstern, and J. E. Russel, *Phys. Rev.* **103,** 727 (1956), included in this volume, page 189; G. Breit and J. P. Lazarus, *ibid.*, **100,** 942 (1955).

[23] A. C. Douglas, AWRE Rept. No. NR/P-2/62, Aldermaston, 1962, and *Nucl. Phys.* **42,** 428 (1963); A. Winther, to be published.

[24] The first observation of double excitation was made by J. O. Newton and F. S. Stephens, *Phys. Rev. Letters* **1,** 63 (1958), included in this volume, page 201.

[25] D. Eccleshall, B. M. Hinds, M. J. L. Yates, and N. MacDonald, *Nucl. Phys.* **37,** 377 (1962).

[26] P. H. Stelson and F. K. McGowan, *Phys. Rev.* **121,** 209 (1961); **122,** 1274 (1961).

ϑ and ξ_{ab}. The parameter $P_{ab}(\vartheta,\xi_{ab})$ does not, however, measure the rate of transition between a and b but rather the total transition. Thus, if the transition probability is a small number, this could be due to a large value of ξ, i.e. it could be small because the transition was adiabatic. The strength with which the state a is coupled to the state b through the electromagnetic interaction with the projectile during the collision is rather measured by the parameter $P_{ab}(\vartheta,0)$ with $\xi_{ab} = 0$. We shall define the dimensionless coupling parameter $\chi_{ab}(\vartheta)$ as the square root of this number, i.e.,

$$\chi_{ab} = \pm[P_{ab}(\vartheta, 0)]^{1/2} \tag{8}$$

where P_{ab} is determined by Eqs. (3) and (4) with $\Delta E = 0$ and the sign is fixed by the nuclear matrix element.

One can obtain a qualitative picture of the excitation process by arranging the parameters χ_{ab} connecting any pair of nuclear states in a matrix. This χ-matrix is then resolved in submatrices in such a way that states within the submatrix are connected by large χ's, while all the matrix elements between the submatrices are small. In the subgroups the states will mix during the collision and multiple Coulomb excitation will occur. The transitions between the groups are weak and may be treated by a perturbation theory. From this picture it is seen that the above-discussed simple perturbation theory applies only if all χ's connecting to the ground state as well as to the final state are small.

The order of magnitude of the parameter $\chi(\vartheta)$ can be estimated quite easily from Eq. (4) as the matrix element of the average interaction energy multiplied by the collision time divided by $\hbar$. For the excitation of the lowest state in a rotational band with intrinsic quadrupole moment Q_0 one finds for backward scattering

$$\chi(\pi) \approx \frac{Z_1 e Q_0}{(2a)^3 \hbar} \frac{2a}{v} \tag{9}$$

The parameter $\chi(\pi)$ measures the higher multipole (e.g. quadrupole) interactions in exactly the same way as the parameter η is a measure of the strength of the monopole interaction.

From Eq. (9) one sees explicitly that χ may become essentially larger than 1. When this occurs it is in general quite complicated to compute the excitation probabilities. However, large values of χ occur mostly for collective nuclear states with low excitation energies, thus usually leading to small values of the parameter ξ. This suggests that as a first approximation one could neglect the energy differences between all the states involved in the multiple excitation. In this sudden approximation it is possible to avoid the perturbation expansion and give an explicit expression for the excitation amplitude,

$$b_{if} = \left\langle f \left| \exp\left[-\frac{i}{\hbar} \int_{-\infty}^{+\infty} H_{\text{int}}(t)\, dt \right] \right| i \right\rangle \tag{10}$$

It is noted that the series development of the exponential function leads to the perturbation expansion for $\xi = 0$ of which the first term is given by Eq. (4).

In Eq. (10) the excitation amplitude is expressed as a matrix element of a complicated but well-known operator which depends on the scattering parameters such as projectile energy, deflection angle, etc. If the nuclear wave functions of the initial and final states are known, the evaluation of b_{if} is simply a matter of evaluating definite integrals. For the excitation of rotational states in deformed nuclei and for surface vibrational states one finds that (10) can be evaluated in terms of rather elementary functions[27] by inserting the interaction energy in terms of the collective nuclear degrees of freedom.

As was stressed earlier, the only nuclear properties which could enter into the calculation of the Coulomb excitation amplitude are the ordinary electric multipole matrix elements. Since some of these may already be known from gamma spectroscopy, it may be convenient to express the excitation amplitude directly in terms of these parameters. For a limited number of nuclear states this can be done,[27] expressing the matrix element of the exponential function in Eq. (10) in terms of the matrix elements of the exponent.

The sudden approximation gives a qualitative picture of the multipole excitation process. It is, however, only for the excitation of rotational states in deformed nuclei that the results can be compared directly with the experiments. Even for rotational spectra corrections for finite ξ are important if the parameter ξ for the lowest state becomes larger than approximately 0.05. For ξ of this order of magnitude the deviations can be treated in a series expansion in ξ, in which Eq. (10) is the first term.[25] One should also consider that there are deviations from the pure rotational model. Such deviations have been treated in a perturbation theory.[27–29] However, the results are rather complicated, depending on ξ as well as on several nuclear parameters, and an exhaustive tabulation is impracticable.

A more convenient approach for the analysis of multiple excitation experiments should be based on model-independent calculations, in which the multipole matrix elements are free parameters. Model-independent calculations for finite ξ have been performed in two different ways. The first method relies on a generalization of Eq. (10) for finite values of ξ which is accurate for much larger values of ξ ($\xi \lesssim 0.2$) than the expansion mentioned above.[29] Another approach is the direct numerical integration of the coupled differential equations which describe the amplitudes of the nuclear levels during the collision.

For the actual execution of these general methods one must rely on large high-speed electronic computers. A computer program by which one may calculate cross sections and angular distributions for multiple quadrupole Coulomb excitation in an arbitrary nucleus is included in this volume.[30]

[27] K. Alder and A. Winther, *Kgl. Danske Videnskab. Selskab, Mat.-Fys. Medd.* **32,** No. 8 (1960), included in this volume, page 209.

[28] H. Lütken and A. Winther, *Kgl. Danske Videnskab. Selskab, Mat.-Fys. Skr.* **2,** No. 6 (1964).

[29] K. Alder, *Proc. Conf. on Reactions between Complex Nuclei, Asilomar*, 1963.

[30] A. Winther and J. de Boer, California Institute of Technology Technical Report, November 18, 1965, included in this volume, page 303.

Experiments on multiple Coulomb excitation became feasible with the construction of high energy accelerators for heavy ions (Hilacs and Tandems). The first experiments were made with deformed nuclei[31] and a qualitative agreement with the results of the sudden approximation was established. In later experiments the multiple Coulomb excitation process was used mostly as a tool for populating states in deformed nuclei.[32,33] More detailed investigation of the population of the nuclear levels confirmed the theory of multiple Coulomb excitation for deformed nuclei, including corrections for finite ξ, and gave indications that it is also possible to extract the nuclear matrix elements.[34,35]

It is especially this aspect of the Coulomb excitation process which may be important for future experiments. As was stressed earlier the excitation amplitude can depend only on the electromagnetic multipole matrix elements, and any observable quantity can theoretically be determined from these quantities by means of a computer program. By a combination of various experimental techniques, observing excitation functions, angular distribution, dependence on charge and mass of projective, etc., it is in principle possible to determine uniquely all the electromagnetic matrix elements involved.[36]

[31] B. S. Stephens, R. M. Diamond, and I. Perlman, *Phys. Rev. Letters* **3,** 435 (1959), included in this volume, page 205.

[32] J. S. Greenberg, D. A. Bromley, E. Bishop, and G. Seaman, *Intern. Symp. on Direct Interaction and Nuclear Reaction Mechanisms, Padua, September 3–8, 1962*; J. S. Greenberg *et al.*, *Phys. Rev. Letters* **11,** 211 (1963).

[33] R. M. Diamond, B. Elbek, and F. S. Stephens, *Nucl. Phys.* **43,** 560 (1963).

[34] R. Graetzer and E. M. Bernstein, *Phys. Rev.* **129,** 1772 (1963).

[35] J. de Boer, G. Goldring, and H. Winkler, *Phys. Rev.* **134B,** 1032 (1964), included in this volume, page 287.

[36] J. de Boer, R. G. Stokstad, G. D. Symons, and A. Winther, *Phys. Rev. Letters*, **14,** 564 (1965).

International Physics Conference, Copenhagen, June, 1952

*Collective Motions in Nuclei**

B. Mottelson (Copenhagen)

The single-particle model has been successful in accounting for spins and parities of nuclear states. However, more detailed calculations with this model have lead to discrepancies. Among the more conspicuous of these are:

(1) large nuclear quadrupole moments
(2) fast E2 (electric 2^2 pole) transitions
(3) shifts of magnetic moments from Schmidt lines
(4) smallness of many nuclear matrix elements (M4 transitions, β-decay etc.).

Especially (1) and (2) are strong evidence that collective modes are playing an important part in nuclear phenomena.

A model has therefore been investigated (in collaboration with A. Bohr [1]) in which the nuclear wave function depends explicitly on certain parameters α describing the shape of the nuclear field. The single particle model is obtained by replacing the dynamical variables α by their mean value. The equations of motion correspond to a coupled system of surface oscillators and easily excitable single particle modes. Various consequences of this model have been examined:

(1) *Nuclear spins.* For a $(j^{\pm 1})$ configuration one obtains $I = j$ for the ground state spin. For $(j^{\pm 3})$ configurations, one obtains $I = j - 1$ for sufficiently strong particle-surface coupling. For odd-odd nuclei, certain rules can be given which are found to agree with observed spins.

(2) *Magnetic moment.* A variety of effects influence the μ-value: For $I \geq \frac{3}{2}$ the coupling to the core shifts μ inward by about 1 magneton for $j = l + \frac{1}{2}$. Presence of nearlying single particle levels produces further shifts as for example in the case of $p_{3/2}$-$f_{5/2}$, $d_{3/2}$-$d_{5/2}$ interactions, etc. In this way an interpretation is also obtained for μ of $I = \frac{1}{2}$ nuclei.

It is however significant that the nondeformable Bi^{209} shows a large shift, which has not been explained.

(3) *Quadrupole moments.* By assuming a rather strong coupling, we can account for the observed magnitude and correlation with magnetic moment shifts. Measurements of the Q's for the K isotopes may provide a test since the theory here predicts opposite correlation with $\Delta\mu$ compared to the usual theory.

(4) *Isomeric transitions.* The fast E2 transitions can be explained by assuming similar deformations (in even-even nuclei) as are inferred from empirical

* From Conference report edited by O. Kofoed-Hansen *et al.*, Copenhagen, 1952.

quadrupole moments and anomalies in isotope shifts. Also, higher order electric transitions may proceed principally through surface deformations, which accounts for equality of lifetimes of E3 transitions in odd-neutron and odd-proton nuclei. The smallness of the M4 matrix elements follows from a kind of Franck-Condon principle applying to the differently shaped cores of the initial and final state.

(5) *β-decay.* The results obtained by Kofoed-Hansen and Winther [2] for mirror nuclei were discussed. Improved agreement with the empirical data is obtained by including the effect of collective motions. The reduction of matrix elements for the allowed unfavored transitions is similar to the case of isomeric transitions.

(6) *Scattering processes.* The present model may provide a description of the compound nucleus formation within the scope of the single particle picture. Attention was called to the information about nuclear collective properties which one can obtain from nuclear scattering of charged heavy particles with energies less than the Coulomb barrier.

The various nuclear properties which may be interpreted on the basis of the present model throw further light on the properties of the collective types of motion. In particular the experimental data in nearly all cases indicate a rather strong coupling, implying that the surface oscillation frequencies are somewhat smaller than those obtained from the simple liquid drop model of the nucleus.

References

1. A. Bohr and B. Mottelson, *Kgl. Danske Videnskab. Selskab Mat.-Fys. Medd.* **27,** No. 16 (1953).
2. O. Kofoed-Hansen and A. Winther, *Phys. Rev.* **86,** 428 (1952); *Kgl. Danske Videnskab. Selskab Mat.-Fys. Medd.* **28,** No. 3 (1953).

Discussion

Jensen (Heidelberg) remarked that the spin $I = j - 1$ in $(j^{\pm 3})$ configurations could also be obtained by pure *j-j* coupling.
Ref. Kurath, *Phys. Rev.* **80,** 98 (1950); Flowers, private communication.

Rosenfeld (Manchester) said that Flowers used gaussian potentials for the interaction between the nucleons and found that for large values of the range (compatible with other data) the $\frac{5}{2}$ state in $(d^3_{3/2})$ lies energetically above the $\frac{3}{2}$ state whereas for short range forces (M. Goeppert-Mayer) $\frac{5}{2}$ is lowest. For Yukawa potentials it seems impossible for reasonable ranges to get a crossover.

Mottelson quoted a paper by Talmi [*Helv. Phys. Acta* **25,** 85 (1952)] who shows that for any reasonable potential it is very unlikely that the $\frac{3}{2}$ state is lowest.

Jensen (Heidelberg) asked if it is possible to explain the spread in the matrix elements for the E3 isomeric transitions compared with the approximate constancy of those for M4 transitions.

A. Bohr (Copenhagen) answered that the somewhat larger spread for the E3 transitions may be due to the fact that for the $(g^3_{9/2})_{7/2} \rightarrow p_{1/2}$ transitions, the matrix elements depend on the admixture of $g_{7/2}$ orbits. Such admixtures are expected on the present model but will depend sensitively on the nuclear deformation.

Jensen (Heidelberg) said that Maria Goeppert-Mayer has calculated isomeric transitions with nuclear wave functions built up by single particle wave functions giving the right quadrupole moment, but it was found that the reduction factor of the matrix elements was too small.

Mottelson then remarked that the reduction factor depends essentially on the zero point oscillations of the surface. For the assumed surface properties large reduction factors may be obtained.

Rosenfeld (Manchester) asked how far this model is based on first principles.

N. Bohr (Copenhagen) answered that it appeared difficult to define what one should understand by first principles in a field of knowledge where our starting point is empirical evidence of different kinds which is not directly combinable. So far it has not been easy to deduce the conditions for the binding of the individual nucleons in an atomic nucleus from the evidence obtained by collisions between free nucleons. Moreover, as regards the problems of deformation of the nuclear shape it is to be remembered that quite apart from the evidence discussed here the discovery of nuclear fission clearly shows the possibility of even large deformations of nuclei. The present attempt of developing the shell model to take into account deformations of the nuclear field associated with collective types of motions would seem to indicate a way for a rational description of the various types of nuclear phenomena.

Rosenfeld (Manchester) asked if there is an unambiguous distinction between particle degrees of freedom and those of the core.

A. Bohr (Copenhagen) answered that the criterion is whether the excitation frequencies of a given particle structure are smaller or larger than the collective frequencies.

Hund (Frankfurt) described some calculations of Pfirsch [*Z. Physik* (**132**, 409 1952)] who has estimated quadrupole moments for many particle configurations by filling the lowest levels available in a deformed core and agreement with empirical variation of quadrupole moments has been found.

A. Bohr (Copenhagen) gave in this connection an argument due to Swiatecki for the fact that the nuclear deformability is very small near closed shells.

Mottelson told (called upon by *Casimir* (Eindhoven)) about two informative scattering experiments below the Coulomb barrier which could test the theory (point 6 in the lecture):

(1) Elastic scattering could give information on quadrupole moments which could be compared with atomic data (Sternheimer effect could be tested).

(2) Inelastic scattering could give information on the excitation of the collective motions.

It was emphasized that no detailed estimates of effects to be expected have been carried out so far.

Salvetti (Milano) said that the configuration $(j^{\pm 5})$ would lead to $I = j - 2$.

A. Bohr (Copenhagen) agreed that this might happen in a few cases with extreme strong coupling but remarked that no such effect has been observed.

Casimir (Eindhoven) asked about the descrepancy in Bi^{209}.

Jensen (Heidelberg) remarked that Bi is a difficulty for all shell models but that it might be explained by change in nuclear moments in bound configurations (quenching).

A. Bohr (Copenhagen) remarked that Bi may not be an isolated case. There is some indication of a general trend of magnetic moment shifts increasing with the spin value for the nuclei with $j = l - \frac{1}{2}$.

Jensen (Heidelberg) called attention to an experiment carried out at Minnesota on the scattering of polarized protons by He^4 (double scattering). The compound state of Li^5 thus formed showed a resonance at a proton energy of $\sim$2.4 MeV. This state was demonstrated to be a $p_{3/2}$ state and from other experiments it could be concluded that the $p_{1/2}$ state lies at least 4 MeV higher. The excited state of 0.47 MeV in Li^7 could thus not be the $p_{1/2}$ level but must be attributed to the configuration $((p^2_{3/2})_2 p_{3/2})_{1/2}$.

*Excitation of Nuclei by the Coulomb Field of Charged Particles**

K. A. Ter-Martirosyan

Leningrad Physico-Technical Institute, Academy of Sciences, Leningrad, USSR

(Received 10 February 1951)

A semiclassical calculation of the probability of excitation of a nucleus under the action of an electric field of a heavy charged particle passing by it is presented.

A heavy charged particle (proton, α-particle, etc.) colliding with an atomic nucleus can excite it by its electrical field. If the energy of collision E_1 is less than the height B of the coulomb barrier of the nucleus, then the probability of penetration by the particle into the region of the nucleus is exponentially small. Nuclear forces do not enter into collisions of such a type, and excitation of the nucleus—if it does take place—is entirely dependent upon the action of the electric field of the particle. Excitation of the nucleus In^{115} by protons and α-particles has been experimentally observed under these conditions (Refs. 1-2); the cross section is estimated to be approximately 10^{-28} to 10^{-29} cm^2.

A calculation of the cross section for excitation of the nucleus by a charged particle is presented without consideration of the nuclear interaction. This means, in particular, that resonance effects are not considered; it is assumed that the energy of collision E_1 is not near one of its resonant values $E_1^{(0)}$, $E_1^{(1)}$, $E_1^{(2)}$, . . . , etc., near which the probability of penetration of the particle into the nucleus rises sharply.

1. Semiclassical Calculation

Let us select as our origin the coordinates of the center of gravity of the excited nucleus. Let the vector $\mathbf{R}_i = (R_i, \Theta_i, \Phi_i)$ determine the position of the ith proton of the nucleus ($i = 1, 2, 3, \ldots, Z$) and the vector $\mathbf{r} = (r, \vartheta, \varphi)$ the position of particle P colliding with the nucleus. If we do not take into account the nuclear interaction between particle P and the nucleons of the nucleus, then we should consider that during the collision the transition of the nucleus from ground state A to excited state B is caused by the interaction of charge $Z'e$ of particle P with the electrical potential of the nucleus:

$$V(\mathbf{r}) = \int \Psi_A{}^* \sum_{i=1}^{Z} \frac{e}{|\mathbf{r} - \mathbf{R}_i|} \Psi_B \, d\tau = Ze \int \frac{\Psi_A{}^*\Psi_B}{|\mathbf{r} - \mathbf{R}_i|} \, d\tau. \tag{1}$$

* Article from *Zh. Eksperim. i Teor. Fiz.* **22,** 284 (1952). Translated by J. W. Butler, Naval Research Laboratory, Washington, D. C.

Here Ψ_A and Ψ_B are the normalized eigenfunctions of the nuclear levels A and B depending on the coordinates of all the nucleons, $d\tau$ being the product of the differentials of these coordinates. The summation over i in (1) is replaced by multiplication by Z inasmuch as all members of the sum over i are equal to each other.

In order to determine the probability of excitation, let us make use of the method of impact parameters. The problem examined above appears to be quasi classical*:

$$\alpha_1 = ZZ'e^2/\hbar v_1 > 1. \tag{2}$$

Here v_1 is the relative velocity of the particle P and the nucleus before the collision. Therefore, we are correct in examining the movement of the particle P in the field Ze/r of the nucleus according to the classical trajectory, considering $\mathbf{r} = \mathbf{r}(t)$. Let us assume that the energy $\Delta E = E_B - E_A$ transferred to the nucleus during the collision is small compared to the bombarding energy E_1:

$$\Delta E < E_1 \tag{3}$$

The case $\Delta E \sim E_1$ will be examined later. Condition (3) permits one to neglect the change of the orbit of particle P during the transfer of energy to the nucleus. Under these circumstances the interaction energy $Z'eV(\mathbf{r})$ of the charge of particle P with potential (1) will be a specific function of time. We shall consider it as a time-dependent perturbation which causes the transition of the nucleus. The probability of transition from A to B of the nucleus will be determined as usual by the Fourier component of the perturbation:

$$W = \left[(Z'e/\hbar) \int_{-\infty}^{\infty} e^{i\omega t} V[\mathbf{r}(t)]\, dt \right]^2, \tag{4}$$

where $\omega = \Delta E/\hbar$. The differential cross section of excitation corresponding to the given impact parameter ρ will be expressed through this probability in the following manner:

$$d\sigma_{AB} = \frac{2\pi\rho\, d\rho}{2J_A + 1} \sum_{M_A=-J_A}^{J_A} \sum_{M_B=-J_B}^{J_B} W. \tag{5}$$

Here the cross section is summed over all final orientations of the nucleus (i.e., according to the magnetic quantum number M_B of level B) and averaged over the $2J + 1$ degenerate states of level A.

For the calculation of (4) and (5) we will first set down the potential (1) in convenient form. Let us expand in (1) $\Psi_A{}^*\Psi_B$ in a Clebsch-Gordon series:

$$\Psi_A{}^*\Psi_B = \sum_{n=|J_A-J_B|}^{J_A+J_B} (-1)^{M_A} C^{nm}_{J_A,-M_A;J_B,M_B} \Psi_{nm}. \tag{6}$$

* It is known that condition (2) results from the inequality $E_1 = \mu v_1{}^2/2 < B$, $B = ZZ'e^2/(R_0 + R_0')$, if the reduced mass μ of the system of nucleus and particle P is sufficiently large: $\mu > 2\hbar^2/(R_0 + R_0')ZZ'e^2 \approx (10^5/A^{\frac{1}{3}}ZZ')m_0$, where R_0 and R_0' are radii of the nucleus and of particle P, A being the atomic weight and m_0 the electron mass.

Here Ψ_{nm} is a function of the coordinates of the nucleons of the nucleus, which transforms under rotation of the system of coordinates like the spherical harmonic Y_{nm}; $m = M_B - M_A$; $C^{nm}_{J_A, -M_A; J_B, M_B}$ are the normalized coefficients of the Clebsch-Gordon series (an obvious form of this is given for example in ref. 3, p. 407). Substituting (6) into (1) we note that

$$\int \frac{\Psi_{nm}}{|\mathbf{r} - \mathbf{R}_i|} \, d\tau = \left(\frac{4\pi}{2n+1}\right)^{1/2} Q^{(n)}_{AB} \frac{Y_{nm}(\vartheta, \varphi)}{r^{n+1}}, \tag{7}$$

inasmuch as

$$\frac{1}{|\mathbf{r} - \mathbf{R}_i|} = \sum_{l=0}^{\infty} \sum_{m'=-l}^{l} \frac{4\pi}{2l+1} R_i{}^l Y^*_{lm'}(\Theta_i \Phi_i) \frac{Y_{lm'}(\vartheta, \varphi)}{r^{l+1}} \qquad (r > R_i)$$

$$\left(\frac{4\pi}{2l+1}\right)^{1/2} \int \Psi_{nm} R_i{}^l Y^*_{lm'}(\Theta_i \Phi_i) \, d\tau = Q^{(n)}_{AB} \, \delta_{nl} \, \delta_{mm'}.$$

The last equation follows from the properties mentioned above of the transformation of Ψ_{nm} under rotations; $Q^{(n)}_{AB}$ is a quantity, which according to (6) is determined by the internal structure of the nucleus:

$$(-1)^{M_A} C^{nm}_{J_A, -M_A; J_B, M_B} Q^{(n)}_{AB} = \left(\frac{4\pi}{2n+1}\right)^{1/2} \int \Psi_A{}^* R_i{}^n Y^*_{nm}(\Theta_i \Phi_i) \Psi_B \, d\tau, \tag{8}$$

and is usually called the electric 2^n-pole transition moment of the nucleus.

According to (1), (6), and (7) we have as a result

$$V(\mathbf{r}) = \sum_{n=|J_A - J_B|}^{J_A + J_B} (-1)^{M_A} C^{nm}_{J_A, -M_A; J_B, M_B} V^{(nm)}(\mathbf{r}), \tag{9}$$

where

$$V^{(nm)}(\mathbf{r}) = Ze Q^{(n)}_{AB} \frac{[4\pi/(2n+1)]^{1/2} Y_{nm}(\vartheta, \varphi)}{r^{n+1}}. \tag{10}$$

Substituting (9) into (4) and (5), and taking into account the known relationship

$$\sum_{M_A - M_B = m} C^{nm}_{J_A, -M_A; J_B, M_B} C^{n'm}_{J_A, -M_A; J_B, M_B} = \delta_{nn'},$$

we obtain

$$d\sigma_{AB} = \frac{1}{2J_A + 1} \sum_{n=|J_A - J_B|}^{J_A + J_B} d\sigma_n, \tag{11}$$

$$d\sigma_n = 2\pi\rho \, d\rho \left[\frac{ZZ'e^2}{\hbar} Q^{(n)}_{AB}\right]^2 \sum_{m=-n}^{n} \left[\int_{-\infty}^{\infty} \frac{[4\pi/(2n+1)]^{1/2} Y_{nm}(\vartheta\varphi)}{r^{n+1}} e^{i\omega t} \, dt\right]^2. \tag{12}$$

In sum (11) the index n runs through either even or odd values only (depending on the parity of the levels A and B of the nucleus) in conformity with the values of n in sum (6).

Let us select the plane xOy of the system of coordinates to be the plane of motion of particle P, and the axis Ox to be directed along the axis of symmetry of the trajectory. As a result we will have $\mathbf{r} = (r, \pi/2, \varphi) = (x, y, 0)$; moreover, $\tan \varphi = y/x$, and $\varphi = y = 0$, when $t = 0$. The equation of motion of particle P in parametric representation will have the form

$$\begin{aligned} r &= a(\epsilon \cosh w + 1), & x &= a(\cosh w + \epsilon), \\ y &= a(\epsilon^2 - 1)^{1/2} \sinh w, & t &= (a/v_1)(\epsilon \sinh w + w), \end{aligned} \tag{13}$$

where $a = ZZ'e^2/\mu v_1^2$, $\epsilon = [1 + (\rho/a)^2]^{1/2}$ is the eccentricity of the orbit. In order to calculate the integral in Eq. (12) we note that

$$\left(\frac{4\pi}{2n+1}\right)^{1/2} Y_{nm}\left(\frac{\pi}{2}, \varphi\right) = \begin{cases} (-1)^{(n+m)/2}(c_{nm})^{1/2} e^{im\varphi}, & \text{when} \quad n + m \text{ is even,} \\ 0 & \text{when} \quad n + m \text{ is odd,} \end{cases}$$

where

$$c_{nm} = \frac{(n-m)!(n+m)!}{\left[2^n \dfrac{n-m}{2}!\, \dfrac{n+m}{2}!\right]^2}.$$

According to (13), we express r, t, and $e^{im\varphi} = ((x + iy)/r)^m$ with the variable w. As a result we obtain

$$d\sigma_n = 2\pi a_1{}^2 \left[\frac{Q_{AB}^{(n)}}{a^{n-1}}\right]^2 \sum_{m=-n}^{n}{}' c_{nm} I_{nm}^2(\epsilon, \beta) \epsilon \, d\epsilon, \tag{14}$$

where

$$I_{nm}(\epsilon, \beta) = \int_{-\infty}^{\infty} \exp\left[i\beta(\epsilon \sinh w + w)\right] \frac{(\cosh w + \epsilon + i(\epsilon^2 - 1)^{1/2} \sinh w)^m}{(\cosh w + 1)^{n+m}} \, dw \tag{15}$$

depends on the eccentricity ϵ and the dimensionless parameter

$$\beta = \frac{a\omega}{v_1} = \frac{\Delta E}{2E_1} \frac{ZZ'e^2}{\hbar v_1}.$$

The prime on the summation sign in (14) means that the index m in the summation takes on those values only where $n + m$ is even.

The perturbation theory used above is not applicable if the total probability

$$W_{\text{tot}} = \frac{d\sigma_n}{2\pi\rho \, d\rho} = \left[\frac{a_1 Q_{AB}^{(n)}}{a^n}\right]^2 \sum_{m=-n}^{n}{}' c_{nm} I_{nm}^2(\epsilon, \beta)$$

of excitation of the nucleus under the given parameter of collision exceeds unity. This probability has its greatest value during head-on collisions, when $\epsilon = 1$. Taking into account that $I_{nm}(1, \beta) = I_{n0}(1, \beta) \leqslant I_{n0}(1, 0) < 2^{2-n}$, and $\Sigma' c_{nm} = 1$, we can conclude that $W_{\text{tot}} < 1$ for any collision parameter, provided that

$$4\alpha_1 Q_{AB}^{(n)}/(2a)^n < 1. \tag{16}$$

From experimental data it is known that the estimate $Q_{AB}^{(n)} \sim R_0{}^n$ or $Q_{AB}^{(n)}/(2a)^n \sim (R_0/2a)^n \approx (E_1/B)^n$, which follows from the definition in (8) of $Q_{AB}^{(n)}$, is too large (especially for $n = 1$). Therefore, there is a basis to suppose that (16) is well satisfied not only with $E_1/B < 1/(4\alpha_1)^{1/n}$, but also when $E_1/B < 1$. Therefore, with $E_1 < B$, application of perturbation theory should not be subject to large error.

Integrating (11) with respect to the angle of scattering (or with respect to ϵ) we determine the total cross section for excitation:

$$\sigma_{AB} = \sum_{n=|J_A - J_B|}^{J_A+J_B} \frac{\sigma_n}{2J_A + 1}, \tag{17}$$

$$\sigma_n = 2\pi \left[\alpha_1 \frac{Q_{AB}^{(n)}}{a^{n-1}} \right]^2 f_n(\beta) e^{-2\pi\beta}, \tag{18}$$

where

$$f_n(\beta) = e^{2\pi\beta} \sum_{m=-n}^{n}{}' c_{nm} \int_1^\infty I_{nm}^2(\epsilon, \beta)\epsilon \, d\epsilon \tag{19}$$

is a slowly changing function of the parameter β.*

With $\beta < 1$, the first term in the sum (17), i.e., $\sigma_{|J_A - J_B|}$ or $\sigma_{|J_A - J_B|+1}$ depending on the parity of the levels A and B of the nucleus has the largest value. It follows from (18) if we use the rough estimate $Q_{AB}^{(n)} \sim R_0{}^n$ that

$$\frac{\sigma_{n+2}}{\sigma_n} \approx \left(\frac{E_1}{B}\right)^4 \frac{16 f_{n+2}(\beta)}{f_n(\beta)},$$

and the above-mentioned statement then follows from $E_1 < B$, and the fact that $f_n(\beta)$ and $16 f_{n+2}(\beta)$ (for $\beta < 1$, see below) are of the same order of magnitude. We note that when $\Delta E/E_1 \ll 1$, we have approximately

$$\beta \approx (ZZ'e^2/\hbar)(\Delta v/v_1{}^2) \approx \alpha_2 - \alpha_1,$$

where $\alpha_2 = ZZ'e^2/\hbar v_2$, v_2 being the relative velocity of particle P and the nucleus after collision. The cross section, therefore, is proportional to the exponential $\exp[-2\pi(\alpha_2 - \alpha_1)]$ which was to be expected [5].

The values of the parameter β which is the argument of $f_n(\beta)$ are not limited by conditions of applicability of calculations (2) and (3). It is possible that $\beta < 1$

* Expressions (18) and (19) were already utilized in the work of the author [4] for the determination of the cross section of disintegration of Be^9 and the deuteron in the field of another nucleus.

as well as $\beta > 1$. Calculation of the variation of $f_n(\beta)$ according to (15) and (19) is a purely mathematical problem. It is presented in the appendix; here we will extract the result.

For $n = 1$ (dipole transition) we have

$$f_1(\beta) = i\pi^2\beta H_{i\beta}^{(1)}(i\beta)H_{i\beta}^{(1)\prime}(i\beta),$$

where $H_p^{(1)}(z)$ is the Hankel function of the first order and $H_p^{(1)\prime}(z)$ is its derivative with respect to z. The curve $f_1(\beta)$ is shown in Fig. 1. Its asymptotic limit for small and large values of β is of the form

$$\beta < 1, \qquad f_1(\beta) \approx 4(1 + \pi\beta)\log(2/\xi\beta),$$

where $\xi = 1.781 \ldots$, and

$$\beta > 1, \qquad f_1(\beta) \approx (1 + 0.218\beta^{-2/3})(4\pi/\sqrt{3}).$$

$f_1(\beta)$ differs only by the factor $4\pi/\sqrt{3}$ from the function G' given by Kramers [6], based on the classical theory of Bremsstrahlung in the Coulomb field.

If $n \neq 1$, then $f_n(\beta)$ is not expressed by any type of known function. One can obtain the following limiting values:

$$\left.\begin{array}{ll} \text{when} \quad \beta = 0, & \begin{cases} f_2(0) = \dfrac{\pi^2}{16} - \dfrac{1}{3} = 0.284, \\[2ex] f_3(0) = \dfrac{8}{45} - \dfrac{\pi^2}{64} = 0.023; \end{cases} \\[4ex] \text{when} \quad \beta \gg 1, & f_n(\beta) \approx \eta_n \beta^{4(n-1)/3}, \end{array}\right\} \tag{20}$$

where

$$\eta_n = 2^{2n} \sum_{m=-n}^{n}{}' c_{nm} \int_0^\infty \exp\left(\tfrac{2}{3}y^2\right)\Phi_{nm}^2(y)y\,dy.$$

Here the integral

$$\Phi_{nm}(y) = \int_C \exp\left(-\frac{yx^2}{2} - \frac{ix^3}{6}\right)\frac{dx}{x^{n-m}(x - 2iy)^{n+m}}$$

is taken over the complex plane in the variable x along the real axis bypassing the pole below the origin. An exact value of the constant η_n can be obtained only as a result of double numerical integration.*

A rough estimate of this value can be obtained by expanding

$$\Phi_{nm}(y) = \Phi_{nm}(0) + y\Phi'_{nm}(y) + \ldots$$

* The asymptotic value of (20) is useful also when $n = 1$. In this case the calculation of $\Phi_{1,\pm1}(y)$ can be accomplished by integration by parts and gives for η_1 the same value $4\pi/\sqrt{3}$ which follows from the expression presented above for $f_1(\beta)$.

into a series, and restricting oneself to the first term. $\Phi_{nm}^{(0)}$ is calculated without difficulty with the aid of the well-known expression for the inverse gamma function in the form of a contour integral, and for η_n we obtain the approximation

$$\eta_n = \frac{4^{n+1/3}\pi^2\Gamma(\frac{2}{3})}{27[\Gamma((2n+2)/3)]^2 6^{4(n-1)/3}}.$$

The values of η_n determined on the basis of this are equal to $\eta_2 = 1.15$, $\eta_3 = 0.186$, etc. These values of η_n are apparently less than the true values. Thus, when $n = 1$, the exact value of $\eta_1 = 4\pi/\sqrt{3} = 4\pi \cdot 0.577$, whereas the above approximation gives the value $4\pi \cdot 0.31$.

In Fig. 1 are given the curves $f_2(\beta)$ and $f_3(\beta)$ constructed according to the limiting values. The dotted line shows the most probable path for the curves in the region where $\beta \sim 1$. For a numerical approximation (considering that according to (8)

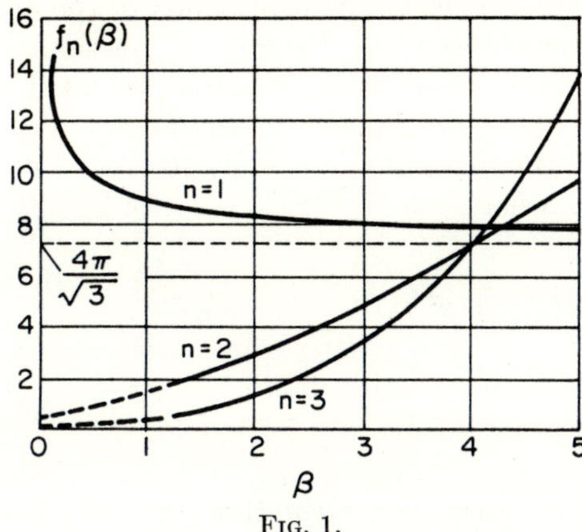

Fig. 1.

$Q_{AB}^{(n)} \sim R_0^n$) we get from (18) that σ_n is of the order of magnitude 10^{-26} to 10^{-29} cm^2 (depending on the multipole order and energy of excitation ΔE) provided E_1 is selected to be near the height of the Coulomb barrier B. The approximation obtained for the total cross section lies within the limits of the experimental values [1, 2] mentioned above.

Weisskopf [7] gives an expression for the cross section σ_n (for $n \geq 2$). His approximation has the form of (18), where in place of $f_n(\beta)e^{-2\pi\beta}$ is the expression

$$\frac{n}{2(2n-1)(2n-2)}\frac{v_2}{v_1}\left(\frac{E_2}{2E_1}\right)^{2n-4}.$$

The manner of obtaining this expression is not understood. In addition, it is known to be incorrect for large values of β inasmuch as it does not contain the exponentially decreasing factor.

2. Large Energies of Excitation $\Delta E \sim E_1$

If $\Delta E \sim E_1$, then the above-presented calculation is not applicable because the trajectory of the particle P becomes indeterminate. For the calculation of $d\sigma_{AB}$ in this instance also, we introduce the wave functions $\psi_{\mathbf{k}_1}(\mathbf{r})$ and $\psi_{\mathbf{k}_2}^{(-)}(\mathbf{r})$ describing the motion of the particle P in the field $Ze^2/\mathbf{r}$ of the nucleus according to the momenta $p_1 = \hbar\mathbf{k}_1$ and $p. = \hbar\mathbf{k}_2$ before and after collision. Applying the usual theory of excitation we have

$$d\sigma_{AB} = \left(\frac{4\pi^2}{k_1}\right)^2 \frac{1}{2J_A + 1} \sum_{M_A, M_B} \left| \int \psi_{\mathbf{k}_1}^* Z'eV(1)\psi_{\mathbf{k}_2}^{(-)}\, d\mathbf{r} \right|^2 d\Omega. \tag{21}$$

The factor $(4\pi^2/k_1)^2$ normalizes $\psi_{\mathbf{k}_1}(\mathbf{r})$ and $\psi_{\mathbf{k}_2}^{(-)}(\mathbf{r})$ according to the energy. For this normalization $V_{\mathbf{k}_1\mathbf{k}_2} = \int \psi_{\mathbf{k}_1}^* Z'eV\psi_{\mathbf{k}_2}\, d\mathbf{r}$ is a dimensionless quantity. The application here of the theory of perturbation corresponds to the expansion of $d\sigma_{AB}$ into a power series of the quantities $V_{\mathbf{k}_1\mathbf{k}_2}$. Usually it is assumed that the mentioned series converges if $V_{\mathbf{k}_1\mathbf{k}_2} < 1$. From the calculation presented below it follows that this condition is actually always fulfilled in a quasi-classical sense.*

Substituting in (21) the expression (9) and (10) for $V(\mathbf{r})$, we determine $d\sigma_{AB}$ in the form of sum (11), where

$$d\sigma_n = (4\pi^2/k_1)^2 \sum_{m=-n}^{n} |V_{\mathbf{k}_1\mathbf{k}_2}^{(nm)}|^2\, d\Omega, \tag{22}$$

$$V_{\mathbf{k}_1\mathbf{k}_2}^{(nm)} = ZZ'e^2Q_{AB}^{(n)} \int \psi_{\mathbf{k}_1}^*(\mathbf{r}) \frac{[4\pi/(2n+1)]^{1/2}Y_{nm}(\vartheta, \varphi)}{r^{n+1}} \psi_{\mathbf{k}_2}^{(-)}(\mathbf{r})\, d\mathbf{r}. \tag{23}$$

With the help of these formulas, the calculation for $n = 1$ is easy. Designating in this case $V_{\mathbf{k}_1\mathbf{k}_2}^{(1m)} = [\mathbf{V}_{\mathbf{k}_1\mathbf{k}_2}^{(1)}]_m$, $m = 0, \pm 1$, we have $\mathbf{V}_{\mathbf{k}_1\mathbf{k}_2}^{(1)} = Q_{AB}^{(n)}(ZZ'e^2\mathbf{r}/r^3)_{\mathbf{k}_1\mathbf{k}_2}$. The index $\mathbf{k}_1\mathbf{k}_2$ following the parentheses denotes the matrix element between the coresponding states. According to Ehrenfest's theory the matrix element of the force $ZZ'e^2\mathbf{r}/r^3 = \mu\ddot{\mathbf{r}}$ can be expressed in terms of the matrix element of the radius vector $\mathbf{r}$. One finds

$$\left(\frac{ZZ'e^2\mathbf{r}}{r^3}\right)_{\mathbf{k}_1\mathbf{k}_2} = \mu(\ddot{\mathbf{r}})_{\mathbf{k}_1\mathbf{k}_2} = -(\mathbf{r})_{\mathbf{k}_1\mathbf{k}_2}\mu\omega^2 \tag{24}$$

* However, in order to be certain that the first term of this series ($V_{\mathbf{k}_1\mathbf{k}_2}$) in the perturbation theory used in (21) is a good approximation, it should be compared at least with the second term

$$\int \frac{V_{\mathbf{k}_1\mathbf{k}_2}V_{\mathbf{k}_3\mathbf{k}_2}}{E_1 - E_3}\, d\Omega_3\, dE_3.$$

We failed to do this.

$[\omega = (\hbar/2)\mu(k_1^2 - k_2^2) = \Delta E/\hbar]$ which is an expression known from the theory of bremsstrahlung. Substituting in (24) the symbol $\mathbf{M}$ for $(\mathbf{r})_{\mathbf{k}_1\mathbf{k}_2}$, [8]

$$(\mathbf{r})_{\mathbf{k}_1\mathbf{k}_2} = \frac{\mu(k_1\alpha_1k_2\alpha_2)^{1/2}}{(2\pi\hbar)^2}\, e^{-\pi(\alpha_1+\alpha_2)}\mathbf{M}.$$

Substituting (24) into (22) and (23), we obtain

$$d\sigma_1 = \left[\frac{\mu^2\omega^2}{\hbar^2}\,\alpha_1 Q_{AB}^{(1)}\right]^2 e^{-2\pi(\alpha_1+\alpha_2)}|\mathbf{M}|^2\, d\Omega. \tag{25}$$

The vector $\mathbf{M}$ is calculated accurately by Sommerfeld [8]; however, we shall use its limiting value for $\alpha_1 \gg 1$ (and consequently for $\alpha_2 - \alpha_1 > 1$ inasmuch as $\Delta E \sim E_1$). One finds

$$|\mathbf{M}|^2\, d\Omega = 2\pi(\hbar^2/\mu^2\omega^2)^2 e^{4\pi\alpha_1}\Phi(s)\, ds, \tag{26}$$

where

$$\Phi(s) = \pi^2\{[e^{5\pi i/6}H_{2/3}^{(1)}(is)]^2 + [e^{2\pi i/3}H_{1/3}^{(1)}(is)]^2\}s$$

is a real function* of the parameter s depending on the angle of scattering χ:

$$s = \frac{(\pi-\chi)^3}{6}\,\frac{\alpha_1\alpha_2(\alpha_2-\alpha_1)}{(\alpha_2+\alpha_1)^2}.$$

Substituting (26) into (25), we obtain

$$d\sigma_1 = 2\pi\alpha_1^2[Q_{AB}^{(1)}]^2 e^{-2\pi(\alpha_2-\alpha_1)}\Phi(s)\, ds. \tag{27}$$

The function $\Phi(s)$ has the following limiting values:

$$\begin{aligned} s < 1, &\qquad \Phi(s) \approx [2^{2/3}\Gamma(\tfrac{2}{3})]^2 s^{-1/3}, \\ s > 1, &\qquad \Phi(s) \approx 4\pi e^{-2s}. \end{aligned}$$

Introducing the angle of scattering χ in place of s, we obtain for the angular distribution of the inelastic scattering of particle P

$$\begin{aligned} &\Phi(s)\, ds = F(\chi)\, d\Omega/4\pi, \\ \pi - \chi \ll 1, \quad &F(\chi) \approx 4\Gamma^2(\tfrac{2}{3})(\tfrac{3}{2}\tau^2)^{1/3} = \text{const}, \\ \pi - \chi \sim 1, \quad &F(\chi) \approx 4\pi\tau\exp[-(\tfrac{1}{3}\tau)(\pi-\chi)^2](\pi-\chi), \end{aligned} \tag{28}$$

where

$$\tau = \frac{\alpha_1\alpha_2}{(\alpha_1+\alpha_2)^2}\,(\alpha_2 - \alpha_1) > 1.$$

* See Sommerfeld [8], p. 559, formula (3). The appearance of the factor $e^{4\pi\alpha_1}$ in (26) follows from formula (41), p. 807 and (2a) p. 558. Here the charge of the particle was assumed to be positive.

From (27) and (28) it follows that the cross section decreases rapidly for increasing values of $\pi - \chi$; i.e., in practice, excitation takes place only for head-on collisions.

The total cross section σ_1 is determined by integration of (27) with respect to s. In view of the rapid decrease of $\Phi(s)$ this integration can be extended to infinity. Taking into account† that

$$\int_0^\infty \Phi(s)\, ds = 4\pi/\sqrt{3},$$

we get

$$\sigma_1 = 2\pi\alpha_1{}^2[Q_{AB}^{(1)}]^2 e^{-2\pi(\alpha_2 - \alpha_1)}(4\pi/\sqrt{3}). \tag{29}$$

This expression agrees with (18) if in the latter β is replaced by $\alpha_2 - \alpha_1$ inasmuch as $f_1(\alpha_2 - \alpha_1) \approx 4\pi\sqrt{3}$ for $\alpha_2 - \alpha_1 > 1$. Thus, for $\Delta E < E_1$ and when $\beta \approx \alpha_2 - \alpha_1$, expressions (29) and (28) agree.

The calculation of $d\sigma_n$ for $n \neq 1$ by formulas (22) and (23) involves mathematical difficulties and is omitted here. One can only suppose that the result is analogous to expression (29) and will not be different from (18), if in the latter β is replaced by $\alpha_2 - \alpha_1 > 1$ and the asymptotic value (20) of the function $f_n(\alpha_2 - \alpha_1)$ is used.

3. Distant Collisions for Large Energies E_1

If the energy E_1 of collision is large, $E_1 > B$, then excitation of the nucleus by means of the electric field takes place only in those instances when the trajectory of the particles P does not traverse the region of the nucleus. According to (13) this takes place when

$$r_{\min} = a(\epsilon + 1) \geq R_0 + R_0'$$

or

$$\epsilon \geq \frac{R_0 + R_0'}{a} - 1 = \frac{2E_1}{B} - 1.$$

If $2E_1 \gg B$, then the one in the latter formula can be neglected. The angle of deflection χ of particle P is related to ϵ according to the relationship $\sin(\chi/2) = 1/\epsilon$. Therefore, it follows that inelastic scattering takes place without involving nuclear forces, if the angle χ is small: $\chi < \chi_0$ where $\chi_0 = B/E_1$.

We shall estimate the total cross section S_n of the inelastic scattering of the particle into a cone bounded by the angle χ_0. We shall here assume that the energy $E_1 \gg B_1/2$ of the particle is not yet relativistic, and that condition (2) is fulfilled; i.e., the differential cross section is determined by the general formula (14).

The cross section S_n will be determined obviously by integration of (14) with respect to ϵ within the limits from $\epsilon_0 = 2E_1/B \gg 1$ to ∞. A simple calculation gives

$$S_n = 2\pi(R_0 + R_0')^2[\alpha_1 Q_{AB}^{(n)}/(R_0 + R_0')^n]^2 e^{-2\gamma}\varphi_n(\gamma), \tag{30}$$

† See, for example Sommerfeld [8], p. 560.

where

$$\varphi_n(\gamma) = e^{2\gamma} \sum_{m=-n}^{n}{}' c_{nm} \int_1^\infty \left[\epsilon_0{}^n I_{nm}\left(\epsilon, \frac{\gamma}{\epsilon_0}\right)\right]^2 \frac{\epsilon}{\epsilon_0}\, d\left(\frac{\epsilon}{\epsilon_0}\right) \tag{31}$$

is a slowly (not exponentially) changing function of the parameter $\gamma = \epsilon_0\beta = (\Delta E/\hbar v_1)(R_0 + R_0')$.

The calculation of $\varphi_n(\gamma)$, presented in the Appendix, gives us the following result:

$$\varphi_1(\gamma) = i\pi^2\gamma H^{(1)}_{i\gamma/\epsilon_0}(i\gamma)H^{(1)\prime}_{i\gamma/\epsilon_0}(i\gamma)e^{(2-\pi/\epsilon_0)\gamma},$$

or in limiting cases:

$$\gamma < 1, \qquad \varphi_1(\gamma) \approx 4\log(2/\xi\gamma), \qquad \xi = 1.781\ \ldots,$$
$$\gamma > 1, \qquad \varphi_1(\gamma) \approx 2\pi.$$

In general for arbitrary $n \neq 1$ we have

$$\gamma = 0, \qquad \varphi_n(\gamma) = 2^{2n}\frac{(n-1)!(n-2)!}{(2n)!},$$

$$\gamma > 1, \qquad \varphi_n(\gamma) \approx \pi\frac{(2n)!}{[n!(2n-1)!!]^2}\gamma^{2(n-1)}e^{-\pi\gamma/\epsilon_0}.$$

After a numerical evaluation (assuming $Q_{AB}^{(n)} \sim R_0{}^n$) the cross section (30) is found to be of the order of magnitude 10^{-27} to 10^{-29} cm^2 depending on the parameters characterizing the collision.

We will note also that in the cone $\chi \leqq \chi_0$ there also are particles inelastically scattered by the nucleus as a result of nuclear interaction. However, the appropriate cross section $\Delta\sigma_{\text{nuc}} = \sigma_{\text{nuc}}\,\Delta\Omega$ (for $\vartheta = 0$) can be considerably less than the cross section S_n inasmuch as $\sigma_{\text{nuc}} \sim 10^{-26}$ cm^2 and $\Delta\Omega = 2\pi\chi_0{}^2 = 2\pi(B/E_1)^2 \ll 1$, if $E_1 \gg B$.

Acknowledgment

The author expresses deep appreciation to academician L. D. Landau under whose guidance the present work was carried out.

Appendix. The Calculation of $f_n(\beta)$ and $\varphi_n(\gamma)$

1. The calculation of $f_n(\beta)$ will begin with the case where $n = 1$, where the index m in the sum of Eq. (14) takes on the values of ± 1. Since $c_{1,1} = c_{1,-1} = \frac{1}{2}$, then

$$f_1(\beta) = \tfrac{1}{2}e^{-2\pi\beta}\int_1^\infty [I^2_{1,1}(\epsilon, \beta) + I^2_{1,-1}(\epsilon, \beta)]\epsilon\, d\epsilon. \tag{a}$$

For the calculation of

$$I_{1,1}(\epsilon, \beta) = \int_{-\infty}^{\infty} e^{i\beta(\epsilon \sinh w + w)}\frac{\cosh w + \epsilon + i(\epsilon^2 - 1)^{1/2}\sinh w}{(\epsilon\cosh w + 1)^2}\, dw$$

we note that

$$\frac{\cosh w + \epsilon + i(\epsilon^2 - 1)^{1/2} \sinh w}{(\epsilon \cosh w + 1)^2} dw$$
$$= \frac{1}{i(\epsilon^2 - 1)^{1/2}} d \left\{ \frac{\cosh w + \epsilon + i(\epsilon^2 - 1)^{1/2} \sinh w}{\epsilon \cosh w + 1} \right\},$$

and, therefore, the integration can be carried out by parts. This gives us

$$I_{1,1}(\epsilon, \beta) = -\frac{\beta}{(\epsilon^2 - 1)^{1/2}} \int_{-\infty}^{\infty} e^{i\beta(\epsilon \sinh w + w)} (\cosh w + \epsilon + i(\epsilon^2 - 1)^{1/2} \sinh w)\, dw. \quad \text{(b)}$$

For $I_{1,-1}(\epsilon, \beta)$ we obtain an analogous expression with a minus sign in front of $i(\epsilon^2 - 1)^{1/2} \sinh w$.

From the parametric representation (13) of the orbit of particle P, it follows that $I_{1,1}(\epsilon, \beta)$ and $I_{1,-1}(\epsilon, \beta)$ will agree (up to the constant factor) with the Fourier components of frequency $\omega = \Delta E/\hbar$, of the coordinates $x + iy$ and $x - iy$ of particle P. Exactly the same Fourier components occur [6, 9] in the classical theory of Bremsstrahlung of a charged particle in a Coulomb field; the total intensity of emission is precisely proportional to the square of sum of Fourier components of the form $I_{1,1}^2(\epsilon, \beta) + I_{1,-1}^2(\epsilon, \beta)$. Therefore, the calculation of the integral (b) and the double integration in (a) agree accurately with the well-known calculations [9] of this theory. The result for $f_1(\beta)$ shown in the text therefore differs from the function G', introduced in the work of Kramers [6], only by the factor $4\pi/\sqrt{3}$.

2. When $n \neq 1$, $f_n(\beta)$ cannot be expressed by any known function. We obtain asymptotic values $f_n(\beta)$ for $\beta \to 0$ and $\beta > 1$. When $\beta = 0$, it is convenient for the calculation of $I_{nm}(\epsilon, 0)$ in integral (15) to introduce the variable φ instead of the variable w. Using the polar equation of the hyperbolic trajectory (13),

$$r = a(\epsilon^2 - 1)/(\epsilon \cos \varphi - 1) = a(\epsilon \sinh w + 1),$$

we have

$$\frac{dw}{(\epsilon \cosh w + 1)^n} = \frac{(\epsilon \cos \varphi - 1)^n}{[(\epsilon^2 - 1)^{1/2}]^{2n-1}} d\varphi.$$

It therefore follows that

$$I_{nm}(\epsilon, 0) = \frac{1}{[(\epsilon^2 - 1)^{1/2}]^{2n-1}} \int_{-\varphi_0}^{\varphi_0} e^{im\varphi} (\epsilon \cos \varphi - 1)^{n-1}\, d\varphi, \quad \text{(c)}$$

where φ_0 is the limiting value of the angle φ for $r(\varphi_0) \to \infty$, i.e., $\tan \varphi_0 = (\epsilon^2 - 1)^{1/2} = \zeta$. With the specific values $n = 2$, $n = 3$, etc., the integration in (c) is carried out in a simple manner and gives us:

$$(1) \quad n = 2: \quad \sideset{}{'}\sum_{m=-2}^{2} c_{2m} I_{2m}^2(\epsilon, 0) = \frac{1}{\zeta^4} - \frac{2 \arctan \zeta}{\zeta^5} + \frac{\arctan^2 \zeta}{\zeta^6} + \frac{1}{3(1 + \zeta^2)^2};$$

$$(2)\quad n = 3: \qquad \sum_{m=-3}^{3}{}' \, c_{3m} I_{3m}^2(\epsilon, 0) = \left(\frac{3}{2\zeta^8} + \frac{2}{\zeta^6} + \frac{2}{\zeta^4}\right)\frac{1}{1+\zeta^2} - \frac{3+2\zeta^2}{\zeta^9}\arctan\zeta$$

$$+ \frac{3}{2}\frac{1+\zeta^2}{\zeta^{10}}\arctan^2\zeta + \frac{2}{45}\frac{1}{(1+\zeta^2)^3}.$$

Hence, after integration over ζ (we have $\zeta\, d\zeta = \epsilon\, d\epsilon$),

$$\int_0^\infty \sum_{m=-n}^{n}{}' \, c_{nm} I_{nm}^2(\epsilon, 0)\zeta\, d\zeta = f_n(0),$$

we obtain the values $f_2(0)$, $f_3(0)$, etc., presented in the text.

3. When $\beta \gg 1$, the exponential $\exp[i\beta(\epsilon \sinh w + w)] = e^{L(w)}$ is a rapidly fluctuating function of w. For this reason, it is convenient for calculation of (15) to use the method of the saddle point or of the "steepest descent." Considering $w = u + iv$ as a complex variable, we move the path of integration from the real axis to curve C given by the equation

$$\operatorname{Im}\{L(w)\} = 0, \qquad \text{that is,} \qquad \epsilon \cos v = -u/\sinh u.$$

In order that the value of integral (15) remains unchanged by this, let us agree to pass below the singular point $w_0 = i(\pi - \varphi_0)$ of the integrand function, which lies exactly on the curve C. The exponential $e^{L(w)}$, remaining real on C, forms a sharp peak at the point $w = w_0$, inasmuch as $[dL(w)/dw]_{w=w_0} = 0$. For this reason the value of the integral (15) is determined by the values of the integrand function at points close to $w = w_0$. Expanding the integrand in (15) into a series in $\xi = w - w_0$, we have, with an accuracy up to the third order, with respect to ξ:

$$L(w) = -\pi\beta - \beta(\zeta - \arctan\zeta) - \frac{\beta\zeta}{2}\xi^2 - \frac{i\beta}{6}\xi^3,$$

$$\frac{(\cosh w + \epsilon + i\zeta \sinh w)^m}{(\epsilon \cosh w + 1)^{n+m}}\, dw = (-2)^n \frac{d\xi}{\xi^{n-m}(\xi - 2i\zeta)^{n+m}},$$

where, as before, $\zeta = (\epsilon^2 - 1)^{1/2}$. Substituting this expansion into (15), we obtain

$$I_{nm}(\epsilon, \beta) = \exp[-\pi\beta - \beta(\zeta - \arctan\zeta)] \int_C \frac{\exp\left\{-\frac{\beta\zeta}{2}\xi^2 - \frac{i\beta}{6}\xi^3\right\}}{\xi^{n-m}(\xi - 2i\zeta)^{n+m}}\, d\xi.$$

This value should be substituted into (19), and since $\beta > 1$, it is possible for us to expand $\zeta - \arctan\zeta$ in the exponent, considering $\zeta - \arctan\zeta = -\zeta^3/3$.

If we now introduce the variables $x = (\beta\xi)^{1/3}$ and $y = (\beta\zeta)^{1/3}$, then the integrals would no longer be dependent on β as a parameter, and for $f_n(\beta)$ we obtain the value (20) mentioned in the text.

4. The calculation of $\varphi_n(\gamma)$ is analogous to that of $f_n(\beta)$. With $n = 1$, it reduces to the calculation of the integral

$$\varphi_1(\gamma) = e^{2\gamma} \frac{\pi^2\gamma^2}{2} \int_1^\infty \left[H^{(1)^2}_{i\beta-1}\left(i\gamma \frac{\epsilon}{\epsilon_0}\right) + H^{(1)^2}_{i\beta+1}\left(i\gamma \frac{\epsilon}{\epsilon_0}\right) - 2H^{(1)^2}_{i\beta}\left(i\gamma \frac{\epsilon}{\epsilon_0}\right)\right] \frac{\epsilon}{\epsilon_0}\, d\left(\frac{\epsilon}{\epsilon_0}\right).$$

No difficulties are involved with this calculation; it is similar to the case of $f_1(\beta)$. The result is given in the text.

5. When $n \neq 1$ and $\gamma = 0$, it follows from (c) that

$$\epsilon_0{}^n I_{nm}(\epsilon, 0) = \frac{1}{(\epsilon/\epsilon_0)^n} \int_{-\pi/2}^{\pi/2} e^{im\varphi} \cos^{n-1}\varphi \, d\varphi \approx \frac{1}{(\epsilon/\epsilon_0)^n} \frac{2(n-1)!}{[(n+m)!(n-m)!c_{nm}]^{1/2}},$$

since $\epsilon \geqslant \epsilon_0 \gg 1$, $\zeta = (\epsilon^2 - 1)^{1/2} \approx \epsilon$, and $\cos\varphi = 1/\epsilon \ll 1$. Substitution into (31) gives us

$$\varphi_n(0) = 2(n-1)!(n-2)! \sum_{m=-n}^{n}{}' \frac{1}{(n-m)!(n+m)!}.$$

Since

$$\sum_{m=-n}^{n}{}' \frac{1}{(n-m)!(n+m)!} = \frac{2^2 n - 1}{(2n)!},$$

one obtains for $\varphi_n(0)$ the value presented in the text.

6. The calculation of $\varphi_n(\gamma)$, when $\gamma > 1$, is quite analogous to that shown above for $f_n(\beta)$, when $\beta > 1$. The only difference is that, in expanding the integrand function in (15), it is sufficient to account for only the squared terms with respect to $\xi = w - w_0$:

$$L(w) = i\beta(\epsilon \sinh w + w) \approx -\beta\zeta - \frac{\beta\zeta}{2}\xi^2 - \frac{\pi\beta}{2},$$

$$\frac{(\cosh w + \epsilon + i\zeta \sinh w)^m}{(\epsilon \cosh w + 1)^{n+m}} dw = (-1)^{(n-m)/2} \frac{d\xi}{2^m \zeta^n \xi^{n-m}}.$$

This follows from the fact that here only large values of $\zeta \approx \epsilon > 1$ are of interest, whereas in $f_n(\beta)$ the values of $I_{nm}(\epsilon, \beta)$ were significant for small ζ also, i.e., of the order of $1/\beta^{1/3} < 1$.

Substitution of these expansions into (15) gives us

$$\epsilon_0{}^n I_{nm}(\epsilon, \beta) = (-1)^{(n-m)/2} \frac{\exp[-(\pi\beta/2) - (\gamma\epsilon/\epsilon_0)]}{2^m(\epsilon/\epsilon_0)^n} \int_C \exp[-(\gamma/2)(\epsilon/\epsilon_0)\xi^2] \frac{d\xi}{\xi^{n-m}}.$$

Integration along the contour C is easily carried out with the use of the well-known representation of the inverse Γ-function by the contour integral:

$$\int_C \exp[-(\gamma/2)(\epsilon/\epsilon_0)\xi^2] \frac{d\xi}{\xi^{n-m}} = (-1)^{(n-m)/2} \frac{\pi(\gamma\epsilon/2\epsilon_0)^{(n-m-1)/2}}{\Gamma(n - m + \frac{1}{2})}.$$

Therefore,

$$[\epsilon_0{}^n I_{nm}(\epsilon, \beta)]^2 = \frac{\pi^2(\gamma/2)^{n-m-1}}{2^{2m}[\Gamma(n - m + \frac{1}{2})]^2} \exp(-\pi\beta - 2\gamma\epsilon/\epsilon_0) \frac{1}{(\epsilon/\epsilon_0)^{n+m+1}}.$$

Substituting this expression into (31) and retaining only the term with $m = -n$, containing $\gamma > 1$ in the highest power, we obtain

$$\varphi_n(\gamma) = \frac{\pi^2 c_{n,-n}}{[\Gamma(n + \frac{1}{2})]^2} \gamma^{2(n-1)} e^{-\pi\gamma/\epsilon_0}$$

($\beta = \gamma/\epsilon_0$). Thus, one finds for $\varphi_n(\gamma)$ the value introduced in the text, inasmuch as

$$c_{n,-n} = \frac{(2n)!}{2^{2n}(n!)^2},$$

and

$$[\Gamma(n + \tfrac{1}{2})]^2 = \frac{\pi}{2^{2n}} [(2n - 1)!!]^2.$$

References

1. S. W. Barnes and P. W. Aradine, *Phys. Rev.* **55,** 50 (1939).
2. K. Lark-Horowitz *et al.*, *Phys. Rev.* **55,** 878 (1939).
3. L. D. Landau and E. M. Lifshitz, "Quantum Mechanics," Part I. GITTL. Moscow, 1948. (Translator's note: "GITTL" is Russian abbreviation for Government Publication of Technical-Theoretical Literature.)
4. K. Ter-Martirosyan, *Zh. Eksper. i Teor. Fiz.* **20,** 937 (1950).
5. L. D. Landau, *Phys. Z. Sowjetun.* **1,** 88 (1932).
6. H. A. Kramers, *Phil. Mag.* **46,** 836 (1923).
7. V. F. Weisskopf, *Phys. Rev.* **53,** 1018 (1938).
8. A. Sommerfeld, "Atombau und Spektrallinien," Vol. II, Vieweg-Verlag, Braunschweig 1939.
9. G. Wentzel, *Z. Physik.* **27,** 257 (1924).

Office of Naval Research, European Scientific Notes, No. 7–9, May 1, 1953

*Proposed Coulomb Excitation of Nuclei**

At a recent colloquium in the Institute for Theoretical Physics, Copenhagen, Dr. B. R. Mottelson discussed a novel proposal for producing excitation of nuclear states. Mottelson's proposal was to bombard atomic nuclei with charged particles whose energy is well below the barrier. The probability of penetrating the barrier would consequently not be appreciable, and the nucleus would experience only the Coulomb field. Thus the excitation resulting from the pulse of electromagnetic energy would reflect the properties of the target nucleus free from complication by the nuclear force between target and projectile.

Principle of the Method

Mottelson outlined a classical treatment of the problem. A criterion for the validity of such an approach is that the parameter $\kappa = 2Z_1Z_2e^2/\hbar v \gg 1$, a condition which is easily realized, particularly for heavy ions bombarding all but the lightest nuclei. The Coulomb field at the nucleus is then treated as a time-dependent perturbation with its value as a function of time and position determined from the known Kepler orbit of the bombarding particle. The Coulomb potential may be decomposed into electric multipoles, and a treatment much like that of Fermi for the similar problem in atomic excitation may be carried out.

The Coulomb field will contain a very large number of electric multipoles of comparable intensity; however, because the speed of the bombarding particle is small compared to that of light, magnetic effects may be completely neglected. In this respect the method of excitation is analogous to a nuclear photoeffect produced by a hypothetical photon which had an electric vector but no magnetic vector. This treatment yields a differential cross section for each electric multipole order which contains the nuclear matrix element for the isomeric transition between the excited nuclear state and the ground state. The virtue of the type of excitation suggested is that the other factors which appear are completely susceptible to calculation. Consequently, in principle, the matrix element may be determined from a measurement of the differential cross section.

Possible Applications

The method suggested may provide an alternative technique for measuring transition probabilities of gamma-ray transitions which are too short-lived for study by direct time measurement. In principle, the differential cross section for the

* Reprinted by permission of Dr. Mottelson and the Office of Naval Research, London.

process may be determined by counting the inelastically scattered particles. Measurement of the angular distribution of the gamma rays could give information on spins, and an absolute measurement of the number of gamma rays emitted in a specified direction would also give the nuclear matrix element.

The size of the effect predicted and some of the experimental difficulties can be seen by considering a specific possible experiment, namely, the bombardment of Er^{166} with α-particles. An α-energy of 12 Mev, half the barrier height, is chosen to prevent barrier penetration. Mottelson finds, in this case, that the 80-kev state of erbium will be excited by the E2 multipole, the differential cross section at 180° for the process being about one fortieth of the Rutherford cross section. In this case the experimentalist is required to separate the inelastically scattered α-particles which have lost 80 kev from the elastically scattered α-particles and/or to distinguish 80-kev gamma rays from X rays.

Det Kongelige Danske Videnskabernes Selskab
Matematisk-fysiske Meddelelser, bind **28**, nr. 1

Dan. Mat. Fys. Medd. **28**, no. 1 (1953)

EXCITATION OF NUCLEAR ROTATIONAL STATES BY THE ELECTRIC FIELD OF IMPINGING PARTICLES

BY

TORBEN HUUS AND ČRTOMIR ZUPANČIČ

København
i kommission hos Ejnar Munksgaard
1953

1. Introduction.

When heavy nuclei are bombarded with charged particles with an energy appreciably below the Coulomb barrier the short range nuclear forces cannot operate. The collisions may, however, still give rise to nuclear excitations produced by the electric field of the impinging particles. Such reactions are especially simple to interpret, and the cross-sections can be expressed in terms of the same nuclear properties that determine the transition probabilities for electromagnetic radiative processes.*†§

It has been suggested** that the method of Coulomb excitation should provide a powerful tool for the study of nuclear collective properties. According to the nuclear model which describes the dynamics of the nucleus in terms of the coupled motion of individual particles and surface oscillations***, low-lying excited states of collective type are expected for nuclei possessing large deformations. These states exhibit a rotational spectrum, and should reveal themselves by their especially large cross-sections for Coulomb excitation.

In order to test some of these predictions we have undertaken an investigation of the Coulomb excitation effects produced

* The mechanism of nuclear excitation by the Coulomb field of bombarding particles has been discussed by several authors: Weisskopf (1938); Ramsey (1951); Mullin and Guth (1951); Huby and Newns (1951); Breit, Hull, and Gluckstern (1952); Ter-Martirosyan (1952); Bohr and Mottelson (1953).

† Nuclear excitations with cross-sections too large to be explained by barrier penetration have been observed on a number of occasions, and the possibility of attributing the effects to Coulomb excitation has been discussed. Cf., e. g., Barnes and Aradine (1939); Risser, Lark-Horowitz and Smith (1940).

§ While the present work was being prepared for publication, we have learned that γ-rays resulting from Coulomb excitation of heavy nuclei have recently been observed by C. L. McClelland and C. Goodman. We are indebted to Professor Goodman for sending us a manuscript of their work in advance of publication.

** A. Bohr and B. Mottelson. (Cf., e. g., Report of the International Physics Conference, Copenhagen, June 1952).

*** A. Bohr (1952); Bohr and Mottelson (1953); the latter paper will be referred to in the following as B.-M.

1*

by the protons from a 2 MeV electrostatic accelerator*. We here report the results obtained by the bombardment of tantalum and tungsten.

In § 2, a brief summary is given of some of the relevant aspects of the theory of Coulomb excitation. A description of the nuclear rotational spectrum appears in § 3. The experimental arrangement is described in § 4. In § 5, the excitation cross-sections for the first excited state in Ta^{181} are given and compared with the theory of Coulomb excitation; the nuclear data obtained are discussed in relation to the theory of rotational states. The excitation of the second rotational state in Ta is described in § 6. In § 7 the Coulomb excitation of the first excited states of the even-A isotopes of W is reported. The yields of the characteristic K X-rays, which are also excited by the protons, are discussed in § 8. A survey of main conclusions is contained in § 9.

2. Theory of Coulomb Excitation**.

When the energy of the bombarding particles is low enough to exclude penetration through the Coulomb barrier, the parameter

$$\varkappa = 2 \frac{Z_1 Z_2 e^2}{\hbar v} \tag{1}$$

is large compared to unity and the trajectory may be described by means of classical mechanics (N. Bohr, 1948). In expression (1), the charge numbers of the projectile and the target nucleus are denoted by Z_1 and Z_2, respectively, while v is the relative velocity.

* In previous experiments performed with the electrostatic accelerator of the California Institute of Technology (Day and Huus, 1952), a strong γ-ray had been observed to arise from the Ta target backing. The present experiments were undertaken after the recognition that this γ-ray of 137 keV resulted from Coulomb excitation.

** The present formulation is based on the work of Ter-Martirosyan (1952). Cf. also B.-M., Appendix VI, who, in connection with a review of this work, have especially discussed the relationship between Coulomb excitation and electromagnetic transitions, and the applications to the study of nuclear collective properties. We here follow the presentation given in the latter reference.

The effect of the projectile on the target nucleus can then be described in terms of the time varying electric potential, given by

$$V(t) = \sum_{p=1}^{Z_2} \frac{Z_1 e^2}{|\vec{r}_p - \vec{r}(t)|}, \tag{2}$$

where $\vec{r}_p$ are the coordinates of the protons in the target nucleus, and where $\vec{r}(t)$ gives the classical trajectory of the projectile considered as a point charge. Since the probability for exciting the nucleus in any single collision is small, the excitation process may be treated by quantum-mechanical perturbation theory.

The collective nuclear excitations are produced by the electric quadrupole component of (2) and, for the total cross-section for excitation of a given level, one obtains

$$\sigma = \frac{2\pi^2}{25} \frac{1}{Z_2^2 e^2} \left(\frac{mv}{\hbar}\right)^2 B_e(2)\, g_2(\xi), \tag{3}$$

where m is the reduced mass. The quantity $B_e(2)$ is a constant containing the nuclear matrix element; this quantity also determines the electric quadrupole ($E2$) decay probability for the inverse transition. Theoretical values of B for rotational excitations are quoted in the following paragraph. The last factor $g_2(\xi)$ can be expressed in terms of integrals over the trajectories of the bombarding particles and depends on the parameter

$$\xi = \frac{\Delta E}{2E} \frac{Z_1 Z_2 e^2}{\hbar v}, \tag{4}$$

where E is the bombarding energy and ΔE the nuclear excitation energy. The parameter ξ thus represents the ratio of the collision time to the nuclear period; for small ξ, the function g_2 approaches the constant value 1.13 while, for large ξ, the collisions become adiabatic, resulting in an exponentially decreasing g_2. The function $g_2(\xi)$ has been evaluated numerically (cf. forthcoming publication by A. Winther, whose results for g_2 are also given in B.-M., Appendix VI).

The angular distribution of the emitted radiation can also be expressed in terms of integrals over the trajectories, which have been calculated numerically (cf. forthcoming publication by K. Alder and A. Winther).

3. Nuclear Rotational States.

The coupled particle-surface model predicts the occurrence of low-lying collective excitations for the strongly deformed nuclei encountered in regions removed from closed shells (B.-M., Chapter VI). These states exhibit a spectrum of rotational character, and also reveal themselves by their very large $E2$ transition probabilities.

In even-even nuclei, the spectrum is given by

$$E = \frac{\hbar^2}{2\mathfrak{J}} I(I+1) \qquad \underset{\text{even parity}}{I = 0, 2, 4 \cdots} \tag{5}$$

where the effective moment of inertia $\mathfrak{J}$ is proportional to the square of the nuclear deformation, and is expected to vary slowly with the atomic number A. States of this type have recently been identified by the regularity of the spectrum, the systematic dependence of the energies on A, and the lifetimes which are often more than a hundred times shorter than expected for single-particle transitions (Bohr and Mottelson, 1952, 1953a, 1953b; cf. also Ford, 1953 and Asaro and Perlman, 1953).

In odd-A nuclei, the rotational spectrum is given by

$$E = \frac{\hbar^2}{2\mathfrak{J}} (I(I+1) - I_0(I_0+1)) \qquad \underset{\text{same parity as ground state}}{I = I_0, I_0+1, I_0+2 \cdots} \tag{6}$$

where I_0 is the ground state spin*. Since $\mathfrak{J}$ is expected in general to vary slowly with A, the rotational excitation energies in odd-A nuclei can be related to those in even-even nuclei.

Due to the very large $E2$ transition probabilities for these rotational states, the method of Coulomb excitation is especially suited for their identification. The excitation cross-section (3) depends on the reduced transition probability $B_e(2)$ which for the rotational excitations $I_0 \rightarrow I_0 + 1$ and $I_0 \rightarrow I_0 + 2$ is given by (B.-M., § VIIc.ii and Appendix VI)

* An additional term in the rotational energy, resulting in a less regular spectrum, occurs when the angular momentum of the particles along the nuclear axis equals $1/2\ \hbar$ (cf. B.-M., § VIc.iii).

$$B_e(2) = \frac{15}{16\pi} e^2 Q_0^2 \frac{I_0}{(I_0+1)(I_0+2)} \qquad I_0 \rightarrow I_0+1 \tag{7}$$

and

$$B_e(2) = \frac{15}{8\pi} e^2 Q_0^2 \frac{1}{(2I_0+3)(I_0+2)} \qquad I_0 \rightarrow I_0+2. \tag{8}$$

The latter formula also refers to the excitation of the (2+) first excited state in even-even nuclei.

The transition probabilities (7) and (8) are expressed in terms of the intrinsic nuclear quadrupole moment Q_0, measured with respect to the nuclear axis (B.-M., Chapter V). The spectroscopically measured quadrupole moment Q is related to Q_0 by

$$Q = \frac{I_0}{I_0+1} \frac{2I_0-1}{2I_0+3} Q_0. \tag{9}$$

Thus, the measurement of the excitation cross-sections, just as of the corresponding $E2$ decay probability, provides a measure of the nuclear deformation which can be directly compared with the spectroscopic data (cf. B.-M., Table XXVII).

The rotational character of the collective excitation spectrum represents a limiting situation realized when the zero-point oscillations of the nuclear surface are negligible compared with the total deformation. These zero-point oscillations give rise to deviations from the expressions given in this section. In the region of closed shells, where the deformations are small, an entirely different spectrum results (B.-M., § VIc.i and ii).

4. Experimental Arrangement.

The 2 MeV electrostatic generator was used to produce a separated beam of protons, which passed through the system of stops shown in Fig. 1. The guard ring (G in the figure) was kept at a negative voltage of 100 volts in order to avoid the influence of secondary electrons on the current measurements.

The target holder was mounted in a horizontal tube in such a way that a number of different targets could be inserted into the beam and set at an arbitrary angle. An aluminum target holder and target tube were employed to make the absorption small. Aluminum, however, radiates strongly when bombarded

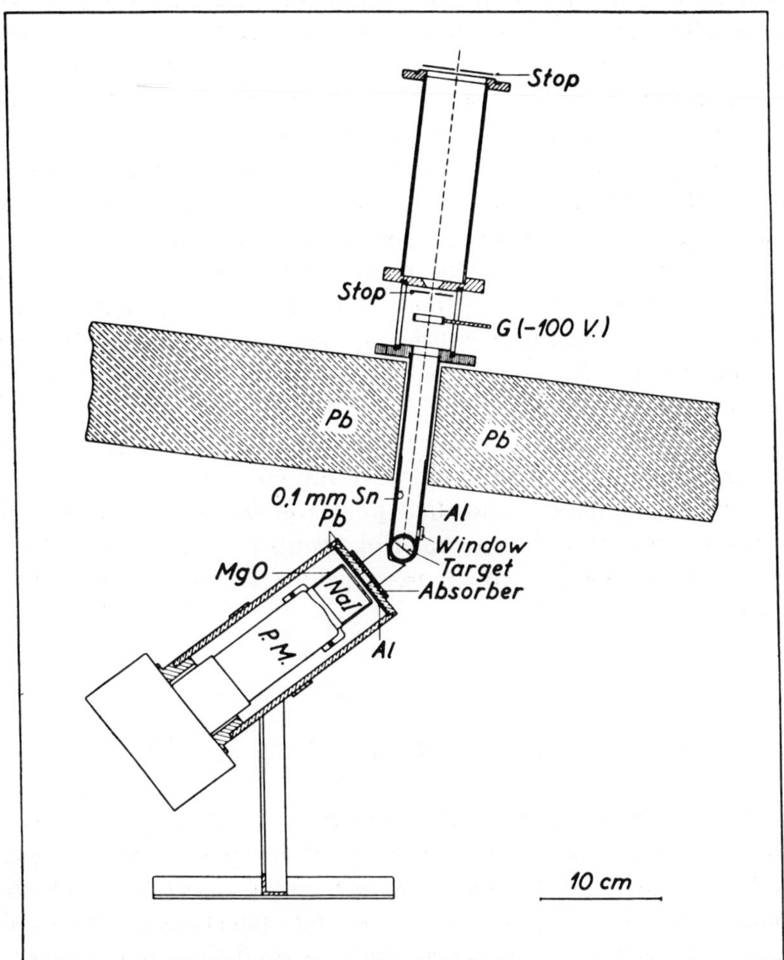

Fig. 1. Experimental arrangement.

with protons, and the target tube was, therefore, lined with a 0.1 mm tin foil in order to reduce the effect from the scattered protons. Tin is good for this purpose because its closed proton shell structure implies a low cross-section for Coulomb excitation; moreover, its nuclear charge is great enough that capture radiation is unimportant and small enough that the characteristic X-radiation does not interfere with the present experiments.

A 10 cm thick layer of lead, 50 cm × 50 cm, was placed above the counter in order to shield it from the hard X-rays from the acceleration tube. These X-rays would produce a background of soft, secondary radiation. The counter was further mounted in a 5 mm lead tube to absorb the scattered radiation. On the top of this tube various absorbers could be placed for the purpose of filtering the radiation from the target. The distance between the counter and the target tube could be fixed by means of two lucite rods when measuring angular distributions.

The γ-rays were detected by a cylindrical NaI (Tl) crystal, 4 cm in diameter and 3 cm high. The crystal was mounted on a RCA 5819 photo multiplier tube by means of a glass cap which was filled with Nujol mineral oil and covered on the outside with a layer of MgO.

The pulse spectrum was measured with a single-channel analyzer. The linearity of the equipment was checked by a relay-pulser connected to the preamplifier, and energy calibrations were made by means of the annihilation radiation from Na^{22} and by means of the characteristic *K* X-rays emitted by the bombarded targets.

5. Excitation of First State in Ta.

For a detailed study of the Coulomb excitation process, Ta is a particularly suited target, since it is known to have a large nuclear quadrupole moment, and occurs in a region of the elements where the trends of the rotational states seem well established. Moreover, Ta possesses only a single isotope ($_{73}Ta^{181}$).

A piece of Ta metal of 0.1 mm thickness (i. e. thick to protons of 2 MeV) was used as a target. The γ-ray spectrum shown in Fig. 2a was obtained with 1.75 MeV protons; the measurements were made at an angle of 80° with respect to the beam. The strong peak is due to the *K* X-radiation from the Ta atoms, which results mainly from the ejection of *K*-electrons by the bombarding particles (cf. Chadwick, 1913; Bothe and Fränz,

1928; HENNEBERG, 1933 LIVINGSTON, GENEVESE and KONOPINSKI, 1937), and partly also from the internal conversion of the nuclear excitation. The weak peak on the low energy side corresponds to the *K*-radiation from the Sn-lining of the target tube.

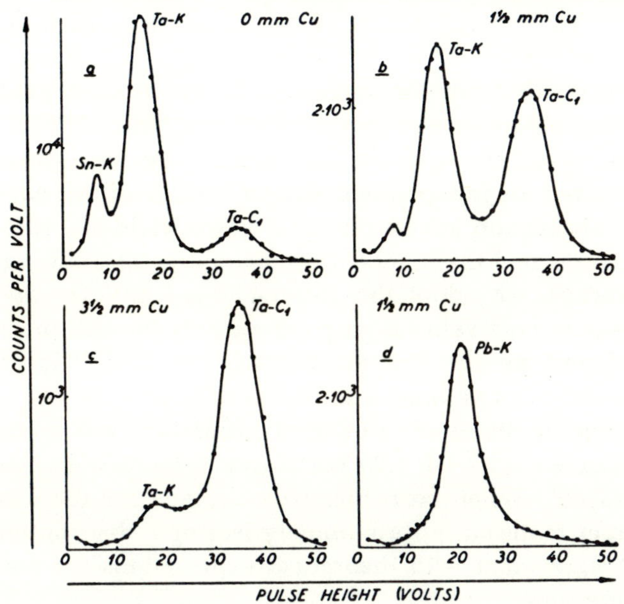

Fig. 2. Pulse spectra measured at a proton energy of 1.75 MeV. The Ta-C_1 peak is due to Coulomb excitation of the first excited state of Ta, and the *K*-peaks to the characteristic X-radiation from the atoms. Figs. *a*, *b*, and *c* show the spectra which are obtained for tantalum when various absorbers are used; Fig. *d* shows the spectrum for lead with a 1.5 mm Cu absorber.

The high energy peak (C_1 in the figure) is due to the Coulomb excitation of the Ta-nuclei. The relative intensity of the C_1-peak with respect to the Ta *K*-peak can be increased by the insertion of absorbers, since the absorption coefficients are strongly energy dependent in this region. This is illustrated in Figs. 2 *a*, *b* and *c*. It was convenient, for most of the quantitative measurements, to use an absorber of 1.5 mm Cu, which makes the two peaks about equally strong.

The nuclear origin of the C_1-peak was checked by measurements on Pb (Fig. 2 *d*), which has a similar X-ray spectrum to that of Ta, but whose closed-shell nuclear structure implies high excitation energies and correspondingly small cross-sections for Coulomb excitation under the present conditions.

The cross-section for formation of a compound nucleus in the bombardment of Ta by 1.75 MeV protons is expected to be of the order of 10^{-38} cm^2 (cf., e.g., Blatt and Weisskopf, 1952, p. 352), which is many orders of magnitude too small to explain the observed yield. It is therefore strongly indicated that the excitations result from the influence of the electric field of the protons.

The energy of the C_1-peak was found to be 135 ± 5 keV from the pulse size. An independent determination of the energy was made by absorption measurements which yielded 140 ± 7 keV, and also confirmed the monochromatic character of the radiation. As an average, we adopt the value 137 ± 5 keV for the energy of the C_1-line. This value is in good agreement with the value of 136 keV found for the first excited level of ${}_{73}Ta^{181}$ in other experiments (cf. Goldhaber and Hill, 1952).

The angular distribution of the C_1-radiation was measured at bombarding energies of 1.25 MeV and 1.75 MeV in the region from 0° to 80°. When corrections were made for the absorption in the target material, approximately isotropic distributions were found, in agreement with theoretical expectations for the transitions in question.

The thick target yield for the C_1-radiation was measured at 80° with respect to the beam for bombarding energies from 1 to 2.2 MeV. The total yield of γ-quanta per proton, obtained by assuming isotropic distribution, is shown in Fig. 3. The solid curve represents the theoretical energy dependence for $E2$ Coulomb excitation obtained from (3), using the numerically calculated g_2 function. Since the energy loss ΔE in the nuclear excitation process is not quite negligible compared with the bombarding energies used, the theoretical curve has been calculated for an effective energy equal to the incident energy minus $1/2\ \Delta E$. The stopping power for Ta is taken from the semi-empirical relation given by Lindhard and Scharff (1952).

The close agreement between the energy dependence of the experimental yield of the C_1-radiation and that given by the theory confirms the interpretation of the observed effects in terms of Coulomb excitation by the electric quadrupole field of the protons.

For a bombarding energy of 2 MeV, a cross-section for γ-emission of approximately 0.5 millibarns (mb) was obtained. In order to derive the total excitation cross-section, it is necessary to take into account the de-excitation by internal conversion. The decay following the $E2$ excitation is expected to be $E2$ or $M1$, both of which yield total conversion coefficients of about 2 (ROSE et al., 1951; GOLDHABER and SUNYAR, 1951), which is also consistent with the experimentally measured electron yield (FAN,

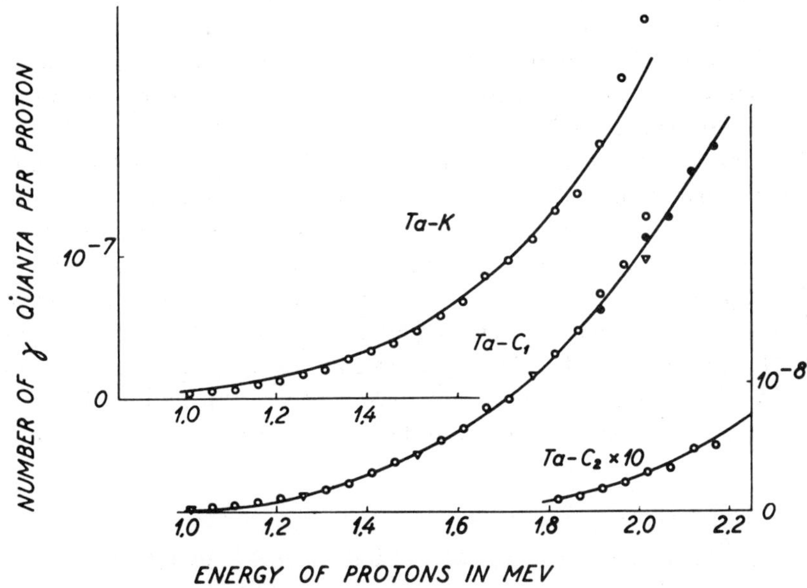

Fig. 3. Total yields for the Ta-K, Ta-C_1, and Ta-C_2 peaks. The theoretical curves are adjusted to fit the measured yields at a proton energy of about 1.8 MeV.

1952). Employing the value 2 for the total conversion coefficient, an excitation cross-section of approximately 1.6 mb is found for protons of 2 MeV.

From the measured cross-section, one derives from (3) the reduced transition probability $B_e(2)$, and the value obtained is more than a hundred times that expected for the excitation of a single particle in the nucleus (cf. B.-M. § VIIb.i for estimates of $B_e(2)$ for single-particle transitions). Thus, the observed large cross-section directly indicates a collective excitation and suggests a rotational interpretation.

The rotational spectrum (6) for an odd-A nucleus depends on the ground state spin, which for $_{73}Ta^{181}$ is known to be 7/2 (cf. MACK, 1950). The first rotational excitation should thus have $I = 9/2$ and an energy of $9\hbar^2/2\mathfrak{J}$ which is 3/2 times the rotational energy expected for the first excited (2+) state in an even-even nucleus with the same value of $\mathfrak{J}$ (cf. (5)). The energy of 93 keV for the first excited state in $_{72}Hf^{180}$ (cf. SCHARFF-GOLDHABER, 1953) thus implies an energy of about 140 keV for the 9/2 state in Ta, in good agreement with the 137 keV γ-ray produced by Coulomb excitation*.

Assuming the rotational interpretation of the 137 keV state, one derives from the empirically determined $B_e(2)$, by means of (7), the intrinsic nuclear quadrupole moment $Q_0 \simeq 7 \times 10^{-24}\,\text{cm}^2$. This may be compared with the value $Q_0 \simeq 14 \times 10^{-24}\,\text{cm}^2$. derived by means of (9) from the quadrupole moment Q determined from the atomic hyperfine structure (BROWN and TOMBOULIAN, 1952). Another estimate of deformations for nuclei in this region may be obtained from measured lifetimes of rotational states in even-even nuclei, which yield $Q_0 \simeq 8 \times 10^{-24}\,\text{cm}^2$ (cf. B.-M., Table XXVII). The three estimates of Q_0 are of the same order of magnitude; it does not seem excluded that the differences may be attributed to experimental uncertainties.

6. Excitation of Second Rotational State in Ta.

The rotational interpretation of the first excited state in Ta implies the existence of a second rotational state with spin 11/2, and an energy of 20/9 times that of the first state. Since this state may also be excited in an $E2$ transition of collective type, it is expected to have an appreciable cross-section for Coulomb excitation. However, the yield of the higher energy γ-ray is expected to be considerably smaller than that for the 137 keV γ-line for several reasons: partly, the value of $B_e(2)$ for the former transition is about four times smaller than for the latter (cf. (7) and

* A level scheme for Ta^{181} has been given (GOLDHABER and HILL, 1952) according to which the first excited state of 136 keV is assigned a spin of 7/2. However, it does not seem inconsistent with the data to identify this level with the one observed in the Coulomb excitation process, and to assign it a spin of 9/2.

(8)); partly, the higher value of ξ implies smaller values of $g_2(\xi)$; finally, the de-excitation of the higher state may proceed either by a cascade through the 137 keV state or by a cross-over transition to the ground state, of which only the latter could be detected with the available resolution.

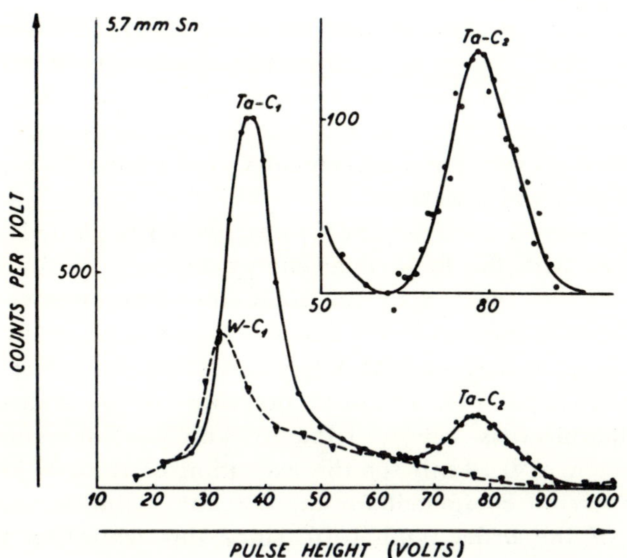

Fig. 4. Pulse spectra for tantalum and tungsten, measured at a proton energy of 2.1 MeV and with a 5.7 mm Sn-absorber. In the insert, the difference between the yields for Ta and *W* is plotted in the region of the Ta-C_2 line, which corresponds to the second excited state in Ta. No other γ-lines were found in Ta.

The importance of this level as a test of the rotational spectrum led us to undertake a careful search for higher energy γ-rays. In order to determine the background radiation, measurements were also made with targets of the neighbouring element *W*, since, in the relevant energy range, *W* is not expected to give rise to appreciable γ-radiation resulting from Coulomb excitation (cf. the following paragraph).

The comparison between the γ-spectra obtained with Ta and *W* targets, for a proton energy of 2.1 MeV, is shown in Fig. 4 and clearly reveals a higher energy line in Ta (the Ta-C_2 peak).

In order to exclude the possibility of effects arising from the coincidence of two C_1-quanta or from the distortion of the background due to absorbers, the spectrum was measured with a number of different strong absorbers. The interpretation of the yield from W-targets as representing the background for higher energies was further supported by the observation that the Ta and W yields in all cases coincided at a point somewhat above the Ta-C_1 peak.

Subtracting the background from the observed Ta-spectrum, the Ta-C_2 peak shown in the insert to Fig. 4 was obtained. The energy of the Ta-C_2 line was determined to be 300 ± 10 keV, which corresponds to a ratio of 2.2 for the C_2 and C_1 energies. This ratio is in very good agreement with the ratio 20/9 predicted for the rotational spectrum.

The observed yield for the C_2-line, which is plotted in Fig. 3, also agrees with the theoretical energy-dependence for $E2$ Coulomb excitation, shown by the solid line. The cross-section for the emission of the 300 keV γ-ray was found to be about 0.02 mb for a proton energy of 2 MeV.

From (3) and (8), assuming the value of Q_0 deduced from the excitation cross-section of the 137 keV line, one calculates a cross-section of 0.10 mb for the excitation of the 300 keV level. The conversion coefficient for the 300 keV γ-line is expected to be only of the order 0.05—0.10, since this transition is of $E2$ type. The comparison between the excitation cross-section and the γ-yield therefore indicates that about 20 $^0/_0$ of the de-excitations take place via the cross-over transition to the ground state.

The relatively large probability for cross-over transitions is characteristic of the rotational spectrum, arising from the strong enhancement of the $E2$ radiation (cf. B.-M. § VIIc.ii).

From the branching ratio one can obtain an estimate of the $M1$ transition probability for the $\Delta I = 1$ radiative transition within the rotational family. This transition probability can also be estimated from the static magnetic moment of the ground state (B.-M. § VIIc.ii). The value obtained in this way is somewhat smaller than that indicated by the branching ratio, but the estimate is very sensitive to the value of the magnetic moment. A precision measurement of this moment would thus be of interest.

7. Coulomb Excitation of *W*.

In even-even nuclei with large deformations, the first excited (2+) states should show up strongly in the Coulomb excitation process. As a first example, $_{74}W$ was selected since it consists predominantly of even isotopes, and since the excitation energies in this region indicated that it would be possible to resolve the nuclear radiation from the *K* X-rays.

The *W*-spectrum is shown in Fig. 5 and clearly exhibits the well separated peaks resulting from the X-radiation (*W-K*) and the Coulomb excitation (W-C_1). The W-C_1 peak was observed

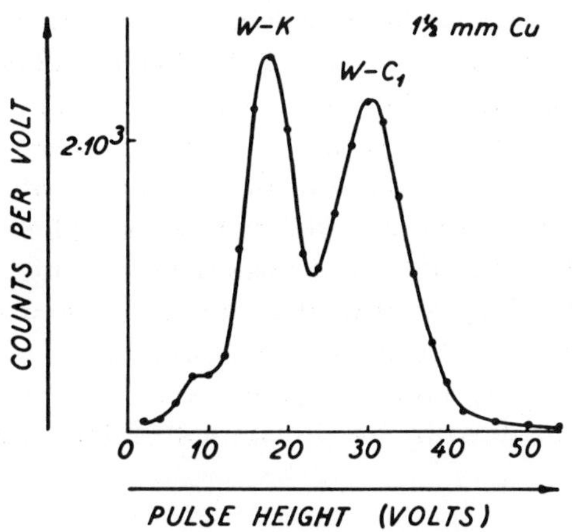

Fig. 5. Pulse spectrum for tungsten, measured at a proton energy of 1.75 MeV and with a 1.5 mm Cu absorber.

to be composite, presumably arising from the different isotopes, since its position and shape could be altered by the use of absorbers.

As an average energy of the W-C_1 lines, we obtained about 115 keV, which is consistent with the known excitation energies of 102 keV for W^{180} and 123 keV for W^{186} (cf. Scharff-Goldhaber, 1953).

The cross-section for γ-emission at a proton energy of 2 MeV is found to be about 1.5 mb, assuming the theoretical energy dependence (3). Using an average total internal conversion coefficient of 2 for these $E2$ transitions, one obtains from (8) a deformation of $Q_0 \simeq 7 \times 10^{-24}$ cm². which is just of the same magnitude as that observed in the excitation of the neighbouring element Ta. This value of Q_0 is also in agreement with that derived from the measured lifetimes of excited states of even-even nuclei in this region.

8. Yield of *K* X-rays.

As a by-product of these investigations, we obtained the yield of X-rays resulting from the ejection of *K*-electrons by protons on Ta. The results are shown in Fig. 3. The X-ray yield has been corrected for internal conversion of the Ta-C_1 line, which amounts to about 20 % of the total X-ray yield. The solid line represents the theoretical energy dependence (HENNEBERG, 1933), which is seen to agree rather well with the measurements. The absolute yield is found to be about a factor of two smaller than predicted by the theory, but it is not yet clear to what extent this discrepancy is to be attributed to experimental uncertainties or to the approximations involved in the theory. A comparison of the Ta-*K* and Pb-*K* yields for a proton energy of 1.75 MeV was found to be in good agreement with the theoretical *Z*-dependence.

9. Conclusions.

One may summarize as the main conclusions of these investigations:

1) The feasibility of nuclear excitation by the electric field of bombarding particles has been confirmed.

2) Over the energy region investigated the theoretical energy dependence of the cross-sections has been verified.

3) The first two rotational levels in Ta have been found at the predicted energies.

4) The large cross-sections for excitation of the levels in Ta and *W* confirm the collective character of the excitation.

5) The nuclear deformations deduced from the measured

cross-sections are of the same magnitude as those derived from spectroscopic evidence and measured lifetimes.

The use of the electric field of charged particles to excite nuclear levels provides the possibility of studying the spectra of a wide variety of nuclei. Preliminary experiments have also exhibited the effect in other elements, and further investigations of the Coulomb excitation processes are in progress.

This work has been performed at the Institute for Theoretical Physics, University of Copenhagen. We wish to thank Professor Niels Bohr for his continued interest in the experiments and for the good working facilities provided at the Institute. We are also very grateful for the many stimulating discussions on the subject, which we have had with A. Bohr and B. Mottelson, in close co-operation with whom the theoretical parts of this paper have been written. To A. Winther and K. Alder we are indebted for making results of their calculations available to us in advance of publication. Our thanks are further due B. Madsen for developing the electronic equipment used in the investigations, and J. Bjerregaard for aid in the experiments.

One of us (Č. Z.), on leave from "J. Stefan" Institute of Physics, Ljubljana, Yugoslavia, is indebted to the Rask-Oersted Foundation for a grant enabling him to take part in the work in Copenhagen, and to the Ljubljana Institute for financial support.

References.

F. Asaro and I. Perkman (1953); submitted to the Physical Review.
S. W. Barnes and P. W. Aradine (1939); Phys. Rev. **55**, 50.
J. M. Blatt and V. F. Weisskopf (1952). Theoretical Nuclear Physics. J. Wiley and Sons, New York and London.
A. Bohr (1952); Dan. Mat. Fys. Medd. **26**, no. 14.
A. Bohr and B. R. Mottelson (1952); Physica **18**, 1066.
A. Bohr and B. R. Mottelson (1953); Dan. Mat. Fys. Medd. **27**, no. 16.
A. Bohr and B. R. Mottelson (1953a); Phys. Rev. **89**, 316.
A. Bohr and B. R. Mottelson (1953b); Phys. Rev. **90**, 717.
N. Bohr (1948); Dan. Mat. Fys. Medd. **18**, no. 8.
W. Bothe and H. Fränz (1928); ZS. f. Phys. **52**, 466.
G. Breit, M. H. Hull and R. L. Gluckstern (1952); Phys. Rev. **87**, 74.

B. M. Brown and D. H. Tomboulian (1952); Phys. Rev. **88**, 1158.
J. Chadwick (1913); Phil. Mag. **25**, 193.
R. B. Day and T. Huus (1952); Phys. Rev. **85**, 761.
C. Y. Fan (1952); Phys. Rev. **87**, 252.
K. Ford (1953); Phys. Rev. **90**, 29.
M. Goldhaber and A. W. Sunyar (1951); Phys. Rev. **83**, 906.
M. Goldhaber and R. D. Hill (1952); Revs. Mod. Phys. **24**, 179.
W. Henneberg (1933); ZS. f. Phys. **86**, 592.
R. Huby and H. C. Newns (1951); Proc. Phys. Soc. Lond. A **64**, 619.
J. Lindhard and M. Scharff (1953); Dan. Mat. Fys. Medd. **27**, no. 15.
M. Livingston, F. Genevese and E. J. Konopinski (1947); Phys. Rev. **51**, 835.
J. E. Mack (1950); Revs. Mod. Phys. **22**, 64.
C. J. Mullin and E. Guth (1951); Phys. Rev. **82**, 141.
N. F. Ramsey (1951); Phys. Rev. **83**, 659.
J. R. Risser, K. Lark-Horowitz and R. N. Smith (1940); Phys. Rev. **57**, 355.
M. E. Rose, G. H. Goertzel, B. I. Spinrad, J. Horr and P. Strong (1951); Phys. Rev. **83**, 79.
G. Scharff-Goldhaber (1953); Phys. Rev. **90**, 587.
K. A. Ter-Martirosyan (1952); Journ. Exp. and Theor. Phys. (U.S.S.R.) **22**, 284.
V. F. Weisskopf (1938); Phys. Rev. **53**, 1018.

Indleveret til selskabet den 23. juni 1953.
Færdig fra trykkeriet den 9. juli 1953.

Excitation of Heavy Nuclei by the Electric Field of Low-Energy Protons*,†

Clyde L. McClelland and Clark Goodman
Department of Physics and Laboratory for Nuclear Science, Massachusetts Institute of Technology, Cambridge, Massachusetts
(Received June 16, 1953)

SEVERAL theoretical predictions[1] have been made that heavy charged particles may excite nuclei even though the collision energy is much too small to allow appreciable wave penetration of the Coulomb barrier. Nuclear forces do not take part in this type of inelastic scattering—the excitation of the nucleus results solely from the interaction of the Coulomb fields.

Barnes and Aradine[2] observed the reaction $In^{115}(p,p')In^{115*}$ (4.1 hr) with protons of energy 5.8 Mev and estimated a cross section of about 10^{-29} cm². However, there is some question as to whether this reaction involved barrier penetration rather than purely Coulomb excitation.

During the summer of 1952 we observed what appeared to be excitation of a low-lying level in tantalum[3] using protons of such low energy (1.4 to 1.8 Mev) that formation of the compound nucleus is entirely negligible. We observed the prompt mono-energetic gamma rays rather than induced radioactivity. More recently we have examined[4] Ta, Pt, Au, Hg, Tl, and Bi using bombarding energies in the range of proton energies, $E_p=1.4$ to 2.6 Mev.

Our experimental setup is quite simple. Monoenergetic protons from the Rockefeller electrostatic generator strike a metallic target (thick to protons) adjacent to which is the crystal of a NaI(Tl) spectrometer using an RCA 5819 and single channel discriminator. Pulse-height distribution curves are obtained in the usual manner. Co^{57}, Na^{22}, Cs^{137}, Hg^{203}, and Pb (K x-rays) were used for energy calibration.

Figure 1 summarizes the results for three different tantalum targets and a single gold target at proton energies of 1.42 and 1.48 Mev. Each Ta target shows a strong photopeak at 138±5 kev, well resolved from the K x-rays to the left (not plotted). To determine whether this gamma ray was due to impurities in the target or to reactions with scattered protons in the target assembly rather than to the 136-kev first excited state[5] in ${}_{73}Ta^{181}$, the following steps were taken:

(1) four targets of Ta from two sources were tested;

(2) the targets were cleaned successively with steel wool, emery, acid, water, and alcohol;

(3) iron, vanadium, and carbon targets were tested, since these are the most common bulk impurities in commercial Ta;

(4) an Au target was tested in a similar manner.

The pulse-height distributions from the Fe, V, C, and Au (see Fig. 1) targets were such that these substances could not account for the 138-kev peak in Ta. Our spectrometer would not reveal the 77-kev gammas from ${}_{79}Au^{197}$ (assuming they were produced) in the presence of the K x-rays. The possibility that the 138-kev line is the result of the pile-up of pulses in the NaI(Tl) crystal is also eliminated by the absence of a similar line in Au at about twice the K x-ray energy.

Another possibility suggested by Professor H. Bethe[4] is that the 138-kev peak is the result of a superposition of K x-rays from the double ionization of the K shell in Ta. We have measured the absorption of the 138-kev line relative to the K x-rays in Pb (~20 mils) and find that the x-rays are attenuated much more markedly than the 138-kev peak. Hence, the 138-kev line does not appear to be composed of lower energy quanta. What is more significant, however, is that the shape of the 138-kev peak was retained after attenuation in the Pb—the effect to be expected for a line spectrum.

The tentative conclusion is that these are $M1$ gammas from the transition $g_{7/2}$, first excited state, to $g_{7/2}$ ground state of ${}_{73}Ta^{181}$ induced by Coulombian excitation of the nucleus. The basis for discarding barrier penetration at the proton energies used is that the cross section for formation of the compound nucleus alone is

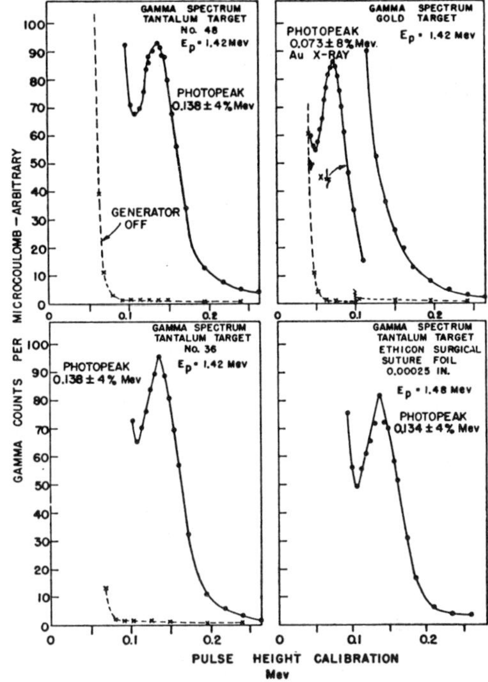

FIG. 1. Pulse-height spectra for Ta and Au. The pulse height has been calibrated with 511-kev annihilation radiation from Na^{22}, and the energy of the gamma ray may be read directly from the position of its photopeak.

only about 10^{-39} cm² at $E_p=1.5$ Mev. It would be even smaller for $\sigma(p,p')$.

Because the yield of gammas has been measured only for thick Ta targets, only rough estimates can be made of the observed cross section. We calculate this to be about 0.4 millibarn and essentially constant for $E_p=1.4$ to 1.6 Mev. However, assuming 80 percent internal conversion of the 138-kev gamma, the cross section would be about 8 millibarns.

Naturally we were reluctant to accept the above conclusions without additional experimental evidence. Dr. de-Shalit suggested[6] that we bombard platinum since there are two low-lying levels:[5] a $p_{1/2}$ at 97 kev and an $f_{5/2}$ at 126 kev above the $p_{1/2}$ ground state in ${}_{78}Pt^{195}$. We have done this, with the results shown in Fig. 2. Clearly there is a peak at about 127 kev for $E_p=1.64$, 1.78, and 2.0 Mev. There is also evidence at these energies, and at 1.51 Mev as well, for some radiation between 90 and 100 kev, since otherwise the curve should dip much more in this region (see corresponding section of Au data in Fig. 1). We, therefore, tentatively conclude that nuclear gammas from ${}_{78}Pt^{195}$ have been activated by the purely Coulombian process. We are unable at this time to quote a cross section for Pt, but it certainly appears to be substantially less than for Ta for the same proton energy.

The results with Au, Hg, Tl, and Bi are more preliminary than those for Ta and Pt. The first measurements with Au (Fig. 1) show no evidence of the 191-kev radiation[5] from the second to the first excited state. More recent measurements with an improved resolution crystal (Harshaw) show two small peaks which we very tentatively would attribute to 250- and 450-kev gammas. The improved resolution also revealed another higher peak in Ta (not evident in Fig. 1) which increased in height as the proton energy was increased. We would tentatively assign this an energy of about 500 kev. Conceivably this radiation may be from the next

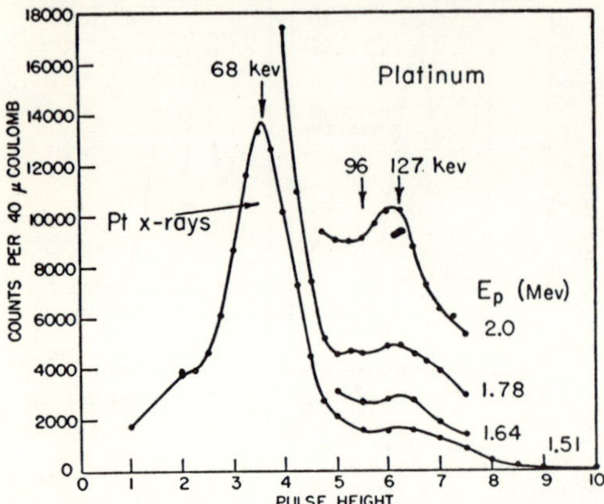

FIG. 2. Pulse-height spectrum of Pt at four proton energies E_p. A small adjustment has been made in the horizontal position of the data taken at $E_p = 2.0$ Mev in order to correct for electronic drift in the spectrometer.

higher level in $_{73}Ta^{181}$ decaying to the ground state. This would imply the presence of 345-kev radiation as well, and some evidence for this has been observed.

After studying the Au radiation, we made an amalgam and examined the increment due to Hg. There is clearly a reproducible break in the curve at an energy of about 200 kev which increases with proton energy. At least two isotopes of Hg are reported to have levels which could account for radiation of this energy.

We examined Tl (cp) and observed a well-resolved peak at 380 ± 10 kev. This radiation does not correspond to transitions between known levels to the best of our knowledge. The 280-kev radiation, from Tl^{203}, if present, would have been obscured by the Compton peak from the 380 kev and the high-energy tail of the x-rays. Preliminary tests on Bi as yet have yielded no significant results.

In the future we plan to use only thin targets since these are not only essential for accurate cross-section measurements, but also result in a greatly reduced x-ray background. We can also further reduce the x-ray background by critical absorption foils, a technique which we have already used successfully to a limited extent.

We wish to express our sincere thanks to Professors Deutsch, Feshbach, and Weisskopf, and Dr. de-Shalit and Dr. Ajzenberg for their advice and encouragement.

* This work was supported by the Bureau of Ships and the U. S. Office of Naval Research.

Reprinted with corrections by the authors.

[1] V. F. Weisskopf, Phys. Rev. **53**, 1018 (1938); C. J. Mullin and E. Guth, Phys. Rev. **82**, 141–155 (1951); R. Huby and H. C. Newns, Proc. Phys. Soc. (London) **A64**, 619–632 (1951); K. A. Ter-Martirosian, Zhur. Eksptl. i Teort. Fiz. **22**, 284 (1952); A. Bohr and B. R. Mottelson [see U. S. Office of Naval Research Scientific Notes, May 1, 1953, p. 93 (unpublished)].

[2] S. W. Barnes and P. W. Aradine, Phys. Rev. **55**, 50–52 (1939).

[3] C. L. McClelland, S.M. Thesis, Massachusetts Institute of Technology, August 15, 1952 (unpublished).

[4] These results were reported at the Medium Energy Conference held at the University of Pittsburgh, June 4–6, 1953 (unpublished).

[5] M. Goldhaber and R. D. Hill, Revs. Modern Phys. **24**, 225 (1952).

[6] A. de-Shalit (private communication).

Reprinted from THE PHYSICAL REVIEW, Vol. 96, No. 1, 237, October 1, 1954
Printed in U. S. A.

Theory of Coulomb Excitation

K. ALDER* AND A. WINTHER
Institute for Theoretical Physics, University of Copenhagen, Copenhagen, Denmark

(Received August 11, 1954)

THE theory of Coulomb excitation has been considered earlier by many authors. Especially helpful for the interpretation of the experiments is a semiclassical calculation by Ter-Martirosyan.[1,2] In this theory the trajectory of the impinging particle is described by a classical hyperbolic orbit.

We have obtained a somewhat more accurate description of the process by approximating the Coulomb wave function, entering in the quantum-mechanical treatment, by means of the WBK method. The results of this calculation are very similar to the results of the semiclassical theory. The cross sections for electric dipole and electric quadrupole excitation are:

$$\sigma(E1)=\frac{2\pi^2 Z_1^2 e^2}{9\hbar^2 v_i^2}B(E1)g_{E1}(\xi), \tag{1}$$

$$\sigma(E2)=\frac{2\pi^2 m_1^2 v_f^2}{25 Z_2^2 e^2 \hbar^2}B(E2)g_{E2}(\xi), \tag{2}$$

with the following new definition of the parameter:

$$\xi=\alpha_f-\alpha_i=\frac{Z_1Z_2e^2}{\hbar}\left(\frac{1}{v_f}-\frac{1}{v_i}\right), \tag{3}$$

where Z_1e and Z_2e are the charges of the impinging projectile and the nucleus, respectively. The initial and final relative velocities are denoted by v_i and v_f, while m is the reduced mass. $B(E\lambda)$ is the reduced transition probability for electric 2^λ-pole radiation in the notation of Bohr and Mottelson.[3] The functions $g_{E\lambda}(\xi)$ are the same as those entering in the semiclassical expression.[1,2] For small excitation energies ΔE, the formulas (1), (2), and (3) reduce to the corresponding semiclassical formulas. We have then $v_f\sim v_i\sim v$, and

$$\xi=\frac{Z_1Z_2e^2}{\hbar v}\frac{\Delta E}{2E}. \tag{4}$$

The expression (1) for electric dipole excitation has been compared with the exact quantum mechanical formula[4]

$$\sigma(E1)=\frac{128\pi^4}{9}\frac{Z_1^2e^2}{\hbar^2 v_i^2}B(E1)\left(\frac{\alpha_f}{\alpha_i}\right)^2 \times\frac{e^{2\pi\alpha_i}}{(e^{2\pi\alpha_i}-1)(e^{2\pi\alpha_f}-1)}\frac{d}{dx_0}|F(x_0)|^2,$$

where

$$x_0=-4\alpha_i\alpha_f/(\alpha_i-\alpha_f)^2, \tag{5}$$

and

$$F(x_0)={}_2F_1(-i\alpha_i,-i\alpha_f,1,x_0)$$

in terms of ${}_2F_1$, the ordinary hypergeometric function. In the range $\alpha_i\geqslant 1$ and $0\leqslant\alpha_f-\alpha_i=\xi\leqslant 1$, the WBK expression (1) agrees within 2 percent with the quantum mechanical calculation, which represents a considerable improvement over the semiclassical treatment.

In the case of the quadrupole excitation, no exact quantum mechanical treatment has been given, but the adequacy of the WBK treatment for the $E1$ case, suggests that (2) should also represent a good approximation when the effects resulting from penetration of the bombarding particle into the nucleus itself can be neglected.

Indeed, it is found that (2) satisfactorily represents all the experimental yield curves available to us at the present time.

Other kinds of electromagnetic excitations have also been considered and a detailed account of all the calculations will be published in the proceedings of the Danish Academy.[5,6]

* Theoretical Study Group, European Council for Nuclear Research. On leave from Physikalisches Institut der Eidgenössischen Technischen Hochschule, Zürich, Switzerland.

[1] K. A. Ter-Martirosyan, J. Exptl. Theoret. Phys. (U.S.S.R.) 22, 284 (1952).

[2] K. Alder and A. Winther, Phys. Rev. **91**, 1578 (1953).

[3] A. Bohr and B. Mottelson, Kgl. Danske Videnskab. Selskab, Mat.-fys. Medd. **27**, No. 16 (1953).

[4] R. Huby and H. C. Newns, Proc. Phys. Soc. (London) **A64**, 619 (1951); C. T. Mullin and E. Guth, Phys. Rev. **82**, 141 (1951).

[5] K. Alder and A. Winther, Kgl. Danske Videnskab. Selskab (to be published).

[6] A review article on the theory of Coulomb excitation, containing also an analysis of experimental data, is being prepared in collaboration with A. Bohr, T. Huus, B. Mottelson, and C. Zupančič.

PHYSICAL REVIEW VOLUME 96, NUMBER 2 OCTOBER 15, 1954

Coulomb Excitation Process in the Lighter Odd-Mass Nuclei*

G. M. TEMMER AND N. P. HEYDENBURG
Department of Terrestrial Magnetism, Carnegie Institution of Washington, Washington, D. C.
(Received June 21, 1954)

We have studied the Coulomb excitation functions for thin targets of F^{19}, Na^{23}, Ti^{47}, Mn^{55}, and Ge^{73}, and for thick targets of V^{51} and Fe^{57}, with alpha particles up to 3.5 Mev; the energy levels excited in these nuclei are at 113 and 196 kev, 446 kev, 160 kev, 128 kev, 68 kev, 320 kev, and 137 kev, respectively. The de-excitation gamma rays from these levels to the ground states were detected except for Fe^{57}, where a 123-kev gamma ray is predominantly emitted. In addition, we have excited the 182-kev level in Zn^{67}, whose de-excitation takes place partly by cascade through the 92-kev first excited state. In the cases of F^{19} and Na^{23} we were able to compare directly the relative contributions of Coulomb excitation and compound nucleus formation by means of the $(\alpha,p\gamma)$ reactions taking place via the same compound nuclei. In all cases the excitation curves are in fair agreement with the theoretical $E2$ curves at the lower energies, but show definite deviations in the direction of too much excitation at the higher energies, pointing to some resonant compound-inelastic contribution as well as possible penetration effects not accounted for by the classical theory. The transition probabilities of all transitions are about one-tenth of those in the rare earth region.

A. INTRODUCTION

ONE of the advantages of using alpha particles for the Coulomb excitation of low-lying nuclear energy levels[1–4] which became immediately apparent was the possibility of studying elements of low atomic number; our early results with Mn^{55} [1] encouraged us to pursue the present investigation. There are several reasons why ions heavier than protons are much better suited for the excitation of low-Z elements, and it might be useful to list them:

(a) The relation $2\eta=2Z_1Z_2e^2/\hbar v\gg1$, which must hold in order to justify the classical orbit calculations,[5,6] is well satisfied to much lower values of Z_2, since Z_1/v (projectile charge over incident velocity) is larger for a given bombarding energy; for alpha particles this amounts to a factor of 4 over protons. As an example, $2\eta=8.00$ for 3-Mev alphas on ${}_{11}Na^{23}$.

(b) Because of the higher charge, the Coulomb barrier is higher and prevents appreciable interference of compound nucleus formation with the process of interest here until one goes to higher bombarding energies; we are thus allowed a range of energies over which Coulomb excitation is essentially the only mechanism contributing to the excitation of nuclear energy levels. This is important because (aside from lifetime determinations) Coulomb excitation is probably the best understood process today from which to extract nuclear transition matrix elements directly.

(c) Troublesome high-energy gamma radiation from light targets is almost entirely absent with alpha particles, whereas protons produce appreciable capture radiation in the (p,γ) process (E_γ of the order of 7 Mev) up to $Z\sim40$, thus making the detection of low-energy gamma rays from the deexcitation of low-lying levels difficult if not impossible.

(d) Many of the nuclei between $Z=20$ and $Z=50$ have (p,n) thresholds lying between 1 and 2 Mev[7]; neutron background and induced positron activities then complicate the problem considerably. In some cases we have found the gamma radiation from the $(p,n\gamma)$ reaction, i.e., from the first excited state of the nucleus $(A, Z+1)$ when bombarding with protons. This is an interesting approach in itself but does not concern us in the present investigation.

(e) Targets which are available only as compounds, such as oxides and chlorides, can be used without difficulty when bombarding with alpha particles for the reasons discussed under (c) and (d); furthermore, the problem of finding a suitably inert backing material for thin target studies is minimized. Nickel turned out to be satisfactory in this respect, but not for protons (see Mn^{55} below).

(f) Finally, general background gamma radiation from the electrostatic generator is many times higher with protons than it is with alpha particles, mainly for the same reasons as listed under (c) and (d) above as applied to the walls of the accelerator tube.

It is for these reasons that we have experienced great difficulties in the few instances where we have attempted to measure Coulomb excitation cross sections with protons; this turns out to be necessary in order to determine the multipolarity of the transitions involved, when making use of the method suggested by Bjerregaard and Huus.[8] The lightest element so far where proton bombardment yielded useful results for us was rhodium ($Z=45$).

* A preliminary account of some of the results in this paper were presented at the Washington meeting of the American Physical Society [Phys. Rev. 95, 629 (1954)].

1 G. M. Temmer and N. P. Heydenburg, Phys. Rev. **94**, 351 (1954).
2 N. P. Heydenburg and G. M. Temmer, Phys. Rev. **94**, 906 (1954).
3 G. M. Temmer and N. P. Heydenburg, Phys. Rev. **94**, 1399 (1954).
4 N. P. Heydenburg and G. M. Temmer, Phys. Rev. **95**, 861 (1954).
5 K. A. Ter-Martirosyan, J. Exptl. Theoret. Phys. (U.S.S.R.) **22**, 284 (1952).
6 K. Alder and A. Winther, Phys. Rev. **91**, 1578 (1953).
7 C. C. Trail and C. H. Johnson, Phys. Rev. **91**, 474 (1953).
8 J. H. Bjerregaard and T. Huus, Phys. Rev. **94**, 204 (1954).

The largest cross sections for Coulomb excitation were found in F^{19} (196 kev), Na^{23} (446 kev), Ti^{47} (160 kev), Mn^{55} (128 kev), and Ge^{73} (68 kev). For these we were able to perform thin-target experiments successfully. Large cross sections do not necessarily imply large matrix elements for the transitions involved; in fact, the latter turn out to be small compared to the 137-kev transition in Ta^{181}. The cross sections are large only for strictly kinematical reasons having to do with the Coulomb excitation mechanism and not with the intrinsic excitability of nuclei under study.

B. EXPERIMENTAL DETAILS

The main features of our experimental setup have already been described.[9] In some cases we have used a 2-in.×2-in. well-type NaI(Tl) crystal with a $\frac{3}{4}$-in. diameter hole $1\frac{1}{2}$ in. deep, the target being located at the bottom of the well.[10] In this arrangement, we approach 100-percent efficiency (including geometry) for radiations up to about 200 kev (see Fig. 1, reference 9). We were thus able to measure cross sections down to a fraction of a microbarn.

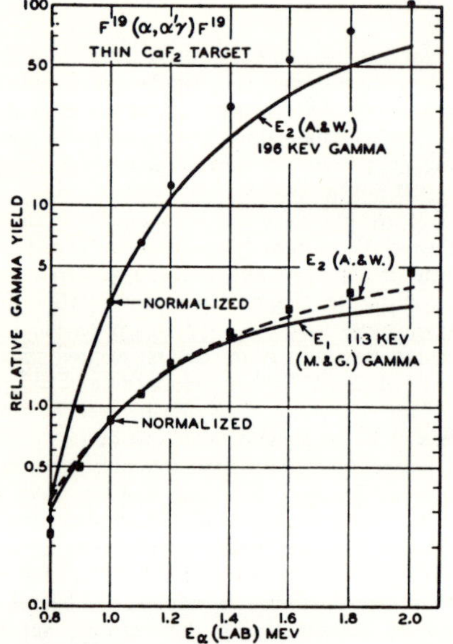

FIG. 1. Coulomb excitation by alpha particles of 113-kev and 196-kev levels in thin F^{19} target. Solid curves are theoretical $E1$ (113-kev) and $E2$ (196-kev) curves, according to Mullin and Guth (reference 13) and Alder and Winther (reference 6), respectively. Dashed curve is theoretical $E2$ curve for comparison. Experimental points are normalized to the curves at 1 Mev. For higher-energy region, see Figs. 3 and 4 of reference 9.

Note added in proof.—Recent calculations by K. Alder and A. Winther (to be published) using the WKB approximation for the $E1$ case have produced two major modifications: (a) the new definition of the parameter ξ (as discussed above); (b) the introduction of v_f (final relative velocity) in the place of v (unspecified) in their expression for the Coulomb excitation cross section (see reference 6). Modification (a) has been incorporated into this paper; modification (b) was not available in time. Suffice it to state that when this correction is applied to our results, essentially *all* the discrepancies between theory and experiment at the higher energies, seen in Figs. 3, 5, 6, 7, and 10, are removed. This correction is of course the larger, the greater ΔE. Only ~40 percent of the discrepancy disappears in the case of the 196-kev level in F^{19}.

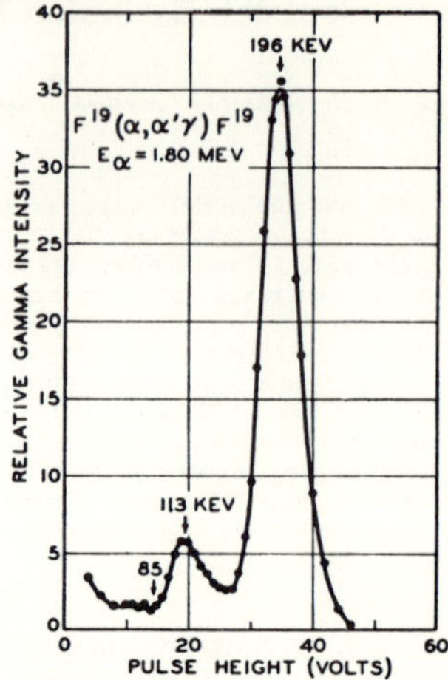

FIG. 2. Pulse-height distribution of gamma radiation from thin F^{19} target under 1.8-Mev alpha bombardment, using a well crystal. Note the absence of 83-kev cascade radiation. Note also that the peak below the 113-kev gamma ray (Fig. 5 of reference 9) has been eliminated.

We prepared our thin targets by vacuum evaporation onto nickel backing. In the cases of Ti^{47}, Fe^{57}, Ge^{73}, and Zn^{67} we were able to obtain enriched isotopes.[11] We have no good measure of the target thicknesses for the thin targets, but we ascertained that they were thin enough for our purposes (~30 kev) by preparing targets yielding fewer gamma rays and comparing the shapes of the excitation curves. (Targets which are too thick produce characteristically steeper curves.)

C. EXCITATION FUNCTIONS

(a) F^{19}

We have previously reported some thin-target excitation functions for both the 113-kev and 196-kev levels of F^{19} under alpha particle bombardment.[9] In this case, we know the target to be no thicker than the width of the narrowest resonance observed in the $(\alpha,p\gamma)$ reaction (see Fig. 6, reference 9). Some additional work, extending the energy range to lower values by using our high-efficiency well crystal, is plotted in Fig. 1. These curves are in essential agreement with the work of the group

[9] N. P. Heydenburg and G. M. Temmer, Phys. Rev. 94, 1252 (1954).

[10] We are indebted to Dr. P. H. Abelson for the loan of this crystal.

[11] From the Oak Ridge National Laboratory, Oak Ridge, Tennessee.

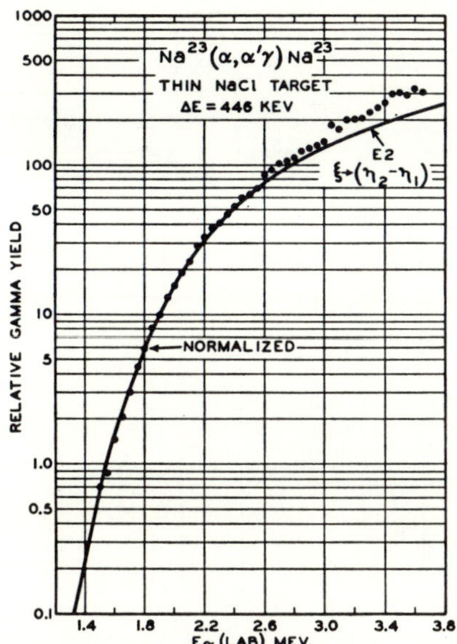

FIG. 3. Coulomb excitation function for thin NaCl target. Level energy is 446 kev. Solid curve is the theoretical $E2$ function (reference 6). Experimental points normalized at 1.8 Mev (see note added in proof, Fig. 1).

at the California Institute of Technology.[12] The theoretical curve for the electric dipole ($E1$) case is calculated according to Mullin and Guth;[13] the curves for the electric quadrupole ($E2$) case are calculated according to the expressions given by Alder and Winther;[6] it should be noted that the quantity $\eta_2-\eta_1$ ($\eta_1=Z_1Z_2e^2/\hbar v_1$; $\eta_2=Z_1Z_2e^2/\hbar v_2$, v_1 and v_2 being initial and final projectile velocities, respectively) rather than

$$\xi=(\Delta E/2E)(Z_1Z_2e^2/\hbar v)$$

[which is the limiting value of $(\eta_2-\eta_1)$ for $\Delta E/E\ll 1$, ΔE being the excitation energy, E the bombarding energy] has been used in all excitation curves, in accordance with prevailing theoretical preference. This is justified mainly by the results of the exact calculations for the electric dipole case;[13] furthermore, the Born approximation calculation for $E2$ employing $(\eta_2-\eta_1)$ as parameter[14] seems to agree with the exact (numerical) $E2$ calculations.[6] Incidentally, the fit with experiment is considerably improved when using the latter parameter. (Note, however, the discrepancy from theory at the higher energies.) Figure 2 shows the pulse-height distribution obtained from a thin CaF_2 target at 1.8-Mev alpha-particle energy, using the well crystal described above. Note that the satellite peak (see Fig. 5, reference 9) to the left of the 113-kev peak has disappeared. This peak seems to have also troubled Cal Tech[15] and Cavendish workers[16] when studying this reaction. The absence of cascade radiation from the 196-kev level to the 113-kev level is now very evident.

[12] Sherr, Li, and Christy, Phys. Rev. **94**, 1076 (1954).
[13] C. J. Mullin and E. Guth, Phys. Rev. **82**, 141 (1951).
[14] C. J. Mullin (private communication).

(b) Na^{23}

The excitation curve for the yield of the 446-kev gamma radiation from a thin NaCl target is shown in Fig. 3. The experimental points have arbitrarily been normalized to the theoretical $E2$ curve at 1.8 Mev. The agreement over a factor of 100 in cross section is seen to be excellent, leaving little doubt as to the $E2$ nature of the excitation process.† The disagreement with theory at the higher energies can be attributed to some compound nucleus formation (see note added in proof, Fig. 1). With the same target we were able to obtain the excitation function for the $Na^{23}(\alpha,p\gamma)Mg^{26}$ reaction as measured by the yield of the 1.83-Mev gamma radiation resulting from the deexcitation of the first excited state of Mg^{26} (see note added in proof, Fig. 1). This

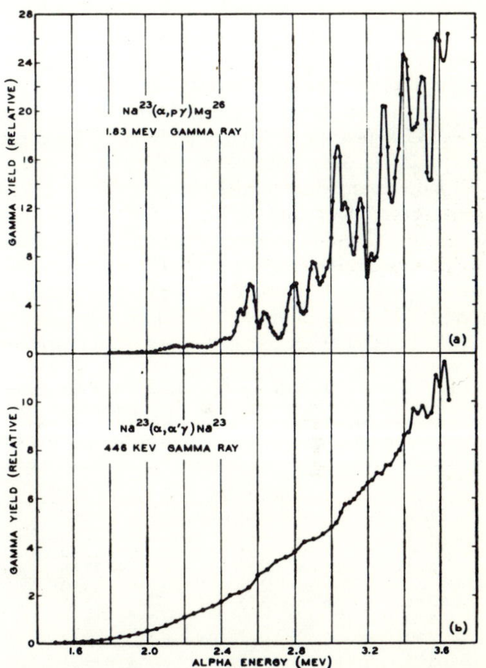

FIG. 4. (a) Excitation function for the reaction $Na^{23}(\alpha,p\gamma)Mg^{26}$, as detected by the 1.83-Mev gamma ray from Mg^{26}. Resonances are levels in the compound nucleus Al^{27}. (b) Excitation function for the reaction $Na^{23}(\alpha,\alpha'\gamma)Na^{23}$. Level energy is 446 kev. Same data as in Fig. 3. Thin NaCl target.

[15] Peterson, Barnes, Fowler, and Lauritsen, Phys. Rev. **93**, 951 (1954).
[16] G. A. Jones and D. H. Wilkinson, Phil. Mag. **45**, 230 (1954).
† *Note added in proof.*—The possibility that some of these transitions are of the $E1$ type cannot be completely ruled out; e.g., in the case of Na^{23} our data fall within ±5 percent of the theoretical $E1$ curve.

reaction takes place via the compound nucleus Al^{27}. The many peaks in Fig. 4(a) correspond to levels in that nucleus, and also show that we are dealing with a thin target. In Table I we list the resonances of $Na^{23}+\alpha$ and the corresponding levels in Al^{27}. The comparison with the $Na^{23}(\alpha,\alpha'\gamma)Na^{23}$ reaction plotted in Fig. 4(b) strikingly demonstrates the interplay of Coulomb excitation and compound nucleus formation. At the high-energy end the latter excitation curve begins to show irregularities which can be correlated with levels in the compound nucleus Al^{27} as indicated in the upper curve. One can say that 3.6 Mev are about 90 percent Coulomb excitation and 10 percent compound-inelastic excitation of the 446-kev state. Also, there is undoubtedly some nonresonant excitation over and above the classically expected Coulomb excitation at the higher energies because of the gradual breakdown of the assumption $2Z_1Z_2e^2/\hbar v \gg 1$ used in the derivation of the classical expressions.[6]

(c) Ti^{47}

The 160-kev gamma ray we reported earlier[1] was shown to belong to Ti^{47}. This gamma ray is presumably the one also seen in the beta decay of Sc^{47}.[17] The thin target excitation function for an enriched target of $Ti^{47}O_2$ (82.05 percent, natural abundance 7.75 percent) is shown in Fig. 5. In this case we know that the target is no more than 50 kev thick, because we observe a resonance of about that width in the $O^{18}(\alpha,n\gamma)Ne^{21}$ reaction by means of a 342-kev gamma ray[18] coming from the first excited state of Ne^{21}. The experimentally observed yields are normalized to the theoretical $E2$ curve[6] at 2 Mev. Again, the agreement with the theory is found to be good at the lower energies, with characteristic departures occurring around 2.5 Mev and above (see note added in proof, Fig. 1).

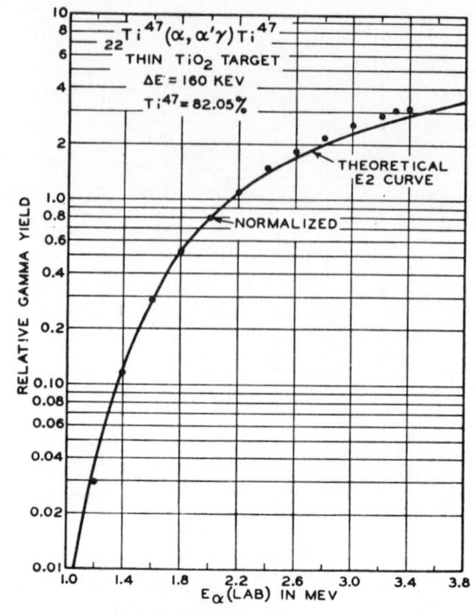

FIG. 5. Coulomb excitation function for thin enriched $Ti^{47}O_2$ target. Level energy is 160 kev. Solid curve is the theoretical $E2$ function (reference 6). Experimental points normalized at 2 Mev (see note added in proof, Fig. 1).

TABLE I. Levels in the compound nucleus Al^{27} as obtained from $Na^{23}(\alpha,p\gamma)Mg^{26}$, observing the 1.83-Mev gamma-ray yield. E_r = resonance energy in the laboratory system; $E_{c.m.}$ = resonance energy in the center-of-mass system; E^* = excitation energy of Al^{27}.

E_r (Mev)	$E_{c.m.}$ (Mev)	E^* (Mev)
1.95	1.66	12.08
2.15	1.83	12.24
2.43	1.98	12.40
2.51	2.14	12.56
2.56	2.18	13.60
2.64	2.25	12.67
2.80	2.39	12.79
2.90	2.47	12.87
3.04	2.59	13.01
3.07	2.62	13.04
3.16	2.69	13.11
3.23	2.75	13.17
3.29	2.80	13.20
3.40	2.90	13.32
3.50	2.98	13.40
3.58	3.05	13.47

(d) Mn^{55}

We have already reported our thick-target results for the excitation of the 128-kev line from this nucleus.[1] The main reason for investigating the thin-target yield was to establish how reliable our thick-target calculations were and how much detail is generally lost when we are forced, because of intensity considerations, to confine ourselves to thick-target excitation. Figure 6 shows the thin-target curve, along with the theoretical $E2$ function.[6] The agreement is again seen to be quite good at the lower energies, with deviations (excess over theory) becoming apparent at higher energies (see note added in proof, Fig. 1). The target used was electrolytic manganese metal evaporated on nickel foil. Generally speaking, not too much detail is lost when using thick targets, if the main objective is to measure absolute cross sections. It must be remembered that thin targets introduce the difficulty of having to know the target thickness absolutely.

In Mn^{55} we were able to make use of the comparison method[8] already mentioned. It turns out that if the thin-target yield of the gamma radiation is measured for protons and alpha particles having the same value of the parameter ξ, the ratio of the yields will always be either 16 for $E3$, 10 for $E2$, or 6.4 for $E1$ (numerical values approximate). In the case of Mn, 1.86-Mev

[17] Cork, LeBlanc, Brice, and Nester, Phys. Rev. 92, 367 (1953).
[18] This gamma ray is always present when we use oxide targets and represents the only observable effect ascribable to oxygen under alpha-particle bombardment.

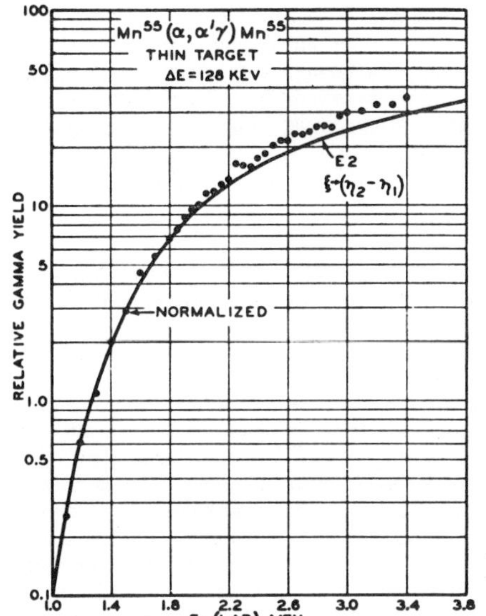

FIG. 6. Coulomb excitation function for thin Mn^{55} target. Level energy is 128 kev. Solid curve is the theoretical *E*2 function (reference 6). Experimental points normalized at 1.5 Mev. For thick-target excitation curve, see Fig. 2 of reference 1 (see note added in proof, Fig. 1).

alpha particles and 0.75-Mev protons have the same value of ξ. The experimentally observed ratio of alpha to proton yields at these energies was 10.8±1.0, thus once again confirming the *E*2 nature of the excitation process. The thin target of Mn was deposited on niobium foil, using a 0.032-in. copper absorber to reduce the Nb x-rays. Niobium was found to have a lower background than nickel for proton excitation.

(e) Fe^{57}

Even with our enriched target of Fe^{57}(59.3 percent, natural abundance 2.25 percent) in the form of Fe_2O_3 and our 4π geometry we did not have enough intensity for thin target work. The previously reported 123-kev gamma ray[1] was shown to belong to Fe^{57} as expected. The oxide was reduced at about 1000°F in a hydrogen atmosphere, the resulting iron powder was compressed into a pure Fe^{57} foil 0.002 in. thick. This step facilitated the theoretical thick-target yield calculation. The excitation curve for a thick target of pure Fe^{57} for the 123-kev gamma ray is shown in Fig. 7. The insert in Fig. 7 shows the decay scheme as recently given by Alburger and Grace.[19] We are evidently exciting the second excited state of Fe^{57} at 137 kev. In order to show the sensitivity of the Coulomb excitation process to the value of the excitation energy ΔE, we have plotted the theoretical thick target yields for both $\Delta E=123$ kev and $\Delta E=137$ kev, both normalized to the experimental points at 2.0 Mev. The better agreement with the 137-kev curve is very evident. Although there is no doubt about the state of affairs in this particular instance, it is interesting to note that the excitation curve can in general decide whether a given gamma ray represents a transition to the ground state, even if the energy difference between excitation energy and gamma-ray energy were only about 10 percent.

In the gamma-ray spectrum of Fe^{57} we found evidence for gamma radiation at 14 kev, which is undoubtedly produced by cascade only, since that transition is known to be too slow ($t_{\frac{1}{2}}=1.1\times10^{-7}$ sec) and of the wrong multipolarity (*M*1) to be Coulomb excited. We also see some slight indication for the 137-kev crossover transition, which is known to be weak compared to the 123-kev radiation (ratio of 123/137=15±7 as found by proportional counter measurement[19]). Our excitation of the 137-kev level confirms the electric quadrupole nature of the transition.

We also compared the thick-target yields for Fe^{57} and $Fe_2{}^{57}O_3$ at 3.0 Mev in order to have some empirical information on the stopping power effect of oxygen for cases (such as the rare earths) where only oxides are available. The ratio for the yields of the 123-kev radiation was 2.07. A similar thick-target comparison for Ta and Ta_2O_5 yielded a ratio of 2.00. We shall therefore use a factor of 2 to put oxides and pure substances on a correct relative basis.

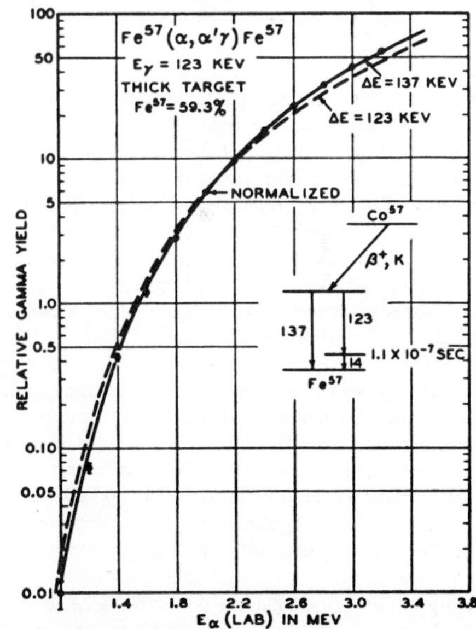

FIG. 7. Coulomb excitation function for thick enriched Fe^{57} target. Level energy is 137 kev, gamma ray detected is 123 kev. Solid curve is the theoretical *E*2 function for $\Delta E=137$ kev, dashed curve for $\Delta E=123$ kev. Experimental points normalized at 2 Mev. Insert shows level scheme as given in reference 19 (see note added in proof, Fig. 1).

[19] D. E. Alburger and M. A. Grace, Proc. Phys. Soc. **A67**, 280 (1954).

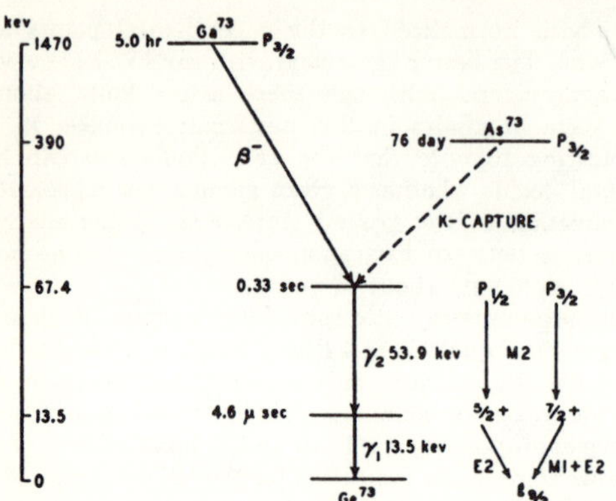

FIG. 8. Level scheme of Ge^{73} as given in reference 21. Level at 67.4 kev is not the one we excite (see text). Crossover transition to ground is less than 2×10^{-3} of the unconverted 54-kev radiation (reference 21).

(f) Ge^{72}

Our early work with ordinary GeO_2 revealed a strong gamma-ray line at 68 kev.[1] Since the only odd-A isotope of germanium is Ge^{73}, we suspected the line to belong to this isotope, since the even-even isotopes are either known or expected to have first excited states lying considerably higher (around 600 kev).[20] An enriched target of $Ge^{73}O_2$ (78.04 percent, natural abundance 7.67 percent) confirmed our expectation. Figure 8 shows the latest information given by Welker *et al.*,[21] on the level scheme of Ge^{73} as known from the decay of Ga^{73} and As^{73}. We see that a level exists at 67.4 kev; however, the crossover gamma ray has never been seen, with an upper limit on the intensity of 2×10^{-3} compared to the unconverted 53.9-kev radiation. Furthermore, the lifetime of that level is known to be 0.33 second, which implies a Coulomb excitation cross section (if it were an $E2$ transition) about 10^{-7} times that of the 137-kev transition in tantalum. On the other hand, we were unable to detect a trace of either the 53.9-kev or the 13.5-kev radiation. Figure 9 gives the pulse-height distribution as obtained with enriched $Ge^{73}O_2$. Three points locating the peak of the 67.8-kev gamma ray of ionium (Th^{230}), which is one of our calibration points, are also shown. We see that within the accuracy of our measurements the energy of the Ge^{73} line and the 68-kev ionium line is indistinguishable. We are forced to the conclusion that we are observing an energy level in Ge^{73} lying within less than 500 electron volts of a known level, but evidently having entirely different properties (spin, parity, lifetime).

Our thin-target excitation curve for this level is shown in Fig. 10. Again the agreement with the theoretical $E2$ curve[6] is very good. This also confirms the fact that the 68-kev transition leads to the ground state (see remarks under Fe^{67}). The case of Ge^{73} points up the necessity for some caution when we try to identify gamma rays observed in Coulomb excitation with "known" gamma rays from beta- and gamma-ray spectroscopy.

[20] G. Scharff-Goldhaber, Phys. Rev. **90**, 587 (1953).
[21] Welker, Schardt, Friedlander, and Howland, Phys. Rev. **92**, 401 (1953).

D. OTHER RESULTS

(a) Zn^{67}

The pulse-height distribution of the gamma radiation from an enriched thick $Zn^{67}O$ target (60.46 percent, natural abundance 4.11 percent) is shown in Fig. 11, along with a partial decay scheme of this nucleus as it is known from the decay of Ga^{67} [22] and Cu^{67}.[23] Two gamma rays, one at around 90 kev and one at 182 kev, were observed. Since the cascade from the 182-kev level involves a 90-kev and a 92-kev gamma ray (not resolvable by scintillation counter), it is of interest to see if the entire peak at around 90 kev can be accounted for without having to invoke direct Coulomb excitation of the first excited state. Using the best available in-information on internal conversion coefficients and branching ratios,[22] we find that this is indeed the case. That is to say, within the accuracy of our measurements all gamma radiation originated from the Coulomb excitation of the 182-kev level in Zn^{67}. We have further confirmation of this from the observation that the intensity ratio of (90+92)-kev gamma radiation to 182-kev gamma radiation is unchanged at 6 Mev (He^{++})

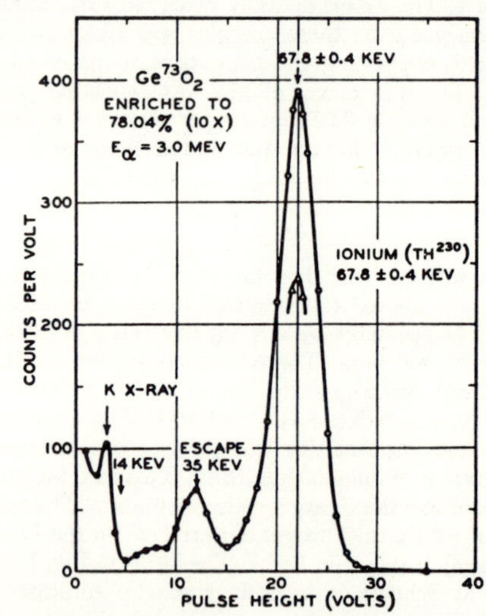

FIG. 9. Pulse-height distribution of gamma radiation from enriched $Ge^{73}O_2$ target, obtained with 3-Mev alpha particles. Three points under main peak locate the 67.8-kev known line from a Th^{230} calibration source. Note absence of 14-kev or 54-kev radiation.

[22] Meyerhof, Mann, and West, Phys. Rev. **92**, 758 (1953).
[23] H. T. Easterday, Phys. Rev. **91**, 653 (1953).

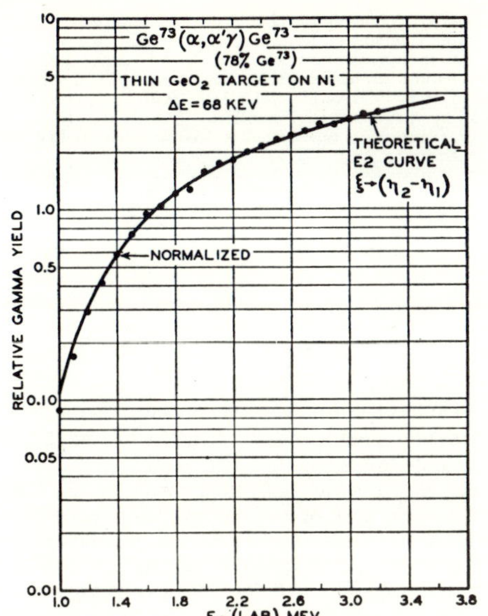

FIG. 10. Coulomb excitation function for thin enriched $Ge^{73}O_2$ target. Level energy is 68 kev. Solid curve is the theoretical $E2$ function (reference 6). Experimental points normalized at 1.4 Mev (see note added in proof, Fig. 1).

bombarding energy; the ratio would have been altered in favor of the 182-kev line, had some direct excitation of the 92-kev level taken place. This is as it should be, since the first excited state at 92 kev is known to be isomeric with a lifetime of about 10^{-5} second; this corresponds to a very small cross section for Coulomb excitation (~1000 times smaller than the cross section for the 137-kev level in Ta^{181}). The spin and parity assignments which have been made for this level[22] are compatible with $E2$ excitation, although the return to the ground state is presumably by $M1$ radiation.

(b) V^{51}

The 320-kev gamma ray associated with the first excited state of this nucleus, known from the decay of Ti^{51} [24] and Cr^{51},[25] as well as from inelastic proton scattering,[26] was observed, with an observed intensity of only about 5 percent of the other radiations discussed in this paper. We have obtained a thick-target excitation curve up to 3.4 Mev (not illustrated). Because of the relatively high location of the level, the thick-target yield varied by a factor of about 50 000 between 1.6 Mev and 3.4 Mev. The data show greater departures from the theoretical $E2$ curves than the other cases described in this paper, again in a direction so as to exceed the theoretical yield. Some, but not all, of this discrepancy may be ascribed to incorrectly known stopping power in this case (see note added in proof, Fig. 1).

(c) As^{75} and Se^{77}

As previously reported,[1] gamma rays of 68, 199, and 283 kev were observed in As^{75}, and of 237 and 445 kev in Se^{77}. A more detailed study of these radiations will be presented in a future publication.

We observed no other gamma radiation with 3-Mev alpha particles on nuclei with $22<Z<34$. We shall re-examine most of these nuclei with our 6-Mev He^{++} beam[4] with which we should be able to excite energy levels up to about 1 Mev in the lighter nuclei, and hence several of the even-even excited states in this region.[27]

E. CONCLUSIONS

In all cases we have studied we have verified the electric quadrupole nature of the transition involved.† We believe that on the strength of the Coulomb excitation function alone no choice can be made between the possible $E1$ or $E2$ character of the transition to the first excited state of F^{19} at 113 kev; in fact, the $E2$ curve is seen to fit somewhat better. However, the life-

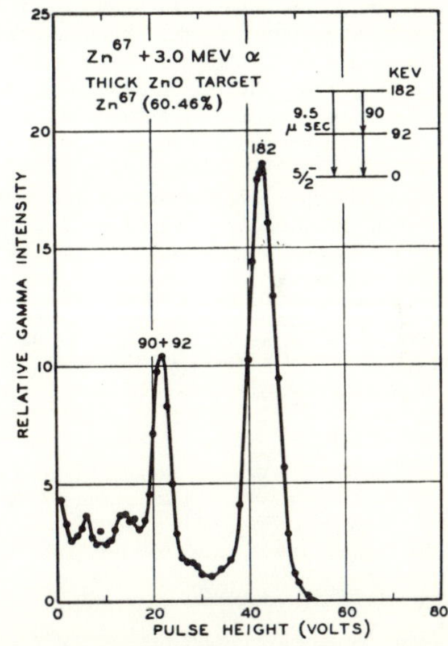

FIG. 11. Pulse-height distribution of gamma radiation from enriched $Zn^{67}O$ target, obtained with 3-Mev alpha particles. Peak marked 90+92 is entirely accounted for by cascade transitions from the 182-kev state, without direct excitation of the 92-kev level. Partial level scheme shown as insert is taken from reference 22.

[24] Koester, Maier-Leibnitz, Mayer-Kuckuk, Schmeiser, and Schulze-Pillot, Z. Physik 133, 319 (1952).

[25] W. S. Lyon, Phys. Rev. 87, 1126 (1952).

[26] Hausman, Allen, Arthur, Bender, and McDole, Phys. Rev. 88, 1296 (1952).

[27] We have already succeeded in detecting the gamma rays from the first excited states of the even-even isotopes of selenium.

time measurement[28] combined with the absolute value of the Coulomb excitation cross section and the isotropic angular distribution of the 113-kev radiation[12] seem to settle the question in favor of an electric dipole transition,[29] the only such case we have encountered among some 72 nuclei we have studied (see note added in proof, Fig. 1). Although the Coulomb excitation process favors $E1$ over $E2$ transitions by a factor of about 300 for 3-Mev alpha particles on F^{19}, the intrinsic nuclear matrix element for $E1$ is depressed about 1000-fold in this case.

Na^{23} provides a clear illustration of the interplay between Coulomb excitation and compound nucleus formation, since we have a comparison reaction proceeding via the compound nucleus Al^{27} at all energies. This case is similar to F^{19}, where we also have the $(\alpha,p\gamma)$ reaction taking place via the compound nucleus Na^{23}, as previously described.[9]

The deviations from the simple theory for $E2$ excitation evident at the high-energy end of most of these curves are presumably of two different origins: (a) compound nucleus contributions, as identified in certain favorable instances by the existence of resonances which agree with resonances of the respective compound nuclei; (b) barrier penetration effects due to the breakdown of "geometrical optics," and the existence of a finite nuclear radius (see note added in proof, Fig. 1). In view of the fact that the region of excitation in the compound nuclei involved in this investigation is completely unexplored (except for $F^{19}+\alpha$ and $Na^{23}+\alpha$) there is considerably difficulty in separating the two contributions experimentally. A more complete theory of the process can presumably hope to cope with (b) but not with (a).

In the way of spin assignments for the levels we excite, we can, of course, state that because of the $E2$ character of the transitions induced, the parities of the excited states are equal to the respective ground-state parities, and their spins differ by 0, ±1, or ±2 from the ground-state spins.

Because of the good agreement with the $E2$ theory[6] over at least part of the energy range covered (see note added in proof, Fig. 1), we are justified in using the simple theory to estimate the relative sizes of the transition probabilities $B_e(2)$. Without making detailed corrections for variations in stopping power over the range of Z under investigation ($22 \leqslant Z \leqslant 32$; Na^{23} corrected, however) nor for internal conversion, we list in Table II the approximate transition probabilities as well as F, the ratio of observed to single-particle transition probability (using $r_0 = 1.2\times10^{-13}$ cm). We note that in all cases the transition probabilities are at least an order of magnitude smaller than the (rotational) transition in Ta^{181} (or most rare-earth nuclei). Since the latter is about 100 times what is predicted from an independent-particle model estimate, our observed values are still very much larger than is expected for single-particle transitions (with the possible exception of V^{51}). This is *a fortiori* true when compared to empirical values of transition probabilities, which usually fall short of the theoretical ones. There also seems to be no systematic difference between odd-proton and odd-neutron nuclei.

TABLE II. Transition probabilities for some odd-A nuclei as found from Coulomb excitation. I_0 = ground-state spin and parity; I^* = spin of excited state (if known); ΔE = transition energy; $B_e(2)$ = reduced transition probability as defined in reference 13, obtained from thick-target yield Y_α at 3 Mev; F = ratio of observed to single-particle transition probability (using $r_0 = 1.2\times10^{-13}$ cm). Y_α is normalized to 1.00 for the 137-kev transition in Ta^{181}.

Nucleus	I_0	I^*	ΔE(kev)	$B_e(2)$ (10^{-48} cm^4)	Y_α	F
$_{11}Na^{23}$	$3/2^+$	$?^+$	446	0.041	1.6	71
$_{22}Ti^{47}$	$5/2^-$	$?^-$	160	0.047	4.5	31
$_{23}V^{51}$	$7/2^-$	$?^-$	320	0.013	0.41	8
$_{25}Mn^{55}$	$5/2^-$	$?^-$	128	0.070	10	38
$_{26}Fe^{57}$	$(3/2^-)$	$?^-$	137	0.051	7.5	26
$_{30}Zn^{67}$	$(5/2^-)$	$(5/2^-)$	182	0.043	3.7	18
$_{32}Ge^{73}$	$9/2^+$	$?^+$	68	0.042	5.8	16

[28] Thirion, Barnes, and Lauritsen, Phys. Rev. **94**, 1076 (1954).

[29] D. H. Wilkinson (private communication) has obtained some independent evidence for the $1/2^-$ assignment for the 113-kev state from a study of the O^{19} beta decay.

Reprinted from THE PHYSICAL REVIEW, Vol. 99, No. 1, 127–134, July 1, 1955
Printed in U. S. A.

Angular Distribution of Gamma Rays from Coulomb Excitation

F. K. MCGOWAN AND P. H. STELSON
Oak Ridge National Laboratory, Oak Ridge, Tennessee
(Received March 30, 1955)

The angular distributions of gamma rays with respect to the incident proton beam on a thick target have been measured for gamma rays following Coulomb excitation in $Pt^{194,196}$, Au^{197}, Ta^{181}, $Ag^{107,109}$, Pd^{106}, Pd^{108}, Pd^{110}, and Rh^{103}. The observed angular distributions deviate considerably from the semiclassical theory of angular distributions of gamma rays following Coulomb excitation given by Alder and Winther. Empirical curves of energy-dependent coefficients $a_\nu(\xi)$ for a thick target are obtained from the results for $Pt^{194,196}$ and Pd^{106}. With these empirical coefficients, information on the spin sequences and the character of the gamma transitions are deduced from the angular distribution measurements in the odd mass nuclei. The spin sequences are as follows: $7/2(E2)3/2$ and $5/2(E2+M1)3/2$ with $\delta_\gamma=-0.75$ (where $\delta_\gamma^2=E2/M1$) for the 550- and 277-kev transitions, respectively, in Au^{197}; $11/2(E2)7/2$ and $11/2(E2+M1)9/2$ with $\delta_\gamma=0.51$ for the 303- and 166-kev transitions, respectively, in Ta^{181}; $5/2(E2)1/2$ and $3/2(E2+M1)1/2$ with $\delta_\gamma=-0.19$ or -1.14 for the 427- and 325-kev transitions, respectively, in $Ag^{107,109}$; and $5/2(E2)1/2$ and $3/2(E2+M1)1/2$ with $\delta_\gamma=-0.18$ or -1.17 for the 365- and 305-kev transitions, respectively, in Rh^{103}.

Drs. McGowan and Stelson note that the paragraphs bracketed on pp.68-69 concerning multiple scattering of protons be replaced by the appended material from a subsequent paper reproduced on pp.70-71.

1. INTRODUCTION

ALDER and Winther,[1] using a semiclassical treatment, have derived explicit expressions for the angular distribution of the gamma radiation following Coulomb excitation. They find that the angular distribution of the gamma radiation with respect to the incident particles is similar to the angular correlation between two gamma rays in cascade. The distribution function is

$$W(\theta)=1+\sum_\nu A_\nu a_\nu(\xi)P_\nu(\cos\theta), \qquad (1)$$

where the coefficients A_ν are the gamma-gamma directional angular correlation coefficients tabulated by Biedenharn and Rose[2] for the spin sequence $j_1(E2)j(L_2)j_2$ and the j's are the spins of the target nucleus, the Coulomb excited state, and the final state after gamma-ray emission, respectively. The coefficients $a_\nu(\xi)$ which depend on the excitation process through the parameter ξ have been evaluated by numerical methods for electric quadrupole excitation by Alder and Winther.

Several workers[3,4] have reported agreement between theory and experiment for the angular distribution of the 303-kev gamma ray of Ta^{181} and the 550-kev gamma ray of Au^{197}. However, our measurements have shown significant deviations from theory. For instance, the energy coefficient $a_2(\xi)$ was observed to be 17 percent smaller than theoretically expected at $E_p=4.0$ Mev for the 303-kev gamma ray in Ta^{181} on the assumption that the spin sequence is $7/2(E2)11/2(E2)7/2$. The results given in this paper suggest that the apparent agreement between theory and experiment found by the other workers is in part the result of the choice of the incident proton energy used in their experiments.

To further test the theory, the angular distributions of the gamma rays from Coulomb excited states of spin 2 have been examined. In these cases the spin of the excited state is known from gamma-gamma directional angular correlation measurements whereas the spins for the states of the odd-mass nuclei mentioned above were not known with certainty. The spin sequence $0(E2)2(E2)0$ is particularly suitable because the coefficients A_ν are large. In addition, for the cases that have been examined in these experiments, the gamma-gamma directional angular correlation measurements have shown no observable influence of extranuclear fields. This point is important for proton-gamma angular distribution measurements where a target in the solid state is necessary.

A number of other proton-gamma ray angular distributions have been measured and the results are presented. In cases for which the spin of the Coulomb excited state is known from other measurements, the energy dependent coefficients $a_\nu(\xi)$ for a thick target are tabulated. The observed deviations from theory are rather large. Finally, an interpretation of the results, in combination with the empirically determined energy-dependent coefficients, is discussed.

2. APPARATUS

The ORNL 5.5-Mv Van de Graaff accelerator was used to produce a separated beam of protons. Metallic targets which were thick to protons (the range of the protons being ≤ 100 mg/cm²) but thin for the gamma rays were oriented at 45° with respect to the incident protons. The targets (≤ 100 mg/cm²) were prepared from thin foils or were electrodeposited onto 0.005-inch nickel. For the detection of the gamma rays, a scintillation spectrometer employing a NaI crystal 1.5 inches in diameter and 1 inch thick mounted on a DuMont 6292 photomultiplier was used. In all angular distribution experiments the front face of the crystal was located at distances 10.0 or 13.5 cm from the target. To suppress the characteristic K x-rays from the target produced by the impinging protons by factors of 10^2

[1] K. Alder and A. Winther, Phys. Rev. **91**, 1578 (1953).
[2] L. C. Biedenharn and M. E. Rose, Revs. Modern Phys. **25**, 729 (1953).
[3] Eisinger, Cook, and Class, Phys. Rev. **94**, 735 (1954); **95**, 628(A) (1954); **96**, 658 (1954).
[4] W. I. Goldburg and R. M. Williamson, Phys. Rev. **95**, 767 (1954).

to 10^3, a graded shield was placed in front of the NaI detector. The graded shields were as follows: for targets with $Z \geq 78$ the shield consisted of 0.010 inch of Ta plus 0.030 inch of Sn plus 0.005 inch of Cu; for a Ta target the shield consisted of 0.040 inch of Sn plus 0.005 inch of Cu; and for targets with $45 \leq Z \leq 53$ the shield consisted of 0.0035 inch of Mo plus 0.005 inch of Cu. Most of the data in these experiments were recorded automatically at 10° increments from 0° to 90° and from 210° to 270°. The time for the collection of a fixed number of counts was printed on a paper tape by a printing timer and the integrated current was recorded by a traffic counter. The angular positions were changed manually. To assure that the axis of rotation of the detector passed through the target, the following alignment procedure was used. The position at which the proton beam impinged on the target was located. A source of Cs^{137} of the same area as the beam was placed on the target at this position. The axis of rotation was adjusted until counting rates showed that the variation in the solid angle subtended by the detector at the target as a function of angular position was less than 0.5 percent. The position of the beam on the target was observed to remain fixed as a function of the beam energy.

3. DISCUSSION OF METHOD

Angular distribution measurements of the gamma rays were carried out with either a single-channel or a multichannel pulse-height analyzer of the ORNL design. In the measurements with a single channel the window of the analyzer was always operated to include only the full energy pulse spectrum peak of the gamma ray. In the case of the multichannel measurements, the full energy pulse spectrum peak was observed and the area of the peak was taken as a measure of the intensity. After each measurement of the intensity at an angle θ_i the intensity was measured either at $\theta=90°$ or at $\theta=0°$. In this way the intensities could be corrected for changes in gain of the detector or fractional acceptance of the window of the single-channel analyzer. In all cases, the intensities have been corrected for the bremsstrahlung continuum by measuring the intensity of the bremsstrahlung as a function of θ_i. For angular distribution measurements involving $Z \geq 73$ a Bi target electrodeposited onto nickel was used and for measurements involving $Z \sim 50$ a tin target was used. Bi and Sn targets are well suited for this purpose because of the absence of gamma rays from Coulomb excitation. We believe the extrapolation to neighboring Z will give little error since our investigations of the bremsstrahlung process show relatively little change in character with a small change in Z. In general for the measurements to be discussed below the intensity of the bremsstrahlung in the angular distributions was never more than a few percent of the gamma-ray intensity from Coulomb excitation. Finally, a correction for the attenuation of the gamma rays in the target and target backing as a function of θ was applied to the observed intensities. If this correction was no larger than 5 percent, a computed attenuation using the absorption cross sections taken from NBS-1003[5] was applied. If the correction was larger, the attenuation was measured directly by placing a source of gamma radiation of the same energy on the target. A least squares fit of the corrected intensities (with the appropriate weight factors) in terms of a series of Legendre polynomials,

$$W(\theta) = \mathfrak{a}_0' + \mathfrak{a}_2' P_2(\cos\theta) + \mathfrak{a}_4' P_4(\cos\theta), \qquad (2)$$

was carried out on an I.B.M. calculator. The standard deviations quoted in Table I have been obtained from Eq. (30) in a paper by Rose.[6] The values of ϵ^2, defined by Eq. (27),[6] clustered about unity indicating that nonstatistical errors were not large. A least-squares fit of each set of data in terms of a series of $\cos^{2n}\theta$ was carried out to serve as a check on the I.B.M. calculations. In Table I we tabulate $(a_\nu A_\nu)_{\text{exp}}$, which have been corrected for finite angular resolution[6] and are defined as

$$(a_\nu A_\nu)_{\text{exp}} = \mathfrak{a}_\nu / \mathfrak{a}_0.$$

4. MEASUREMENTS AND RESULTS

A. $Pt^{194,196}$

From the β decay of Ir^{194} and Au^{196}, the first excited states in Pt^{194} and Pt^{196} are known to exist at 330 and 358 kev, respectively. Directional angular correlation measurements[7–9] have verified the spin assignment of 2 for these excited states.

The differential pulse-height spectrum of the gamma radiation from platinum for $E_p = 5.0$ Mev is shown in an accompanying paper.[10] The proton-gamma angular distribution of the 330- and 358-kev gamma rays taken together has been measured for $E_p = 2.5$ to 5.0 Mev. The results are tabulated in Table I. All entries in Table I represent the mean of several determinations of the angular distribution coefficients. A comparison between theory and experiment is shown in Fig. 1. The solid curves labelled "theory" are the thick target energy-dependent coefficients $(a_\nu)_t$ deduced from the excitation cross section and the thin target coefficients a_ν given by Alder and Winther. A procedure for obtaining $(a_\nu)_t$ will be discussed in Sec. 5 The observed energy dependence of the coefficients for a thick target deviates considerably from the theory.

The second excited states in Pt^{194} and Pt^{196} are known to have spin 2.[8,9] If these states were appreciably excited by Coulomb excitation, then the observed proton-gamma angular distribution would be a com-

[5] G. R. White, National Bureau of Standards Circular NBS-1003, 1952 (unpublished).

[6] M. E. Rose, Phys. Rev. **91**, 610 (1953).

[7] J. J. Kraushaar and M. Goldhaber, Phys. Rev. **89**, 1081 (1953).

[8] R. M. Steffen, Phys. Rev. **89**, 665 (1953).

[9] Mandeville, Varma, and Saraf, Phys. Rev. **98**, 94 (1955).

[10] P. H. Stelson and F. K. McGowan. preceding paper [Phys. Rev. **99**, 112 (1955)]

TABLE I. Proton-gamma angular distribution coefficients of the terms in the expansion of the correlation function in Legendre polynomials for a thick target.

Nucleus	E_γ (kev)	E_p (Mev)	Spin sequence	$(A_2)_{\gamma\gamma}$	$(a_2A_2)_{exp}$	$(a_2)_{exp}$	$(A_4)_{\gamma\gamma}$	$(a_4A_4)_{exp}$	$(a_4)_{exp}$	$(A_2)_{exp}$	δ_γ
$_{78}Pt^{194,196}$	330 and 358	2.5	$0(E2)2(E2)0$	0.3571	0.286±0.005	0.800±0.015	1.143	−(0.078±0.010)	−(0.068±0.010)		
		3.0			0.253±0.004	0.709±0.010		−(0.052±0.006)	−(0.054±0.005)		
		3.5			0.227±0.003	0.636±0.008		−(0.026±0.008)	−(0.023±0.007)		
		4.0			0.211±0.003	0.591±0.007		−(0.013±0.004)	−(0.011±0.004)		
		4.5			0.191±0.005	0.535±0.015		−(0.007±0.008)	−(0.006±0.007)		
		5.0			0.171±0.003	0.479±0.008		0.001±0.003	0.001±0.003		
Au^{197}	550	3.5[a]	$3/2(E2)7/2(E2)3/2$	0.2186	0.176±0.006	0.80 ±0.03	0.1282	−(0.004±0.008)			
		4.0[a]			0.161±0.006	0.74 ±0.03		−(0.010±0.008)			
		5.0[a]			0.123±0.006	0.58 ±0.03		0.010±0.008			
		4.0			0.164±0.004	0.75 ±0.02		−(0.018±0.005)			
Au^{197}	277	4.0	$3/2(E2)5/2(E2+M1)3/2$		−(0.115±0.004)			0.003±0.005		−(0.21±0.01)	−(0.75±0.20)
Ta^{181}	303	4.0	$7/2(E2)11/2(E2)7/2$	0.1688	0.089±0.004	0.53 ±0.02	0.0500	−(0.006±0.004)			
Ta^{181}	166	4.0	$7/2(E2)11/2(E2+M1)9/2$		0.112±0.006			0.020±0.008		0.21±0.01	0.51 or 3.0
$Ag^{107,109}$	325	2.5	$1/2(E2)3/2(E2+M1)1/2$		−(0.282±0.004)			0.007±0.007		−(0.39±0.02)	−0.19 or −1.14
$Ag^{107,109}$	427	2.5	$1/2(E2)5/2(E2)1/2$	0.2857	0.248±0.004	0.868±0.015	0.3809	−(0.039±0.009)	−(0.102±0.024)		
Pd^{110}	380	2.1	$0(E2)2(E2)0$	0.3571	0.285±0.008	0.80 ±0.02	1.143	−(0.088±0.011)	−(0.077±0.010)		
		2.5			0.243±0.006	0.68 ±0.02		−(0.032±0.008)	−(0.028±0.007)		
		2.5			0.270±0.004	0.75 ±0.01		−(0.050±0.007)	−(0.044±0.007)		
		2.9			0.211±0.006	0.59 ±0.02		−(0.005±0.010)	−(0.005±0.009)		
Pd^{108}	445	2.5	$0(E2)2(E2)0$	0.3571	0.255±0.009	0.72 ±0.03	1.143	−(0.071±0.008)	−(0.062±0.007)		
Pd^{106}	520	2.1	$0(E2)2(E2)0$	0.3571	0.343±0.007	0.96 ±0.02	1.143	−(0.143±0.010)	−(0.125±0.009)		
		2.5			0.315±0.007	0.88 ±0.02		−(0.092±0.010)	−(0.080±0.009)		
		2.5			0.340±0.005	0.95 ±0.02		−(0.107±0.018)	−(0.094±0.017)		
		2.9			0.279±0.006	0.78 ±0.02		−(0.065±0.008)	−(0.057±0.008)		
Rh^{103}	305	1.7	$1/2(E2)3/2(E2+M1)1/2$		−(0.346±0.006)			0.006±0.007		−0.380	
		2.1			−(0.307±0.005)			0.024±0.006		−0.394	
		2.5			−(0.274±0.004)			0.019±0.006		−0.400	−0.18 or −1.17
		2.5			−(0.255±0.004)			0.007±0.004		−0.372	
		2.9			−(0.234±0.003)			0.012±0.004		−0.377	
Rh^{103}	365	1.7	$1/2(E2)5/2(E2)1/2$	0.2857	0.288±0.007	1.01 ±0.02	0.3809	−(0.046±0.009)	−(0.121±0.024)		
		2.1			0.247±0.005	0.87 ±0.02		−(0.021±0.006)	−(0.055±0.016)		
		2.5			0.224±0.004	0.78 ±0.01		−(0.027±0.005)	−(0.071±0.013)		
		2.5			0.227±0.003	0.79 ±0.01		−(0.023±0.004)	−(0.060±0.011)		
					0.200±0.002	0.70 ±0.01		−(0.012±0.003)	−(0.032±0.007)		

[a] Thin target (10.5 mg/cm²).

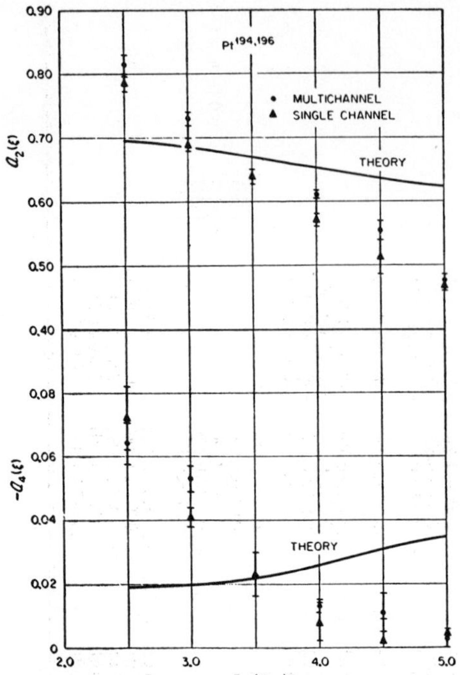

FIG. 1. Thick-target energy-dependent coefficients deduced from the angular distribution measurements of the 330-kev and 358-kev gamma rays from Pt^{194} and Pt^{196} as a function of the incident proton energy. The curves labelled "theory" are the thick-target energy-dependent coefficients deduced from the total cross section for excitation and the the thin-target coefficients given by Alder and Winther.

posite angular distribution function. Any significant excitation of the second excited states can be excluded in these isotopes for the following reasons. Recent measurements by Johns and Nablo[11] indicate that the second excited state is at 620 kev in Pt^{194}. The intensity ratio of the cascade gamma ray to the cross-over gamma ray is 3. From the differential pulse-height distribution of the gamma radiation resulting from 4.5-Mev protons incident on a thick platinum target we find that the excitation of the 620-kev state in Pt^{194} relative to the direct excitation of the 330-kev state is less than 3 percent. The contribution of 330-kev gamma rays resulting from excitation of the second excited state to the observed angular distribution function is less than 2 percent.

In Pt^{196}, the second excited state is at 688 kev and decays 99 percent of the time by a cascade gamma ray of 330 kev which is 95 percent $E2$ radiation and 5 percent $M1$ radiation.[8] Assuming the reduced transition probability for excitation of the 688-kev and 358-kev levels are equal, the yield of 330-kev gamma rays by the cascade transition in Pt^{196} relative to 330- and 358-kev gamma rays would be 5, 7, 10, and 13 percent at $E_p=3.5$, 4.0, 4.5, and 5.0 Mev, respectively. However, the reduced transition probability for the 358-kev transition in Pt^{196} is 20 times larger than independent particle theoretical estimate [see Eq. (VII. 9) given by Bohr and Mottelson].[12] Consequently, the assumption made above, i.e., that the reduced transition probability for excitation of the 682-kev state is equal to that for the 358-kev state, requires the reduced transition probability for the 330-kev cascade to be 4×10^3 times larger than the independent particle estimate and this seems rather unlikely.

B. Au^{197}

The proton-gamma angular distributions of the 550-kev and 277-kev gamma rays in Au^{197} have been measured by several workers.[3,4] Spins of 7/2 and 5/2 for the 550-kev and 277-kev levels, respectively, were deduced from their measurements. For the 277-kev transition Goldburg and Williamson,[4] and Eisinger *et al.*[3] obtained $E2/M1=0.07$ and 0.59, respectively, by applying the energy-dependent coefficients a_ν given by Alder and Winther to their angular distribution measurements. Measurements of the proton-gamma angular distributions have been carried out for the 550-kev gamma ray in Au^{197} using thick and thin targets and for the 277-kev gamma ray in Au^{197} using a thick target. The thin target was prepared by spot-welding a thin Au foil on a Bi backing which was electrodeposited on nickel. The Bi backing served as a catcher for the protons emerging from the Au target. The results of the measurements are tabulated in Table I. The measurements for the 277-kev radiation have been treated as entirely the result of direct excitation of 277-kev level. The cascade radiation (273-, 277-kev γ rays) from the 550-kev level contributes about 4 percent to the intensity of 277-kev gamma radiation[10] at $E_p=4.0$ Mev.

C. Ta^{181}

The proton-gamma angular distribution of the 303-kev gamma ray in Ta^{181} has been measured by several workers[3,4] and a spin of 11/2 for the 303-kev level was deduced from their measurements. In addition to measurements of the proton-gamma angular distribution of the 303-kev gamma ray we have measured the proton-gamma angular distribution of the 166-kev cascade gamma ray from the 303-kev state. For the latter measurements, the window of the analyzer was set on the high energy edge of the 166-kev full pulse spectrum peak in order to exclude the detection of 137-kev gamma radiation. The results of the measurements are tabulated in Table I.

[11] M. W. Johns and S. V. Nablo, Phys. Rev. **96**, 1599 (1954).

[12] A. Bohr and B. R. Mottelson, Kgl. Danske Videnskab. Selskab, Mat.-fys. Medd. **27**, No. 16 (1953).

D. $Ag^{107,109}$ and Rh^{103}

Proton-gamma angular distribution measurements have been carried out for the gamma rays in $Ag^{107,109}$ at $E_p = 2.5$ Mev using thick targets of normal silver and for two gamma rays in Rh^{103} at $E_p = 1.7$ to 2.9 Mev using thick targets. Since the completion of our distribution measurements, Heydenburg and Temmer[13] have reported that the two corresponding excited states in Ag^{107} and Ag^{109} differ by only a few kev. As a result the measurements for $Ag^{107,109}$ tabulated in Table I represent angular distribution coefficients of a composite angular distribution function. A comparison of the results with the observed angular distributions for the two gamma rays of Rh^{103} indicates that the angular distribution coefficients for $Ag^{107,109}$ are meaningful, i.e., the spins of the corresponding Coulomb excited states in Rh^{103}, Ag^{107}, and Ag^{109} and the character of the gamma radiation are the same.

E. Pd

Proton-gamma angular distributions have been carried out for the 445-kev gamma ray at $E_p = 2.5$ Mev and for the 380-kev and 520-kev gamma rays at $E_p = 2.1$ to 2.9 Mev using thick targets of palladium. A differential pulse-height spectrum of the gamma radiation from palladium produced by Coulomb excitation is shown in the accompanying paper.[10] The results of measurements are tabulated in Table I. The 520-kev gamma ray is attributed to Pd^{106} resulting from the decay of the well-known 513-kev level observed in the β-decay of Rh^{106}. From gamma-gamma directional angular correlation measurements the spins of the first and second excited states of Pd^{106} are known to be 2 and 0, respectively, and the angular correlation measurements are not disturbed by extranuclear fields.[14] The very short lifetime[10] of the intermediate state deduced from the reduced transition probability for excitation of the first excited state in Pd^{106} lends support to this latter statement. The second excited state in Pd^{106} is, of course, not excited by Coulomb excitation. As a result the angular distribution measurements of the 520-kev gamma ray serve to test the angular distribution theory for medium weight nuclei.

5. DISCUSSION OF RESULTS

In order to compare experimental angular distribution coefficients for a thick target with theory, we must evaluate the expected thick target coefficients. Now, Alder and Winther[1,15] have given both the total excitation cross section and the angular distribution of the gamma rays with respect to the incident protons for excitation by the electric quadrupole field. The differential cross section at a given energy E is

$$d\sigma(E)/d\Omega = \sigma(E)W(\theta), \tag{3}$$

where

$$\sigma(E) = \frac{4\pi^2 mB(E2)}{25Z_2^2e^2\hbar^2}E_f g_2(\xi)$$

and $W(\theta)$ is given by Eq. (1). For the differential cross section from a thick target for an incident proton energy E_i in the laboratory system, we have

$$\frac{d\sigma(E_i)}{d\Omega} \propto \int_0^{E_i} \frac{\sigma(E)W(\theta)dE}{dE/d\rho x} \tag{4}$$

We therefore have that the expected coefficient of the Legendre polynomial P_ν for a thick target is given by

$$\frac{\mathcal{A}_\nu(E_i)}{\mathcal{A}_0(E_i)} = A_\nu \int_0^{E_i} \frac{\sigma(E)a_\nu(E)dE}{dE/d\rho x} \Big/ \int_0^{E_i} \frac{\sigma(E)dE}{dE/d\rho x}, \tag{5}$$

or the thick target energy-dependent coefficient is

$$[a_\nu(E_i)]_t = \frac{1}{A_\nu}\frac{\mathcal{A}_\nu(E_i)}{\mathcal{A}_0(E_i)}. \tag{6}$$

Now let us change from E to the variable ξ, where

$$\xi = \frac{Z_1Z_2e^2}{\hbar}\left(\frac{1}{v_f} - \frac{1}{v_i}\right). \tag{7}$$

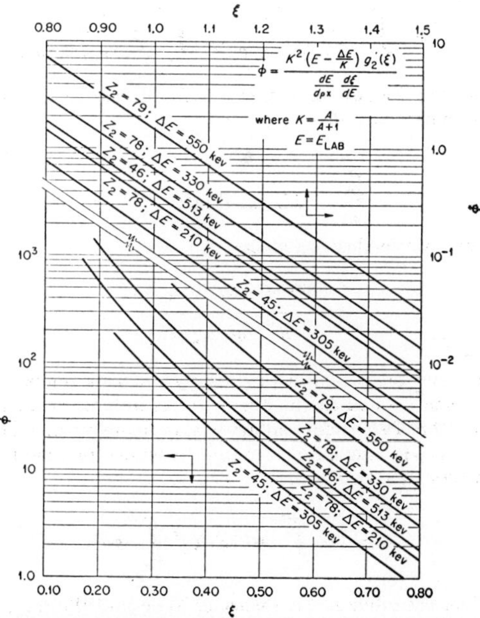

FIG. 2. The function ϕ for representative cases plotted as a function of ξ.

---

[13] N. P. Heydenburg and G. M. Temmer, Phys. Rev. **95**, 861 (1954).

[14] E. D. Klema and F. K. McGowan, Phys. Rev. **92**, 1469 (1953).

[15] K. Alder and A. Winther, Phys. Rev. **96**, 237 (1954).

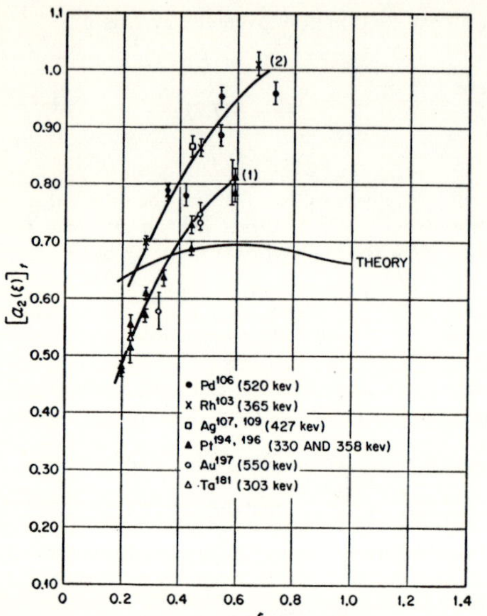

FIG. 3. Thick-target energy-dependent coefficient $a_2(\xi)$ deduced from angular distribution measurements plotted at a value of ξ corresponding to the incident proton energy on the thick target. The solid curve labelled "theory" represents the thick-target energy-dependent coefficient $[a_2(\xi_i)]_t$ as given by Eq. (9).

For any given case we have

$$[a_\nu(\xi_i)]_t = \int_\infty^{\xi_i} \frac{\sigma a_\nu d\xi}{dE/d\rho x \; d\xi/dE} \Big/ \int_\infty^{\xi_i} \frac{\sigma d\xi}{dE/d\rho x \; d\xi/dE}. \quad (8)$$

Now for the cases of interest it is found that the energy-dependent part of

$$\frac{\sigma}{dE/d\rho x \; d\xi/dE}$$

which we shall call ϕ, where

$$\phi = \frac{K^2(E-\Delta E/K)g_2(\xi)}{dE/d\rho x \; d\xi/dE}$$

as a function of ξ has very nearly the same shape for different Z_2 and ΔE. In Fig. 2, ϕ has been plotted as a function of ξ for representative cases, namely; $\Delta E = 200$ to 550 kev and incident proton energies of practical interest. We then have

$$[a_\nu(\xi_i)]_t = \int_\infty^{\xi_i} \phi a_\nu(\xi) d\xi \Big/ \int_\infty^{\xi_i} \phi d\xi. \quad (9)$$

Thus, according to the theory of Alder and Winther, it follows that $[a_\nu(\xi_i)]_t$ *vs* ξ_i will be nearly an unique function. Or to put it differently, if many different cases are measured one would expect the resulting points $a_\nu(\xi_i)_{\text{exp}}$ to fall on a smooth curve as a function of ξ_i.

A comparison between theory and experiment is shown in Figs. 3 and 4 for the cases of $Pt^{194,196}$ and Pd^{106} where the spin of the Coulomb excited state is known with reasonable certainty. The solid curves labelled "theory" represent the expected thick target energy-dependent coefficient $[a_\nu(\xi_i)]_t$ given by Eq. (9) using for a_ν the numerical calculations of Alder and Winther. For Pd^{106}, the deviations of the observed energy-dependent coefficients from theory for a thick target are even larger than they are for $Pt^{194,196}$. In addition the coefficient $a_2(\xi_i)_{\text{exp}}$ seems to have a Z_2 dependence over and above that contained in the parameter ξ.

The effect of multiple scattering of the protons by Rutherford scattering as they traverse a thick target on the angular distribution coefficient $[a_\nu(\xi_i)]_t$ should be discussed. Let us consider a proton-gamma angular distribution experiment, where $W(\theta_1)$ is the correlation function and $F(\theta')$ is the multiple scattering function due to Rutherford scattering. The probability that any proton from the collimated incident beam is multiple scattered through an angle θ' in the target prior to a nuclear excitation and that the resulting gamma ray is correlated to the multiple scattered proton by $W(\theta_1)$ is

$$P = \int d\Omega_1 W(\theta_1) F(\theta'). \quad (10)$$

Goudsmit and Saunderson[16] have already expressed the multiple Rutherford scattering function as a series in Legendre polynomials for the case of electrons. Their treatment is exact if one considers electrons with the same total path length in the scatterer. Thus, the coef-

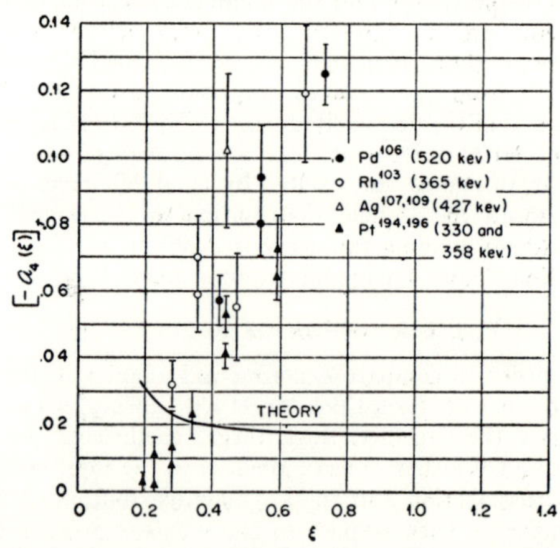

FIG. 4. Thick-target energy-dependent coefficient $a_4(\xi)$ deduced from angular distribution measurements plotted at a value of ξ corresponding to the incident proton energy on the thick target. The solid curve labelled "theory" represents the thick-target energy-dependent coefficient $[a_4(\xi_i)]_t$ as given by Eq. (9).

[16] S. Goudsmit and J. L. Saunderson, Phys. Rev. **57**, 24 (1940).

ficients G_k (coefficients of the Legendre polynomials representing the multiple Rutherford scattering function) given by Goudsmit and Saunderson should be applicable to the case of protons. A solution of this problem on the effect of multiple scattering on the proton-gamma angular distribution is similar to the effect of scattering of electrons in a conversion electron-gamma angular distribution.[17] The form of the correlation function is unchanged and each coefficient $a_\nu A_\nu$ becomes multiplied by an attenuation factor G_ν. We have evaluated the effect of multiple scattering on the thick target angular distribution coefficients by replacing $a_\nu(\xi)$ in Eq. (9) by $a_\nu(\xi)G_\nu$ and considering the case $a_\nu(\xi)$ is constant. The results are tabulated in Table II. The effect of multiple Rutherford scattering on the observed thick target energy-dependent coefficients is in the right direction but is not nearly large enough to account for the difference between curves (1) and (2) of Fig. 3. In any case the attenuation coefficients for multiple scattering are significant and should be included in the analysis of thick target angular distribution measurements. The coefficients in Table I have not been corrected for multiple Rutherford Scattering because we have no satisfactory thin target a_ν from which we could evaluate these attenuation coefficients for a thick target.

TABLE II. Attenuation coefficients for multiple Rutherford scattering for proton-gamma angular distributions from thick targets for the case $a_\nu(\xi)$ is constant.

Pt^{194}	$\Delta E = 330$ kev		Rh^{103}	$\Delta E = 365$ kev	
E_p(Mev)	$\bar{G}_2$	$\bar{G}_4$	E_p(Mev)	$\bar{G}_2$	$\bar{G}_4$
5.0	0.955	0.841	2.9	0.980	0.940
4.0	0.961	0.872	2.5	0.982	0.946
3.0	0.971	0.916	2.1	0.985	0.956
2.0	0.988	0.961	1.7	0.989	0.969

Since the deviations between theory and experiment appears to be rather large, we have chosen to analyze the remainder of the data in Table I using empirically determined energy dependent coefficients from a plot of $a_\nu(\xi_i)_{\text{exp}}$ *vs* ξ_i. For neighboring nuclei this method of presenting the data appears to be useful. Of the possible spin sequences that need to be considered for the 550-kev and 303-kev gamma rays in Au^{197} and Ta^{181}, respectively, on the assumption that the radiation is E2, only the spin sequences suggested originally by Eisinger *et al.*[3] and by Goldburg and Williamson[4] yield energy dependent coefficients $a_2(\xi_i)_{\text{exp}}$ that agree with the empirical $[a_2(\xi_i)]_t$ given in Fig. 3. These additional coefficients determined from the Au^{197} and Ta^{181} data are tabulated in Table I and are plotted in Fig. 3.

In general, for odd mass nuclei, the radiation from the first Coulomb excited state and the cascade radiation from the second Coulomb excited state will be $E2+M1$ radiation. In addition to inferring the spin of the excited states, information on the ratio of the quadrupole to dipole intensity is obtained from angular distribution measurements. We have applied the empirical $[a_2(\xi_i)]_t$ to our data for Au^{197} and Ta^{181} and for the indicated spin sequence in Table I. From the tabulated $(A_2)_{\text{exp}}$ for the 277-kev transition in Au^{197} one finds $\delta_\gamma = -(0.75 \pm 0.20)$ where δ_γ^2 is the ratio of the squares of the reduced matrix elements[2] and is defined as the intensity ratio (in this case) of quadrupole to dipole radiation in the gamma-ray transition. The large uncertainty in δ_γ results from the fact that A_2 has a broad maximum at $\delta = -0.75$. This value of δ_γ^2 is in good agreement with that deduced from K-shell internal conversion coefficient measurements.[18] The data for the 166-kev transition in Ta^{181} lead to $\delta_\gamma = 0.51$ or 3.0. The former value agrees with the δ_γ^2 deduced from a K/L ratio measurements.[19] This sign and magnitude of δ_{166} is identical to δ_{137} of the 137-kev transition in Ta^{181}.[20]

Huus and Lunden[21] have suggested spin assignments for the Coulomb excited states in $Ag^{107,109}$ from the position of the levels using the nuclear model of Bohr and Mottelson.[12] Heydenburg and Temmer[13] have reached similar conclusions for $Ag^{107,109}$ and Rh^{103}. The signs of the angular distribution coefficients $(a_2A_2)_{\text{exp}}$ in Table I fix the level order as 1/2, 3/2, and 5/2 independent of a nuclear model. The angular distribution data for the 427-kev and 365-kev transitions in $Ag^{107,109}$ and Rh^{103}, respectively, yield additional determinations of $a_\nu(\xi_i)$ for a thick target and are plotted in Figs. 3 and 4. These additional data are in good agreement with the empirical curve deduced from the proton-gamma angular distribution measurements of Pd^{106}. Using these empirical data the spins of the 325-kev and 305-kev states in $Ag^{107,109}$ and Rh^{103}, respectively, and the character of the radiation are deduced from the angular distribution data. Actually, the observed angular distributions are composite functions, namely: proton-gamma angular distribution function for the direct excitation of the 325-kev or 305-kev levels and the proton-gamma angular distribution function for 325-kev or 305-kev gamma rays by excitation of the 427-kev and 365-kev levels and not observing the cascade transitions. This latter distribution function is not known explicitly, although we do know approximately what fraction of the observed 325-kev and 305-kev gamma rays result from direct Coulomb excitation of the 325- and 305-kev levels.[10] For the 325-kev transition in $Ag^{107,109}$ about 96 percent of the gamma rays observed are the result of direct Coulomb excitation of the 325-kev level. If we assume that the reduced $E2$ transition probabilities for the 65- and 365-kev transitions in Rh^{103} are equal, then the cascade transition must be predominantly $M1(E2/M1 = 0.002)$ or 86 to 92 percent of the 305-kev

[17] S. Frankel, Phys. Rev. 83, 673 (1951).

[18] Huber, Halter, Joly, Maeder, and Brunner, Helv. Phys. Acta 26, 591(A) (1953).

[19] T. Huus and J. H. Bjerregaard, Phys. Rev. 92, 1579 (1953).

[20] F. K. McGowan, Phys. Rev. 93, 471 (1954).

[21] T. Huus and A. Lunden, Phil. Mag. 45, 966 (1954).

gamma rays result from direct Coulomb excitation of the 305-kev level. In the absence of any additional information, we have treated the data as resulting from the direct Coulomb excitation of the 305-kev level. The constancy of $(A_2)_{exp}$ for different E_p for the 305-kev transition in Rh^{103} probably indicates that the distribution functions entering into the composite function are nearly alike.

In the treatment of the angular distribution data for the 380-kev and 445-kev gamma rays in Pd we assume that the transitions are in ${}_{46}Pd^{110}$ and ${}_{46}Pd^{108}$, respectively, as was done in the analysis of the yield data.[10] The trend of $B(E2)/e^2$ with neutron number indicated that the transitions are predominantly in these isotopes. The angular distributions indicate that transitions are predominantly of the type $2(E2)0$. The empirical coefficients $(a_2)_{exp}$ for these transitions are slightly smaller (the coefficients $(a_2)_{exp}$ for the 445-kev transition are more so than for the 380-kev transition) than those in Fig. 3 for Pd^{106}, $Ag^{107,109}$, and Rh^{103}. This is not surprising since we have not attributed any of the gamma rays to Coulomb excitation of Pd^{105} (22.23 percent).

6. CONCLUSIONS

The coefficients $a_\nu(\xi)$ are found to deviate considerably from those given by the semiclassical theory of the process. The fact that the points for heavy (or for medium weight) nuclei fall on a smooth curve indicates that the parameter ξ correctly takes into account the excitation energy and the exciting proton energy. However, from the fact that two distinct curves are obtained we conclude that the dependence on Z_2 is not correctly accounted for. Recently, Biedenharn and Class[22] have obtained an exact result for the particular case of no energy loss ($\xi=0$). The behavior of a_2 for this special case as a function of the parameter $\eta=Z_1Z_2e^2/\hbar v$ shows appreciable deviations from the classical limit. It is clear that more accurate calculations of the angular distribution coefficients a_ν are needed. Inferring spins of Coulomb excited states using a nuclear model is not very satisfying. One would prefer to arrive at the spins more directly from angular distribution measurements.

For pure multipole radiation from heavy nuclei, one might deduce the correct spin assignment from the existing calculations by Alder and Winther. However, for the medium weight nuclei the deviations between experiment and theory are larger and the deduction of the correct spin assignment would be less certain. In odd-mass nuclei, the radiation from the first Coulomb excited state and the cascade radiation from the second Coulomb excited state will in general be $E2+M1$ radiation. In addition to inferring the spin of the excited states, information on the ratio $E2/M1$ can be determined from the angular distribution measurements. A determination of this ratio demands an accurate knowledge of the coefficients a_ν.

From the results discussed in this paper, it appears possible to use empirically determined energy-dependent coefficients a_ν for a thick target to interpret angular distribution measurements involving pure or mixed multipole transitions in neighboring odd-mass nuclei.

[22] L. C. Biedenharn and C. M. Class, Phys. Rev. 98, 691 (1955).

ACKNOWLEDGMENTS

It is a pleasure to express our appreciation to Dr. M. E. Rose and Dr. L. C. Biedenharn for several stimulating discussions concerning the results. We should like to express appreciation to Buford Carter, who carried out the I.B.M. calculations and to Dr. M. H. Lietzke of the Chemistry Division for advice on the electrodeposition of bismuth.

Reprinted from The Physical Review, Vol. 106, No. 3, 522-529, May 1, 1957
Printed in U. S. A.

Angular Distribution of Gamma Rays Following Coulomb Excitation in Even-Even Nuclei

F. K. McGowan and P. H. Stelson
Oak Ridge National Laboratory, Oak Ridge, Tennessee
(Received January 7, 1957)

The angular distributions of gamma radiation from Coulomb excitation with respect to the incident proton beam on thick targets have been measured in the even-even nuclei Ru^{104}, Pd^{108}, Pd^{110}, Cd^{110}, Cd^{112}, Cd^{114}, Cd^{116}, $Os^{190,192}$, and Pt^{194}. The particle parameters a_ν for a thick target deduced from the measurements on Ru^{104}, Pd^{110}, Cd^{114}, and Pt^{194} agree to within the accuracy of the experiments (± 2 to 5% for a_2) with the numerical results from the quantum-mechanical treatment of the Coulomb excitation process.

The slight attenuation of the angular distribution of the gamma rays by the multiple scattering of the protons by Rutherford scattering as they traverse a thick target has been discussed in a previous paper.[4] The form of the correlation function is unchanged and

[4] F. K. McGowan and P. H. Stelson, Phys. Rev. 99, 127 (1955).

each coefficient $a_\nu A_\nu$ becomes multiplied by an attenuation factor G_ν. Following the notation of Goudsmit and Saunderson,[13] who have expressed the multiple scattering function as a series in Legendre polynomials, the G_ν are given by

$$G_\nu = e^{-x_\nu}, \tag{4}$$

where

$$x_\nu = 2\pi K^2 N t \nu(\nu+1)[\log\beta - (\tfrac{1}{2}+\tfrac{1}{3}+\cdots 1/\nu)].$$

The reader is referred to their treatment for a discussion of the assumptions inherent in Eq. (4). In our previous paper[4] we chose for the effect of screening by the atomic electrons a cutoff in the Rutherford scattering law which was not appropriate for the scattering of protons.[14] The effect of the cutoff comes into the attenuation coefficient through the factor $\log\beta$. According to Williams[15] the Born approximation is not valid for $\eta_i \equiv Ze^2/\hbar v_i \gg 1$ but instead one must use the classical method of orbits in the treatment of multiple scattering with shielding. For our experiments we have $4 \leq \eta_i \leq 11$ and we have chosen a cutoff which Williams found appropriate to fit some data from multiple scattering of α particles. In addition the values of $dE/d\rho x$ have been revised upward by 8 to 10%. The net result of these changes is to reduce the x_ν about 40%. The effective attenuation coefficients $[G_\nu]_t$ for a thick target have been evaluated for a few cases, where $[G_\nu]_t$ is defined by

$$[G_\nu]_t = \int_0^{E_i} \frac{\sigma(E) a_\nu G_\nu dE}{dE/d\rho x} \Big/ \int_0^{E_i} \frac{\sigma(E) a_\nu dE}{dE/d\rho x}, \tag{5}$$

and G_ν is given by Eq. (4). These results are listed in Table IV under columns 7 and 8. Since the effect of multiple Rutherford scattering on the angular distribution coefficients is relatively small, we have not computed the $[G_\nu]_t$ for every entry in Table IV. For neighboring nuclei we find that $[G_\nu]_t$ fall on a smooth curve as a function of ξ_i, where

$$\xi = \frac{Z_1 Z_2 e^2}{\hbar}\left(\frac{1}{v_f} - \frac{1}{v_i}\right)$$

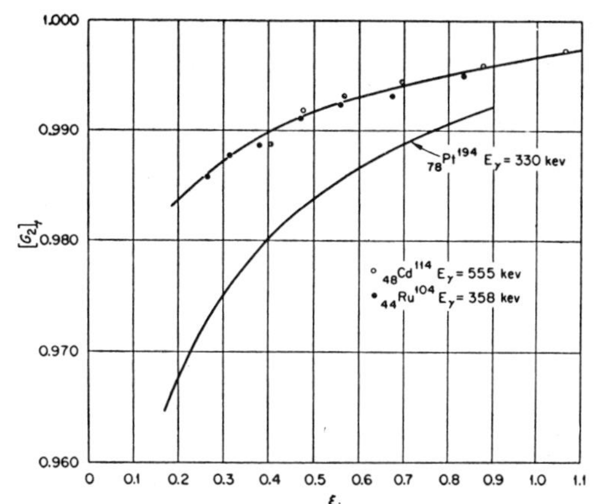

FIG. 4. The effective attenuation coefficient $[G_2]_t$ for multiple Rutherford scattering as a function of ξ_i for a few representative cases.

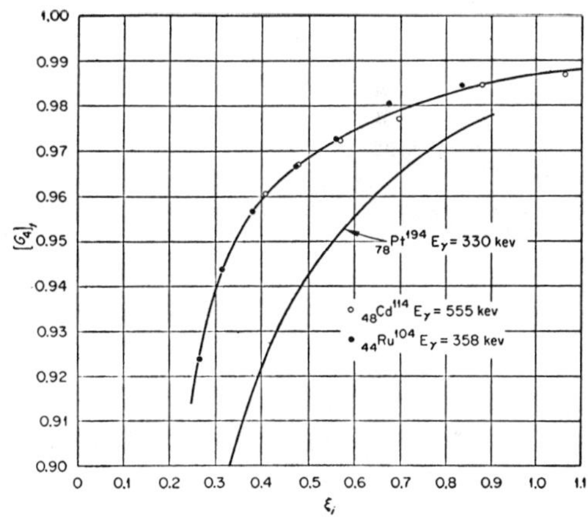

FIG. 5. The effective attenuation coefficient $[G_4]_t$ for multiple Rutherford scattering as a function of ξ_i for a few representative cases.

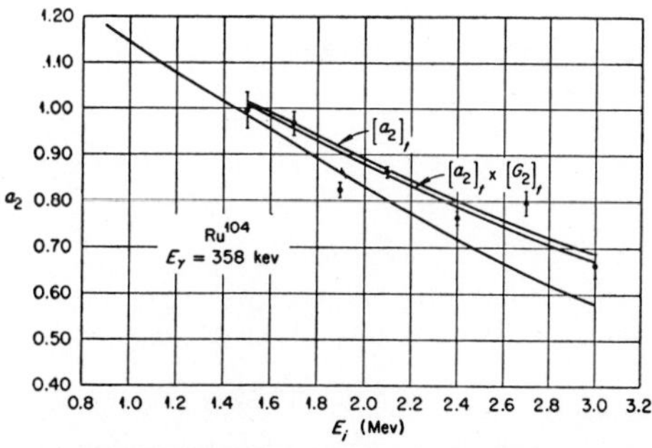

FIG. 8. Particle parameter a_2 as a function of the incident proton energy for Ru^{104}.

[13] S. Goudsmit and J. L. Saunderson, Phys. Rev. 57, 24 (1940).
[14] We are indebted to Dr. T. Huus (private communication) for pointing out that the attenuation appeared to be too large in our previous paper.
[15] E. J. Williams, Phys. Rev. 58, 292 (1940).

Reprinted from THE PHYSICAL REVIEW, Vol. 98, No. 3, 691–693, May 1, 1955 *
Printed in U. S. A.

Soluble Problem in the Theory of Coulomb Excitation

L. C. BIEDENHARN AND C. M. CLASS
The Rice Institute, Houston, Texas
(Received January 6, 1955)

The quantum treatment of Coulomb excitation in the limiting case of no energy loss involves radial Coulomb integrals that can be expressed in simple terms. The excitation function, G, and particle parameter for the directional correlation, a_2, are evaluated for this case and compared to the classical limit. The deviation from the classical limit is found to be negligible for the excitation function in the region of experimental interest, but for the particle parameter a_2 the deviation is sizeable.

THE Coulomb excitation of nuclear levels has been customarily treated[1,2] as an interaction involving the electric field of impinging particles traveling in definite Kepler orbits. This use of classical trajectories is valid, according to Bohr,[3] if the parameter $\eta \equiv Z_1Z_2e^2/\hbar v$ is large compared to unity. There have been two types of problem considered: the total cross section for the excitation process[1] as measured, say, by the γ quanta produced, and the directional correlation[2] of these quanta with the incident particle beam. Numerous experiments in the range $\eta \sim 3$ to 10 have shown generally good agreement with the approximate theory for the total cross section, but recent data on the correlation has suggested appreciable deviations.[4] It is therefore of some interest to examine more critically the validity of the classical approximation.[5] An essentially exact quantum mechanical treatment of both problems has been carried out,[6] reducing the problem to integrals over the radial Coulomb wave functions. (Evaluation of these integrals, a rather formidable task, is in progress using electronic computers.) The usual classical approximation results from this exact quantum mechanical treatment by a simultaneous limiting process $\eta \rightarrow \infty$, $1-\rho \rightarrow 0$, $\eta(1-\rho) \equiv \xi \rightarrow$finite, where ρ is the ratio of the emergent to incident wave numbers for the impinging particle. This limit process is, in general, difficult, if not impossible, to carry through explicitly for the case of arbitrary energy losses.[7] For the particular case of no energy loss ($\xi=0$), however, the two limits may be carried out separately, letting first $\rho=1$ and then $\eta \rightarrow \infty$. Fortunately the relevant Coulomb integrals may be integrated exactly in this case,[8] and it is thus possible to treat the approach to the classical limit in detail. It is obvious from Bohr's considerations that the calculations must agree precisely in the limit $\eta \rightarrow \infty$, and we may therefore use the classical limit to normalize our results and simplify the discussion.

The analogy between the classical calculation as given in reference 2, and the quantum mechanical calculations given in reference 6, is remarkably close. Let us restrict attention to the quadrupole transitions in the following. The excitation function then is defined[2] as:

$$g_2(0) = \int_1^\infty \epsilon d\epsilon \sum_{\mu=0,\pm 2} |S_\mu^{(2)}(0)|^2, \tag{1}$$

while the quantum result[6] is, normalized as discussed above,

$$G(\eta,1) = 32\eta^2 \sum_{L=0}^{\infty} \left[\frac{L(L+1)(2L+1)}{(2L-1)(2L+3)} I_{L,L}^2 + \frac{3L(L-1)}{2(2L-1)} \times I_{L,L-2}^2 + \frac{3}{2} \frac{(L+1)(L+2)}{(2L+3)} I_{L,L+2}^2 \right], \tag{2a}$$

$$I_{L,L'} \equiv \int_0^\infty r^{-3} dr F_L(\eta,r) F_{L'}(\eta,r). \tag{2b}$$

(The $F_L(\eta,r)$ are the radial Coulomb wave functions, and the $I_{L,L'}$ is taken to be zero if $L+L' \leqslant 0$.)

Noting that the eccentricity ϵ is related to the angular momentum L by the equation $\epsilon^2 = 1 + L^2/\eta^2$, one sees already the close formal connection between the two calculations. The Coulomb integrals can be evaluated, with the result that:

$$I_{L,L} = [2L(L+1)(2L+1)]^{-1}[2L+1-\pi\eta - i\eta\psi(L+1+i\eta) + i\eta\psi(L+1-i\eta)], \tag{3a}$$

$$I_{L,L+2} = I_{L+2,L} = \tfrac{1}{6} |L+1+i\eta|^{-1} |L+2+i\eta|^{-1}. \tag{3b}$$

The function $\psi(z)$ is the logarithmic derivative of the gamma function.

For large values of L these integrals approach the

*Corrections given in Phys. Rev. 100, 1790 (1955) have been incorporated in the article reproduced here.

[1] K. A. Ter-Martirosyan, J. Exptl. Theoret. Phys. (U.S.S.R.) **22**, 284 (1952); see also V. Weisskopf, Phys. Rev. **53**, 1018 (1938); C. Mullin and E. Guth, Phys. Rev. **82**, 141 (1951); R. Huby and H. C. Newns, Proc. Phys. Soc. (London) **A64**, 619 (1951).

[2] K. Alder and A. Winther, Phys. Rev. **91**, 1518 (1953).

[3] N. Bohr, Kgl. Danske Videnskab. Selskab, Mat.-fys. Medd. **18**, No. 8 (1948).

[4] P. H. Stelson and F. K. McGowan, Bull. Am. Phys. Soc. **29**, No. 7, 34 (1954).

[5] K. Alder and A. Winther, Phys. Rev. **96**, 237 (1954), have obtained somewhat more accurate results using a WKB approximation. G. Breit and P. B. Daitch, Phys. Rev. **96**, 1447 (1954); Daitch, Lazarus, Hull, Benedict, and Breit, Phys. Rev. **96**, 1449 (1954).

[6] L. C. Biedenharn and M. E. Rose, Oak Ridge National Laboratory Report ORNL-1789, September, 1954.

[7] An exception in the exact Sommerfeld dipole case.

[8] This result is a special case taken from a paper on radial Coulomb matrix elements, in preparation.

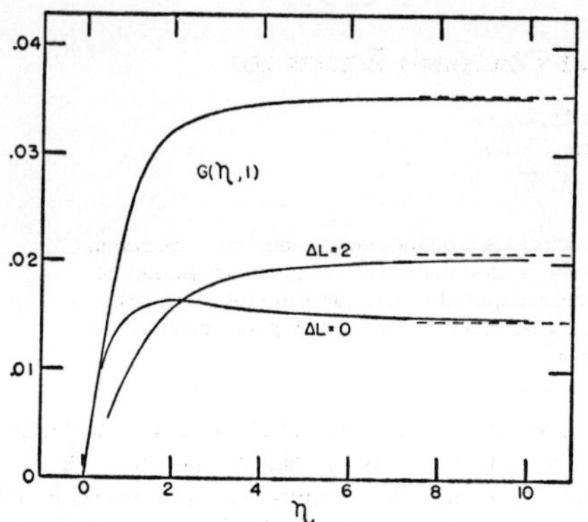

FIG. 1. The excitation function G for no energy loss ($\xi=0$) *versus* η. The lower curves are the $\Delta L=0,2$ components. The dotted lines are the classical limits.

functions $S_\mu^{(2)}$ given by Alder and Winther, and the correspondence between Eqs. (1) and (2a) is complete.

Figure 1 shows the behavior of the exact $G(\eta,1)$ function *versus* η. The classical limit is approached very quickly, the difference being only 9 percent at $\eta=2$. On the other hand, this behavior is to some extent fortuitous, as shown by the curves of the $\Delta L=0,2$ components, also plotted in Fig. 1. These depart in opposite directions from their classical limits by amounts which nearly cancel.

If one measures only the angle of emission of the γ-rays with respect to the incident particles, one obtains an angular distribution which is very similar to the angular correlation between two γ's in cascade. It can be shown that the distribution function is

$$W(\theta)=1+B_2a_2(\eta,\rho)P_2(\cos\theta)+B_4a_4(\eta,\rho)P_4(\cos\theta), \quad (4)$$

which is tc be compared with the angular correlation

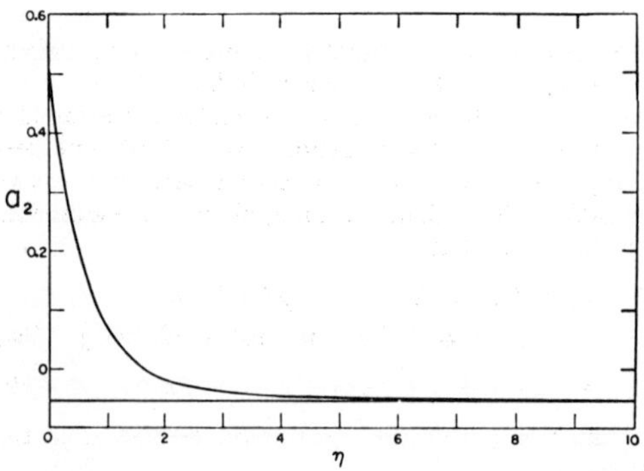

FIG. 2. The particle parameter a_2 *versus* η for no energy loss. The classical limit (solid line) is -0.05425. The intercept at $\eta=0$ is $+0.5002$.

in the $\gamma-\gamma$ cascade,

$$J_i \xrightarrow{E2} J_{ce} \xrightarrow{\gamma} J_f,$$

given by

$$W'(\theta)=1+B_2P_2(\cos\theta)+B_4P_4(\cos\theta). \quad (5)$$

The J's are the spins of the initial, Coulomb-excited, and the final state after the γ emission, respectively. The first γ transition, being an electric quadrupole radiation, corresponds to the electric quadrupole excitation process. The B_k are tabulated in reference 9, among others.

Restricting attention to the coefficient of $P_2(\cos\theta)$, one finds in the quantum calculation that:

$$a_2(\eta,1)=[G(\eta,1)]^{-1}(32\eta^2)\sum_{L=0}^{\infty}\left[\frac{3(L-2)(L-1)L}{(2L-1)^2}I_{L,L-2}{}^2 - \frac{L(L+1)(2L+1)(2L-3)(2L+5)}{(2L-1)^2(2L+3)^2}I_{L,L}{}^2 + \frac{3(L+1)(L+2)(L+3)}{(2L+3)^2}I_{L,L+2}{}^2 - 6\cos(\sigma_L-\sigma_{L-2})\cdot\frac{(L-1)(L)(L+1)}{(2L-1)^2}I_{L,L-2}I_{L,L} - 6\cos(\sigma_L-\sigma_{L+2})\cdot\frac{L(L+1)(L+2)}{(2L+3)^2}I_{L,L+2}I_{L,L}\right]. \quad (6)$$

(The σ_L are the Coulomb phase shifts for angular momentum L.)

In the limit of large L this is seen to correspond precisely term by term to the classical results given by Alder and Winther.

The behavior of the exact a_2 as a function of η is shown in Fig. 2. It is immediately apparent that the deviations from the classical limit are more significant than for the total cross section. This is not unexpected since the particle parameter a_2 is "phase-sensitive," and a classical approximation is generally inadequate. For example, the results differ by a factor of 2 at $\eta=2$, whereas the total cross section here differed by only a few percent.

The calculations at $\xi=0$ show great simplifications, and the origin of this may be found in the fact that this parameter measures the ratio of the collision time to the period of the emitted radiation. For $\xi=0$, therefore, the process is insensitive to all but the grossest details of the motion. The Coulomb integrals, in particular, show that the turning point radius, $r_t\sim(L^2+\eta^2)^{\frac{1}{2}}$, is effectively the only significant feature of the motion.

[9] L. C. Biedenharn and M. E. Rose, Revs. Modern Phys. **25**, 729 (1953).

Since the major contribution to the sums comes from $L \sim \eta$, one sees that large η implies large L. Here the effects of quantization are small, and in this way the calculation becomes classical.

The particular case $\xi = 0$ is not of too much intrinsic importance, but it does serve to illustrate several typical features of the problem and the importance of more accurate calculations for the general case. Such calculations are in progress.

We would like to express our appreciation to the Oak Ridge National Laboratory where this work was begun while one of the authors was a summer visitor. In particular, we would like to thank Dr. M. E. Rose and Dr. R. A. Charpie for their help and encouragement.

Reprinted from Reviews of Modern Physics, Vol. 28, No. 4, 432–542, October, 1956
Printed in U. S. A.

Study of Nuclear Structure by Electromagnetic Excitation with Accelerated Ions*

K. Alder, A. Bohr, T. Huus, B. Mottelson, and A. Winther

CERN Theoretical Study Division and Institute for Theoretical Physics, University of Copenhagen, Copenhagen, Denmark

* Reprinted with corrections by the authors.

TABLE OF CONTENTS

CHAPTER I. INTRODUCTION

ALREADY in the early stages of the study of nuclear reactions, the possibility was discussed[1] of producing nuclear excitations by the long-range electric interactions with bombarding particles.[2] Particularly for incident energies so low that the Coulomb repulsion prevents the particles from penetrating into the nucleus, such excitation processes can be studied without interference from the more complicated nuclear interactions. Following these early theoretical suggestions, the possibility was discussed that an isomeric activity in indium, observed in charged particle bombardment, might have been produced by processes of this kind.[3]

In subsequent years, the theoretical description of the "Coulomb excitation" reactions was considerably developed.[4,5] In particular, it was found that in many cases of interest a classical treatment of the trajectory of the bombarding particle is justified and leads to simple quantitative expressions for the excitation cross sections.[5]

At the same time, it was recognized that such experiments were a particularly appropriate means for investigating certain features of the nuclear structure. The nuclear states most strongly produced in the Coulomb excitation reactions are the low-lying collective excitations which are induced by the electric quadrupole field of the impinging particles. Such experiments, thus, appeared as an especially promising tool for the exploration of the nuclear rotational and vibrational spectra.[6,7]

Nuclear gamma rays which were later identified as resulting from Coulomb excitation of tantalum were first seen as a background radiation in experiments on proton induced reactions in light nuclei, in which target backings of tantalum were employed.[8] About a year after these first observations, the origin of the gamma rays and the mechanism of their production were clearly established.[9,10] It was also shown[10] that the yield of this radiation as a function of the proton energy was in agreement with the theoretical expression for electric quadrupole Coulomb excitation[5,11] and that the absolute cross sections and excitation energies could be interpreted in terms of rotational excitations.

Since these first investigations, the Coulomb excitation reaction has been extensively employed in many laboratories for the study of nuclear levels. Apart from furnishing new information on previously known nuclear states, a large number of new levels have been identified.[12]

The scope of the experiments has been extended by the utilization of alternative methods of detection. Thus, additional information has been obtained from measurements of the internal conversion electrons[13] and of the inelastically scattered proton groups.[14] While most of the experiments performed so far have employed protons, deuterons, or α particles as projectiles, the use of still heavier ions[15] may in certain respects provide additional advantages. The Coulomb excitation reaction has so far been used for the study of rather low-lying states (excitation energies up to about one Mev), but with the use of higher bombarding energies it should be possible also to explore excitations of somewhat higher energy.

Extensive experimental investigations of the Coulomb excitation process itself have also been performed and have tested the adequacy of the theoretical description. Thus, the excitation cross section as a function of the energy, charge, and mass of the projectile has been found to be in good agreement with the classical theory, when the latter is appropriately modified to take into account the energy loss of the projectile.[16] On the other hand, the accurately measured angular distribution of the γ radiation from Coulomb excitation revealed[17] significant deviations from the theory, and stimulated the development of a complete quantum-mechanical treatment of the process.[18]

---

[1] See, e.g., the discussion in Rutherford, Chadwick, and Ellis, *Radiations from Radioactive Substances* (Cambridge University Press, Cambridge, England, 1930), p. 247 ff. and the later work by L. Landau, Physik. Z. Sowjetunion **1**, 88 (1932) and V. F. Weisskopf, Phys. Rev. **53**, 1018 (1938).

[2] We shall, in the present article, confine our attention mainly to electric and magnetic excitations produced by bombardment with nuclear particles. A brief review of the theory of inelastic electron scattering is given in Sec. IIE.3.

[3] S. W. Barnes and P. W. Aradine, Phys. Rev. **55**, 50 (1939); Risser, Lark-Horovitz, and Smith, Phys. Rev. **57**, 355 (1940).

[4] C. J. Mullin and E. Guth, Phys. Rev. **82**, 141 (1951); R. Huby and H. C. Newns, Proc. Phys. Soc. (London) A **64**, 619 (1951).

[5] K. A. Ter-Martirosyan, J. Exptl. Theoret. Phys. (U.S.S.R.) **22**, 284 (1952).

[6] A. Bohr and B. Mottelson, Report of the International Physics Conference, Copenhagen, June, 1952; Kgl. Danske Videnskab. Selskab Mat. fys. Medd. **27**, No. 16 (1953).

[7] It has also been suggested [N. F. Ramsey, Phys. Rev. **83**, 659 (1951); Breit, Hull, and Gluckstern. Phys. Rev. **87** 74 (1952); Malenka, Kruse, and Ramsey, Phys. Rev. **91**, 1165 (1953)] that the polarization of the nucleus as a whole, as well as of the projectile, could be studied in reactions where the energy of the bombarding particle was insufficient to enable it to surmount the Coulomb barrier (see Sec. II D.3).

[8] R. B. Day and T. Huus, Phys. Rev. **85**, 761 (1952); C. L. McClelland, S. M. thesis, Massachusetts Institute of Technology, August, 1952.

[9] C. L. McClelland and C. Goodman, Phys. kev. **91**, 760 (1953).

[10] T. Huus and Č. Zupančič, Kgl. Danske Videnskab. Selskab. Mat. fys. Medd. **28**, No. 1 (1953).

[11] K. Alder and A. Winther, Phys. Rev. **91**, 1578 (1953).

[12] A special reference should be made to the extensive and systematic survey performed by N. Heydenburg and G. Temmer, Phys. Rev. **93**, 351 and 906 (1954); **94**, 1399 (1954); **95**, 861 (1954); **96**, 426 (1954); **98**, 1308 (1955); **100**, 150 (1955).

[13] T. Huus and J. H. Bjerregaard, Phys. Rev. **92**, 1579 (1953).

[14] B. Elbek and C. K. Bockelman (submitted for publication).

[15] Recently, Coulomb excitation studies have been made employing cyclotron accelerated nitrogen ions [Alkhazov, Andreyev, Greenberg, and Lemberg, Nuclear Phys. 2, 65 (1956)].

[16] K. Alder and A. Winther, Phys. Rev. **96**, 237 (1954).

[17] F. K. McGowan and P. H. Stelson, Phys. Rev. **99**, 127 (1955).

[18] Biedenharn, McHale, and Thaler, Phys. Rev. **100**, 376 (1955); K. Alder and A. Winther, Kgl. Danske Videnskab. Selskab Mat. fys. Medd. **29**, No. 19 (1955). Numerical results have been given by K. Alder and A. Winther, reference 11; Biedenharn, Goldstein, McHale, and Thaler, Phys. Rev. **101**, 662 (1956), and **102**, 1567 (1956). A WKB approximation which yields results in essential agreement with the detailed quantal treatment has been given

In the present review article, we begin in Chapter II with a discussion of the theory of Coulomb excitation, and give in tables and figures the numerical results necessary for the analysis of the experiments. Chapter III deals with the experimental techniques which have been employed in Coulomb excitation measurements, and also contains a discussion of the main background effects in these experiments. Chapter IV contains a compilation of results obtained in Coulomb excitation experiments and also a comparison with the theory of Chapter II. Finally, in Chapter V, we outline the theory of collective nuclear excitations, which makes possible an interpretation of many of the observed transitions.

We wish to acknowledge the benefit we have derived from contacts with experimental and theoretical physicists working in the field of Coulomb excitation, many of whom have kindly communicated to us the results of their investigations prior to publication. We are also indebted to the members of the Institute for Theoretical Physics, Copenhagen, as well as to Dr. N. P. Heydenburg, Dr. G. M. Temmer, and Dr. G. Breit for valuable discussions.

CHAPTER II. THEORY OF ELECTROMAGNETIC EXCITATIONS

The excitation of nuclei by impinging nuclear particles with energies well below the Coulomb barrier proceeds only through the electromagnetic interaction between the projectile and the nucleus.

The motion of the projectile in the Coulomb field of the nucleus is essentially characterized[19] by the dimensionless quantity η defined by

$$\eta = Z_1 Z_2 e^2/\hbar v, \tag{II A.1}$$

where Z_1 and Z_2 are the charge numbers of the projectile and the nucleus, while v is the velocity of the incident particle. The parameter η measures the effective strength of the interaction. Thus, for $\eta \ll 1$, the Coulomb field produces only a small distortion of the incident wave, and the collision process can be treated by Born approximation. For the particle velocities involved in Coulomb excitation, however, the interaction must be strong to prevent the projectiles from entering the nucleus. Under such conditions, we always have $\eta \gg 1$, and the collision may then be approximately described by considering the particle as moving along a classical trajectory. For inelastic collisions, it is a further condition for the application of a classical description that the energy loss of the particle is small compared to the bombarding energy, so that the effect of the excitation on the particle motion can be neglected.

In such a treatment, the nuclear excitation is a result of the time dependent electromagnetic field of the projectile acting on the nucleus. In most cases, the effect of this field is small and may be treated by first-order quantum-mechanical perturbation theory. The excitation probability can be expressed in terms of the same nuclear matrix elements as determine the radiative transitions between the nuclear states.

In the following we shall first consider, in Sec. A, the Coulomb excitation process in terms of such a classical treatment of the projectile. We shall describe this simplified method in some detail, since it illustrates the main physical features of the process, without involving the more complex mathematical formalism of the quantum-mechanical theory.

The more rigorous treatment of the excitation process, in which the particles are described by the Coulomb wave functions, is given in Sec. B. At the end of this Section, we also consider the application of the WKB approximation, which is intermediate between the classical and the exact quantum-mechanical treatment.

In Sec. C, the final formulas for the excitation cross sections and the angular distribution of the emitted nuclear radiation are collected. These depend on the collision parameters through certain functions which have been evaluated numerically and are tabulated and given in figures.

In Sec. D, we briefly consider some of the effects associated with the higher order excitation processes, while Sec. E has the form of an appendix, which contains partly results appropriate to certain limiting cases and partly a discussion of certain processes related to Coulomb excitation, such as bremsstrahlung and nuclear excitation produced by fast electrons.

II A. Classical Theory

In the classical treatment of the Coulomb excitation process,[5] we consider the projectile as moving along a hyperbolic orbit in the repulsive Coulomb field of the target nucleus (see Fig. II.1). The differential scattering

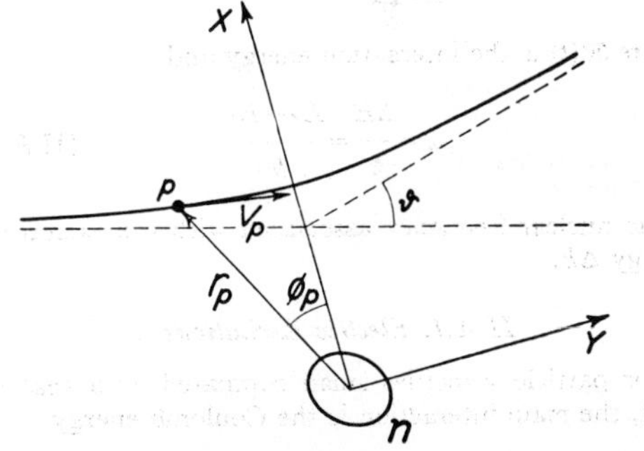

FIG. II.1. Classical picture of the projectile orbit in the Coulomb field of the nucleus. The hyperbolic orbit of the projectile, P, is shown in the frame of reference in which the nuclear mass center is at rest. The focal coordinate system employed in the evaluation of the orbital integrals (II A.24) is indicated. The position and velocity of the projectile are denoted by ϕ_p, r_p, and v_p, respectively, and the total deflection angle by ϑ.

by Benedict, Daitch, and Breit, Phys. Rev. **101**, 171 (1956); Gluckstern, Lazarus, and Breit, *ibid.* **101**, 175 (1956); F. D. Benedict, *ibid.* **101**, 178 (1956).

[19] N. Bohr, Kgl. Danske Videnskab. Selskab Mat. fys. Medd. **18**, No. 8 (1948).

cross section is given by the Rutherford law

$$d\sigma_R = \tfrac{1}{4}a^2 \sin^{-4}(\vartheta/2)d\Omega, \qquad \text{(II A.2)}$$

where ϑ is the scattering angle in the center-of-mass system, and

$$a = \frac{Z_1 Z_2 e^2}{m_0 v^2} \qquad \text{(II A.3)}$$

is half the distance of closest approach in a head-on collision. The reduced mass of the projectile and the nucleus is denoted by m_0.

Since we have assumed that the orbit of the particle is not appreciably affected by the excitation, the differential excitation cross section is given by

$$d\sigma = P d\sigma_R, \qquad \text{(II A.4)}$$

where P is the probability that the nucleus is excited in a collision in which the particle is scattered into the solid angle $d\Omega$.

The probability P can be expressed in terms of the amplitudes b_{if} for a transition from the initial nuclear state i to the various final states f. If we ask for the probability for excitation of a given energy level, irrespective of the orientation of the initial or final nuclear state, we have

$$P = (2I_i+1)^{-1} \sum_{M_i M_f} |b_{if}|^2, \qquad \text{(II A.5)}$$

where I_i is the spin of the initial nuclear state, and where M_i and M_f are the magnetic quantum numbers of the initial and final states.

Under most experimental conditions the probability for excitation in a single encounter is very small. Thus, by first-order time dependent perturbation theory,[20] we obtain

$$b_{if} = \frac{1}{i\hbar}\int_{-\infty}^{\infty} \langle f|\mathcal{H}(t)|i\rangle e^{i\omega t} dt, \qquad \text{(II A.6)}$$

where $\mathcal{H}(t)$ is the interaction energy and

$$\omega = \frac{\Delta E}{\hbar} = \frac{E_f - E_i}{\hbar} \qquad \text{(II A.7)}$$

is the nuclear frequency associated with the excitation energy ΔE.

II A.1. Electric Excitations

For particle velocities small compared with that of light, the main interaction is the Coulomb energy

$$\mathcal{H}_E(t) = \int \rho_n(\mathbf{r})\varphi(\mathbf{r},t)d\tau, \qquad \text{(II A.8)}$$

[20] P. A. M. Dirac, *The Principles of Quantum Mechanics* (Oxford University Press, New York, 1947), third edition, p. 172.

where

$$\varphi(\mathbf{r},t) = \frac{Z_1 e}{|\mathbf{r}-\mathbf{r}_p(t)|} - \frac{Z_1 e}{r_p(t)} \qquad \text{(II A.9)}$$

and $\rho_n(\mathbf{r})$ is the nuclear charge density operator. The projectile is considered as a point charge and its position vector $\mathbf{r}_p(t)$ is measured from the nuclear center of mass. In (9)* we have subtracted the interaction between the mass centers, which is responsible for the scattering and does not contribute to the excitation.

In order to evaluate the matrix element in (6) we expand the potential (9) in multipole components, whereby one obtains

$$\mathcal{H}_E(t) = 4\pi Z_1 e \sum_{\lambda=1}^{\infty} \sum_{\mu=-\lambda}^{\lambda} \frac{1}{2\lambda+1} r_p^{-\lambda-1} \times Y_{\lambda\mu}(\theta_p,\phi_p)\mathfrak{M}^*(E\lambda,\mu), \qquad \text{(II A.10)}$$

which holds if the projectile remains outside the nucleus. The electric multipole moments of the nucleus are defined by

$$\mathfrak{M}(E\lambda,\mu) = \int r^\lambda Y_{\lambda\mu}(\theta,\phi)\rho_n(\mathbf{r})d\tau. \qquad \text{(II A.11)}$$

The polar coordinates are referred to a coordinate system with origin in the nuclear center of mass and with a fixed direction of the polar axis. The $Y_{\lambda\mu}(\theta,\phi)$ are the normalized spherical harmonics.[21]

The multipole operators (11) are the same as those responsible for the emission of electric multipole radiation with wavelength large compared with the nuclear radius.[22] If we assume that the nuclear charge density can be described in terms of point charge protons we have

$$\rho_n(\mathbf{r}) = \sum_k e_k \delta(\mathbf{r}-\mathbf{r}_k), \qquad \text{(II A.12)}$$

where e_k and $\mathbf{r}_k$ are the charge and the position vector of the kth nucleon. The multipole moment can then be written in the familiar form

$$\mathfrak{M}(E\lambda,\mu) = \sum_k e_k r_k^\lambda Y_{\lambda\mu}(\theta_k,\phi_k). \qquad \text{(II A.13)}$$

Inserting (10) into (6) we get for the transition amplitude

$$b_{if} = \frac{4\pi Z_1 e}{i\hbar} \sum_{\lambda\mu} \frac{1}{2\lambda+1} \times \langle I_i M_i|\mathfrak{M}(E\lambda,\mu)|I_f M_f\rangle S_{E\lambda,\mu}, \qquad \text{(II A.14)}$$

* In each chapter we have referred to the equations in that chapter without adding the chapter or section designation. For example, this reference is to Eq. (II A.9)

[21] We use the phases employed by E. U. Condon and G. H. Shortley, *Theory of Atomic Spectra* (Cambridge University Press, New York, 1935).

[22] See, e.g., J. M. Blatt and V. F. Weisskopf, *Theoretical Nuclear Physics* (John Wiley & Sons, Inc., New York, 1952).

where we have introduced the notation

$$S_{E\lambda,\mu}=\int_{-\infty}^{\infty} e^{i\omega t}Y_{\lambda\mu}(\theta_p(t),\phi_p(t))[r_p(t)]^{-\lambda-1}dt \quad \text{(II A.15)}$$

for the orbital integrals, and where we have specified the nuclear states by their total angular momentum I and magnetic quantum number M.

Since the multipole moments are tensor operators, we may write[23]

$$\langle I_iM_i|\mathfrak{M}(\lambda,\mu)|I_fM_f\rangle$$
$$=(-1)^{I_i-M_i}\begin{pmatrix} I_i & \lambda & I_f \\ -M_i & \mu & M_f\end{pmatrix}\langle I_i\|\mathfrak{M}(\lambda)\|I_f\rangle, \quad \text{(II A.16)}$$

where the last factor is the reduced matrix element. We use the Wigner notation for the vector addition coefficients,[24] which is related to the notation employed by Condon and Shortley[21] by

$$\begin{pmatrix} j_1 & j_2 & j_3 \\ m_1 & m_2 & m_3\end{pmatrix}$$
$$=\frac{(-1)^{j_1-j_2-m_3}}{(2j_3+1)^{\frac{1}{2}}}\langle j_1j_2m_1m_2|j_1j_2j_3-m_3\rangle. \quad \text{(II A.17)}$$

Further, we introduce

$$B(E\lambda;I_i\rightarrow I_f)$$
$$=\sum_{M_f\mu}|\langle I_iM_i|\mathfrak{M}(E\lambda,\mu)|I_fM_f\rangle|^2$$
$$=(2I_i+1)^{-1}|\langle I_i\|\mathfrak{M}(E\lambda)\|I_f\rangle|^2, \quad \text{(II A.18)}$$

which represents the reduced transition probability associated with a radiative transition of multipole order $E\lambda$.

By inserting into (4) Eqs. (2), (5), and (14), and using the orthogonality relation,[24]

$$\sum_{M_iM_f}\begin{pmatrix} I_i & \lambda & I_f \\ -M_i & \mu & M_f\end{pmatrix}\begin{pmatrix} I_i & \lambda' & I_f \\ -M_i & \mu' & M_f\end{pmatrix}$$
$$=(2\lambda+1)^{-1}\delta_{\lambda\lambda'}\delta_{\mu\mu'} \quad \text{(II A.19)}$$

for the vector addition coefficients, we get for the differential excitation cross section

$$d\sigma_E=\sum_{\lambda=1}^{\infty} d\sigma_{E\lambda}, \quad \text{(II A.20)}$$

with

$$d\sigma_{E\lambda}=\frac{4\pi^2Z_1^2e^2}{\hbar^2}a^2$$
$$\times\sin^{-4}\frac{\vartheta}{2}\,\frac{B(E\lambda)}{(2\lambda+1)^3}\sum_\mu|S_{E\lambda,\mu}|^2d\Omega. \quad \text{(II A.21)}$$

The evaluation of the integrals $S_{E\lambda,\mu}$ is most easily performed if the coordinates (r_p,θ_p,ϕ_p) are given in the focal system of the hyperbolic orbit (see Fig. II.1). In this system, a convenient parametric representation is

$$\begin{aligned} x_p&=a(\cosh w+\epsilon),\\ y_p&=a(\epsilon^2-1)^{\frac{1}{2}}\sinh w,\\ z_p&=0, \quad \text{(II A.22)}\\ r_p&=a(\epsilon\cosh w+1),\\ t&=\frac{a}{v}(\epsilon\sinh w+w).\end{aligned}$$

The eccentricity ϵ is related to the deflection angle ϑ by

$$\epsilon=\frac{1}{\sin(\vartheta/2)}. \quad \text{(II A.23)}$$

Since $\theta_p=\pi/2$, the $S_{E\lambda,\mu}$ take the form

$$S_{E\lambda,\mu}=Y_{\lambda\mu}\left(\frac{\pi}{2},0\right)\int_{-\infty}^{\infty}\frac{(x_p+iy_p)^\mu}{r_p^{\lambda+\mu+1}}e^{i\omega t}dt$$
$$=v^{-1}a^{-\lambda}Y_{\lambda\mu}\left(\frac{\pi}{2},0\right)I_{\lambda\mu}(\vartheta,\xi), \quad \text{(II A.24)}$$

where[25]

$$Y_{\lambda\mu}\left(\frac{\pi}{2},0\right)=\begin{cases}\left(\dfrac{2\lambda+1}{4\pi}\right)^{\frac{1}{2}}\dfrac{[(\lambda-\mu)!(\lambda+\mu)!]^{\frac{1}{2}}}{(\lambda-\mu)!!(\lambda+\mu)!!}(-1)^{(\lambda+\mu)/2} & (\lambda+\mu\text{ even})\\ 0 & (\lambda+\mu\text{ odd})\end{cases} \quad \text{(II A.25)}$$

and where

$$I_{\lambda\mu}(\vartheta,\xi)=\int_{-\infty}^{\infty}e^{i\xi(\epsilon\sinh w+w)}$$
$$\times\frac{[\cosh w+\epsilon+i(\epsilon^2-1)^{\frac{1}{2}}\sinh w]^\mu}{[\epsilon\cosh w+1]^{\lambda+\mu}}dw. \quad \text{(II A.26)}$$

The dimensionless quantity ξ is defined by

$$\xi=\frac{a\Delta E}{\hbar v}=\frac{Z_1Z_2e^2}{\hbar v}\frac{\Delta E}{2E} \quad \text{(II A.27)}$$

---

[23] G. Racah, Phys. Rev. **62**, 438 (1942). In the following we assume the phases of the nuclear wave functions to be chosen in such a manner that the matrix element (II A.16) is real. In this case, one finds $\langle I_i\|\mathfrak{M}(\lambda)\|I_f\rangle=(-1)^{I_f-I_i}\langle I_f\|\mathfrak{M}(\lambda)\|I_i\rangle$.

[24] A. R. Edmonds, Angular Momentum in Quantum Mechanics, CERN 55-26, Geneva, 1955. The notation is also employed by A. de Shalit, Phys. Rev. **91**, 1479 (1953).

[25] We have used the notation $(2n)!!=2\cdot4\cdot6\cdots2n$ and $(2n+1)!!=1\cdot3\cdot5\cdots(2n+1)$.

with $E=\frac{1}{2}m_0v^2$. The product $\xi\epsilon$ represents the ratio between the collision time and the nuclear period, and is thus a measure of the extent to which the process is adiabatic.

The properties of the integrals $I_{\lambda\mu}$ are discussed in Secs. II E.4–7.

The differential excitation cross section (21) may thus be written

$$d\sigma_{E\lambda}=\left(\frac{Z_1e}{\hbar v}\right)^2 a^{-2\lambda+2}B(E\lambda)df_{E\lambda}(\vartheta,\xi), \quad \text{(II A.28)}$$

with

$$df_{E\lambda}(\vartheta,\xi)=\frac{4\pi^2}{(2\lambda+1)^3}\sum_\mu \left|Y_{\lambda\mu}\left(\frac{\pi}{2},0\right)\right|^2 \times|I_{\lambda\mu}(\vartheta,\xi)|^2\sin^{-4}\frac{\vartheta}{2}d\Omega. \quad \text{(II A.29)}$$

The total excitation cross section of order $E\lambda$, obtained by integration over all scattering directions, is given by

$$\sigma_{E\lambda}=\left(\frac{Z_1e}{\hbar v}\right)^2 a^{-2\lambda+2}B(E\lambda)f_{E\lambda}(\xi), \quad \text{(II A.30)}$$

where

$$f_{E\lambda}(\xi)=\int\frac{df_{E\lambda}(\vartheta,\xi)}{d\Omega}d\Omega$$

$$=\frac{16\pi^3}{(2\lambda+1)^3}\sum_\mu\left|Y_{\lambda\mu}\left(\frac{\pi}{2},0\right)\right|^2 \times\int_0^\pi |I_{\lambda\mu}(\vartheta,\xi)|^2\frac{\cos\frac{\vartheta}{2}}{\sin^3\frac{\vartheta}{2}}d\vartheta. \quad \text{(II A.31)}$$

The excitation processes considered so far, which are produced by the electrostatic interaction (8), are subject to the usual parity selection rule for electric multipole radiation. Thus, an excitation of order λ involves a parity change of $(-1)^\lambda$.

Excitations of opposite parity can be produced by the magnetic field from the projectile.[26] Such magnetic excitations usually have very small cross sections, since for bombarding energies below the Coulomb barrier the projectile velocity is small compared with that of light. Still, in cases where electric transitions are forbidden, or in the case of γ-ray angular distributions where there are interference terms between electric and magnetic excitations, it may be possible to observe the magnetic effects.

[26] The classical treatment of the magnetic excitations was first given by M. Jean and J. Prentki, Compt. rend. **238**, 2290 (1954).

II A.2. Magnetic Excitations

To lowest order in the particle velocity, the magnetic interaction is contained in the expression

$$\mathcal{H}(t)=-\frac{1}{c}\int\mathbf{j}_n(\mathbf{r})\cdot\mathbf{A}(\mathbf{r},t)d\tau, \quad \text{(II A.32)}$$

where $\mathbf{j}_n(\mathbf{r})$ is the nuclear current density and

$$\mathbf{A}(\mathbf{r},t)=\frac{Z_1e}{c}\frac{\mathbf{v}_p(t)}{|\mathbf{r}-\mathbf{r}_p(t)|}, \quad \text{(II A.33)}$$

is the unretarded vector potential produced by the projectile. We measure $\mathbf{j}_n$ and $\mathbf{A}$ in the nuclear rest system, and thus $\mathbf{v}_p(t)$ is the instantaneous relative velocity of projectile and nucleus.

In (33) we have neglected the contribution from a possible magnetic moment of the projectile. This effect, however, is usually small compared to the magnetic effect of the orbital motion, since, for $\eta\gg1$, the main contribution to the excitation arises from collisions with large orbital angular momenta l. For $\xi\lesssim1$, the order of magnitude of the effective l is given by $l\hbar\gtrsim m_0av=\eta\hbar\gg\hbar$.

Expanding in spherical harmonics we obtain for the vector potential (33)

$$\mathbf{A}(\mathbf{r})=\sum_{\lambda\mu}\frac{4\pi}{2\lambda+1}\frac{Z_1e}{c}\mathbf{v}_p r_p^{-\lambda-1} \times Y_{\lambda\mu}(\theta_p,\phi_p)r^\lambda Y_{\lambda\mu}^*(\theta,\phi). \quad \text{(II A.34)}$$

In this potential the terms involving r^λ contain, besides the magnetic multipole component of order λ, also electric multipole components of order $\lambda\pm1$. These contribute a small relativistic correction to the electric excitations, and will here be disregarded.[27] In order to extract the magnetic part of (34) we take the component of $\mathbf{A}$ along the direction of $\mathbf{L}$ (see expression (II B.6) below), where

$$\mathbf{L}=-i[\mathbf{r}\times\nabla]. \quad \text{(II A.35)}$$

Thus, one obtains

$$\mathbf{A}_M(\mathbf{r})=\frac{4\pi Z_1e}{c}\sum_\lambda r^\lambda r_p^{-\lambda-1}\frac{\mathbf{L}(\mathbf{L}\cdot\mathbf{v}_p)}{(2\lambda+1)\lambda(\lambda+1)} \times\sum_\mu Y_{\lambda\mu}(\theta_p,\phi_p)Y_{\lambda\mu}^*(\theta,\phi), \quad \text{(II A.36)}$$

since

$$(\mathbf{L}^2-\lambda(\lambda+1))Y_{\lambda\mu}(\theta,\phi)=0. \quad \text{(II A.37)}$$

The sum over μ in (36) depends only on the relative angle of the vectors $\mathbf{r}$ and $\mathbf{r}_p$, and the operator $\mathbf{L}$ acting on this sum can therefore be replaced by $\mathbf{L}_p$, where $\mathbf{L}_p$ acts on the projectile coordinates.

For the magnetic multipole part of (32) we thus

[27] The complete relativistic interaction is derived in Sec. II B.1.

obtain

$$\mathcal{H}_M(t)=\frac{4\pi i Z_1 e}{c}\sum_{\lambda\mu}\frac{1}{\lambda(2\lambda+1)}(\mathbf{L}_p\cdot\mathbf{v}_p)r_p^{-\lambda-1}\times Y_{\lambda\mu}(\theta_p,\phi_p)\mathfrak{M}^*(M\lambda,\mu), \quad \text{(II A.38)}$$

where

$$\mathfrak{M}(M\lambda,\mu)=-\frac{i}{c(\lambda+1)}\int(\mathbf{j}_n\cdot\mathbf{L})r^\lambda Y_{\lambda\mu}(\theta,\phi)d\tau \quad \text{(II A.39)}$$

is the nuclear magnetic multipole moment, which is associated also with radiative transitions of order $M\lambda$.[22]

If we describe the nuclear current in terms of a convection current of point charge protons and a magnetization current associated with point dipole moments of the nucleons, we have

$$\mathbf{j}_n(\mathbf{r})=\sum_k e_k(\mathbf{v}_k\delta(\mathbf{r}-\mathbf{r}_k))_{\text{sym}}+\frac{e\hbar}{2M}g_{sk}\nabla\times\mathbf{s}_k\delta(\mathbf{r}-\mathbf{r}_k), \quad \text{(II A.40)}$$

where $\mathbf{s}_k$ and g_{sk} are the spin-vector and the spin-gyromagnetic ratio for the kth nucleon, while M is the proton mass. The subscript "sym" indicates a symmetrization of the factors in the parenthesis. With the expression (40) for $\mathbf{j}_n$ we obtain the multipole moment (39) in the familiar form

$$\mathfrak{M}(M\lambda,\mu)=\frac{e\hbar}{2Mc}\sum_k\left(g_{sk}\mathbf{s}_k+\frac{2}{\lambda+1}g_{lk}\mathbf{l}_k\right)\cdot\nabla(r_k^\lambda Y_{\lambda\mu}(\theta_k,\phi_k)), \quad \text{(II A.41)}$$

where g_{lk} is the orbital g factor for the kth nucleon.

Inserting (38) we get for the transition amplitude (6)

$$b_{if}=\frac{4\pi Z_1 e}{i\hbar}\sum_{\lambda\mu}\frac{1}{2\lambda+1}\times\langle I_iM_i|\mathfrak{M}(M\lambda,\mu)|I_fM_f\rangle S_{M\lambda,\mu}, \quad \text{(II A.42)}$$

where

$$S_{M\lambda,\mu}=-\frac{1}{\lambda}\frac{\hbar}{m_0c}\mathbf{l}_p\cdot\int\nabla_p r_p^{-\lambda-1}\times Y_{\lambda\mu}(\theta_p,\phi_p)e^{i\omega t}dt. \quad \text{(II A.43)}$$

We have used the relation [see (35)]

$$\mathbf{L}_p\cdot\mathbf{v}_p=\frac{i\hbar}{m_0}\mathbf{l}_p\cdot\mathbf{\Delta}_p, \quad \text{(II A.44)}$$

where $\hbar\mathbf{l}_p$ is the relative orbital angular momentum, which is a constant of the motion. This vector is perpendicular to the plane of the orbit and its magnitude is related to the deflection angle ϑ by

$$\hbar l_p=am_0v\cot\frac{\vartheta}{2}. \quad \text{(II A.45)}$$

It is convenient, as for the electric excitations, to evaluate the orbital integrals in the focal system (22). In this coordinate system we have

$$\mathbf{l}_p\cdot\nabla_p\bigg|_{\theta_p=\pi/2}=l_pr_p^{-1}\frac{\partial}{\partial\theta_p}\bigg|_{\theta_p=\pi/2} \quad \text{(II A.46)}$$

and, by employing the formula

$$\frac{\partial}{\partial\theta_p}Y_{\lambda\mu}(\theta_p,\phi_p)\bigg|_{\theta_p=\pi/2}=\left(\frac{2\lambda+1}{2\lambda+3}\right)^{\frac{1}{2}}[(\lambda+1)^2-\mu^2]^{\frac{1}{2}}Y_{\lambda+1,\mu}\left(\frac{\pi}{2},\phi_p\right), \quad \text{(II A.47)}$$

we may express the orbital integrals (43) in terms of those involved in the electric excitations (15). By means of (45) and (24), one obtains

$$S_{M\lambda,\mu}=-c^{-1}a^{-\lambda}\frac{1}{\lambda}\left(\frac{2\lambda+1}{2\lambda+3}\right)^{\frac{1}{2}}[(\lambda+1)^2-\mu^2]^{\frac{1}{2}}\times Y_{\lambda+1,\mu}\left(\frac{\pi}{2},0\right)I_{\lambda+1,\mu}(\vartheta,\xi)\cot\frac{\vartheta}{2}. \quad \text{(II A.48)}$$

In complete analogy with the derivation of the cross sections for electric excitations we thus obtain

$$d\sigma_M=\sum_{\lambda=1}^{\infty}d\sigma_{M\lambda}, \quad \text{(II A.49)}$$

with

$$d\sigma_{M\lambda}=\left(\frac{Z_1e}{\hbar c}\right)^2a^{-2\lambda+2}B(M\lambda)df_{M\lambda}(\vartheta,\xi), \quad \text{(II A.50)}$$

and

$$df_{M\lambda}(\vartheta,\xi)=\frac{4\pi^2}{(2\lambda+1)^2}\sum_\mu\frac{(\lambda+1)^2-\mu^2}{\lambda^2(2\lambda+3)}\left|Y_{\lambda+1,\mu}\left(\frac{\pi}{2},0\right)\right|^2\times|I_{\lambda+1,\mu}(\vartheta,\xi)|^2\cot^2\frac{\vartheta}{2}\sin^{-4}\frac{\vartheta}{2}d\Omega. \quad \text{(II A.51)}$$

We have here introduced

$$B(M\lambda)=\sum_{\mu M_f}|\langle I_iM_i|\mathfrak{M}(M\lambda,\mu)|I_fM_f\rangle|^2=(2I_i+1)^{-1}|\langle I_i\|\mathfrak{M}(M\lambda)\|I_f\rangle|^2 \quad \text{(II A.52)}$$

in analogy to (18) and have employed the relation (16).

The total excitation cross section of order $M\lambda$ is given by

$$\sigma_{M\lambda}=\left(\frac{Z_1e}{\hbar c}\right)^2a^{-2\lambda+2}B(M\lambda)f_{M\lambda}(\xi), \quad \text{(II A.53)}$$

with

$$f_{M\lambda}(\xi)=\frac{16\pi^3}{(2\lambda+1)^2}\sum_{\mu}\frac{(\lambda+1)^2-\mu^2}{\lambda^2(2\lambda+3)}\left|Y_{\lambda+1,\mu}\left(\frac{\pi}{2},0\right)\right|^2$$

$$\times\int_0^{\pi}|I_{\lambda+1,\mu}(\vartheta,\xi)|^2\cot^2\frac{\vartheta}{2}\cdot\frac{\cos\frac{\vartheta}{2}}{\sin^3\frac{\vartheta}{2}}d\vartheta. \quad \text{(II A.54)}$$

II A.3. Discussion of Cross Sections

The electromagnetic excitation cross sections, derived above, are expressed in terms of the reduced nuclear transition probabilities and the functions $f(\xi)$ and $df(\vartheta,\xi)$. The orbital integrals $I_{\lambda\mu}$ entering in these functions are defined by Eq. (26) and can be expressed in terms of confluent hypergeometric functions of two variables (see Sec. II E.4). In the special cases of $\lambda=1$ or $\xi=0$, the $I_{\lambda\mu}$ reduce to simpler functions (see Secs. II E.5 and 6). The integrals have also been evaluated numerically, and the results are given, for $\lambda=2$, in Sec. II E.4. The numerically evaluated f and df functions are given in Sec. II C for $E1$, $E2$, $E3$, $E4$, $M1$, and $M2$ excitations.

An important feature of the functions $f(\xi)$ is the exponential decrease for large values of ξ (see, e.g., Fig. II.4). This is a consequence of the approximately adiabatic character of the collisions for $\xi>1$, for which the collision time is large compared to the nuclear period [see (27)]. In the opposite limit of $\xi\to 0$, all the $f(\xi)$ approach a finite value except for the $E1$ and $M1$ excitations. The functions $f_{E1}(\xi)$ and $f_{M1}(\xi)$ increase logarithmically for small ξ (see Sec. II E.5) in analogy to the well-known logarithmic dependence of the atomic stopping power on the atomic excitation frequencies.

As already mentioned, the electromagnetic excitation involves the same nuclear matrix elements as the radiative transition of corresponding multipole order. Thus, the excitation process is subject to the usual selection rules

$$|I_i-I_f|\leqslant\lambda\leqslant I_i+I_f,$$

$$\pi_i\pi_f=\begin{cases}(-1)^{\lambda} & \text{for } E\lambda,\\ (-1)^{\lambda+1} & \text{for } M\lambda,\end{cases} \quad \text{(II A.55)}$$

where π_i and π_f are the parities of the initial and final nuclear states.

There is therefore also a simple relation between the excitation cross section and the lifetime for the radiative decay of the excited state by the corresponding multipole transition. The probability per unit time for such a transition is given by[22]

$$T=\frac{8\pi(\lambda+1)}{\lambda[(2\lambda+1)!!]^2}\frac{1}{\hbar}\left(\frac{\omega}{c}\right)^{2\lambda+1}B(\lambda;I_f\to I_i), \quad \text{(II A.56)}$$

where the reduced transition probability $B(\lambda;I_f\to I_i)$ for the decay is related by

$$B(\lambda;I_f\to I_i)=\frac{2I_i+1}{2I_f+1}B(\lambda;I_i\to I_f) \quad \text{(II A.57)}$$

to the reduced transition probability $B(\lambda;I_i\to I_f)$ entering into the expression for the excitation cross section. The relation (57) is equivalent to the fact that the magnitude of the reduced matrix elements $\langle I_i\|\mathfrak{M}(\lambda)\|I_f\rangle$ is symmetric with respect to interchange of initial and final state[23] (see (18)).

The electromagnetic field acting on the nucleus in a collision with a charged particle differs, however, in various respects from that involved in the emission or absorption of a photon, and this implies certain essential differences between the two processes as regards the relative contributions of the various multipole components. Thus, while in the radiative field the electric and magnetic field strengths are of equal magnitude, the magnetic field of the bombarding particle is only of order v/c as compared with the electric field. Magnetic excitations therefore are reduced, with respect to electric ones, by a factor $(v/c)^2$, apart from differences in the nuclear matrix elements. Moreover, while, in radiative processes, the relative intensities of consecutive multipole orders involve a factor $[(\omega/c)R_0]^2$, where R_0 is the nuclear radius, the corresponding factor in Coulomb excitation is $(R_0/a)^2$ [see (30) and (53)]. The latter factor is much larger than the former since, according to (27), we have $(\omega/c)a=(v/c)\xi$. Therefore, the cross section for Coulomb excitation does not decrease as rapidly with increasing multipole order as does the intensity of radiative processes.

A convenient unit in which to measure the nuclear transition probabilities $B(\lambda)$ is the "single-particle unit" defined by[22]

$$B_{\mathrm{sp}}(\lambda)=(2\lambda+1)\frac{e^2}{4\pi}\left(\frac{3}{3+\lambda}\right)^2R_0^{2\lambda}\times\begin{cases}1 & \text{for } E\lambda\\ 10\left(\frac{\hbar}{McR_0}\right)^2 & \text{for } M\lambda,\end{cases} \quad \text{(II A.58)}$$

where M is the proton mass. We have included, somewhat arbitrarily, a statistical factor $2\lambda+1$, since the Coulomb excitation usually, and always in even-even nuclei, involves an increase in the nuclear spin.

Figure II.1a gives the excitation cross sections for proton bombardment of a medium heavy nucleus ($Z_2=50$, $A_2=120$), assuming $B(\lambda)$ equal to the unit (58) with $R_0=1.2\cdot A^{\frac{1}{3}}\cdot 10^{-13}$ cm.[28] The excitation energy is taken to be 200 kev.

[28] This value for the nuclear radius seems the most appropriate in connection with the interpretation of evidence regarding the nuclear charge distribution (see Chapter V).

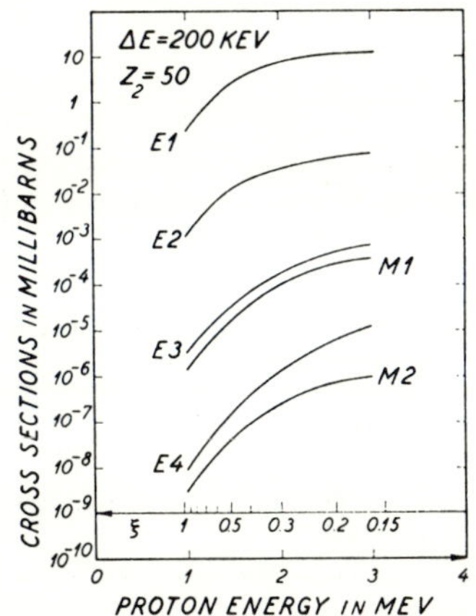

FIG. II.1a. Excitation cross sections for nuclear transitions of single particle strength. The curves give the total Coulomb excitation cross sections of various multipole orders for proton bombardment of a nucleus with $Z_2=50$ [see (II A.30) and (II A.53), and Fig. II.4]. The excitation energy is taken to be 200 kev, and the reduced nuclear transition probabilities to be given by the single particle units (II A.58) with $R_0=5.9\ 10^{-13}$ cm.

The empirical values of the nuclear transition probabilities obtained from lifetime determinations of γ transitions show major departures from the single-particle unit (58).[29] Thus, the relatively few electric dipole transitions, which have been observed in the low-energy nuclear spectra, have in most cases transition probabilities many orders of magnitude smaller than (58). In contrast low-energy electric quadrupole transitions, which occur with great frequency, are often found to be strongly enhanced as compared with single-particle estimates. Thus, $E2$ transitions with a strength of 10–100 single-particle units occur systematically in most regions of elements (see Chapter V).

For these reasons, the electric quadrupole transitions are of special importance in the Coulomb excitation, and in fact it appears that the overwhelming majority of the excitations so far observed are of $E2$ type (see Chapter IV).

As seen from Fig. II.1a, the cross sections for magnetic excitation are very much smaller than for electric excitations; thus, even in cases where the radiative de-excitation process takes place by a mixed $M1+E2$ transition, the excitation will almost always be of rather pure $E2$ type.

II A.4. Angular Distribution of De-Excitation γ Rays

The nuclear states populated by Coulomb excitation decay by emission of γ radiation or conversion electrons. The angular distribution of this radiation can be obtained from the excitation amplitudes b_{if} given above.

[29] For a survey of these data, see M. Goldhaber and A. W. Sunyar, Chapter XVI of *Beta- and Gamma-Ray Spectroscopy*, edited by K. Siegbahn (North Holland Publishing Company, Amsterdam, 1955).

Denoting the nuclear state to which the de-excitation takes place by ff, the angular distribution of the emitted γ radiation is given by[30]

$$W_{\vartheta,\varphi}(\Omega_\gamma)=\sum_{\sigma M_i M_{ff}}\left|\sum_{M_f} b_{if}\langle I_{ff}M_{ff}|H_\gamma(\Omega_\gamma,\sigma)|I_fM_f\rangle\right|^2, \tag{II A.59}$$

where $H_\gamma(\Omega_\gamma,\sigma)$ is the interaction Hamiltonian for emission of a γ quantum in the direction Ω_γ, and with polarization σ. We have assumed unpolarized target nuclei and have summed over the polarizations of the γ quantum. The distribution (59) refers to a definite orbit of the projectile characterized by the polar angles ϑ, φ of the scattered particle.

We first consider the case in which the excitation takes place by a transition of pure multipole order λ which may be either electric or magnetic. Using the expressions (14) and (42), and the relation (16), we get from (59)

$$\begin{aligned}W_{\vartheta,\varphi}(\Omega_\gamma)=\sum &\begin{pmatrix} I_i & \lambda & I_f\\ -M_i & \mu & M_f\end{pmatrix}\\ &\times\begin{pmatrix} I_i & \lambda & I_f\\ -M_i & \mu' & M_f'\end{pmatrix}S_{\lambda\mu}S_{\lambda\mu'}^*\\ &\times\langle I_{ff}M_{ff}|H_\gamma(\Omega_\gamma,\sigma)|I_fM_f\rangle\\ &\times\langle I_{ff}M_{ff}|H_\gamma(\Omega_\gamma,\sigma)|I_fM_f'\rangle^*,\end{aligned} \tag{II A.60}$$

where we have left out constant factors. The summation in (60) is to be extended over M_i, M_f, M_f', μ, μ', M_{ff}, and σ.

The distribution (60) may conveniently be expressed in terms of the correlation function for a hypothetical γ-γ cascade in which the first transition is a pure 2^λ-pole radiation[31] (see Fig. II.2). This latter correlation func-

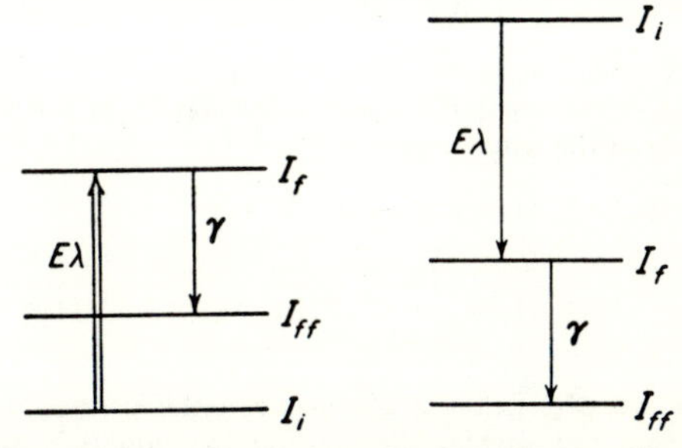

FIG. II.2. Hypothetical transitions involved in describing angular distribution of gamma rays following Coulomb excitation. The ground-state spins, the spin of the state excited by Coulomb excitation, and the spin of the final state populated by the gamma ray are denoted by I_i, I_f, and I_{ff}, respectively. The figure on the right then gives the hypothetical $\gamma-\gamma$ cascade employed in obtaining the angular distribution of the gamma rays following Coulomb excitation.

[30] D. L. Falkoff and G. E. Uhlenbeck, Phys. Rev. **79**, 323 (1950).
[31] L. C. Biedenharn and M. E. Rose, Revs. Modern Phys. **25**, 729 (1953).

tion differs from (60) only in the replacement of $S_{\lambda\mu}$ by the rotation matrix $D_{\mu\sigma'}{}^{\lambda}(\mathcal{R})$, where $\mathcal{R}$ denotes the rotation from the fixed coordinate system to a system whose z axis points in the direction Ω_γ' of the first γ quantum. The polarization index σ' refers to circular polarization.

As in the usual treatment of angular correlation, we employ the relation[24]

$$\sum_{M_i}\begin{pmatrix} I_i & \lambda & I_f \\ -M_i & \mu & M_f\end{pmatrix}\begin{pmatrix} I_i & \lambda' & I_f' \\ -M_i & \mu' & M_f'\end{pmatrix}$$

$$=\sum_{k\kappa}(-1)^{2\lambda'+k-I_i+\mu+M_f'}(2k+1)\begin{pmatrix}\lambda & \lambda' & k\\ \mu & -\mu' & \kappa\end{pmatrix}$$

$$\times\begin{pmatrix} I_f & I_f' & k\\ -M_f & M_f' & \kappa\end{pmatrix}\begin{Bmatrix}\lambda & \lambda' & k\\ I_f' & I_f & I_i\end{Bmatrix}, \quad \text{(II A.61)}$$

where we have introduced the Wigner notation for the Racah coefficient $W(\lambda\lambda' I_f I_f|kI_i)$ through the definition

$$\begin{Bmatrix} j_1 & j_2 & j_3\\ l_1 & l_2 & l_3\end{Bmatrix}$$

$$=(-1)^{j_1+j_2+l_2+l_1}W(j_1j_2l_2l_1|j_3l_3). \quad \text{(II A.62)}$$

In this manner, we obtain an expression for $W(\Omega_\gamma)$ which involves the $S_{\lambda\mu}$ only in the combination

$$\sum_{\mu\mu'}(-1)^{\mu}\begin{pmatrix}\lambda & \lambda & k\\ \mu & -\mu' & \kappa\end{pmatrix}S_{\lambda\mu}S_{\lambda\mu'}{}^{*}. \quad \text{(II A.63)}$$

In the γ-γ correlation the corresponding expression reduces to

$$\begin{pmatrix}\lambda & \lambda & k\\ 1 & -1 & 0\end{pmatrix}\left(\frac{4\pi}{2k+1}\right)^{\frac{1}{2}}Y_{k\kappa}{}^{*}(\Omega_\gamma') \quad \text{(II A.64)}$$

after summation over the polarization index σ'.

Thus, if we write the angular correlation in the γ-γ cascade in the usual way

$$W_{\Omega_\gamma'}(\Omega_\gamma)=\sum_k A_k{}^{(\lambda)}P_k(\cos(\Omega_\gamma',\Omega_\gamma))$$

$$=\sum_{k\kappa}A_k{}^{(\lambda)}\frac{4\pi}{2k+1}Y_{k\kappa}{}^{*}(\Omega_\gamma')Y_{k\kappa}(\Omega_\gamma), \quad \text{(II A.65)}$$

where $(\Omega_\gamma',\Omega_\gamma)$ is the angle between the γ rays, it is seen that the angular distribution function (60) may be written

$$W_{\vartheta,\varphi}(\Omega_\gamma)=\sum_{k\kappa}a_{k\kappa}{}^{\lambda}(\vartheta,\varphi,\xi)A_k{}^{(\lambda)}Y_{k\kappa}(\Omega_\gamma). \quad \text{(II A.66)}$$

The coefficients $a_{k\kappa}{}^{\lambda}$, which are independent of the nuclear states involved, and of the de-excitation process, may be expressed in the form

$$a_{k\kappa}{}^{\lambda}(\vartheta,\varphi,\xi)=b_{k\kappa}{}^{\lambda}/b_{00}{}^{\lambda}, \quad \text{(II A.67)}$$

with

$$b_{k\kappa}{}^{\lambda}=-(2k+1)^{-\frac{1}{2}}\begin{pmatrix}\lambda & \lambda & k\\ 1 & -1 & 0\end{pmatrix}^{-1}$$

$$\times\sum_{\mu\mu'}(-1)^{\mu}\begin{pmatrix}\lambda & \lambda & k\\ \mu & -\mu' & \kappa\end{pmatrix}S_{\lambda\mu}S_{\lambda\mu'}{}^{*}. \quad \text{(II A.68)}$$

The normalization of $a_{k\kappa}{}^{\lambda}$ is then such that $a_{00}{}^{\lambda}=1$. As in the cascade, only terms with even k occur in (66). It is noted that $b_{00}{}^{\lambda}$ reduces to

$$b_{00}{}^{\lambda}=\sum_{\mu}|S_{\lambda\mu}|^2, \quad \text{(II A.69)}$$

and is thus directly related to the differential excitation cross section (see (21)).

The coefficients A_k in the γ-γ correlation (65) are given by[31]

$$A_k{}^{(\lambda)}=F_k(\lambda,I_iI_f)\sum_{LL'}\delta_L\delta_{L'}F_k(LL'I_{ff}I_f), \quad \text{(II A.70)}$$

where $\delta_L{}^2$ is the intensity of the 2^L-pole radiation in the γ transition $I_f\rightarrow I_{ff}$. With the present definition of the multipole operators, the relative values of δ_L are given by

$$\delta_L\propto i^{s(L)}\frac{q^L}{(2L+1)!!}\left(\frac{L+1}{L}\right)^{\frac{1}{2}}\langle I_{ff}\|\mathfrak{M}(\pi L)\|I_f\rangle, \quad \text{(II A.71)}$$

with

$$s(L)=\begin{cases}L & \text{for } EL\\ L+1 & \text{for } ML.\end{cases} \quad \text{(II A.71a)}$$

The product $\delta_L\delta_{L'}$ is always real since $(-1)^{s(L)}=\pi$ (the parity). The coefficients F_k are geometrical factors defined by

$$F_k(LL'I_1I_2)$$

$$=(-1)^{I_1+I_2-1}[(2k+1)(2I_2+1)(2L+1)(2L'+1)]^{\frac{1}{2}}$$

$$\times\begin{pmatrix}L & L' & k\\ 1 & -1 & 0\end{pmatrix}\begin{Bmatrix}L & L' & k\\ I_2 & I_2 & I_1\end{Bmatrix}, \quad \text{(II A.72)}$$

and

$$F_k(LI_1I_2)=F_k(LLI_1I_2),$$

and are tabulated[32] in references 31 and 33.

The orbital integrals $S_{\lambda\mu}$ in (68) are most easily evaluated in the focal system (see Fig. II.1) in which they are given by (24) and (48). One thereby obtains the angular distribution coefficients $b_{k\kappa}{}^{\lambda}$ in the focal system. It is, however, often more convenient to express

[32] L. C. Biedenharn and M. E. Rose (reference 31) have given the interference terms (for $L\neq L'$) in the form

$$G_k(LL'I_1I_2)=(-1)^{I_2-I_1-1}[(2I_2+1)(2L+1)(2L'+1)]^{-\frac{1}{2}}F_k(LL'I_1I_2).$$

The coefficients A_k are tabulated in Table II.11 for some cases often encountered in Coulomb excitation.

[33] M. Ferentz and N. Rosenzweig, Argonne National Laboratory report, ANL 5324.

the angular distribution of the γ quanta in a coordinate system with the z axis in the direction of the incident beam of particles. This may be obtained by a simple transformation which, in the case of electric excitations, gives

$$b_{k\kappa}^{E\lambda}(\vartheta,\varphi,\xi) = -(2k+1)^{-\frac{1}{2}}\begin{pmatrix}\lambda & \lambda & k\\ 1 & -1 & 0\end{pmatrix}^{-1} \times \sum_{\mu\mu'\kappa'}(-1)^{\mu}\begin{pmatrix}\lambda & \lambda & k\\ \mu & -\mu' & \kappa'\end{pmatrix} Y_{\lambda\mu}\left(\frac{\pi}{2},0\right) \times Y_{\lambda\mu'}\left(\frac{\pi}{2},0\right) I_{\lambda\mu}(\vartheta,\xi) I_{\lambda\mu'}(\vartheta,\xi) \times D_{\kappa'\kappa}^{k}\left(\frac{\pi}{2}+\frac{\vartheta}{2},\frac{\pi}{2},\varphi\right), \quad \text{(II A.73)}$$

where the Eulerian angles[32a] $(\pi/2+\vartheta/2, \pi/2, \varphi)$ represent the rotation from the coordinate system of the incident beam to the focal system of the orbit in question.

The distribution function (66) applies to a measurement of γ rays in coincidence with particles scattered inelastically in a definite direction. The total angular distribution of the γ's, irrespective of the scattering angle of the projectile, is obtained by multiplying (66) by the differential excitation cross section (28) or (50), and integrating over ϑ and φ. This gives

$$W(\vartheta_{\gamma}) = \sum_{k} a_{k}^{\lambda}(\xi) A_{k}^{(\lambda)} P_{k}(\cos\vartheta_{\gamma}), \quad \text{(II A.74)}$$

where ϑ_{γ} is the angle between the direction of the incident beam and the γ quantum. The coefficients $a_{k}^{\lambda}(\xi)$ are given by

$$a_{k}^{\lambda}(\xi) = b_{k}^{\lambda}/b_{0}^{\lambda}, \quad \text{(II A.75)}$$

with

$$b_{k}^{E\lambda}(\xi) = -(2k+1)^{-\frac{1}{2}}\begin{pmatrix}\lambda & \lambda & k\\ 1 & -1 & 0\end{pmatrix}^{-1}\sum_{\mu\mu'\kappa}(-1)^{\mu}\begin{pmatrix}\lambda & \lambda & k\\ \mu & -\mu' & \kappa\end{pmatrix} \times Y_{\lambda\mu}\left(\frac{\pi}{2},0\right) Y_{\lambda\mu'}\left(\frac{\pi}{2},0\right)\int_{0}^{\pi} I_{\lambda\mu}(\vartheta,\xi) I_{\lambda\mu'}(\vartheta,\xi) Y_{k\kappa}\left(\frac{\pi}{2},\frac{\pi}{2}+\frac{\vartheta}{2}\right)\frac{\cos\frac{\vartheta}{2}}{\sin^{3}\frac{\vartheta}{2}}d\vartheta, \quad \text{(II A.76)}$$

and

$$b_{k}^{M\lambda}(\xi) = -(2k+1)^{-\frac{1}{2}}\begin{pmatrix}\lambda & \lambda & k\\ 1 & -1 & 0\end{pmatrix}^{-1}\sum_{\mu\mu'\kappa}(-1)^{\mu}\begin{pmatrix}\lambda & \lambda & k\\ \mu & -\mu' & \kappa\end{pmatrix}[((\lambda+1)^{2}-\mu^{2})((\lambda+1)^{2}-\mu'^{2})]^{\frac{1}{2}} \times Y_{\lambda+1,\mu}\left(\frac{\pi}{2},0\right) Y_{\lambda+1,\mu'}\left(\frac{\pi}{2},0\right)\int_{0}^{\pi} I_{\lambda+1,\mu}(\vartheta,\xi) I_{\lambda+1,\mu'}(\vartheta,\xi) Y_{k\kappa}\left(\frac{\pi}{2},\frac{\pi}{2}+\frac{\vartheta}{2}\right)\cot^{2}\frac{\vartheta}{2}\,\frac{\cos\frac{\vartheta}{2}}{\sin^{3}\frac{\vartheta}{2}}d\vartheta. \quad \text{(II A.77)}$$

The coefficients $a_{k}^{E\lambda}(\xi)$ have been evaluated numerically for $E1$ and $E2$ excitations. The results are shown in Fig. II.8, where they represent the limiting values for $\nu \to 0$ of the corresponding quantum-mechanical expressions. In the case of $M1$ excitations, where (77) only contains terms with $\mu=\mu'=\kappa=0$ (see (25)), one obtains

$$a_{2}^{M1}(\xi) = 1, \quad \text{(II A.78)}$$

independent of ξ.

If the polarization of the decay γ ray is measured, one may obtain the correlation functions in a similar manner as above by comparing with a γ-γ cascade in which the polarization of the second quantum is measured. Thus, the probability for emission of a γ ray at an angle ϑ_{γ} and with a given direction of polarization is again of the form (74) with the only difference that the functions $P_{k}(\cos\vartheta_{\gamma})$ are to be replaced by[31]

$$\mathcal{P}_{k}(LL';\vartheta_{\gamma}\psi_{\gamma}) = P_{k}(\cos\vartheta_{\gamma}) + (-1)^{p(L')}\left[\frac{(k-2)!}{(k+2)!}\right]^{\frac{1}{2}}\begin{pmatrix}L & L' & k\\ 1 & 1 & -2\end{pmatrix} \times \begin{pmatrix}L & L' & k\\ 1 & -1 & 0\end{pmatrix}^{-1}\cos 2\psi_{\gamma} P_{k}^{2}(\cos\vartheta_{\gamma}), \quad \text{(II A.78a)}$$

where ψ_{γ} is the angle between the electric vector and the plane determined by the direction of the incident projectile and the γ ray. The phase $(-1)^{p(L')}$ is $+1$ if L' is an electric radiation, and -1 if it is magnetic. The functions P_{k}^{2} are the associated Legendre polynomials. If the decay radiation is of mixed multipole

[32a] We have used the same definition of the Eulerian angles as that used in reference 6 (see also reference 24).

type, each term in the coefficient $A_k^{(\lambda)}$ (see (70)) is to be multiplied by the appropriate angular function $\mathcal{P}_k(LL')$.

If the excitation is of mixed multipole type, the angular distribution of the γ rays, in contrast to the excitation cross sections, contains additional interference terms.

In order to derive the general expression for the angular distribution, we write the transition amplitude b_{if} in the following form (see (14), (16), and (42))

$$b_{if}=\frac{4\pi Z_1 e}{i\hbar}\sum_{\pi\lambda\mu}(-1)^{I_i-M_i}(2\lambda+1)^{-1}\begin{pmatrix} I_i & \lambda & I_f \\ -M_i & \mu & M_f \end{pmatrix} \times\langle\pi_i I_i\|\mathfrak{M}(\pi\lambda)\|\pi_f I_f\rangle S_{\pi\lambda\mu}, \quad \text{(II A.79)}$$

where π_i and π_f indicate the parity of the initial and final nuclear state, while π is the parity of the excitation process, i.e., $\pi=\pi_i\pi_f$. The $S_{\pi\lambda\mu}$ are defined by (15) and (43) for electric and magnetic excitations.

By following the same procedure as above, it is readily seen that the angular distribution (66) now takes the form

$$W_{\vartheta,\varphi}(\Omega_\gamma)=\sum_{k\kappa}\sum_{\lambda\pi\lambda'\pi'}c_{k\kappa}^{\pi\lambda,\pi'\lambda'}(\vartheta,\varphi,\xi)F_k(\lambda\lambda' I_i I_f) \times\sum_{LL'}\delta_L\delta_{L'}F_k(LL'I_f I_f)Y_{k\kappa}(\Omega_\gamma), \quad \text{(II A.80)}$$

with

$$c_{k\kappa}^{\pi\lambda,\pi'\lambda'}(\vartheta,\varphi,\xi) =-[(2k+1)(2\lambda+1)^3(2\lambda'+1)^3]^{-\frac{1}{2}}\begin{pmatrix} \lambda & \lambda' & k \\ 1 & -1 & 0 \end{pmatrix}^{-1} \times\langle\pi_i I_i\|\mathfrak{M}(\pi\lambda)\|\pi_f I_f\rangle\langle\pi_i I_i\|\mathfrak{M}(\pi'\lambda')\|\pi_f I_f\rangle \times\sum_{\mu\mu'}(-1)^{\mu}\begin{pmatrix} \lambda & \lambda' & k \\ \mu & -\mu' & \kappa \end{pmatrix}S_{\pi\lambda\mu}S_{\pi'\lambda'\mu'}^{*}. \quad \text{(II A.81)}$$

For $k=\kappa=0$, the $c_{00}^{\pi\lambda,\pi\lambda}$ are proportional to the differential excitation cross sections. For $\lambda\neq\lambda'$, the $c_{00}^{\pi\lambda,\pi'\lambda'}$ vanish.

In order to obtain the total angular distribution of the γ quanta, we multiply (80) by the Rutherford cross section (2) and integrate over ϑ and φ. One thus obtains the angular distribution in the form

$$W(\vartheta_\gamma)=\sum_k\sum_{\lambda\lambda'}a_k^{\pi\lambda\pi'\lambda'}(\xi)(\sigma_{\pi\lambda})^{\frac{1}{2}}(\sigma_{\pi'\lambda'})^{\frac{1}{2}}F_k(\lambda\lambda' I_i I_f) \times\sum_{LL'}\delta_L\delta_{L'}F_k(LL'I_f I_f)P_k(\cos\vartheta_\gamma), \quad \text{(II A.82)}$$

where $\sigma_{\pi\lambda}$ is the total excitation cross section of multipole order $\pi\lambda$ and where the sign of the square root is the same as that of the reduced matrix element, $\langle\pi_i I_i\|\mathfrak{M}(\pi\lambda)\|\pi_f I_f\rangle$. These latter are the same as those occurring in the radiative decay $I_f\rightarrow I_i$ (see (71)). The a coefficients in (82) are given by

$$a_k^{\pi\lambda\pi'\lambda'}(\xi)=b_k^{\pi\lambda\pi'\lambda'}/(b_0^{\pi\lambda})^{\frac{1}{2}}(b_0^{\pi'\lambda'})^{\frac{1}{2}}, \quad \text{(II A.83)}$$

where the $b_0^{\pi\lambda}$ are given by (76) and (77). Furthermore

$$b_k^{\pi\lambda\pi\lambda}=b_k^{\pi\lambda}. \quad \text{(II A.84)}$$

The most important case where interference terms appear is that of a mixed electric and magnetic excitation, for which the $b_k^{\pi\lambda\pi\lambda'}(\xi)$ are given by

$$b_k^{E\lambda M\lambda'}(\xi)=(2k+1)^{-\frac{1}{2}}\begin{pmatrix} \lambda & \lambda' & k \\ 1 & -1 & 0 \end{pmatrix}^{-1} \times\sum_{\mu\mu'\kappa}(-1)^{\mu}\begin{pmatrix} \lambda & \lambda' & k \\ \mu & -\mu' & \kappa \end{pmatrix}[(\lambda'+1)^2-\mu'^2]^{\frac{1}{2}} \times Y_{\lambda\mu}\left(\frac{\pi}{2},0\right)Y_{\lambda'+1,\mu'}\left(\frac{\pi}{2},0\right) \times\int_0^{\pi} I_{\lambda\mu}(\vartheta,\xi)I_{\lambda'+1,\mu'}(\vartheta,\xi) \times Y_{k\kappa}\left(\frac{\pi}{2},\frac{\pi}{2}+\frac{\vartheta}{2}\right)\cot\frac{\vartheta}{2}\,\frac{\cos\frac{\vartheta}{2}}{\sin^3\frac{\vartheta}{2}}\,d\vartheta. \quad \text{(II A.85)}$$

The decay of the excited nuclear level may also take place by emission of internal conversion electrons. The angular distribution of these electrons is given by expressions similar to those applying to the γ distribution, with the only difference that the F_k factors for the decay are to be multiplied by appropriate coefficients depending on the parameters of the conversion process.[34]

II A.5. Symmetrization of Classical Cross Sections

The classical treatment of the excitation process neglects the effect of the energy loss on the motion of the projectile. It may be expected, however, that improved expressions for the excitation cross sections may be obtained by substituting for the particle velocity v entering in these expressions, some mean value of initial and final velocity v_i and v_f, rather than the initial velocity assumed above.

While the choice of v leading to the best approximation for the cross sections cannot be decided within the scope of the classical treatment, it follows immediately from the general character of the quantum-mechanical formalism, considered in the next section, that when the probability for excitation in a single encounter is small the excitation cross section is symmetrical in v_i and v_f, except for a factor v_f/v_i. In fact, the cross section is inversely proportional to the flux of the incident par-

[34] See Biedenharn and Rose, reference 31, which contains tables of the coefficients (denoted in this reference by b) involved in the correlation with K-shell conversion electrons, calculated for a point charge nucleus.

ticles, and thus to v_i, while proportional to the density of final states, i.e., to v_f. In addition, the cross section involves the square of a matrix element which is symmetrical in the initial and final state [see (II B.25 and 26)].

A straightforward way of symmetrizing the classical cross sections is first to introduce symmetrized parameters a and ξ, given by

$$a=\frac{Z_1Z_2e^2}{m_0v_iv_f}, \qquad \text{(II A.86)}$$

and

$$\xi=\frac{Z_1Z_2e^2}{\hbar}\left(\frac{1}{v_f}-\frac{1}{v_i}\right) \qquad \text{(II A.87)}$$

to replace (3) and (27), respectively. It is readily seen that the expressions (86) and (87) for a and ξ are equal to (3) and (27), respectively, to lowest order in $\Delta E/E$.[35] Although (87) changes sign when v_i and v_f are interchanged, the f functions are not affected, since they are even functions of ξ.

Appropriately symmetrized expressions for the excitation cross sections may thus be obtained by replacing (28) and (50) by

$$d\sigma_{E\lambda}=\left(\frac{Z_1e}{\hbar v_i}\right)^2 a^{-2\lambda+2}B(E\lambda)df_{E\lambda}(\vartheta,\xi), \qquad \text{(II A.88)}$$

and

$$d\sigma_{M\lambda}=\left(\frac{Z_1e}{\hbar c}\right)^2 \frac{v_f}{v_i} a^{-2\lambda+2}B(M\lambda)df_{M\lambda}(\vartheta,\xi), \qquad \text{(II A.89)}$$

and similarly for the total cross sections. In these expressions a and ξ are given by (86) and (87). Likewise, symmetrized expressions for the angular distribution of the emitted γ rays are obtained by employing, in the formulas in Sec. II A.4, the symmetrized expression (87) for ξ.

It is found, by comparison with the quantum-mechanical results (see Sec. II B.6), that the symmetrized expressions represent an essential improvement over the unsymmetrized. In fact, the symmetrized total cross sections reproduce the quantum-mechanical to within a few percent, for values of η as low as 3, and even for ξ as large as 2 (see Fig. II.6). This corresponds to a collision in which the particle loses more than half its energy, and for which the unsymmetrized cross sections would be in error by more than a factor hundred.

The angular distribution of the emitted γ radiation is found to be less accurately given by symmetrized classical formulas, except for very large values of η (see Table II.9). Similarly, the differential excitation cross section may be expected to be fairly sensitive to quantum corrections.

[35] Arguments for the special choice (87) for ξ have been given by K. A. Ter-Martirosyan, reference 5, and by Sherr, Li, and Christy, Phys. Rev. **96**, 1258 (1954).

II A.6. Excitation of Projectile

If the projectile is a composite particle, the collision may also lead to the excitation of the projectile. This process is entirely analogous to the excitation of the target nucleus, and corresponds merely to the interchange of the roles of nucleus and projectile. The interaction is now proportional to the nuclear charge and to the projectile transition matrix element, and the excitation cross section is thus obtained from the cross section for target excitation by simply replacing the factor Z_1^2 in (88) and (89) by Z_2^2 and the quantities ΔE, ξ, and $B(\lambda)$ by those appropriate to the projectile excitation.

The angular distribution of the emitted γ rays following projectile excitation is the same as for excitation of the target nucleus. However, the γ energies may be somewhat shifted by the Doppler effect if the stopping time for the projectile is longer than the lifetime of the excited state. To first order in the projectile velocity, the γ energy is given by

$$E_\gamma=\Delta E\left(1+\frac{v_{1f}}{c}\cos u\right), \qquad \text{(II A.90)}$$

where v_{1f} is the velocity of the projectile at the time of emission (measured in the laboratory system) and u the angle between the scattered projectile and the direction of the γ ray. Even if (90) is averaged over the direction of the scattered projectile, there will remain some dependence of the average E_γ on the direction in which the γ ray is observed.

II B. Quantum-Mechanical Theory

In this Section we consider the quantum-mechanical treatment of electromagnetic excitations of nuclei. In the first part (Sec. II B.1) we give a relativistic derivation of the excitation cross section, considering the interaction as arising from the exchange of a photon between projectile and nucleus. This method is equivalent to the use of the retarded Greens functions for the interaction.[36] In most applications it is sufficient to include only the leading term in the projectile velocity in the expressions for the electric and magnetic excitation cross sections (Sec. II B.2).

These cross sections can be expressed as sums of terms referring to the different angular momenta of the incoming and outgoing projectile (Sec. II B.3). Each of the terms involves a radial matrix element which can be evaluated in terms of known functions, and expressed

[36] See M. Jean and J. Prentki, reference 26; Biedenharn, McHale, and Thaler, reference 18.

in a form convenient for numerical computation (Sec. II B.4). Similarly, in Sec. II B.5 the angular distribution of the de-excitation γ rays is expressed in terms of these radial matrix elements.

Approximate values for the radial matrix elements can be obtained by means of the WKB method (Sec. II B.6) which in most cases of interest is found to yield a high degree of accuracy. In this section we also discuss the transition of the quantum-mechanical cross sections to the symmetrized classical expressions for large values of the parameter η.

II B.1. Derivation of Excitation Cross Sections

For the system consisting of projectile, nucleus, and the quantized electromagnetic field, we take as the zero-order Hamiltonian

$$\mathcal{H}_0=\mathcal{H}_p+\mathcal{H}_n+\mathcal{H}_{\text{rad}}+\frac{Z_1Z_2e^2}{r_p}, \qquad \text{(II B.1)}$$

where the three first terms represent the free Hamiltonians of projectile, nucleus, and radiation field, respectively. In (1) we have also included the static point charge interaction between the projectile and the nucleus. In the relativistic treatment we shall neglect the nuclear recoil, so that the nuclear center of mass may be taken as the fixed origin of the coordinate system. The recoil effects may be reintroduced in the nonrelativistic part of the cross sections (see below), and thus the only essential approximation involved here is that of neglecting the effect of the recoil in the relativistic corrections.

It is convenient to divide the electromagnetic field into a transverse part described by a vector potential **A** for which div**A**=0 and a longitudinal part. The latter contributes the instantaneous Coulomb interaction,[37] and the total interaction Hamiltonian is thus

$$\mathcal{H}_{\text{int}}=-\frac{1}{c}\int(\mathbf{j}_p(\mathbf{r})+\mathbf{j}_n(\mathbf{r}))\cdot\mathbf{A}(\mathbf{r})d\tau +\mathcal{H}_{\text{coul}}-\frac{Z_1Z_2e^2}{r_p}, \qquad \text{(II B.2)}$$

where

$$\mathcal{H}_{\text{coul}}=\int\frac{\rho_p(\mathbf{r})\rho_n(\mathbf{r}')}{|\mathbf{r}-\mathbf{r}'|}d\tau d\tau', \qquad \text{(II B.3)}$$

and where ρ_p and $\mathbf{j}_p$ are the charge and current density operators for the projectile.

The vector potential is expanded in multipole components according to[38]

$$\mathbf{A}(\mathbf{r})=\sum_q\sum_{\lambda=1}^{\infty}\sum_{\mu=-\lambda}^{\lambda}\{a(E\lambda,\mu,q)\mathbf{A}(E\lambda,\mu,q) +a(M\lambda,\mu,q)\mathbf{A}(M\lambda,\mu,q)+\text{compl. conj.}\}, \qquad \text{(II B.4)}$$

where the electric and magnetic multipole fields are given by

$$\mathbf{A}(E\lambda,\mu,q)=\left(\frac{8\pi c^2}{\lambda(\lambda+1)}\right)^{\frac{1}{2}}R^{-\frac{1}{2}}\nabla\times\mathbf{L}(j_\lambda(qr)Y_{\lambda\mu}(\theta,\phi)), \qquad \text{(II B.5)}$$

and

$$\mathbf{A}(M\lambda,\mu,q)=i\left(\frac{8\pi c^2q^2}{\lambda(\lambda+1)}\right)^{\frac{1}{2}}R^{-\frac{1}{2}}\mathbf{L}(j_\lambda(qr)Y_{\lambda\mu}(\theta,\phi)). \qquad \text{(II B.6)}$$

The angular momentum operator **L** is defined by (II A.35), and $j_\lambda(qr)$ represents the spherical Bessel function.[39] The multipole fields (5) and (6) are associated with photons of angular momentum λ, magnetic quantum number μ, wave number q, and parity $(-1)^\lambda$ and $(-1)^{\lambda+1}$ for the electric and magnetic multipole fields, respectively. The fields are enclosed in a large sphere of radius R.

The coefficients a in (4) and their conjugates are the photon absorption and the emission operators. With the normalization (5) and (6) the nonvanishing matrix elements of these operators are given by (see reference 37)

$$\langle n|a|n+1\rangle=\langle n+1|a^*|n\rangle=\left(\frac{\hbar(n+1)}{2qc}\right)^{\frac{1}{2}}, \qquad \text{(II B.7)}$$

where $|n\rangle$ represents a state with n photons of the type in question.

The eigenstates of the Hamiltonian (1) are represented by a wave function φ, for the projectile moving in the point Coulomb field of the nucleus multiplied by a nuclear wave function ψ, and is further specified by a number of free photons. We consider a transition from an initial state i with the nucleus in the ground state to a final state f where the projectile has transferred an energy ΔE to the nucleus. In initial and final state no photons are present. To first order in the charge of the projectile this transition receives partly a first-order contribution from the Coulomb term in (2) and partly a second-order contribution from the first term in (2) corresponding to the emission of a photon by the projectile, and its reabsorption by the nucleus (or vice versa).

To this approximation the transition matrix element

[37] See, e.g., W. Heitler, *The Quantum Theory of Radiation* (Oxford University Press, New York, 1944), second edition, Sec. III.10.

[38] See, e.g., W. Franz, Z. Physik **127**, 363 (1950); B. Stech, Z. Naturforsch. **7a**, 401 (1952).

[39] We use the same notation for the spherical cylinder functions as employed in L. I. Schiff, *Quantum Mechanics* (McGraw-Hill Book Company, Inc., New York, 1949).

is thus given by

$$\langle f|\mathcal{H}^{(1)}|i\rangle=\sum_{\lambda\mu q}\frac{\hbar}{2qc^3}\left\{\frac{\left\langle\varphi_f\middle|\int\mathbf{j}_p\cdot\mathbf{A}(E\lambda,\mu,q)d\tau\middle|\varphi_i\right\rangle\left\langle\psi_f\middle|\int\mathbf{j}_n\cdot\mathbf{A}^*(E\lambda,\mu,q)d\tau\middle|\psi_i\right\rangle}{-\Delta E-\hbar cq}\right.$$
$$+\frac{\left\langle\psi_f\middle|\int\mathbf{j}_n\cdot\mathbf{A}(E\lambda,\mu,q)d\tau\middle|\psi_i\right\rangle\left\langle\varphi_f\middle|\int\mathbf{j}_p\cdot\mathbf{A}^*(E\lambda,\mu,q)d\tau\middle|\varphi_i\right\rangle}{\Delta E-\hbar cq}$$
$$\left.+\text{magnetic terms}\right\}+\left\langle\psi_f\varphi_f\middle|\mathcal{H}_{\text{coul}}-\frac{Z_1Z_2e^2}{r_p}\middle|\psi_i\varphi_i\right\rangle.\quad\text{(II B.8)}$$

The summation over q may be replaced by an integral

$$\sum_q\to\int_0^\infty dq\frac{R}{\pi},\quad\text{(II B.9)}$$

in which the path of integration is to circumvent the pole by passing below the real axis.

The integrals over q can be evaluated by using the formulas[40]

$$\int_0^\infty\frac{j_\lambda(qr)j_\lambda(qr')}{q^2-\kappa^2}q^2dq=\frac{i\pi\kappa}{2}j_\lambda(\kappa r_<)h_\lambda^{(1)}(\kappa r_>),\quad\text{(II B.10)}$$

and

$$\int_0^\infty\frac{j_\lambda(qr)j_\lambda(qr')}{q^2-\kappa^2}dq=\frac{i\pi}{2\kappa}j_\lambda(\kappa r_<)h_\lambda^{(1)}(\kappa r_>)-\frac{\pi r_<^{\lambda}r_>^{-\lambda-1}}{2\kappa^2(2\lambda+1)},\quad\text{(II B.11)}$$

where $r_>$ and $r_<$ denote the greater and smaller, respectively, of r and r', while $h_\lambda^{(1)}$ is the spherical Hankel function of first kind. With

$$\kappa=\frac{\Delta E}{\hbar c},\quad\text{(II B.12)}$$

the integrals (10) and (11) occur in the magnetic and electric part of (8), respectively.

It is now seen that the last term in (11) leads to a contribution from the electric multipole photons to the transition matrix element (8) which just cancels the corresponding multipole contribution from the Coulomb terms. This result may be obtained by using the relation

$$\nabla\times\mathbf{L}(r^kY_{\lambda\mu})=i(k+1)\nabla(r^kY_{\lambda\mu})\quad(k=\lambda\text{ or }-\lambda-1),\quad\text{(II B.13)}$$

and the continuity equation

$$\langle f|\text{div}\mathbf{j}|i\rangle=\frac{i}{\hbar}(E_f-E_i)\langle f|\rho|i\rangle.\quad\text{(II B.14)}$$

Since the photon field contains no components with $\lambda=0$, the cancellation is only complete provided the last terms in (8) contain no resulting monopole component, which is the case if the projectile does not penetrate into the nucleus. While this condition is fulfilled in Coulomb excitation with projectile energies below the barrier, there may, for instance in electron scattering, be an important electric monopole interaction causing nuclear excitations.[41]

In the following we shall neglect the effect of penetration so that $r_p=r_>$. Specifying the nuclear states by the quantum numbers I and M, and the scattering state of the projectile by its momentum $\hbar\mathbf{k}$ at infinity, we may express the transition matrix element in the following form

$$\langle f|\mathcal{H}^{(1)}|i\rangle=\sum_{\lambda\mu}\frac{4\pi}{2\lambda+1}(-1)^\mu\{\langle\mathbf{k}_f|\mathfrak{N}(E\lambda,\mu)|\mathbf{k}_i\rangle\times\langle I_fM_f|\mathfrak{M}(E\lambda,-\mu)|I_iM_i\rangle-\langle\mathbf{k}_f|\mathfrak{N}(M\lambda,\mu)|\mathbf{k}_i\rangle\times\langle I_fM_f|\mathfrak{M}(M\lambda,-\mu)|I_iM_i\rangle\},\quad\text{(II B.15)}$$

with the notation

$$\mathfrak{M}(E\lambda,\mu)=\frac{(2\lambda+1)!!}{\kappa^{\lambda+1}c(\lambda+1)}\int\mathbf{j}_n\cdot\nabla\times\mathbf{L}(j_\lambda(\kappa r)Y_{\lambda\mu}(\theta,\phi))d\tau,\quad\text{(II B.16)}$$

$$\mathfrak{M}(M\lambda,\mu)=\frac{-i(2\lambda+1)!!}{\kappa^{\lambda}c(\lambda+1)}\int\mathbf{j}_n\cdot\mathbf{L}(j_\lambda(\kappa r)Y_{\lambda\mu}(\theta,\phi))d\tau,\quad\text{(II B.17)}$$

[40] G. N. Watson, *Theory of Bessel Functions* (Cambridge University Press, New York, 1944), second edition, p. 429.

[41] See L. I. Schiff, Phys. Rev. **98**, 1281 (1955) and Sec. II E.3 following.

and

$$\mathfrak{M}(E\lambda,\mu)=\frac{i\kappa^{\lambda}}{c\lambda(2\lambda-1)!!}\int \mathbf{j}_p\cdot\nabla\times\mathbf{L}(h_\lambda^{(1)}(\kappa r)Y_{\lambda\mu}(\theta,\phi))d\tau, \quad \text{(II B.18)}$$

$$\mathfrak{M}(M\lambda,\mu)=\frac{\kappa^{\lambda+1}}{c\lambda(2\lambda-1)!!}\int \mathbf{j}_p\cdot\mathbf{L}(h_\lambda^{(1)}(\kappa r)Y_{\lambda\mu}(\theta,\phi))d\tau. \quad \text{(II B.19)}$$

It is often convenient to transform the electric multipole transition operators by means of the identity

$$\nabla\times\mathbf{L}(f_\lambda(\kappa r)Y_{\lambda\mu})=i\nabla\left(\frac{\partial}{\partial r}(rf_\lambda(\kappa r))Y_{\lambda\mu}\right)+i\kappa^2\mathbf{r}f_\lambda(\kappa r)Y_{\lambda\mu}, \quad \text{(II B.20)}$$

where $f_\lambda(\kappa r)$ is a spherical Bessel or Hankel function. Performing a partial integration and applying the continuity equation (14) one obtains

$$\mathfrak{M}(E\lambda,\mu)=\frac{(2\lambda+1)!!}{\kappa^{\lambda}(\lambda+1)}\int \rho_n\frac{\partial}{\partial r}(rj_\lambda(\kappa r))Y_{\lambda\mu}(\theta,\phi)d\tau+\frac{i(2\lambda+1)!!}{\kappa^{\lambda-1}c(\lambda+1)}\int \mathbf{j}_n\cdot\mathbf{r}j_\lambda(\kappa r)Y_{\lambda\mu}(\theta,\phi)d\tau, \quad \text{(II B.21)}$$

and

$$\mathfrak{M}(E\lambda,\mu)=-\frac{i\kappa^{\lambda+1}}{\lambda(2\lambda-1)!!}\int \rho_p\frac{\partial}{\partial r}(rh_\lambda^{(1)}(\kappa r)Y_{\lambda\mu}(\theta,\phi)d\tau-\frac{\kappa^{\lambda+2}}{c\lambda(2\lambda-1)!!}\int \mathbf{j}_p\cdot\mathbf{r}h_\lambda^{(1)}(\kappa r)Y_{\lambda\mu}(\theta,\phi)d\tau. \quad \text{(II B.22)}$$

The nuclear transition operators (16) and (17) are precisely the same as those which determine the emission probability for electric and magnetic multipole radiation. If the radiative transition probability is written in the form (II A.56), the reduced transition probabilities $B(\lambda)$ are given in terms of the transition operators through the definition

$$B(\lambda)=\sum_{\mu M_f}|\langle I_fM_f|\mathfrak{M}(\lambda,\mu)|I_iM_i\rangle|^2. \quad \text{(II B.23)}$$

This equation is identical with the definitions (II A.18) and (II A.52), and in fact the transition operators approach the electric and magnetic multipole moments (II A.11) and (II A.39) in the limit $\kappa R_0\ll 1$, in which one may employ the asymptotic expression

$$j_\lambda(\kappa r)\approx\frac{(\kappa r)^\lambda}{(2\lambda+1)!!}. \quad \text{(II B.24)}$$

The differential cross section for excitation with unspecified orientation of the initial and final nuclear state is given by[42]

$$d\sigma=\frac{m_f^2}{4\pi^2\hbar^4}\frac{v_f}{v_i}(2I_i+1)^{-1}\sum_{M_iM_f}|\langle f|\mathfrak{H}|i\rangle|^2d\Omega, \quad \text{(II B.25)}$$

where m_f is the relativistic mass of the outgoing projectile. The scattering states in (15) are eigenstates in the Coulomb field $Z_1Z_2e^2/r_p$, which for large distances behave as distorted plane waves plus spherical waves. While in the initial state these are outgoing, the final state should contain only incoming spherical waves.[43] The scattering states are normalized at infinity to one particle per unit volume. Using (15) and (23) and the relation (II A.16) the cross section may be written

$$d\sigma=\frac{4m_f^2}{\hbar^4}\frac{v_f}{v_i}\sum_\lambda\left\{\frac{B(E\lambda)}{(2\lambda+1)^3}\sum_\mu|\langle\mathbf{k}_f|\mathfrak{M}(E\lambda,\mu)|\mathbf{k}_i\rangle|^2+\frac{B(M\lambda)}{(2\lambda+1)^3}\sum_\mu|\langle\mathbf{k}_f|\mathfrak{M}(M\lambda,\mu)|\mathbf{k}_i\rangle|^2\right\}d\Omega. \quad \text{(II B.26)}$$

II B.2. Nonrelativistic Approximation

For projectile velocities small compared to that of light the product κr_p may be treated as a small quantity so that we may apply the asymptotic expansion

$$h_\lambda^{(1)}(\kappa r)\approx -i\frac{(2\lambda-1)!!}{(\kappa r)^{\lambda+1}}. \quad \text{(II B.27)}$$

In fact, we have

$$\kappa r_p=\frac{v}{c}\cdot\frac{r_p}{v}\cdot\omega\lesssim\frac{v}{c},$$

since for $\omega r_p/v$ larger than unity the interaction becomes almost adiabatic.

If we furthermore consider the projectile as a point particle with charge Z_1e we get to leading order from (22) and (19)

$$\mathfrak{M}(E\lambda,\mu)=Z_1er_p^{-\lambda-1}Y_{\lambda\mu}(\theta_p,\phi_p), \quad \text{(II B.28)}$$

$$\mathfrak{M}(M\lambda,\mu)=\frac{Z_1e\hbar}{mc\lambda}\mathbf{l}_p\cdot\nabla_p(r_p^{-\lambda-1}Y_{\lambda\mu}(\theta_p,\phi_p)), \quad \text{(II B.29)}$$

where we have used the relation (II A.44). The terms neglected in (28) and (29) are at most of the order $(v/c)^2$.

If the projectile possesses a spin with associated magnetic moment, the current density contains a contribution similar to the second term in (II A.40). The magnetic transition operator then becomes

$$\mathfrak{M}(M\lambda,\mu)=\left(\frac{1}{\lambda}\boldsymbol{\mu}_l+\boldsymbol{\mu}_s\right)\cdot\nabla_p(r_p^{-\lambda-1}Y_{\lambda\mu}(\theta_p,\phi_p)), \quad \text{(II B.30)}$$

[42] If the projectile possesses a spin the scattering states must also be specified with respect to spin indices, and the cross sections will involve appropriate averages over these indices.

[43] G. Breit and H. A. Bethe, Phys. Rev. 93, 888 (1954).

where $\mathbf{\mu}_l$ and $\mathbf{\mu}_s$ are the orbital and spin magnetic moments of the projectile. For $\eta \gg 1$, where large angular momenta are involved in the scattering, the effect of the spin moment is expected to be relatively small.[43a]

The magnetic moment also contributes to the electric excitation through the magnetization current and its associated charge density $(c\rho_{\text{spin}} \approx (\mathbf{v}/c)\cdot\mathbf{j}_{\text{spin}})$, but the effect is again at most of the order $(v/c)^2$ as compared with the leading term (28).

As was to be expected, the interaction (15) with the nonrelativistic transition operators (28) and (29) for the projectile is identical with that assumed in the classical treatment [see (II A.10) and (II A.38)]. It is thus also evident that in the nonrelativistic approximation the nuclear recoil may be taken into account as in Sec. II A simply by replacing the projectile mass by the reduced mass m_0 of projectile and nucleus.

The excitation cross section obtained from (26), (28), and (29) may now be written

$$d\sigma = \sum_{\lambda=1}^{\infty} d\sigma_{E\lambda} + d\sigma_{M\lambda}, \qquad \text{(II B.31)}$$

with

$$d\sigma_{E\lambda} = \left(\frac{Z_1 e}{\hbar v_i}\right)^2 a^{-2\lambda+2} B(E\lambda) df_{E\lambda}(\vartheta, \eta_i, \xi), \qquad \text{(II B.32)}$$

and

$$d\sigma_{M\lambda} = \left(\frac{Z_1 e}{\hbar c}\right)^2 \frac{v_f}{v_i} a^{-2\lambda+2} B(M\lambda) df_{M\lambda}(\vartheta, \eta_i, \xi), \qquad \text{(II B.33)}$$

where a is given by the symmetrized expression (II A.86).

We have here introduced the dimensionless functions

$$df_{E\lambda}(\vartheta, \eta_i, \xi) = \frac{4k_i k_f}{(2\lambda+1)^3} a^{2\lambda-2} \times \sum_{\mu} |\langle \mathbf{k}_f | r_p^{-\lambda-1} Y_{\lambda\mu}(\theta_p, \phi_p) | \mathbf{k}_i \rangle|^2 d\Omega, \qquad \text{(II B.34)}$$

and

$$df_{M\lambda}(\vartheta, \eta_i, \xi) = \frac{4a^{2\lambda-2}}{\lambda^2(2\lambda+1)^3} \times \sum_{\mu} |\langle \mathbf{k}_f | \mathbf{l}_p \cdot \nabla_p (r_p^{-\lambda-1} Y_{\lambda\mu}(\theta_p, \phi_p)) | \mathbf{k}_i \rangle|^2 d\Omega \qquad \text{(II B.35)}$$

in analogy with the notation used in the classical treatment [see (II A.88) and (II A.89)]. From dimensional considerations it follows that (34) and (35) for given deflection angle ϑ may be regarded as functions only of η_i and η_f defined by (II A.1) for v equal to v_i and v_f, respectively. To stress the analogy with the classical case we consider (34) and (35) as functions of η_i and the parameter ξ

$$\xi = \eta_f - \eta_i, \qquad \text{(II B.36)}$$

which is identical with (II A.87). As may be expected, the functions $df(\vartheta, \eta_i, \xi)$ approach the classical functions $df(\vartheta, \xi)$ for $\eta_i \to \infty$ (see Sec. II B.6).

For the total excitation cross section one obtains, by integration over the direction of $\mathbf{k}_f$,

$$\sigma_{E\lambda} = \left(\frac{Z_1 e}{\hbar v_i}\right)^2 a^{-2\lambda+2} B(E\lambda) f_{E\lambda}(\eta_i, \xi), \qquad \text{(II B.37)}$$

and

$$\sigma_{M\lambda} = \left(\frac{Z_1 e}{\hbar c}\right)^2 \frac{v_f}{v_i} a^{-2\lambda+2} B(M\lambda) f_{M\lambda}(\eta_i, \xi), \qquad \text{(II B.38)}$$

where

$$f(\eta_i, \xi) = \int \frac{df(\vartheta, \eta_i, \xi)}{d\Omega} d\Omega. \qquad \text{(II B.39)}$$

The scattering states to be used in (34) and (35) are the nonrelativistic Coulomb wave functions, which at large distances behave as distorted plane waves with appropriate in- and outgoing spherical waves. With the normalization employed, these wave functions are given by[44]

$$|\mathbf{k}_i\rangle = e^{-(\pi/2)\eta_i} \Gamma(1+i\eta_i) e^{i\mathbf{k}_i\cdot\mathbf{r}} \times {}_1F_1(-i\eta_i, 1; i(k_i r - \mathbf{k}_i\cdot\mathbf{r})), \qquad \text{(II B.40)}$$

and

$$|\mathbf{k}_f\rangle = e^{-(\pi/2)\eta_f} \Gamma(1-i\eta_f) e^{i\mathbf{k}_f\cdot\mathbf{r}} \times {}_1F_1(i\eta_f, 1; -i(k_f r + \mathbf{k}_f\cdot\mathbf{r})), \qquad \text{(II B.41)}$$

where ${}_1F_1$ is the confluent hypergeometric function.

It may be observed that the approximations involved in the cross sections derived in this paragraph only involve to the neglect of relativistic effects in the motion of the projectile; thus, the nuclear matrix elements entering into the $B(\lambda)$ may be taken to be the fully relativistic expressions [see (16) and (17)] which are identical with those appearing in the radiative transitions.

The matrix elements involving the scattering states of the projectile can be evaluated explicitly in the special case of electric dipole excitations. In fact, these matrix elements are equivalent to those involved in the bremsstrahlung process (see Sec. II E.1), and can be expressed in terms of hypergeometric functions (see Sec. II E.5).

For excitations of higher multipole orders the matrix elements are of essentially more complex character. They may be evaluated,[45,46] however, by expanding the Coulomb wave functions in partial waves; the radial matrix elements may then be expressed in terms of hypergeometric functions of two variables.

[43a] The expression for f_{M1} with the inclusion of spin effects has recently been given by L. C. Biedenharn and R. M. Thaler reference 62a).

[44] See, e.g., A. Sommerfeld, *Atombau und Spektrallinien* (Friedrich Vieweg & Sohn, Braunschweig, Germany, 1939). In the following we leave out the index p for the coordinates of the projectile.

[45] Biedenharn, McHale, and Thaler, Phys. Rev. **100**, 376 (1955).

[46] K. Alder and A. Winther, Kgl. Danske Videnskab. Selskab Mat. fys. Medd. **29**, Nos. 18 and 19 (1955).

II B.3. Reduction to Radial Matrix Elements

The expansions of the Coulomb wave functions (40) and (41) in partial waves are given by[44]

$$|\mathbf{k}_i\rangle=\sum_{lm} 4\pi(-1)^m i^l e^{i\sigma_l(\eta_i)} Y_{l,-m}(\mathbf{k}_i) Y_{lm}(\theta,\phi)(k_i r)^{-1}F_l(k_i r), \tag{II B.42}$$

and

$$|\mathbf{k}_f\rangle=\sum_{lm} 4\pi(-1)^m i^l e^{-i\sigma_l(\eta_f)} Y_{l,-m}(\mathbf{k}_f) Y_{lm}(\theta,\phi)(k_f r)^{-1}F_l(k_f r), \tag{II B.43}$$

where $\sigma_l(\eta)=\arg\Gamma(l+1+i\eta)$ is the Coulomb phase shift, and where $F_l(kr)$ is the regular solution to the radial wave equation for orbital angular momentum l. For large values of r, the function $F_l(kr)$ has the asymptotic form

$$F_l(kr)\sim\sin\left(kr-\frac{\pi}{2}l-\eta\ \ln 2kr+\sigma_l\right). \tag{II B.43a}$$

The angular integrations may now be performed by means of the relation[47]

$$\int Y_{l_1m_1}Y_{l_2m_2}Y_{l_3m_3}d\Omega=\left(\frac{(2l_1+1)(2l_2+1)(2l_3+1)}{4\pi}\right)^{\frac{1}{2}}\begin{pmatrix} l_1 & l_2 & l_3 \\ 0 & 0 & 0\end{pmatrix}\begin{pmatrix} l_1 & l_2 & l_3 \\ m_1 & m_2 & m_3\end{pmatrix}, \tag{II B.44}$$

and one thus obtains, for the matrix elements involved in electric excitations (34),

$$\langle\mathbf{k}_f|r^{-\lambda-1}Y_{\lambda\mu}(\theta,\phi)|\mathbf{k}_i\rangle=(4\pi)^{\frac{1}{2}}\sum_{l_il_fm_im_f} i^{l_i-l_f}(-1)^\mu e^{i(\sigma_i+\sigma_f)}$$
$$\times[(2l_i+1)(2l_f+1)(2\lambda+1)]^{\frac{1}{2}}\begin{pmatrix} l_i & l_f & \lambda \\ 0 & 0 & 0\end{pmatrix}\begin{pmatrix} l_i & l_f & \lambda \\ m_i & -m_f & \mu\end{pmatrix} Y_{l_i,-m_i}(\mathbf{k}_i)Y_{l_fm_f}(\mathbf{k}_f)M_{l_il_f}{}^{-\lambda-1}, \tag{II B.45}$$

where the radial matrix element M is defined by

$$M_{l_il_f}{}^{-\lambda-1}=\frac{1}{k_ik_f}\int_0^\infty F_{l_f}(k_fr)r^{-\lambda-1}F_{l_i}(k_ir)dr. \tag{II B.46}$$

Inserting (45) into (34) one obtains

$$df_{E\lambda}(\vartheta,\eta_i,\xi)=\frac{16\pi}{(2\lambda+1)^2}k_ik_fa^{2\lambda-2}\sum_{l_il_i'l_fl_f'}(2l_i+1)(2l_f+1)(2l_i'+1)(2l_f'+1)i^{l_i-l_f-l_i'+l_f'}(-1)^{\lambda+l_f+l_f'}$$
$$\times\exp i\{\sigma_{l_i}(\eta_i)+\sigma_{l_f}(\eta_f)-\sigma_{l_i'}(\eta_i)-\sigma_{l_f'}(\eta_f)\}\begin{pmatrix} l_i & l_f & \lambda \\ 0 & 0 & 0\end{pmatrix}\begin{pmatrix} l_i' & l_f' & \lambda \\ 0 & 0 & 0\end{pmatrix}M_{l_il_f}{}^{-\lambda-1}M_{l_i'l_f'}{}^{-\lambda-1}$$
$$\times\sum_l(2l+1)\begin{Bmatrix} l_i & l_i' & l \\ l_f' & l_f & \lambda\end{Bmatrix}\begin{pmatrix} l_i & l_i' & l \\ 0 & 0 & 0\end{pmatrix}\begin{pmatrix} l_f & l_f' & l \\ 0 & 0 & 0\end{pmatrix}P_l(\cos\vartheta), \tag{II B.47}$$

where we have used the notation (II A.62) for the Racah coefficient.[47a]

From (47) one obtains by integration over ϑ (see (39))

$$f_{E\lambda}(\eta_i,\xi)=\frac{64\pi^2}{(2\lambda+1)^2}k_ik_fa^{2\lambda-2}\sum_{l_il_f}(2l_i+1)(2l_f+1)\begin{pmatrix} l_i & l_f & \lambda \\ 0 & 0 & 0\end{pmatrix}^2|M_{l_il_f}{}^{-\lambda-1}|^2. \tag{II B.48}$$

Similarly, for the magnetic excitations one obtains from (35)

$$\langle\mathbf{k}_f|\mathbf{l}\cdot\nabla(r^{-\lambda-1}Y_{\lambda\mu}(\theta,\phi))|\mathbf{k}_i\rangle=(4\pi)^{\frac{1}{2}}\sum_{l_il_fm_im_f}(-i)^{l_i-l_f}(-1)^\mu e^{i(\sigma_i+\sigma_f)}2l_i(2\lambda+1)[\lambda(\lambda+1)(l_i+1)(2l_i+3)(2l_f+1)]^{\frac{1}{2}}$$
$$\times\begin{pmatrix} l_i & l_f & \lambda \\ m_i & -m_f & \mu\end{pmatrix}\begin{pmatrix} l_i+1 & l_f & \lambda \\ 0 & 0 & 0\end{pmatrix}\begin{Bmatrix} \lambda & \lambda & 1 \\ l_i & l_i+1 & l_f\end{Bmatrix}Y_{l_fm_f}(\mathbf{k}_f)Y_{l_i-m_i}(\mathbf{k}_i)M_{l_il_f}{}^{-\lambda-2}, \tag{II B.49}$$

---

[47] See, e.g., reference 24 or reference 22, p. 793.

[47a] For the numerical evaluation of df, it would be advantageous to compute (45) and insert afterwards into (34).

and

$$f_{M\lambda}(\eta_i,\xi)=\frac{64\pi^2(\lambda+1)}{\lambda(2\lambda+1)}a^{2\lambda-2}\sum_{l_il_f}(2l_i)^2(2l_i+1)(l_i+1)(2l_f+1)$$

$$\times\begin{pmatrix} l_i+1 & l_f & \lambda \\ 0 & 0 & 0\end{pmatrix}^2\begin{Bmatrix}\lambda & \lambda & 1\\ l_i & l_i+1 & l_f\end{Bmatrix}^2|M_{l_il_f}{}^{-\lambda-2}|^2. \quad \text{(II B.50)}$$

The evaluation of the excitation cross sections is thus reduced to the problem of computing the radial matrix elements and performing the summation over the angular momentum components contained in the scattering states.

II B.4. Evaluation of the Radial Matrix Elements

The radial wave function in (42) and (43) can be expressed in the form[44]

$$F_l(kr)=e^{-(\pi/2)\eta}\frac{|\Gamma(l+1+i\eta)|}{2\Gamma(2l+2)}(2kr)^{l+1}e^{-ikr}{}_1F_1(l+1-i\eta,\,2l+2;\,2ikr), \quad \text{(II B.51)}$$

which may be seen from (II E.90) to be a real function.

The radial matrix element (46) and even the more general matrix element

$$M_{l_il_f}{}^{-\lambda-1,q}=\frac{1}{k_ik_f}\int_0^\infty F_{l_f}(k_fr)r^{-\lambda-1}e^{-qr}F_{l_i}(k_ir)dr \quad \text{(II B.52)}$$

can be evaluated explicitly[48] by employing an integral representation of ${}_1F_1$ and carrying out the integration over r first. The result is [see (II E.91) and (II E.100)]

$$M_{l_il_f}{}^{-\lambda-1,q}=\frac{|\Gamma(l_i+1+i\eta_i)|\,|\Gamma(l_f+1+i\eta_f)|}{(2l_i+1)!(2l_f+1)!}(l_i+l_f-\lambda+1)!\,i^{l_i+l_f-\lambda+2}x^{l_i}(-y)^{l_f}$$

$$\times e^{-(\pi/2)(\eta_i+\eta_f)}(k_i-k_f+iq)^{\lambda-2}F_2(l_i+l_f-\lambda+2,\,l_i+1+i\eta_i,\,l_f+1-i\eta_f,\,2l_i+2,\,2l_f+2;\,x,\,y), \quad \text{(II B.53)}$$

with

$$x=\frac{2\eta_f}{\xi+iq\eta_i/k_f},\quad y=\frac{-2\eta_i}{\xi+iq\eta_i/k_f}. \quad \text{(II B.54)}$$

The function F_2 is a generalized hypergeometric function of two variables, one of the so-called Appell functions, and is defined in the neighborhood of $x=y=0$ by the series expansion (II E.93). The function is multivalued and the branch which is of interest in the present context is determined from (52) in the limit $q\to 0$.

The conservation of angular momentum and parity in the excitation process implies (see (45) and (49)) that the only matrix elements occurring in the cross sections are those for which

$$l_i-l_f=-\lambda,\,-\lambda+2,\,\cdots,\,\lambda. \quad \text{(II B.55)}$$

In the special case of $\lambda=0$, the F_2 function in (53) reduces to an ordinary hypergeometric function according to the reduction relation (II E.96). One thus obtains for the monopole matrix elements[49]

$$M_{ll}{}^{-1}=(k_i-k_f)^{-2}\left(\frac{\xi}{\eta_i+\eta_f}\right)^{i(\eta_i+\eta_f)}\frac{|\Gamma(l+1+i\eta_i)|\,|\Gamma(l+1+i\eta_f)|}{(2l+1)!}$$

$$\times e^{-(\pi/2)\xi}(-x_0)^lF(l+1-i\eta_i,\,l+1-i\eta_f,\,2l+2;\,x_0), \quad \text{(II B.56)}$$

with

$$x_0=-\frac{4\eta_i\eta_f}{\xi^2}. \quad \text{(II B.57)}$$

Although these matrix elements are of no direct importance for Coulomb excitation, they are useful in expressing higher multipole matrix elements by recursion relations (see following).

The series expansion (II E.93) of the function F_2 is valid only for $|x|+|y|<1$ and, since in our case $x+y=2$, an analytic continuation must be employed in the evaluation of (53). This analytic continuation is especially simple

[48] A. Erdélyi, Math. Z. 40, 693 (1936).
[49] W. Gordon, Ann. Physik (5) 2, 1031 (1929).

to perform in the case $l_i=l_f\pm\lambda$, where the F_2 function reduces to an F_1 function [see (II E.97)]. The F_1 function can again be written in terms of the Appell function F_3 [see (II E.98)] for which the analytic continuation is well known [see (II E.99)]. The application of these three formulas leads directly to the following result:

$$M_{l+\lambda,\,l}{}^{-\lambda-1}=e^{(\pi/2)\xi}\left|\frac{\Gamma(l+1+i\eta_f)}{\Gamma(l+1+\lambda+i\eta_i)}\right|\left(\frac{\eta_i}{\eta_f}\right)^l(2k_i)^{\lambda-2}$$
$$\times\left\{\frac{|\Gamma(\lambda+i\xi)|^2}{(2\lambda-1)!}F_2\left(-2\lambda+1,\,l+1-i\eta_f,\,l+1+i\eta_f,\,-\lambda+1-i\xi,\,-\lambda+1+i\xi;\frac{\xi}{2\eta_f},\frac{\xi}{2\eta_f}\right)\right.$$
$$+2\,\Re e\left[\left(e^{i\pi}\frac{\xi}{2\eta_f}\right)^{\lambda+i\xi}\frac{\Gamma(l+\lambda+1-i\eta_i)\Gamma(-\lambda-i\xi)}{\Gamma(l+1-i\eta_f)}\right.$$
$$\left.\left.\times F_2\left(-\lambda+1+i\xi,\,l+\lambda+1-i\eta_i,\,l+1+i\eta_f,\,\lambda+1+i\xi,\,-\lambda+1+i\xi;\frac{\xi}{2\eta_f},\frac{\xi}{2\eta_f}\right)\right]\right\},\tag{II B.58}$$

and

$$M_{l,\,l+\lambda}{}^{-\lambda-1}(\eta_i,\eta_f)=M_{l+\lambda,\,l}{}^{-\lambda-1}(\eta_f,\eta_i)=e^{-\pi\xi}M_{l+\lambda,\,l}{}^{-\lambda-1}(-\eta_f,\,-\eta_i).\tag{II B.59}$$

In (58), the first F_2 function is a polynomial, since the first parameter is a negative integer [see (II E.93)]. Thus, for the first few λ's, one finds

$$F_2\left(-2\lambda+1,\,l+1-i\eta_f,\,l+1+i\eta_f,\,-\lambda+1-i\xi,\,-\lambda+1+i\xi;\frac{\xi}{2\eta_f},\frac{\xi}{2\eta_f}\right)$$
$$=\begin{cases}0, & (\lambda=1)\\[4pt] \dfrac{1}{2(1+\xi^2)}\dfrac{\eta_i(\eta_i+\eta_f)}{\eta_f{}^2}, & (\lambda=2)\\[4pt] \dfrac{1}{2(1+\xi^2)(4+\xi^2)}\dfrac{\eta_i(\eta_i+\eta_f)}{\eta_f{}^4}[5l\xi(\eta_i+\eta_f)+12\eta_f{}^2-8\eta_i{}^2]. & (\lambda=3)\end{cases}\tag{II B.60}$$

If $|l_i-l_f|\neq\lambda$ the F_2 function cannot be reduced to a single F_1 function, but, as shown in Sec. II E.8, it may be written as a finite sum of such functions. The analytic continuation may thus be performed in complete analogy to the case $|l_i-l_f|=\lambda$ and the result can be expressed by two polynomials plus a finite number of F_2 functions of the arguments x^{-1} and y^{-1}. The analytic continuation can, however, be obtained more easily from (II E.99). Identifying one of the F_2 functions of this equation with that involved in (53), one obtains

$$F_2(l_i+l_f-\lambda+2,\,l_i+1-i\eta_i,\,l_f+1+i\eta_f,\,2l_i+2,\,2l_f+2;\,x,\,y)$$
$$=\frac{\Gamma(-l_i-i\eta_i)\Gamma(-l_f+i\eta_f)\Gamma(\lambda-l_i-l_f-1)}{\Gamma(\lambda+1+i\xi)\Gamma(-2l_i-1)\Gamma(-2l_f-1)}(-x)^{-l_i-1+i\eta_i}(-y)^{-l_f-1-i\eta_f}$$
$$\times F_3\left(l_i+1-i\eta_i,\,l_f+1+i\eta_f,\,-l_i-i\eta_i,\,-l_f+i\eta_f,\,\lambda+1+i\xi;\,\frac{1}{x},\frac{1}{y}\right)$$
$$+\frac{\Gamma(2l_f+1)\Gamma(-l_f+i\eta_f)\Gamma(\lambda-l_i-l_f-1)}{\Gamma(-2l_f-1)\Gamma(l_f+1+i\eta_f)\Gamma(\lambda-l_i+l_f)}(-y)^{-2l_f-1}$$
$$\times F_2(l_i-l_f+1-\lambda,\,l_i+1-i\eta_i,\,-l_f+i\eta_f,\,2l_i+2,\,-2l_f;\,x,\,y)$$
$$+\frac{\Gamma(2l_i+1)\Gamma(-l_i-i\eta_i)\Gamma(\lambda-l_i-l_f-1)}{\Gamma(-2l_i-1)\Gamma(l_i+1-i\eta_i)\Gamma(\lambda-l_f+l_i)}(-x)^{-2l_i-1}$$
$$\times F_2(l_f-l_i+1-\lambda,\,-l_i-i\eta_i,\,l_f+1+i\eta_f,\,-2l_i,\,2l_f+2;\,x,\,y)$$
$$+\frac{\Gamma(2l_i+1)\Gamma(2l_f+1)\Gamma(\lambda-l_i-l_f-1)\Gamma(-l_i-i\eta_i)\Gamma(-l_f+i\eta_f)}{\Gamma(-2l_i-1)\Gamma(-2l_f-1)\Gamma(l_i+l_f+\lambda+1)\Gamma(l_i+1-i\eta_i)\Gamma(l_f+1+i\eta_f)}(-x)^{-2l_i-1}(-y)^{-2l_f-1}$$
$$\times F_2(-l_i-l_f-\lambda,\,-l_i-i\eta_i,\,-l_f+i\eta_f,\,-2l_i,\,-2l_f;\,x,\,y).\tag{II B.61}$$

This equation is, however, singular for integer values of l_i and l_f. If one first considers l_i and l_f to have noninteger values while preserving l_i-l_f as an integer, the first and second F_2 functions reduce to polynomials, since the first parameter is in that case a negative integer. The third F_2 function can be eliminated by considering also the complex conjugate equation to (61) which contains the same F_2 functions, according to the transformation (II E.95). After this elimination, the limiting process l_i, l_f approaching integer values and $q\to 0$ can easily be performed and the result gives the following expression for the radial matrix element (46) or (53):

$$M_{l_il_f}{}^{-\lambda-1}=\pi(2k_i)^{\lambda-2}\left(\frac{\eta_i}{\eta_f}\right)^{l_f}\frac{e^{(\pi/2)\xi}}{\sinh\pi\xi}\left|\frac{\Gamma(l_f+1+i\eta_f)}{\Gamma(l_i+1+i\eta_i)}\right|\frac{(2l_i)!}{(2l_f+1)!(l_i-l_f+\lambda-1)!}\left(-\frac{i\xi}{2\eta_f}\right)^{l_i-l_f-1+\lambda}$$
$$\times F_2(l_f-l_i+1-\lambda,\, l_f+1+i\eta_f,\, -l_i-i\eta_i,\, 2l_f+2,\, -2l_i;\, y,\, x)$$
$$+\pi(2k_f)^{\lambda-2}\left(\frac{\eta_f}{\eta_i}\right)^{l_i}\frac{e^{-(\pi/2)\xi}}{\sinh\pi\xi}\left|\frac{\Gamma(l_i+1+i\eta_i)}{\Gamma(l_f+1+i\eta_f)}\right|\frac{(2l_f)!}{(2l_i+1)!(l_f-l_i+\lambda-1)!}\left(\frac{i\xi}{2\eta_i}\right)^{l_f-l_i-1+\lambda}$$
$$\times F_2(l_i-l_f+1-\lambda,\, l_i+1-i\eta_i,\, -l_f+i\eta_f,\, 2l_i+2,\, -2l_f;\, x,\, y)'$$
$$-\frac{\pi}{2}(k_f-k_i)^\lambda k_i^{-1}k_f^{-1}\frac{e^{-(\pi/2)|\xi|}}{\sinh\pi\xi}\left|\frac{\Gamma(l_i+1+i\eta_i)}{\Gamma(l_f+1+i\eta_f)}\right|\mathcal{R}e\left\{\left(\frac{\xi}{2}\right)^{-i\xi}\frac{\Gamma(l_f+1-i\eta_f)}{\Gamma(l_i+1-i\eta_i)\Gamma(\lambda+1-i\xi)}\right.$$
$$\left.\times i^{l_f-l_i-\lambda-1}\eta_i^{i\eta_f}\eta_f^{-i\eta_i}F_3\left(-l_i+i\eta_i,\, -l_f-i\eta_f,\, l_i+1+i\eta_i,\, l_f+1-i\eta_f,\, \lambda+1-i\xi;\, \frac{1}{x},\frac{1}{y}\right)\right\}. \quad \text{(II B.62)}$$

In these equations, x and y represent the limiting values obtained from (54) by setting $q=0$. The F_2 functions in (62) are to be interpreted as the polynomials obtained in the limit of l_i and l_f approaching integer values while l_i-l_f remains an integer. These polynomials are pure imaginary [see (II E.95)] and are for the lowest values of λ given explicitly by

$$F_2(l_i-l_f+1-\lambda,\, l_i+1-i\eta_i,\, -l_f+i\eta_f,\, 2l_i+2,\, -2l_f;\, x,\, y)$$
$$=\begin{cases} 0 & \lambda=1\quad l_f=l_i+1\\ -i\,\dfrac{\eta_i\eta_f}{\xi}\dfrac{1}{l_i(l_i+1)} & \lambda=2\quad l_f=l_i\\ i\,\dfrac{\eta_i\eta_f(\eta_i+\eta_f)}{\xi^2(l_i+1)(l_i+2)(2l_i+3)} & \lambda=2\quad l_f=l_i+2\\ -i\,\dfrac{2\eta_i\eta_f[l_i(l_i+1)(2l_i+1)\xi(\eta_i+\eta_f)-6(l_i+1)^2\eta_i^2-6\eta_i^2\eta_f^2]}{\xi^3l_i(l_i+1)^2(l_i+2)(2l_i+1)(2l_i+3)} & \lambda=3\quad l_f=l_i+1\\ i\,\dfrac{2\eta_i\eta_f(\eta_i+\eta_f)[5l_i\xi(\eta_i+\eta_f)+12\eta_f^2-8\eta_i^2]}{\xi^4(l_i+1)(l_i+2)(l_i+3)(2l_i+3)(2l_i+5)}, & \lambda=3\quad l_f=l_i+3.\end{cases} \quad \text{(II B.63)}$$

As mentioned above, the analytic continuation of the F_2 function in (53) can be written as a finite sum of F_2 functions. This alternative form for the matrix elements may be obtained directly from (62) by expanding the F_3 function in terms of F_2 functions [see (II E.104) and (II E.97)]. The relation (II E.98) shows how (62) reduces to (58) in the special case $l_i-l_f=\pm\lambda$.

In the evaluation of the excitation cross section, it is in general necessary to extend the summation to include large values of l_i and l_f. While the main contribution in most cases arises from terms with $l\sim\eta$, the convergence for large l is rather slow, especially for small values of ξ. The numerical calculations are therefore greatly simplified by the use of recursion formulas connecting matrix elements for different values of l_i, l_f, and λ.

The existence of such recursion formulas is a consequence of simple recursion properties of hypergeometric functions. Thus, five F_2 functions with parameters differing only by integer numbers are always linearly dependent. In special cases, the recursion formulas may of course contain less than five terms. One may derive these formulas either directly from the properties of the hypergeometric functions or from the differential equation for the Coulomb wave functions.[45,46,50] The

[50] See also L. Infeld and T. E. Hull, Revs. Modern Phys. 23, 21 (1951).

recursion relations obtained by the latter method are all contained in the following general formula

$$x_1\frac{|l_f+1+i\eta_f|}{\eta_f(l_f+1)}M_{l_i,l_f+1}{}^{-\lambda-1}+x_2\frac{|l_i+i\eta_i|}{\eta_i l_i}M_{l_i-1,l_f}{}^{-\lambda-1}-x_3\frac{|l_i+1+i\eta_i|}{\eta_i(l_i+1)}M_{l_i+1,l_f}{}^{-\lambda-1}$$
$$-x_4\frac{|l_f+i\eta_f|}{\eta_f l_f}M_{l_i,l_f-1}{}^{-\lambda-1}-\left[\frac{x_1}{l_f+1}+\frac{x_2}{l_i}-\frac{x_3}{l_i+1}-\frac{x_4}{l_f}\right]M_{l_il_f}{}^{-\lambda-1}$$
$$=[k_i\eta_i]^{-1}[x_1(l_f-\lambda)+x_2l_i-x_3(l_i+1)-x_4(l_f+\lambda+1)]M_{l_il_f}{}^{-\lambda-2}$$
$$+(k_i\eta_i)^{-1}(x_1+x_2+x_3+x_4)\int_0^\infty F_{l_f}(k_fr)r^{-\lambda-1}\frac{d}{dr}F_{l_i}(k_ir)dr, \quad \text{(II B.64)}$$

where x_1, x_2, x_3, and x_4 are arbitrary constants. Three independent recursion formulas may be obtained from (64) by giving the factors x_1 to x_4 different values satisfying

$$x_1+x_2+x_3+x_4=0, \quad \text{(II B.65)}$$

whereby the last term of (64) is suppressed.

Additional recursion relations may then be obtained by combining those derived directly from (64) and eliminating the unwanted matrix elements. In the following, we shall give some specific relations which are useful for the numerical evaluation of the radial matrix elements for low multipole orders.

For the monopole ($\lambda=0$) matrix elements (see (56)), one obtains the three term relation

$$y_1M_{l+1,l+1}{}^{-1}+y_2M_{ll}{}^{-1}+y_3M_{l-1,l-1}{}^{-1}=0, \quad \text{(II B.66)}$$

with

$$y_1=2l|l+1+i\eta_i|\,|l+1+i\eta_f|,$$
$$y_2=-(2l+1)\left[\frac{\eta_i^2+\eta_f^2}{\eta_i\eta_f}l(l+1)+2\eta_i\eta_f\right], \quad \text{(II B.67)}$$
$$y_3=(2l+2)|l+i\eta_i|\,|l+i\eta_f|.$$

This relation connects all monopole matrix elements satisfying the condition (55) with the two first ($l=0$ and 1).

For $\lambda=1$, the matrix elements are most easily obtained from the monopole matrix elements by

$$(\lambda+1)M_{l,l+1}{}^{-\lambda-2}=y_1M_{ll}{}^{-\lambda-1}+y_2M_{l+1,l+1}{}^{-\lambda-1}, \quad \text{(II B.68)}$$

with

$$y_1=k_f\frac{|l+1+i\eta_f|}{l+1},$$
$$y_2=-k_i\frac{|l+1+i\eta_i|}{l+1}. \quad \text{(II B.69)}$$

By means of this relation one may obtain an explicit expression for the dipole radial matrix elements in terms of usual hypergeometric functions.[49]

Also in the dipole case the recursion relation which connects different values of l contains only three terms and may be written

$$y_1M_{l-1,l-2}{}^{-2}+y_2M_{l,l-1}{}^{-2}+y_3M_{l+1,l}{}^{-2}=0, \quad \text{(II B.70)}$$

with

$$y_1=2\eta_i\eta_f|l-1+i\eta_f|\,|l+i\eta_i|,$$
$$y_2=-4\eta_i^2\eta_f^2-l(2l+1)\eta_i^2-l(2l-1)\eta_f^2, \quad \text{(II B.71)}$$
$$y_3=2\eta_i\eta_f|l+i\eta_f|\,|l+1+i\eta_i|.$$

The $\lambda=2$ matrix elements cannot be reduced to those with $\lambda=1$, since the recursion relations connecting matrix elements of multipolarity λ with those of multipole order $\lambda+1$ become singular for $\lambda=1$. There are, for $\lambda=2$, two types of matrix elements, namely those for which $l_i-l_f=\pm2$ and those for which $l_i=l_f$. The latter are connected with the former through the relation

$$yM_{ll}{}^{-3}=y_1M_{l,l+2}{}^{-3}+y_2M_{l-1,l+1}{}^{-3}$$
$$+y_3M_{l+2,l}{}^{-3}+y_4M_{l+1,l-1}{}^{-3}, \quad \text{(II B.72)}$$

with

$$y=\frac{l(l+1)}{3}(\eta_f^2-\eta_i^2),$$
$$y_1=-\eta_i^2|l+1+i\eta_f|\,|l+2+i\eta_f|,$$
$$y_2=\eta_i\eta_f\frac{2l+3}{2l+1}|l+i\eta_i|\,|l+1+i\eta_f|, \quad \text{(II B.73)}$$
$$y_3=\eta_f^2|l+1+i\eta_i|\,|l+2+i\eta_i|,$$
$$y_4=-\eta_i\eta_f\frac{2l+3}{2l+1}|l+i\eta_f|\,|l+1+i\eta_i|.$$

For the matrix elements with $|l_i-l_f|=\lambda$ there exist the following four term relations

$$y_1M_{l+\lambda-3,l-3}{}^{-\lambda-1}+y_2M_{l+\lambda-2,l-2}{}^{-\lambda-1}$$
$$+y_3M_{l+\lambda-1,l-1}{}^{-\lambda-1}+y_4M_{l+\lambda,l}{}^{-\lambda-1}=0, \quad \text{(II B.74)}$$

with

$$y_1=2\eta_i\eta_f|l-2+i\eta_i|\,|l-1+i\eta_f|\,|l+\lambda-2+i\eta_i|,$$
$$y_2=-|l-1+i\eta_f|[l^2(2\eta_i^2+4\eta_f^2)+l(4(\lambda-2)(\eta_i^2+\eta_f^2)+\eta_i^2-\eta_f^2)+(\lambda-2)((2\lambda-3)\eta_i^2-3\eta_f^2)+6\eta_i^2\eta_f^2],$$
$$y_3=\frac{\eta_f}{\eta_i}|l+\lambda-1+i\eta_i|[l^2(4\eta_i^2+2\eta_f^2)+l(4(\lambda-2)\eta_i^2+\eta_i^2-\eta_f^2)-2(\lambda-2)\eta_i^2+6\eta_i^2\eta_f^2],\quad \text{(II B.75)}$$
$$y_4=-2\eta_f^2|l+\lambda-1+i\eta_i|\,|l+\lambda+i\eta_i|\,|l+i\eta_f|.$$

For $\lambda=2$, two of these matrix elements are also connected with two monopole matrix elements by

$$y_1M_{l+1,l+3}{}^{-3}+y_2M_{l,l+2}{}^{-3}=y_3M_{l+1,l+1}{}^{-1}+y_4M_{ll}{}^{-1},\quad \text{(II B.76)}$$

with

$$y_1=4(l+1)\eta_i^2|l+2+i\eta_f|\,|l+3+i\eta_f|,$$
$$y_2=-4(l+1)\eta_i\eta_f|l+2+i\eta_f|\,|l+1+i\eta_i|,$$
$$y_3=(k_i^2-k_f^2)[2\eta_i^2\eta_f^2+\eta_f^2(l+1)(2l+3)-\eta_i^2(l+1)],\quad \text{(II B.77)}$$
$$y_4=-(k_i^2-k_f^2)2\eta_i\eta_f|l+1+i\eta_i|\,|l+1+i\eta_f|.$$

By repeated application of this formula one obtains the recurrence relation

$$M_{l,l+2}{}^{-3}=f(l)\left[\frac{M_{l',l'+2}{}^{-3}}{f(l')}+\sum_{j=l'}^{l-1}\frac{A(j)}{f(j+1)}\right],\quad \text{(II B.78)}$$

with $l'<l$ and

$$f(l)=\left(\frac{\eta_f}{\eta_i}\right)^l\left|\frac{\Gamma(l+1+i\eta_i)}{\Gamma(l+3+i\eta_f)}\right|,\quad \text{(II B.79)}$$

$$A(l)=\frac{k_i^2-k_f^2}{4(l+1)|l+2+i\eta_f|\,|l+3+i\eta_f|}\frac{1}{\eta_i^2}\times\{[2\eta_i^2\eta_f^2+\eta_f^2(l+1)(2l+3)-\eta_i^2(l+1)]\times M_{l+1,l+1}{}^{-1}-2\eta_i\eta_f|l+1+i\eta_i|\times|l+1+i\eta_f|M_{ll}{}^{-1}\}.\quad \text{(II B.80)}$$

In the $\lambda=2$ case, one thus needs to calculate directly from (58) either the six matrix elements M_{02}, M_{13}, M_{24}, M_{20}, M_{31}, and M_{42}, from which the remaining ones may be obtained from (72) and (74), or one may use the relations (78) and (72) and thereby obtain all matrix elements from M_{02}, M_{20}, and the monopole matrix elements (see (56) and (66)).

For $\lambda=3$ the matrix elements with $l_i-l_f=\pm1$ may be obtained from the quadrupole matrix elements by means of Eq. (68). The matrix elements with $l_i-l_f=\pm3$ can again be reduced to the six first by means of (74).

II B.5. Angular Distribution of De-Excitation γ Rays

The angular distribution of the γ quanta following an electromagnetic excitation is given by

$$W_{\mathbf{k}_i\mathbf{k}_f}(\Omega_\gamma)=\sum_{M_iM_{ff}\sigma}\left|\sum_{M_f}\langle I_{ff}M_{ff}|H_\gamma(\Omega_\gamma,\sigma)|I_fM_f\rangle\times\langle f|\mathcal{H}^{(1)}|i\rangle\right|^2,\quad \text{(II B.81)}$$

in analogy to the expression (II A.59). The transition matrix element $\langle f|\mathcal{H}^{(1)}|i\rangle$ is given by (15) and is of just the same form as the classical transition amplitude b_{if} [see (II A.79)].

The quantum-mechanical angular distribution (81) can thus be directly obtained from the formulas in Sec. II A.4 by the substitution

$$S_{\pi\lambda\mu}\rightarrow\langle\mathbf{k}_f|\mathfrak{M}(\pi\lambda,\mu)|\mathbf{k}_i\rangle.\quad \text{(II B.82)}$$

While the classical integrals S aside from constant factors depend only on the scattering angles and the parameter ξ [see (II A.24) and (II A.48)] the quantum-mechanical matrix elements depend also on the parameter η_i. Thus, the a and b coefficients involved in the quantum-mechanical angular distributions will also depend on η_i.

In order to obtain the total γ distribution irrespective of scattering angle one simply integrates (81) over the direction of $\mathbf{k}_f$, since the Rutherford cross section is already contained in $|\langle f|\mathcal{H}^{(1)}|i\rangle|^2$. In the most important case of excitations of pure $E\lambda$ type one thus obtains [see (II A.74)][51,45]

$$W(\vartheta_\gamma)=\sum_k a_k^{E\lambda}(\eta_i,\xi)A_k^{(\lambda)}P_k(\cos\vartheta_\gamma),\quad \text{(II B.83)}$$

with

$$a_k^{E\lambda}(\eta_i,\xi)=b_k^{E\lambda}/b_0^{E\lambda},\quad \text{(II B.84)}$$

and, according to (II A.68), (II B.28), and (II B.45),

$$b_k^{E\lambda}(\eta_i\xi)=\begin{pmatrix}\lambda&\lambda&k\\1&-1&0\end{pmatrix}^{-1}\sum_{l_il_i'l_f}(-1)^{l_f+1}(2l_i+1)(2l_i'+1)(2l_f+1)i^{l_i-l_i'}\exp i(\sigma_{l_i}(\eta_i)-\sigma_{l_i'}(\eta_i))\times\begin{Bmatrix}\lambda&\lambda&k\\l_i&l_i'&l_f\end{Bmatrix}\begin{pmatrix}\lambda&l_i&l_f\\0&0&0\end{pmatrix}\begin{pmatrix}\lambda&l_i'&l_f\\0&0&0\end{pmatrix}\begin{pmatrix}l_i&l_i'&k\\0&0&0\end{pmatrix}M_{l_il_f}{}^{-\lambda-1}M_{l_i'l_f}{}^{-\lambda-1}.\quad \text{(II B.85)}$$

[51] L. C. Biedenharn and M. E. Rose, ORNL report 1789 (1954). See also Breit, Ebel, and Russell, Phys. Rev. **101**, 1504 (1956).

We have here employed the relation[24]

$$\sum_{\mu_1\mu_2\mu_3} (-1)^{l_1+l_2+l_3+\mu_1+\mu_2+\mu_3}\begin{pmatrix} j_1 & l_2 & l_3 \\ m_1 & \mu_2 & -\mu_3 \end{pmatrix}\begin{pmatrix} l_1 & j_2 & l_3 \\ -\mu_1 & m_2 & \mu_3 \end{pmatrix}\begin{pmatrix} l_1 & l_2 & j_3 \\ \mu_1 & -\mu_2 & m_3 \end{pmatrix} = \begin{pmatrix} j_1 & j_2 & j_3 \\ m_1 & m_2 & m_3 \end{pmatrix}\begin{Bmatrix} j_1 & j_2 & j_3 \\ l_1 & l_2 & l_3 \end{Bmatrix}, \quad \text{(II B.86)}$$

and have inserted

$$Y_{l_i m_i}(\mathbf{k}_i) = \left(\frac{2l_i+1}{4\pi}\right)^{\frac{1}{2}} \delta_{m_i 0}, \quad \text{(II B.87)}$$

corresponding to the fact that the emission angle ϑ_γ of the γ quantum is measured from the direction of the incident beam.

The coefficient $b_0^{E\lambda}(\eta_i,\xi)$ is related to the total excitation function $f_{E\lambda}(\eta_i,\xi)$ (see (48)) by the equation

$$f_{E\lambda}(\eta_i,\xi) = \frac{64\pi^2}{(2\lambda+1)^2} k_i k_f a^{2\lambda-2} b_0^{E\lambda}(\eta_i,\xi). \quad \text{(II B.88)}$$

II B.6. WKB Approximation and Classical Limit

For $\eta > 1$, a rather accurate approximation to the Coulomb excitation cross sections may be obtained by replacing the radial matrix elements by those derived from the WKB approximation.[52] This treatment is also convenient for the discussion of the transition of the quantum-mechanical expressions to the classical formulas in the limit $\eta \gg 1$.[16,53]

The WKB approximation for the radial wave function is given by

$$F_l(kr) = [f(r)/k^2]^{-\frac{1}{4}} \sin\varphi, \quad \text{(II B.89)}$$

where

$$\varphi = \frac{\pi}{4} + \int_{r_0}^{r} [f(r)]^{\frac{1}{2}} dr, \quad \text{(II B.90)}$$

and

$$f(r) = k^2 - \frac{2k\eta}{r} - \frac{l(l+1)}{r^2}. \quad \text{(II B.91)}$$

The expression (89) holds outside the classical turning point r_0 defined by $f(r_0)=0$. The contribution to the radial matrix element from the region $r<r_0$ is of lower order in η and is neglected in the present approximation.

Inserting (89), one finds that the radial matrix element (46) involves two terms, the first containing the sum of the phases φ_i and φ_f of initial and final wave function and the second containing the difference $\varphi_i - \varphi_f$. The first term may be neglected due to the rapid oscillation of the integrand. In the second term, the phase difference may be expanded as follows

$$\varphi_i - \varphi_f \simeq (k_i - k_f) \int_{r_0}^{r} [f(r)]^{-\frac{1}{2}} k\, dr - [l_i(l_i+1) - l_f(l_f+1)] \int_{r_0}^{r} [f(r)]^{-\frac{1}{2}} \frac{dr}{2r^2}, \quad \text{(II B.92)}$$

since $k_i\eta_i = k_f\eta_f$.

In the integrals in (92) the parameters k, η, and l refer to average values for initial and final state. The differences between the turning points for the initial and final state have been neglected, since these contributions are of higher order in ξ/η.

Evaluating the integrals in (92) and introducing the substitution

$$kr = (\eta^2 + l(l+1))^{\frac{1}{2}} \cosh w + \eta, \quad \text{(II B.93)}$$

one obtains

$$\varphi_i - \varphi_f \simeq \xi(\epsilon \sinh w + w) + \mu \cos^{-1} \frac{\epsilon + \cosh w}{1 + \epsilon \cosh w}, \quad \text{(II B.94)}$$

where

$$\xi = \eta_f - \eta_i, \quad \text{(II B.95)}$$

while

$$\epsilon = \frac{(\eta^2 + l(l+1))^{\frac{1}{2}}}{\eta} = \left\{1 + \frac{(l_i+\mu/2)(l_i+1+\mu/2)}{\eta^2}\right\}^{\frac{1}{2}}, \quad \text{(II B.96)}$$

and

$$\mu = l_f - l_i. \quad \text{(II B.97)}$$

For the radial matrix element (46) we thus obtain

$$M_{l_i, l_i+\mu}^{-\lambda-1} = \frac{k^{\lambda-2}}{4\eta^\lambda} \int_{-\infty}^{\infty} e^{i\xi(\epsilon \sinh w + w)} \times \frac{[\cosh w + \epsilon + i(\epsilon^2-1)^{\frac{1}{2}} \sinh w]^\mu}{(\epsilon \cosh w + 1)^{\lambda+\mu}} dw. \quad \text{(II B.98)}$$

It is seen that this integral is identical with the orbital integral (II A.26) involved in the classical treatment. The quantity ϵ given by (96) just corresponds to the eccentricity of the orbit [see (II A.23 and 45)] and μ represents the transfer of angular momentum in the direction perpendicular to the plane of the orbit. By introducing the deflection angle of this corresponding

[52] Benedict, Daitch, and Breit, Phys. Rev. **101**, 171 (1956); K. Alder and A. Winther, CERN report T/KA-AW-4 (1955).

[53] K. Alder and A. Winther, Phys. Rev. **96**, 237 (1954); G. Breit and P. B. Daitch, Phys. Rev. **96**, 1447 (1954).

TABLE II.1. Comparison between WKB approximation and exact values for the radial matrix elements. The radial matrix elements, M_{l_i,l_f}^{-3}, occurring in $E2$ Coulomb excitation are given for various values of η_i, ξ, l_i, and l_f. The table compares the values obtained from the exact calculation (Sec. II B.4) and from the WKB approximation (II B.100).

		$M_{l+2,l}^{-3}$ Quantal	$M_{l+2,l}^{-3}$ WKB	$M_{l+1,l+1}^{-3}$ Quantal	$M_{l+1,l+1}^{-3}$ WKB	$M_{l,l+2}^{-3}$ Quantal	$M_{l,l+2}^{-3}$ WKB
$\xi=0.2$	0	0.06548	0.06615	0.05993	0.05393	0.02402	0.02338
$\eta_i=1.0$	3	0.01544	0.01528	0.00932	0.00922	0.00171	0.00172
	6	0.00589	0.00583	0.00258	0.00264	0.00032	0.00033
$\xi=0.2$	0	0.008463	0.008577	0.007777	0.007616	0.006137	0.006184
$\eta_i=4.0$	2	0.006717	0.006750	0.005740	0.005669	0.003318	0.003316
	5	0.004052	0.004052	0.003947	0.003918	0.001266	0.001266
$\xi=0.2$	0	0.002219	0.002228	0.002085	0.002073	0.001877	0.001883
$\eta_i=8.0$	1	0.002212	0.002220	0.002007	0.001996	0.001678	0.001681
	3	0.002048	0.002053	0.001901	0.001892	0.001468	0.001470
$\xi=1.0$	0	0.001690	0.001714	0.001155	0.001155	0.000729	0.000727
$\eta_i=4.0$	2	0.001659	0.001681	0.000799	0.000782	0.000273	0.000274
	5	0.001024	0.001032	0.000340	0.000338	0.000061	0.000061
$\xi=1.0$	0	0.0004295	0.0004308	0.0003492	0.0003452	0.0002712	0.0002712
$\eta_i=8.0$	1	0.0004655	0.0004672	0.0003316	0.0003281	0.0002184	0.0002183
	3	0.0004932	0.0004953	0.0002802	0.0002779	0.0001335	0.0001334

classical orbit

$$\vartheta=2\sin^{-1}\frac{1}{\epsilon}, \qquad \text{(II B.99)}$$

where ϵ is given by (96), the matrix element (98) may also be written

$$M_{l_i,l_i+\mu}^{-\lambda-1}=\frac{k^{\lambda-2}}{4\eta^{\lambda}}I_{\lambda\mu}(\vartheta,\xi). \qquad \text{(II B.100)}$$

Since the WKB approximation for the radial wave function is valid not only when $\eta\gg1$, but also when $l\gg1$, the limiting formula (100) holds also for large l irrespective of the magnitude of η.

The formula (100) has also been derived[45] by expressing the exact radial integral (46) in a suitable form and going to the limit $|l+i\eta|\to\infty$ for fixed ξ. In this way one obtains an expansion of the matrix element with (98) as the leading term.

The WKB formula (100) gives very accurate values for the radial matrix elements even for moderate values of η and l; indeed the accuracy is much greater than might have been expected in view of the inaccuracy of the wave function (89) in the neighborhood of the turning point.[54]

An illustration of the accuracy of (100) is given in Table II.1, where exact values of quadrupole matrix elements computed from the formulas in Sec. II B.4 are compared with the WKB values. From this table it is also seen that for not too large values of l_i the particular choice of l in (96) is essential for the close agreement.

If one employs the WKB formula (100) for the radial matrix elements in expressions such as (47), (48), or (85), one obtains cross sections which are valid even for moderate values of η and which represent a major improvement over the classical expressions (see Fig. II.3).

In order to exhibit the transition of the total cross sections to the classical limit one may take advantage of the fact that for large η's the main contribution arises from the large values of l. One may thus use the asymptotic formula

$$\begin{pmatrix} l_1 & l_2 & \lambda \\ m_1 & m_2 & \mu \end{pmatrix} =\frac{(-1)^{l_2-\lambda-m_1}}{(2l_1+1)^{\frac{1}{2}}}D_{\mu,l_1-l_2}{}^{\lambda}(0,\theta,0), \qquad \text{(II B.101}$$

with

$$\cos\theta=\frac{m_2}{[l_2(l_2+1)]^{\frac{1}{2}}},$$

for the vector addition coefficient in the limit of large l

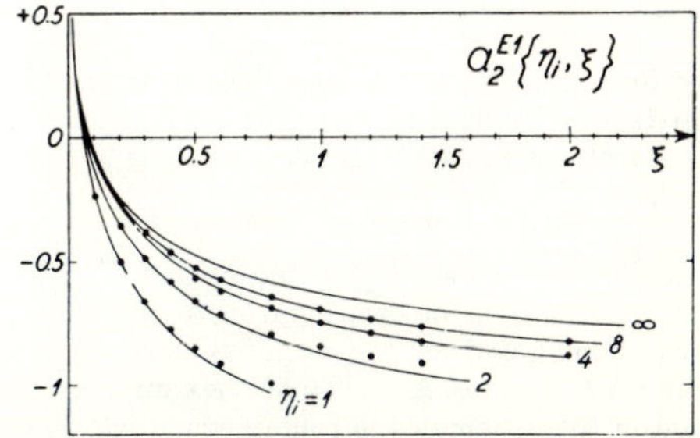

FIG. II.3. Comparison of WKB approximation and exact ca lation for the angular distribution of the gamma rays. coefficient $a_2^{E1}(\eta_i,\xi)$ which describes the angular distributio the gamma rays following $E1$ Coulomb excitation (II C.26 bremsstrahlung (II E.17), is plotted as a function of ξ for diffe values of η_i. The full-drawn curve corresponds to the exact c lation,[61] while the black points have been obtained by the W approximation.[59]

[54] The reasons for the high accuracy of the WKB approximation have been discussed especially by G. Breit and P. B. Daitch, Proc. Natl. Acad. Sci. **41**, 653 (1955); J. P. Lazarus and S. Sack, Phys. Rev. **100**, 370 (1955).

and l_2, for fixed λ. This formula may be derived from the explicit expressions for the Clebsch-Gordon coefficient and the rotation matrix D.[24,55]

If in (48) one further inserts [see (96) and (99)],

$$l=\eta \cot\frac{\vartheta}{2}, \qquad \text{(II B.102)}$$

and replaces the sum over l_i by an integral over ϑ, i.e.,

$$\sum_{l_i l_f} \to \frac{\eta}{2}\sum_{\mu}\int_0^{\pi} d\vartheta\left(\sin\frac{\vartheta}{2}\right)^{-2}, \qquad \text{(II B.103)}$$

one obtains immediately the classical expression (II A.31).

In the case of the magnetic cross section one must in addition employ the asymptotic expression

$$\begin{Bmatrix} \lambda_1 & \lambda_2 & \lambda_3 \\ l_1 & l_2 & l_3 \end{Bmatrix} = \frac{(-1)^{2l_1-2\lambda_1}}{(2l_1+1)^{\frac{1}{2}}}\begin{pmatrix} \lambda_1 & \lambda_2 & \lambda_3 \\ l_3-l_2 & l_1-l_3 & l_2-l_1 \end{pmatrix}, \qquad \text{(II B.104)}$$

for the Racah coefficient in the limit of large l_1, l_2, and l_3 with fixed λ's. The expression (50) then reduces to (II A.54).

In a similar way the classical limit (II A.29) for the differential cross-section function is obtained from (47) by employing the further relation

$$\begin{Bmatrix} l_1 & l_2 & l_3 \\ l_4 & l_5 & \lambda \end{Bmatrix} = \frac{(-1)^{l_1+l_2+l_3}}{[(2l_1+1)(2l_2+1)]^{\frac{1}{2}}} D_{l_4-l_2,l_1-l_5}{}^{\lambda}(0,\theta,0), \qquad \text{(II B.105)}$$

with

$$\cos\theta=\frac{l_3(l_3+1)-l_1(l_1+1)-l_2(l_2+1)}{2[l_1(l_1+1)l_2(l_2+1)]^{\frac{1}{2}}},$$

holding for l_1, l_2, l_3, l_4, and l_5 large. The formulas (104) and (105) may be derived from the explicit formulas for the Racah coefficients.[24]

II C. Numerical Results

In this Section we shall give the results of the numerical computations of excitation cross sections and γ distributions based on the formulas derived in the preceding two Sections. A survey of the approximations involved in these formulas is contained in Sec. II C.5.

II C.1. Collision Parameters

For the application of the theoretical expressions it is convenient to write all parameters involved as functions of the energy of the incident projectile

$$E=\tfrac{1}{2}m_1 v_i^2, \qquad \text{(II C.1)}$$

and the energy independent quantities such as the charge and mass numbers and the excitation energy.

Thus, while the initial projectile velocity is given by

$$v_i=\left(\frac{2E}{A_1 M}\right)^{\frac{1}{2}}, \qquad \text{(II C.2)}$$

where A_1 is the projectile mass in units of the proton mass M, the final relative velocity v_f is obtained from the equation

$$\tfrac{1}{2}m_1 v_f^2=E-\Delta E', \qquad \text{(II C.3)}$$

with

$$\Delta E'=(1+A_1/A_2)\Delta E. \qquad \text{(II C.4)}$$

The nuclear mass number is denoted by A_2, and ΔE represents the excitation energy. Introducing the parameter

$$\zeta=\Delta E'/E, \qquad \text{(II C.5)}$$

we may also write

$$v_f=\left(\frac{2E}{A_1 M}\right)^{\frac{1}{2}}(1-\zeta)^{\frac{1}{2}}. \qquad \text{(II C.6)}$$

For the symmetrized parameter a defined by (II A.86) we thus obtain

$$a=\tfrac{1}{2}(1+A_1/A_2)\frac{Z_1Z_2e^2}{E}(1-\zeta)^{-\frac{1}{2}}$$

$$=0.07199(1+A_1/A_2)\frac{Z_1Z_2(1-\zeta)^{-\frac{1}{2}}}{E_{\mathrm{Mev}}}\cdot 10^{-12}\ \mathrm{cm}, \qquad \text{(II C.7)}$$

where E_{Mev} is the initial energy (1) expressed in Mev.

Furthermore, the parameter η (see (II A.1)), for the initial and final states may be written

$$\eta_i=\frac{Z_1Z_2}{2}\left(\frac{A_1}{10.008\cdot E_{\mathrm{Mev}}}\right)^{\frac{1}{2}}, \qquad \text{(II C.8)}$$

and

$$\eta_f=\eta_i(1-\zeta)^{-\frac{1}{2}}. \qquad \text{(II C.9)}$$

The quantum-mechanical excitation and angular distribution functions are expressed as functions of η_i and $\xi=\eta_f-\eta_i$. Since both these parameters depend on the bombarding energy it is sometimes convenient instead of η_i to use the energy independent parameter ν defined by

$$\nu=2(\eta_i^{-2}-\eta_f^{-2})^{\frac{1}{2}}$$

$$=\frac{4}{Z_1Z_2}\left(\frac{10.008\Delta E'_{\mathrm{Mev}}}{A_1}\right)^{\frac{1}{2}}, \qquad \text{(II C.10)}$$

where the effective energy loss $\Delta E'$ defined by (4) is

[55] E. P. Wigner, *Gruppentheorie* (Friedrich Vieweg & Sohn, Braunschweig, Germany, 1931).

TABLE II.2. List of numerically evaluated functions.

Results for	Method	Given in	
Total cross section			
$E1$, $E2$, $E3$, $E4$, $M1$, $M2$	Classical	Tables II.3, 4	Figs. II.4, 5
$E1$, $E2$	Quantal	Table II.5	Fig. II.6
Differential cross section			
$E1$, $E2$	Classical	Tables II.7, 8	Fig. II.7
$E3$, $E4$, $M1$, $M2$	Classical		Fig. II.7
Angular distribution of γ rays			
$E1$	Quantal		Fig. II.8
$E2$	Quantal	Tables II.9, 10	Fig. II.8
$M1$		Text	

measured in Mev. The parameter ν is related to η_i by

$$\eta_i = 2\zeta^{\frac{1}{2}}/\nu, \tag{II C.11}$$

and for ξ one finds

$$\xi = \frac{2}{\nu}\zeta^{\frac{1}{2}}[(1-\zeta)^{-\frac{1}{2}}-1]. \tag{II C.12}$$

The dependence of $\xi\nu$ on ζ is given graphically in Fig. III.10. A convenient expansion of ξ in powers of the energy loss is given by

$$\xi = \frac{Z_1Z_2A_1^{\frac{1}{2}}\Delta E'_{\text{Mev}}}{12.65(E_{\text{Mev}}-\frac{1}{2}\Delta E'_{\text{Mev}})^{\frac{3}{2}}} \times\left(1+\frac{5}{32}\left(\frac{\Delta E'}{E}\right)^2+\cdots\right). \tag{II C.13}$$

The numerically evaluated excitation functions and angular distribution functions[11,56–61] are given in the series of tables and figures listed in Table II.2.

II C.2. Total Cross Sections

The total cross section for excitation of a given level may be written

$$\sigma = \sum_{\lambda=1}^{\infty}(\sigma_{E\lambda}+\sigma_{M\lambda}), \tag{II C.14}$$

where the partial cross sections are given by (II B.37) and (II B.38). Inserting (2) and (7) one obtains, for the electric excitation,

$$\sigma_{E\lambda} = c_{E\lambda}E_{\text{Mev}}^{\lambda-2}(E_{\text{Mev}}-\Delta E'_{\text{Mev}})^{\lambda-1} \times B(E\lambda)f_{E\lambda}(\eta_i,\xi), \tag{II C.15}$$

with

$$c_{E\lambda} = \frac{Z_1^2A_1}{40.03}[0.07199(1+A_1/A_2Z_1Z_2]^{-2\lambda+2}\text{ barns} \tag{II C.16}$$

i.e.,

$$c_{E\lambda} = \begin{cases} 2.498\cdot10^{-2}Z_1^2A_1\text{ barns} & (\lambda=1)\\ 4.819\cdot(1+A_1/A_2)^{-2}\dfrac{A_1}{Z_2^2}\text{ barns} & (\lambda=2)\\ 9.298\cdot10^{2}(1+A_1/A_2)^{-4}\dfrac{A_1}{Z_1^2Z_2^4}\text{ barns} & (\lambda=3)\\ 1.794\cdot10^{5}(1+A_1/A_2)^{-6}\dfrac{A_1}{Z_1^4Z_2^6}\text{ barns} & (\lambda=4).\end{cases} \tag{II C.17}$$

The reduced nuclear transition probability, $B(E\lambda)$ defined by (II B.21 and 23) [see also (II A.13)] is measured in units of $e^2\cdot(10^{-24}\text{ cm}^2)^\lambda$.

Similarly, for the magnetic excitations, one obtains

$$\sigma_{M\lambda} = c_{M\lambda}E_{\text{Mev}}^{\lambda-\frac{3}{2}}(E_{\text{Mev}}-\Delta E'_{\text{Mev}})^{\lambda-\frac{1}{2}} \times B(M\lambda)f_{M\lambda}(\eta_i,\xi), \tag{II C.18}$$

with

$$c_{M\lambda} = 5.888\cdot10^{-9}Z_1^2 \times[0.07199(1+A_1/A_2)Z_1Z_2]^{-2\lambda+2}\text{ barns} \tag{II C.19}$$

$$= \begin{cases} 5.888\cdot10^{-9}Z_1^2\text{ barns} & (\lambda=1)\\ 1.136\cdot10^{-6}(1+A_1/A_2)^{-2}\dfrac{1}{Z_2^2}\text{ barns} & (\lambda=2).\end{cases} \tag{II C.20}$$

The reduced transition probabilities $B(M\lambda)$ defined by (II B.17) and (II B.23) [see also (II A.41)] are measured in units of $(e\hbar/2Mc)^2\cdot(10^{-24}\text{ cm}^2)^{\lambda-1}$.

TABLE II.3. Classical f functions for $E1$, $E2$, and $M1$ excitations. The entry is given by a number and the power of ten (in parenthesis) by which it should be multiplied. The data are taken from reference 56.

ξ	$f_{E1}(\xi)$	$f_{E2}(\xi)$	$f_{M1}(\xi)$
0.0	∞	0.895 (0)	∞
0.1	0.580 (2)	0.859 (0)	2.230 (1)
0.2	2.721 (1)	0.729 (0)	0.828 (1)
0.3	1.349 (1)	0.561 (0)	3.719 (0)
0.4	0.693 (1)	4.046 (−1)	1.809 (0)
0.5	3.591 (0)	2.781 (−1)	0.905 (0)
0.6	1.872 (0)	1.844 (−1)	4.603 (−1)
0.7	0.980 (0)	1.189 (−1)	2.368 (−1)
0.8	0.514 (0)	0.751 (−1)	1.229 (−1)
0.9	2.707 (−1)	4.663 (−2)	0.642 (−1)
1.0	1.428 (−1)	2.855 (−2)	3.377 (−2)
1.2	3.992 (−2)	1.035 (−2)	0.944 (−2)
1.4	1.121 (−2)	3.628 (−3)	2.664 (−3)
1.6	3.154 (−3)	1.238 (−3)	0.757 (−3)
1.8	0.889 (−3)	4.143 (−4)	2.160 (−4)
2.0	2.511 (−4)	1.363 (−4)	0.618 (−4)
4.0	0.839 (−9)	1.247 (−9)	2.404 (−10)

[56] K. Alder and A. Winther, CERN report T/KA-AW-1 (1954); see also reference 88.

[57] L. C. Biedenharn and C. M. Class, Phys. Rev. **98**, 691 (1955) and **100**, 1790 (1955).

[58] K. Alder and A. Winther, Kgl. Danske Videnskab. Selskab Mat. fys. Medd. **29**, No. 19 (1955).

[59] K. Alder and A. Winther, reference 52.

[60] Biedenharn, Goldstein, McHale, and Thaler, Phys. Rev. **101**, 662 (1956).

[61] Thaler, Goldstein, McHale, and Biedenharn, Phys. Rev. **102**, 1567 (1956).

The classical treatment of the excitation process given in Sec. II A leads, after symmetrization of the cross sections, to expressions of the same form as (15) and (18) [see (II A.88) and (II A.89)]. The entire difference between the classical and quantal cross sections is contained in the f functions, which in the classical case depend only on ξ, and correspond to the limiting values of the quantum-mechanical functions $f(\eta_i,\xi)$ for $\eta_i\rightarrow\infty$.

In most practical cases the quantal f functions differ only slightly from the classical limit. It is thus convenient to write

$$f_\lambda(\eta_i,\xi)=f_\lambda(\xi)\cdot R_\lambda(\eta_i,\xi), \qquad \text{(II C.21)}$$

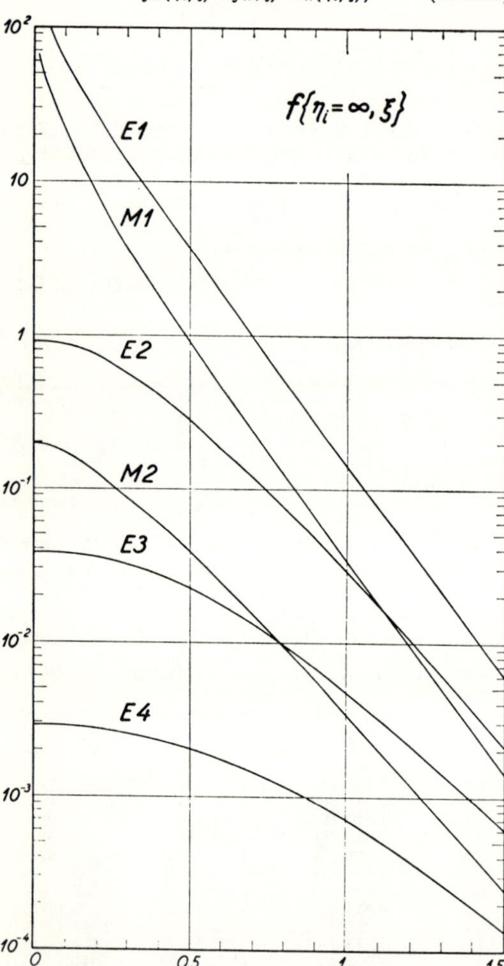

FIG. II.4. The total excitation cross section functions $f_\lambda(\xi)$ in the classical approximation. In the limit of large values for η_i, the f functions approach those obtained from a classical description (see Secs. II A.1 and II A.2). The classical f functions for the lowest electric and magnetic multipole orders are plotted against the parameter ξ. The data are taken from reference 56.

TABLE II.4. Classical f functions for $E3$, $E4$, and $M2$ excitations. The entry is given by a number and the power of ten (in parenthesis) by which it should be multiplied. The data are taken from reference 56.

ξ	$f_{E3}(\xi)$	$f_{E4}(\xi)$	$f_{M2}(\xi)$
0.0	3.797 (−2)	2.862 (−3)	1.936 (−1)
0.2	3.532 (−2)	2.729 (−3)	1.233 (−1)
0.4	2.723 (−2)	2.330 (−3)	0.577 (−1)
0.6	1.736 (−2)	1.760 (−3)	2.391 (−2)
0.8	0.956 (−2)	1.176 (−3)	0.920 (−2)
1.0	4.722 (−3)	0.705 (−3)	3.378 (−3)
1.5	0.593 (−3)	1.370 (−4)	2.433 (−4)
2.0	0.565 (−4)	1.870 (−5)	1.562 (−5)
4.0	1.398 (−9)	1.204 (−9)	1.560 (−10)

where $f_\lambda(\xi)$ is the classical f function and where the quantum effects are contained in the correction factor R_λ.

The classical f functions have been calculated[11,56] for excitations of order $E1$, $E2$, $E3$, $E4$, $M1$, and $M2$ and the results are presented in Tables II.3 and II.4 and Figs. II.4 and II.5.

The functions were obtained from (II A.31) and (II A.54) with the orbital integrals $I_{\lambda\mu}(\vartheta,\xi)$ given by (II A.26). For the evaluation of these integrals, see Secs. II E.4–6.

The quantal f functions have been computed in the case of $E1$[61] and $E2$[57–60,62] excitations.[62a] The function $f_{E2}(\eta_i,\xi)$ is given in Table II.5 and the correction factors $R_{E\lambda}(\eta_i,\xi)$ for $\lambda=1$ and 2 are illustrated in Fig. II.6.

The f_{E1} and f_{E2} functions have been obtained from (II B.48) which for $\lambda=1$ reduces to

$$f_{E1}(\eta_i,\xi)=\frac{64\pi^2}{9}k_ik_fb_0^{E1}, \qquad \text{(II C.22)}$$

with

$$b_0^{E1}=\sum_{l=0}^{\infty}\{l|M_{l-1,l}^{-2}|^2+(l+1)|M_{l+1,l}^{-2}|^2\}. \qquad \text{(II C.23)}$$

For $\lambda=2$ one obtains

$$f_{E2}(\eta_i,\xi)=\frac{64\pi^2}{25}\eta_i\eta_fb_0^{E2}, \qquad \text{(II C.24)}$$

with

$$b_0^{E2}=\sum_{l=0}^{\infty}\left\{\frac{3l(l-1)}{2(2l-1)}|M_{l-2,l}^{-3}|^2+\frac{3(l+1)(l+2)}{2(2l+3)}|M_{l+2,l}^{-3}|^2+\frac{l(l+1)(2l+1)}{(2l-1)(2l+3)}|M_{ll}^{-3}|^2\right\}. \qquad \text{(II C.25)}$$

[62] The quantum-mechanical calculation of the electric quadrupole cross section was first performed for $\xi=0$ and for a particular value of η_i by direct numerical integration of the radial matrix elements by Daitch, Lazarus, Hull, Benedict, and Breit, Phys. Rev. **96**, 1449 (1954).

[62a] *Note added in proof.*—Recently, the quantal f function for $M1$ excitation has been given by L. C. Biedenharn and R. M. Thaler (to be published in Phys. Rev.). This reference also contains the coefficient a_2 describing the angular distribution of the γ rays from an $M1$ or mixed $M1+E2$ excitation process.

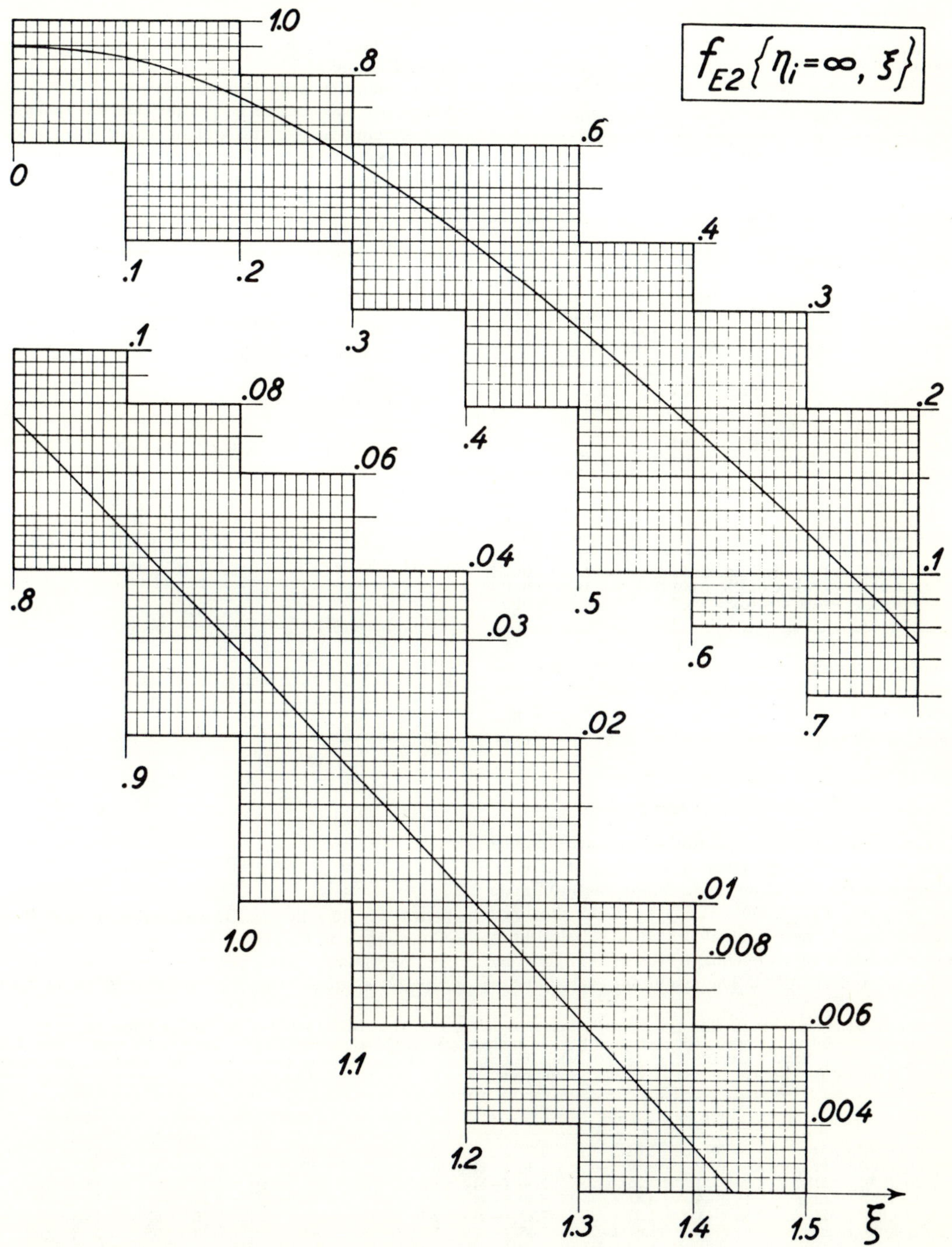

FIG. II.5. The total excitation cross-section function $f_{E2}(\xi)$ in the classical approximation. The data are taken from reference 56.

TABLE II.5. Total f function for $E2$ Coulomb excitation. The total f function for $E2$ excitation, as obtained from the complete quantum-mechanical calculation,[59,60] is listed as a function of η_i and ξ.

$\eta_i \backslash \xi$	0	0.1	0.2	0.3	0.4	0.5	0.6
0.5	0.321	0.344	0.307	0.243			
1	0.620	0.614	0.528	0.409	0.295	0.203	0.1350
1.5	0.754	0.732	0.624	0.480	0.346	0.237	0.1570
2	0.812	0.784	0.666	0.512	0.368	0.253	0.1672
2.5	0.842	0.810	0.688	0.529	0.380	0.261	0.1726
3	0.858	0.825	0.700	0.538	0.387	0.266	0.1759
3.5	0.869	0.834	0.708	0.545	0.392	0.269	0.1779
4	0.875	0.840	0.713	0.548	0.395	0.271	0.1793
5	0.881	0.847	0.719	0.553	0.398	0.273	0.1810
6	0.886	0.851	0.722	0.556	0.400	0.275	0.1819
7	0.888	0.854	0.724	0.558	0.401	0.276	0.1825
8	0.890	0.855	0.726	0.559	0.402	0.276	0.1829
∞	0.895	0.859	0.729	0.561	0.405	0.278	0.1844

$\eta_i \backslash \xi$	0.8	1.0	1.2	1.4	1.6	2.0
1	0.0553					
1.5	0.0640	0.0244	0.00887	0.00312		
2	0.0680	0.0259	0.00939	0.00330	0.001130	
2.5	0.0702	0.0267	0.00968	0.00340	0.001162	0.0001287
3	0.0715	0.0272	0.00986	0.00345	0.001181	0.0001306
3.5	0.0724	0.0275	0.00996	0.00349	0.001194	0.0001319
4	0.0730	0.0277	0.01005	0.00352	0.001203	0.0001328
5	0.0737	0.0280	0.01015	0.00356	0.001214	0.0001339
6	0.0741	0.0282	0.01021	0.00358	0.001221	0.0001345
7	0.0743	0.0283	0.01024	0.00359	0.001225	0.0001350
8	0.0745	0.0283	0.01027	0.00360	0.001228	0.0001353
∞	0.0751	0.0286	0.01035	0.00363	0.001238	0.0001363

The radial matrix elements in (23) and (25) have been evaluated by the methods described in Sec. II B.4. The sum over l must be extended to $l \sim 300$ for large η_i and small ξ, while the convergence for large ξ is much more rapid.

II C.3. Differential Cross Sections

The differential excitation cross sections are obtained from (15) and (18) by replacing $f(\eta_i,\xi)$ by $df(\vartheta,\eta_i,\xi)$, where ϑ is the deflection angle in the center-of-mass system [see (II B.32) and (II B.33)]. These functions have so far only been evaluated[56] in the classical limit $\eta_i \to \infty$ in which they are equal to the $df(\vartheta,\xi)$ given by (II A.29) and (II A.51). The results for excitations of order $E1$, $E2$, $E3$, $E4$, $M1$, and $M2$, are given in Fig. II.7, and for $E1$ and $E2$ excitations also in the Tables II.7 and II.8.

The quantum corrections to the differential cross sections are expected to be greater than for the total cross sections. An indication of the effect of these corrections is provided by a comparison with the results of the Born-approximation treatment which corresponds to the limit $\eta_i = \xi = 0$, and which for $E2$ excitations leads to an isotropic distribution of the inelastically scattered particles[4] (see Sec. II E.2).

II. C.4. Angular Distribution of De-Excitation γ Rays

The angular distribution of the γ rays following electromagnetic excitation is given by (II B.83).

For $E1$ excitations the distribution can be written

$$W(\vartheta_\gamma) = 1 + a_2^{E1}(\eta_i,\xi) A_2^{(1)} P_2(\cos\vartheta_\gamma), \quad \text{(II C.26)}$$

where the $A_2^{(1)}$ coefficients refer to the hypothetical γ-γ correlation of Fig. II.2 and may be obtained from (II A.70). The coefficients a_2^{E1} are given by (II B.84) and (II B.85) which for $\lambda = 1$ reduce to

$$a_2^{E1}(\eta_i,\xi) = (b_0^{E1})^{-1} \sum_l \left\{ -\frac{(l-1)l}{2l+1} |M_{l-1,\,l}^{-2}|^2 - \frac{(l+1)(l+2)}{2l+1} |M_{l+1,\,l}^{-2}|^2 + \frac{6l(l+1)}{2l+1} M_{l+1,\,l}^{-2} M_{l-1,\,l}^{-2} \times \cos(\sigma_{l+1}(\eta_i) - \sigma_{l-1}(\eta_i)) \right\}, \quad \text{(II C.27)}$$

where b_0^{E1} is given by (23). The numerical results[61,63] for a_2^{E1} are shown in Fig. II.8.

In the classical limit $\eta_i \to \infty$ or $\nu \to 0$ the value of $a_2^{E1}(\eta_i,\xi)$ is equal to $a_2^{E1}(\xi)$ which is given by [see

TABLE II.6. Normalization for the angular distributions given in Fig. II.7. The absolute values of the classical differential f functions may be obtained from the relative values given in Fig. II.7 by employing the absolute normalization given in the present table. For electric excitations, the table gives $df/d\Omega$ at $\vartheta = 180°$, while for magnetic excitation the value given is for $\vartheta = 90°$. The entry is given by a number and the power of ten by which it should be multiplied.

	$\xi=0.0$	$\xi=0.2$	$\xi=0.4$	$\xi=0.6$	$\xi=1.0$	$\xi=2.0$	$\xi=4.0$
$df_{E1}(180°,\xi)/d\Omega$	1.40 (0)	5.62 (−1)	1.95 (−1)	6.40 (−2)	6.39 (−3)	1.66 (−5)	8.41 (−11)
$df_{E2}(180°,\xi)/d\Omega$	5.58 (−2)	3.89 (−2)	1.96 (−2)	8.60 (−3)	1.32 (−3)	6.76 (−6)	7.37 (−11)
$df_{E3}(180°,\xi)/d\Omega$	4.56 (−3)	3.76 (−3)	2.36 (−3)	1.26 (−3)	2.71 (−4)	2.51 (−6)	5.58 (−11)
$df_{E4}(180°,\xi)/d\Omega$	5.16 (−4)	4.47 (−4)	3.20 (−4)	1.97 (−4)	5.46 (−5)	8.39 (−7)	3.57 (−11)
$df_{M1}(90°,\xi)/d\Omega$	2.58 (−1)	1.63 (−1)	7.23 (−2)	2.76 (−2)	3.25 (−3)	8.97 (−6)	3.16 (−11)
$df_{M2}(90°,\xi)/d\Omega$	6.94 (−3)	5.64 (−3)	3.45 (−3)	1.77 (−3)	3.44 (−4)	2.30 (−6)	2.36 (−11)

[63] A WKB calculation was given in reference 59.

TABLE II.7. Classical differential cross section function for $E1$ excitation. The table lists $df_{E1}/d\Omega$ as a function of ϑ (in degrees) and ξ.[56] The entry is given by a number and the power of ten by which it should be multiplied.

ϑ \ ξ	0.0	0.1	0.2	0.3	0.4	0.5	0.6	0.7
0	∞	0.000	0.000	0.000	0.000	0.000	0.000	0.000
10	1.838 (2)	0.595 (2)	0.783 (1)	0.827 (0)	0.791 (−1)	0.714 (−2)	0.620 (−3)	0.525 (−4)
20	4.630 (1)	2.792 (1)	1.052 (1)	3.342 (0)	0.976 (0)	2.709 (−1)	0.727 (−1)	1.904 (−2)
30	2.084 (1)	1.420 (1)	0.724 (1)	3.236 (0)	1.346 (0)	0.535 (0)	2.066 (−1)	0.779 (−1)
40	1.194 (1)	0.839 (1)	4.843 (0)	2.520 (0)	1.234 (0)	0.581 (0)	2.664 (−1)	1.196 (−1)
50	0.782 (1)	0.551 (1)	3.373 (0)	1.898 (0)	1.015 (0)	0.524 (0)	2.645 (−1)	1.309 (−1)
60	0.558 (1)	3.920 (0)	2.462 (0)	1.446 (0)	0.813 (0)	4.438 (−1)	2.370 (−1)	1.245 (−1)
70	4.244 (0)	2.954 (0)	1.877 (0)	1.127 (0)	0.653 (0)	3.682 (−1)	2.037 (−1)	1.111 (−1)
80	3.379 (0)	2.331 (0)	1.485 (0)	0.902 (0)	0.531 (0)	3.053 (−1)	1.727 (−1)	0.964 (−1)
90	2.792 (0)	1.909 (0)	1.213 (0)	0.740 (0)	4.391 (−1)	2.554 (−1)	1.463 (−1)	0.828 (−1)
100	2.379 (0)	1.613 (0)	1.021 (0)	0.622 (0)	3.703 (−1)	2.165 (−1)	1.248 (−1)	0.712 (−1)
110	2.081 (0)	1.401 (0)	0.881 (0)	0.536 (0)	3.185 (−1)	1.864 (−1)	1.078 (−1)	0.617 (−1)
120	1.862 (0)	1.245 (0)	0.779 (0)	4.714 (−1)	2.795 (−1)	1.634 (−1)	0.944 (−1)	0.541 (−1)
130	1.700 (0)	1.131 (0)	0.703 (0)	4.236 (−1)	2.503 (−1)	1.459 (−1)	0.841 (−1)	4.814 (−2)
140	1.581 (0)	1.047 (0)	0.648 (0)	3.885 (−1)	2.286 (−1)	1.328 (−1)	0.764 (−1)	4.360 (−2)
150	1.496 (0)	0.988 (0)	0.608 (0)	3.634 (−1)	2.131 (−1)	1.234 (−1)	0.707 (−1)	4.026 (−2)
160	1.440 (0)	0.948 (0)	0.582 (0)	3.466 (−1)	2.026 (−1)	1.170 (−1)	0.669 (−1)	3.799 (−2)
170	1.407 (0)	0.925 (0)	0.567 (0)	3.369 (−1)	1.966 (−1)	1.133 (−1)	0.647 (−1)	3.667 (−2)
180	1.396 (0)	0.918 (0)	0.562 (0)	3.338 (−1)	1.946 (−1)	1.121 (−1)	0.640 (−1)	3.623 (−2)

ϑ \ ξ	0.8	0.9	1.0	1.2	1.4	1.6	1.8	2.0	4.0
0	0.000	0.000	0.000	0.000	0.000	0.000	0.000	0.000	0.000
10	4.365 (−6)	3.573 (−7)	2.890 (−8)	1.840 (−10)	1.140 (−12)	0.693 (−14)	4.144 (−17)	2.449 (−19)	0.894 (−41)
20	4.898 (−3)	1.243 (−3)	3.117 (−4)	1.911 (−5)	1.142 (−6)	0.669 (−7)	3.863 (−9)	2.204 (−10)	0.569 (−22)
30	2.893 (−2)	1.059 (−2)	3.839 (−3)	4.919 (−4)	0.615 (−4)	0.754 (−5)	0.911 (−6)	1.089 (−7)	4.613 (−17)
40	0.529 (−1)	2.311 (−2)	0.999 (−2)	1.826 (−3)	3.256 (−4)	0.571 (−4)	0.986 (−5)	1.685 (−6)	2.566 (−14)
50	0.639 (−1)	3.084 (−2)	1.474 (−2)	3.296 (−3)	0.720 (−3)	1.547 (−4)	3.279 (−5)	0.688 (−5)	0.818 (−12)
60	0.645 (−1)	3.311 (−2)	1.684 (−2)	4.266 (−3)	1.058 (−3)	2.581 (−4)	0.622 (−4)	1.482 (−5)	0.642 (−11)
70	0.598 (−1)	3.192 (−2)	1.690 (−2)	4.643 (−3)	1.251 (−3)	3.317 (−4)	0.869 (−4)	2.255 (−5)	2.301 (−11)
80	0.532 (−1)	2.912 (−2)	1.582 (−2)	4.587 (−3)	1.305 (−3)	3.662 (−4)	1.015 (−4)	2.790 (−5)	0.511 (−10)
90	4.643 (−2)	2.582 (−2)	1.427 (−2)	4.286 (−3)	1.265 (−3)	3.685 (−4)	1.062 (−4)	3.033 (−5)	0.832 (−10)
100	4.030 (−2)	2.263 (−2)	1.264 (−2)	3.880 (−3)	1.173 (−3)	3.502 (−4)	1.035 (−4)	3.034 (−5)	1.097 (−10)
110	3.507 (−2)	1.980 (−2)	1.112 (−2)	3.458 (−3)	1.060 (−3)	3.213 (−4)	0.965 (−4)	2.877 (−5)	1.251 (−10)
120	3.078 (−2)	1.741 (−2)	0.980 (−2)	3.066 (−3)	0.947 (−3)	2.893 (−4)	0.877 (−4)	2.639 (−5)	1.288 (−10)
130	2.737 (−2)	1.547 (−2)	0.871 (−2)	2.727 (−3)	0.844 (−3)	2.588 (−4)	0.787 (−4)	2.380 (−5)	1.236 (−10)
140	2.473 (−2)	1.396 (−2)	0.784 (−2)	2.452 (−3)	0.758 (−3)	2.322 (−4)	0.707 (−4)	2.137 (−5)	1.136 (−10)
150	2.278 (−2)	1.283 (−2)	0.719 (−2)	2.240 (−3)	0.690 (−3)	2.110 (−4)	0.641 (−4)	1.934 (−5)	1.024 (−10)
160	2.145 (−2)	1.205 (−2)	0.674 (−2)	2.092 (−3)	0.642 (−3)	1.957 (−4)	0.592 (−4)	1.784 (−5)	0.927 (−10)
170	2.067 (−2)	1.159 (−2)	0.648 (−2)	2.004 (−3)	0.614 (−3)	1.865 (−4)	0.563 (−4)	1.692 (−5)	0.863 (−10)
180	2.041 (−2)	1.144 (−2)	0.639 (−2)	1.975 (−3)	0.604 (−3)	1.834 (−4)	0.553 (−4)	1.661 (−5)	0.841 (−10)

(II A.75)]

$$a_2^{E1}(\xi) = -\tfrac{1}{2} + \int_0^\pi 3 I_{1,1} I_{1,-1} \cos\vartheta \frac{\cos\frac{\vartheta}{2}}{\sin^3\frac{\vartheta}{2}} d\vartheta \times \left\{ \int_0^\pi [|I_{1,1}|^2+|I_{1,-1}|^2] \frac{\cos\frac{\vartheta}{2}}{\sin^3\frac{\vartheta}{2}} d\vartheta \right\}^{-1}. \quad \text{(II C.28)}$$

For $E2$ excitations the angular distribution of the γ rays is given by

$$W(\vartheta_\gamma) = 1 + a_2^{E2}(\eta_i,\xi) A_2^{(2)} P_2(\cos\vartheta_\gamma) + a_4^{E2}(\eta_i,\xi) A_4^{(2)} P_4(\cos\vartheta_\gamma), \quad \text{(II C.29)}$$

where the $A_k^{(2)}$ coefficients may be obtained from (II A.70). For some of the most frequently occurring transitions these coefficients are given in Table II.11. The a coefficients of (29) are tabulated as functions of

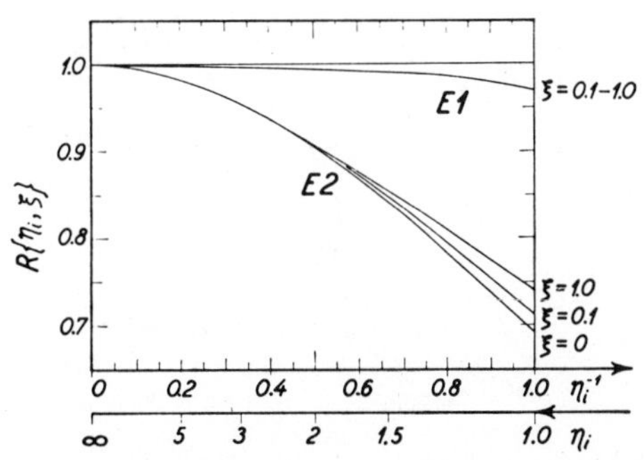

FIG. II.6. Quantum-mechanical corrections to the classical total f functions. The ratio $R_{E\lambda}(\eta_i,\xi)$ between the quantal and the classical total f function for $\lambda=1$ and 2 [see (II C.21)] is plotted as a function of η_i^{-1} for different values of ξ. The data are taken from references 59, 60, and 61.

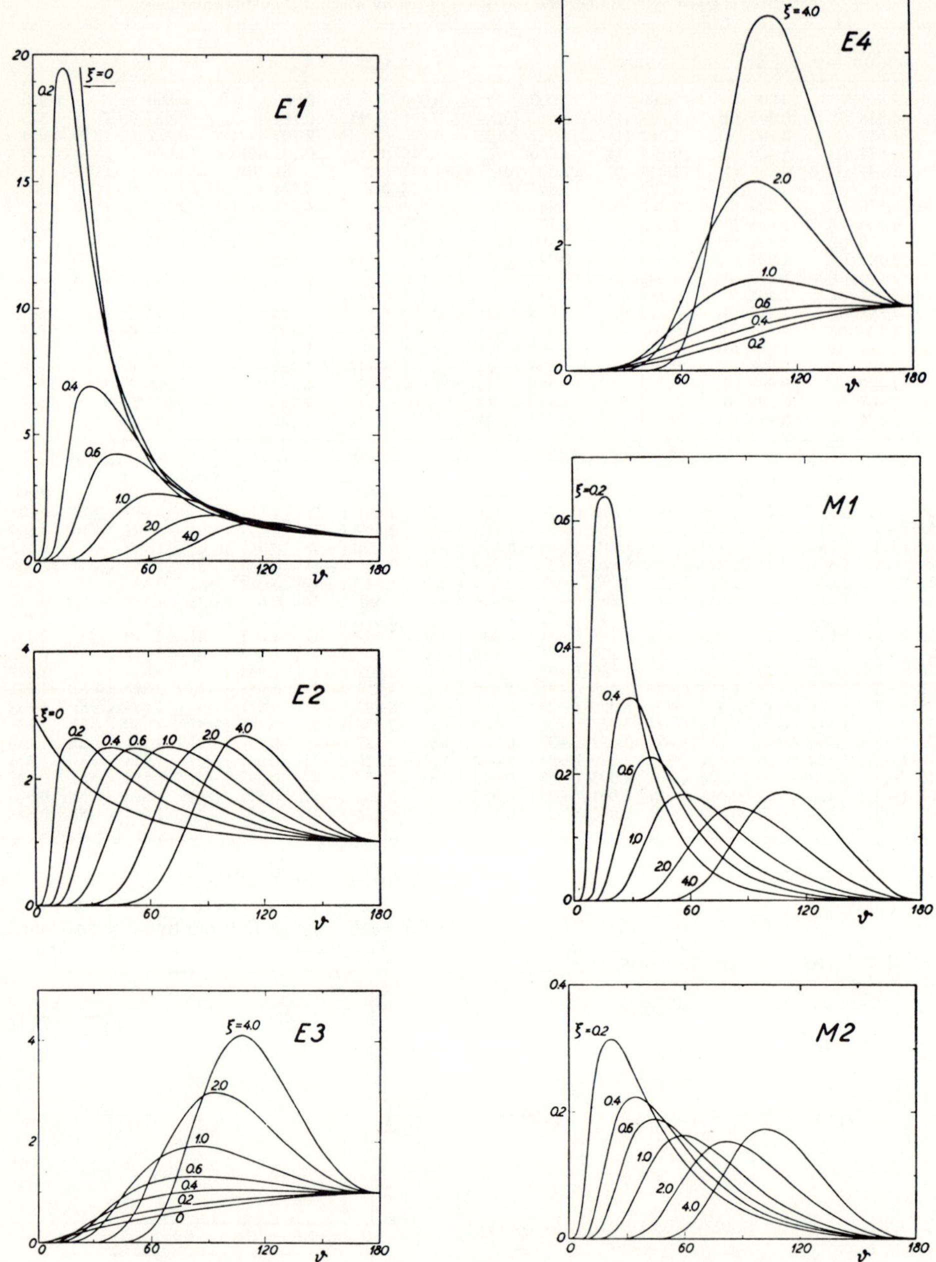

FIG. II.7. Angular distribution of the inelastically scattered particles in classical approximation. The classical differential excitation cross section functions $df_\lambda(\vartheta,\xi)/d\Omega$ are plotted as a function of ϑ for fixed values of ξ. The electric multipole cross sections are normalized to unity at 180°, while the magnetic are normalized to give a total cross section of unity. The absolute values can be obtained by means of Table II.6. The data are taken from reference 56.

TABLE II.8. Classical differential cross section function for $E2$ excitation. The table lists $df_{E2}/d\Omega$ as a function of ϑ (in degrees) and ξ.[56] The entry is given by a number and the power of ten by which it should be multiplied.

$\vartheta \backslash \xi$	0.0	0.1	0.2	0.3	0.4	0.5	0.6	0.7
0	1.676 (−1)	0.000	0.000	0.000	0.000	0.000	0.000	0.000
10	1.385 (−1)	1.403 (−1)	0.521 (−1)	1.110 (−2)	1.788 (−3)	2.443 (−4)	2.996 (−5)	3.403 (−6)
20	1.178 (−1)	1.286 (−1)	1.016 (−1)	0.578 (−1)	2.678 (−2)	1.085 (−2)	4.005 (−3)	1.382 (−3)
30	1.027 (−1)	1.087 (−1)	0.981 (−1)	0.716 (−1)	4.474 (−2)	2.508 (−2)	1.300 (−2)	0.635 (−2)
40	0.916 (−1)	0.943 (−1)	0.873 (−1)	0.694 (−1)	4.895 (−2)	3.154 (−2)	1.899 (−2)	1.085 (−2)
50	0.832 (−1)	0.837 (−1)	0.772 (−1)	0.634 (−1)	4.715 (−2)	3.251 (−2)	2.113 (−2)	1.311 (−2)
60	0.768 (−1)	0.758 (−1)	0.690 (−1)	0.570 (−1)	4.341 (−2)	3.095 (−2)	2.095 (−2)	1.361 (−2)
70	0.719 (−1)	0.697 (−1)	0.624 (−1)	0.513 (−1)	3.932 (−2)	2.846 (−2)	1.967 (−2)	1.311 (−2)
80	0.680 (−1)	0.649 (−1)	0.570 (−1)	4.643 (−2)	3.549 (−2)	2.579 (−2)	1.798 (−2)	1.213 (−2)
90	0.650 (−1)	0.612 (−1)	0.527 (−1)	4.234 (−2)	3.210 (−2)	2.326 (−2)	1.624 (−2)	1.100 (−2)
100	0.627 (−1)	0.583 (−1)	4.930 (−2)	3.895 (−2)	2.920 (−2)	2.100 (−2)	1.460 (−2)	0.988 (−2)
110	0.608 (−1)	0.559 (−1)	4.651 (−2)	3.617 (−2)	2.676 (−2)	1.905 (−2)	1.314 (−2)	0.884 (−2)
120	0.593 (−1)	0.541 (−1)	4.428 (−2)	3.391 (−2)	2.475 (−2)	1.741 (−2)	1.190 (−2)	0.794 (−2)
130	0.582 (−1)	0.526 (−1)	4.252 (−2)	3.210 (−2)	2.311 (−2)	1.606 (−2)	1.086 (−2)	0.718 (−2)
140	0.573 (−1)	0.515 (−1)	4.115 (−2)	3.068 (−2)	2.182 (−2)	1.499 (−2)	1.003 (−2)	0.657 (−2)
150	0.566 (−1)	0.507 (−1)	4.013 (−2)	2.961 (−2)	2.085 (−2)	1.418 (−2)	0.939 (−2)	0.609 (−2)
160	0.562 (−1)	0.501 (−1)	3.942 (−2)	2.887 (−2)	2.017 (−2)	1.361 (−2)	0.895 (−2)	0.576 (−2)
170	0.559 (−1)	4.980 (−2)	3.900 (−2)	2.844 (−2)	1.977 (−2)	1.328 (−2)	0.869 (−2)	0.557 (−2)
180	0.558 (−1)	4.969 (−2)	3.887 (−2)	2.829 (−2)	1.963 (−2)	1.316 (−2)	0.860 (−2)	0.550 (−2)

$\vartheta \backslash \xi$	0.8	0.9	1.0	1.2	1.4	1.6	1.8	2.0	4.0
0	0.000	0.000	0.000	0.000	0.000	0.000	0.000	0.000	0.000
10	3.652 (−7)	3.750 (−8)	3.719 (−9)	3.374 (−11)	2.824 (−13)	2.227 (−15)	1.679 (−17)	1.220 (−19)	1.750 (−41)
20	4.531 (−4)	1.427 (−4)	4.351 (−5)	3.752 (−6)	2.998 (−7)	2.265 (−8)	1.638 (−9)	1.144 (−10)	1.134 (−22)
30	2.960 (−3)	1.331 (−3)	0.581 (−3)	1.032 (−4)	1.708 (−5)	2.680 (−6)	4.032 (−7)	0.587 (−7)	0.929 (−16)
40	0.595 (−2)	3.159 (−3)	1.631 (−3)	4.076 (−4)	0.953 (−4)	2.118 (−5)	4.523 (−6)	0.936 (−6)	0.518 (−13)
50	0.784 (−2)	4.553 (−3)	2.577 (−3)	0.777 (−3)	2.203 (−4)	0.595 (−4)	1.549 (−5)	3.913 (−6)	1.643 (−12)
60	0.855 (−2)	0.523 (−2)	3.123 (−3)	1.054 (−3)	3.356 (−4)	1.022 (−4)	3.001 (−5)	0.857 (−5)	1.269 (−11)
70	0.847 (−2)	0.534 (−2)	3.294 (−3)	1.191 (−3)	4.079 (−4)	1.339 (−4)	4.249 (−5)	1.312 (−5)	4.434 (−11)
80	0.796 (−2)	0.510 (−2)	3.211 (−3)	1.211 (−3)	4.338 (−4)	1.494 (−4)	4.985 (−5)	1.621 (−5)	0.950 (−10)
90	0.727 (−2)	4.700 (−3)	2.986 (−3)	1.153 (−3)	4.245 (−4)	1.506 (−4)	0.519 (−4)	1.743 (−5)	1.477 (−10)
100	0.653 (−2)	4.232 (−3)	2.699 (−3)	1.054 (−3)	3.936 (−4)	1.420 (−4)	4.983 (−5)	1.708 (−5)	1.842 (−10)
110	0.582 (−2)	3.769 (−3)	2.402 (−3)	0.939 (−3)	3.526 (−4)	1.282 (−4)	4.538 (−5)	1.572 (−5)	1.964 (−10)
120	0.520 (−2)	3.345 (−3)	2.123 (−3)	0.826 (−3)	3.094 (−4)	1.125 (−4)	3.990 (−5)	1.387 (−5)	1.869 (−10)
130	4.660 (−3)	2.979 (−3)	1.879 (−3)	0.724 (−3)	2.691 (−4)	0.973 (−4)	3.439 (−5)	1.193 (−5)	1.640 (−10)
140	4.224 (−3)	2.678 (−3)	1.677 (−3)	0.637 (−3)	2.344 (−4)	0.840 (−4)	2.947 (−5)	1.016 (−5)	1.364 (−10)
150	3.888 (−3)	2.445 (−3)	1.519 (−3)	0.569 (−3)	2.069 (−4)	0.733 (−4)	2.546 (−5)	0.870 (−5)	1.106 (−10)
160	3.650 (−3)	2.280 (−3)	1.407 (−3)	0.521 (−3)	1.870 (−4)	0.656 (−4)	2.254 (−5)	0.763 (−5)	0.904 (−10)
170	3.509 (−3)	2.181 (−3)	1.340 (−3)	4.918 (−4)	1.751 (−4)	0.609 (−4)	2.078 (−5)	0.698 (−5)	0.779 (−10)
180	3.462 (−3)	2.148 (−3)	1.318 (−3)	4.822 (−4)	1.712 (−4)	0.593 (−4)	2.019 (−5)	0.676 (−5)	0.737 (−10)

η_i and ξ in Tables II.9 and II.10 and are illustrated as functions of ξ and ν [see (10)] in Fig. II.8.[57–60,64,65]

These coefficients have been computed from the expressions

$$a_k^{E2}(\eta_i,\xi)=b_k^{E2}/b_0^{E2}, \tag{II C.30}$$

where b_0^{E2} is given by (25), while

$$b_2^{E2}=\sum_l \left\{\frac{3l(l-1)(l-2)}{(2l-1)^2}|M_{l-2,l}{}^{-3}|^2+\frac{3(l+1)(l+2)(l+3)}{(2l+3)^2}|M_{l+2,l}{}^{-3}|^2\right.$$

$$-\frac{l(l+1)(2l-3)(2l+1)(2l+5)}{(2l-1)^2(2l+3)^2}|M_{ll}{}^{-3}|^2-\frac{6(l-1)l(l+1)}{(2l-1)^2}M_{l-2,l}{}^{-3}M_{ll}{}^{-3}\cos(\sigma_l(\eta_i)-\sigma_{l-2}(\eta_i))$$

$$\left.-\frac{6l(l+1)(l+2)}{(2l+3)^2}M_{l+2,l}{}^{-3}M_{ll}{}^{-3}\cos(\sigma_l(\eta_i)-\sigma_{l+2}(\eta_i))\right\}, \tag{II C.31}$$

[64] The coefficients a_2^{E2}, given in references 58 and 59, contain a numerical error responsible for the somewhat irregular behavior. We are indebted to Dr. L. C Biedenharn for discussions concerning this point.

[65] A WKB calculation for special values of η_i and ξ has been given by F. D. Benedict, Phys. Rev. **101**, 178 (1956).

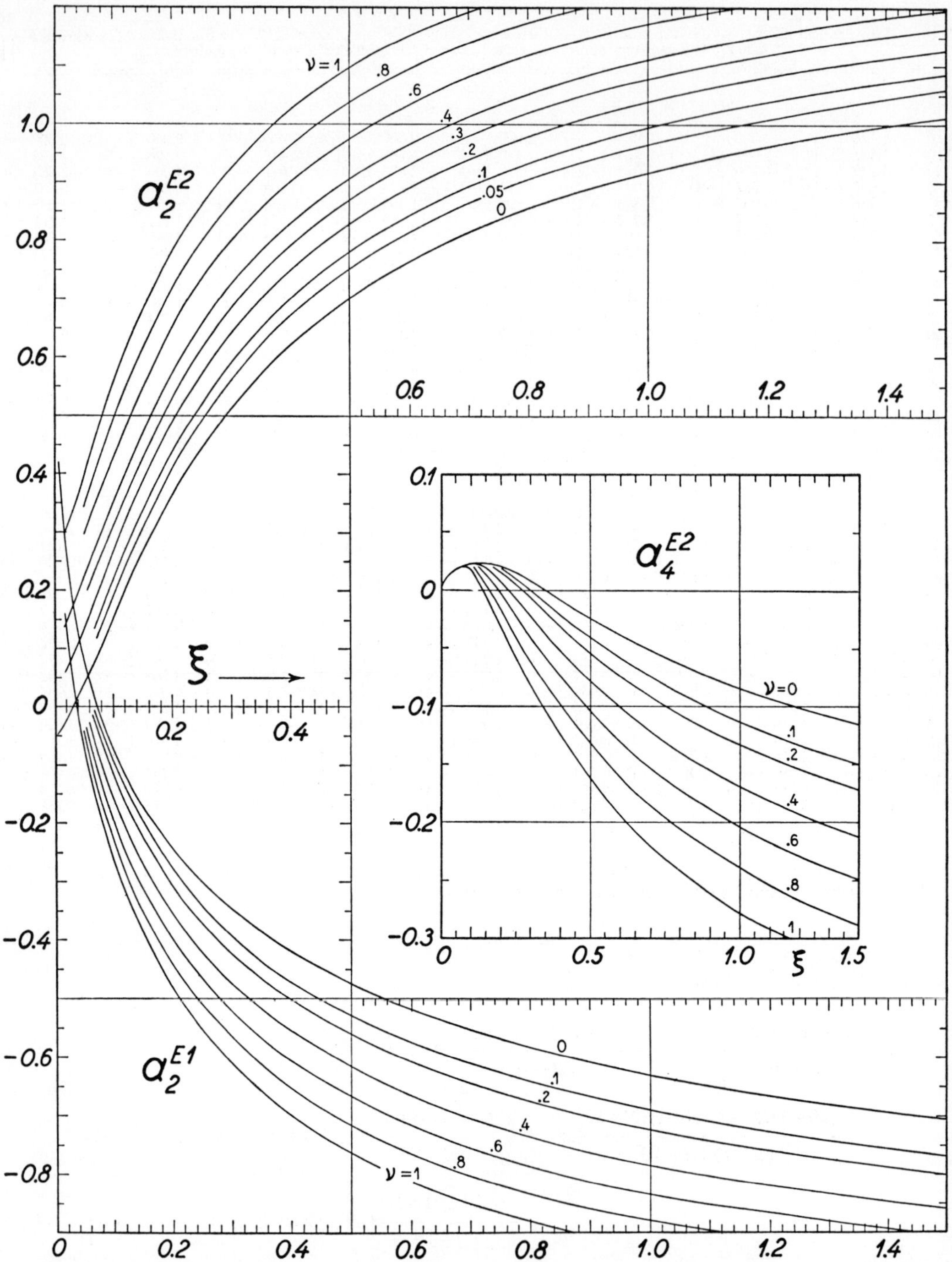

FIG. II.8. Gamma-ray angular distributions following Coulomb excitation. The coefficients $a_k^{E\lambda}(\nu,\xi)$ which describe the angular distribution of the gamma rays following Coulomb excitation (II C.26) and (II C.29) are plotted as a function of ξ for different values of the parameter ν. The classical limit corresponds to $\nu = 0$. The data are taken from references 59, 60, and 61.

TABLE II.9. Gamma-ray angular distribution coefficients $a_2^{E2}(\eta_i,\xi)$. The coefficient a_2^{E2} which describes the angular distribution of the gamma rays following $E2$ Coulomb excitation [see (II C.29)] is given as a function of η_i and ξ. The data are taken from references 59 and 60.

η_i \ ξ	0.0	0.1	0.2	0.3	0.4	0.5	0.6	0.8	1.0	1.2	1.4	1.6	2.0
0.0	0.50	2.000	2.000	2.000	2.000	2.000	2.000	2.000	2.000	2.000	2.000	2.000	2.000
0.5	0.20	0.800	1.142	1.325									
1.0	0.07	0.501	0.809	1.002	1.130	1.223	1.285						
1.5	−0.00	0.372	0.660	0.843	0.970	1.065	1.132	1.228	1.290	1.334	1.360		
2.0	−0.02	0.307	0.577	0.758	0.884	0.971	1.046	1.144	1.210	1.256	1.290		
2.5	−0.03	0.270	0.526	0.707	0.831	0.922	0.993	1.089	1.158	1.206	1.242	1.270	
3.0	−0.04	0.247	0.497	0.673	0.793	0.882	0.956	1.053	1.122	1.171	1.208	1.237	1.279
4.0	−0.04	0.220	0.462	0.633	0.752	0.841	0.909	1.008	1.075	1.125	1.163	1.192	1.235
5.0	−0.04	0.204	0.440	0.606	0.724	0.813	0.878	0.979	1.046	1.095	1.131	1.162	1.207
6.0	−0.05	0.194	0.426	0.589	0.705	0.794	0.859	0.959	1.026	1.075	1.110	1.142	1.185
8.0	−0.05	0.178	0.408	0.569	0.685	0.769	0.838	0.934	1.000	1.048	1.086	1.116	1.159
10.0	−0.05	0.173	0.399	0.557	0.672	0.758	0.823	0.919	0.984	1.030	1.069	1.099	1.143
12.0	−0.05	0.170	0.393	0.549	0.661	0.748	0.813	0.908	0.973	1.019	1.057	1.087	1.130
16.0	−0.05	0.165	0.387	0.540	0.651	0.735	0.803	0.894	0.959	1.005	1.043	1.072	1.115
∞	−0.05	0.150	0.361	0.510	0.619	0.700	0.763	0.853	0.915	0.961	0.996	1.023	1.064

and

$$b_4^{E2}=\sum_l\Bigg\{-\frac{9l(l-1)(l-2)(l-3)}{16(2l-1)^2(2l+1)}|M_{l-2,l}^{-3}|^2-\frac{9(l+1)(l+2)(l+3)(l+4)}{16(2l+1)(2l+3)^2}|M_{l+2,l}^{-3}|^2$$

$$-\frac{9(l-1)l(l+1)(l+2)(2l+1)}{4(2l-1)^2(2l+3)^2}|M_{ll}^{-3}|^2+\frac{15(l-2)(l-1)l(l+1)}{4(2l-1)^2(2l+3)}M_{l-2,l}^{-3}M_{ll}^{-3}\cos(\sigma_l(\eta_i)-\sigma_{l-2}(\eta_i))$$

$$+\frac{15l(l+1)(l+2)(l+3)}{4(2l-1)(2l+3)^2}M_{l+2,l}^{-3}M_{ll}^{-3}\cos(\sigma_l(\eta_i)-\sigma_{l+2}(\eta_i))$$

$$-\frac{105(l-1)l(l+1)(l+2)}{8(2l-1)(2l+1)(2l+3)}M_{l+2,l}^{-3}M_{l-2,l}^{-3}\cos(\sigma_{l+2}(\eta_i)-\sigma_{l-2}(\eta_i))\Bigg\}. \quad \text{(II C.32)}$$

The classical limit ($\eta_i\rightarrow\infty$ or $\nu=0$) is obtained from the formula (II A.76) which in the case $\lambda=2$ reduces to[11,56,66]

$$b_0^{E2}(\xi)=\int_0^\pi(\tfrac{3}{2}|I_{22}|^2+|I_{20}|^2+\tfrac{3}{2}|I_{2,-2}|^2)\frac{\cos\frac{\vartheta}{2}}{\sin^3\frac{\vartheta}{2}}d\vartheta, \quad \text{(II C.33)}$$

$$b_2^{E2}(\xi)=\int_0^\pi(\tfrac{3}{2}|I_{22}|^2-|I_{20}|^2+\tfrac{3}{2}|I_{2,-2}|^2-3I_{20}(I_{22}+I_{2,-2})\cos\vartheta)\frac{\cos\frac{\vartheta}{2}}{\sin^3\frac{\vartheta}{2}}d\vartheta, \quad \text{(II C.34)}$$

$$b_4^{E2}(\xi)=\int_0^\pi\left(-\frac{9}{64}|I_{22}|^2-\frac{9}{16}|I_{20}|^2-\frac{9}{64}|I_{2,-2}|^2+\frac{15}{16}I_{20}(I_{22}+I_{2,-2})\cos\vartheta-\frac{105}{32}I_{22}I_{2,-2}\cos2\vartheta\right)\frac{\cos\frac{\vartheta}{2}}{\sin^3\frac{\vartheta}{2}}d\vartheta. \quad \text{(II C.35)}$$

[66] The γ-ray angular distribution coefficients of references 11 and 56 contain errors of sign as pointed out by Breit, Ebel, and Benedict (Phys. Rev. **100**, 429 (1955)), who have re-evaluated the distributions for certain values of ξ (see also F. D. Benedict and G. Tice, Phys. Rev. **100**, 1545 (1955)).

TABLE II.10. Gamma-ray angular distribution coefficient $a_4^{E2}(\eta_i,\xi)$. The coefficient a_4^{E2} which describes the angular distribution of the gamma rays following $E2$ Coulomb excitation [see (II C.29)] is given as a function of η_i and ξ. The data are taken from references 59 and 60.

η_i \ ξ	0.0	0.1	0.2	0.3	0.4	0.5	0.6	0.8	1.0	1.2	1.4	1.6	2.0
0	+0.0625	−1.500	−1.500	−1.500	−1.500	−1.500	−1.500						
0.5	+0.016	−0.007	−0.179	−0.338	−0.463								
1.0	+0.002	+0.020	−0.040	−0.114	−0.183	−0.241	−0.291	−0.368					
1.5	0.000	+0.026	−0.006	−0.054	−0.101	−0 144	−0.182	−0.242	−0.289	−0.324	−0.352		
2.0	0.000	+0.027	+0.006	−0.031	−0.069	−0.104	−0.136	−0.188	−0.229	−0.262	−0.288	−0.309	
2.5	0.000	+0.027	+0.011	−0.020	−0.053	−0.084	−0.112	−0.159	−0.197	−0.227	−0.251	−0.272	−0.303
3.0	0.000	+0.027	+0.014	−0.013	−0.043	−0.071	−0.097	−0.141	−0.176	−0.205	−0.228	−0.247	−0.278
4.0	0.000	+0.026	+0.017	−0.006	−0.033	−0.058	−0.080	−0.120	−0.151	−0.178	−0.199	−0.218	−0.246
5.0	0.000	+0.025	+0.018	−0.003	−0.027	−0.050	−0.071	−0.108	−0.137	−0.162	−0.182	−0.200	−0.221
6.0	−0.001	+0.025	+0.018	−0.001	−0.023	−0.045	−0.065	−0.100	−0.128	−0.151	−0.171	−0.188	−0.217
8.0	−0.001	+0.024	+0.019	+0.001	−0.019	−0.039	−0.058	−0.090	−0.117	−0.139	−0.157	−0.173	−0.199
10.0	−0.001	+0.024	+0.019	+0.003	−0.016	−0.036	−0.053	−0.084	−0.109	−0.130	−0.147	−0.162	−0.185
12.0	−0.001	+0.023	+0.019	+0.004	−0.013	−0.034	−0.051	−0.081	−0.105	−0.124	−0.142	−0.157	−0.177
16.0	−0.001	+0.023	+0.020	+0.005	−0.010	−0.031	−0.048	−0.076	−0.100	−0.118	−0.135	−0.149	−0.174
∞	−0.001	+0.022	+0.020	+0.007	−0.009	−0.024	−0.039	−0.064	−0.085	−0.101	−0.116	−0.128	−0.146

It is seen from the figures and tables of the a coefficients that the deviations from the classical limit are considerable even for rather large values of η_i.

In the limit of $\xi \to 0$, the curves for $\nu \neq 0$ would exhibit rapid variation and would approach the Born approximation values which are appreciably different from those of the classical approximation (see Sec. II E.2). Since the region of rapid variation corresponds to $\eta \ll 1$ it is outside the domain of Coulomb excitation experiments.

For $M1$ excitations the angular distribution of the γ rays in the classical approximation is given by (see II A.74)

$$W(\vartheta_\gamma)=1+A_2^{(1)}P_2(\cos\vartheta_\gamma), \qquad \text{(II C.36)}$$

where the $A_2^{(1)}$ may be obtained from (II A.70). It is noted that for mixed multipole excitations, the γ-ray angular distribution contains additional interference terms.[66a]

The angular distribution of the γ's following other multipole excitations, as well as the angular distribution for specified direction of the inelastically scattered projectile, may be obtained in the classical approximation from the formulas of Sec. II A.4 using the tabulated values of the orbital integrals (see Table II.12 and reference 88).

The polarization of the decay γ rays is obtained by replacing $A_k^{(\lambda)}P_k(\cos\vartheta_\gamma)$ in (26), (29), and (36) by the appropriate expressions which depend also on the polarization angle ψ_γ, and which are given in Sec. II A [see (II A.70) and (II A.78a)].

II C.5. Survey of Approximations

We here summarize the various effects which have been neglected in the formulas considered in the earlier parts of this Section.

[66a] See reference 62a, where it is also shown that Eq. (36) remains valid in the quantum-mechanical treatment, provided spin effects are neglected.

a. Relativity effects.—The relativistic treatment of the excitation process (see Sec. II B.1) involves a modification of the excitation functions, but the correction terms are at most of order v^2/c^2, where v is the projectile velocity. The effect of the projectile spin, which is not included in (15) and (18), again implies corrections to the electric excitation cross section of order v^2/c^2 or less, while the corrections to the magnetic excitations are of relatively greater significance and may be obtained from (II B.30).[66a]

It is noted that the treatment of the nuclear structure is fully relativistic, provided the expressions (II B.16 and 17) for the multipole operators are employed.

b. Screening by the atomic electrons.—The screening of the nuclear Coulomb field by the atomic electrons gives rise to a minor modification of the projectile orbit. The effect is, however, very small, since the distance of closest approach, $2a$, even in heavy atoms and for proton energies as low as one Mev, is an order of magnitude smaller than the radius of the atomic K shell. The attraction from the electrons implies a small acceleration of the projectiles during their passage to the center of the atom, which results in a small increase in the effective energy in the Coulomb excitation process. This energy increase amounts to about 10 kev in a heavy atom ($Z_2 \sim 80$). An effect of similar magnitude but opposite sign results from the vacuum polarization which implies a small increase of the order of one-half of a percent in the repulsion between the nucleus and the projectile at distances of a few times the nuclear radius.[67]

There are also various processes by which the projectile may lose small amounts of energy in traversing the atom (e.g., ionization of $K, L, \cdots$ shells, or bremsstrahlung (see Sec. III B.3)). These effects are connected with the influence of straggling on the Coulomb excitation yield, and are of minor importance.

c. Attenuation effects in angular distributions.—The atomic electric and magnetic fields may cause a pre-

[67] L. L. Foldy and E. Eriksen, Phys. Rev. 95, 1048 (1954).

TABLE II.11. Some γ-γ angular correlation coefficients. The table gives the γ-γ angular correlation coefficients,[31] $A_k(2)$, for values of the spins I_i, I_f, and I_{ff} which may occur especially often in Coulomb excitation experiments. The excitation process is assumed to be pure $E2$ and the de-excitation is assumed to proceed by mixed $M1$ and $E2$ radiation; δ^2 is the ratio between the intensities of $E2$ and $M1$ gamma radiation, and the sign of δ is defined as in Eq. (II A.70).

I_i	I_f	I_{ff}	$A_2^{(2)}$	$A_4^{(2)}$
0	2	0	0.3571	1.143
1/2	3/2	1/2	$\frac{-0.250+0.8660\delta+0.250\delta^2}{1+\delta^2}$	0
1/2	5/2	1/2	0.2857	0.3810
1/2	5/2	3/2	$\frac{-0.2000+1.014\delta+0.1020\delta^2}{1+\delta^2}$	$\frac{-0.4354\delta^2}{1+\delta^2}$
3/2	5/2	3/2	$\frac{-0.07143+0.3622\delta+0.03644\delta^2}{1+\delta^2}$	$\frac{0.4976\delta^2}{1+\delta^2}$
3/2	7/2	3/2	0.2186	0.1282
3/2	7/2	5/2	$\frac{-0.1530+0.884\delta+0.0364\delta^2}{1+\delta^2}$	$\frac{-0.2280\delta^2}{1+\delta^2}$
5/2	7/2	5/2	$\frac{-0.0255+0.1472\delta+0.00607\delta^2}{1+\delta^2}$	$\frac{0.4054\delta^2}{1+\delta^2}$
5/2	9/2	5/2	0.1870	0.07204
5/2	9/2	7/2	$\frac{-0.1310+0.809\delta+0.00852\delta^2}{1+\delta^2}$	$\frac{-0.1572\delta^2}{1+\delta^2}$
7/2	9/2	7/2	$\frac{-0.00597+0.0368\delta+0.000389\delta^2}{1+\delta^2}$	$\frac{0.3430\delta^2}{1+\delta^2}$
7/2	11/2	7/2	0.1688	0.0500
7/2	11/2	9/2	$\frac{-0.1182+0.762\delta-0.00649\delta^2}{1+\delta^2}$	$\frac{-0.1231\delta^2}{1+\delta^2}$.
9/2	11/2	9/2	$\frac{0.00454-0.0293\delta+0.00025\delta^2}{1+\delta^2}$	$\frac{0.3030\delta^2}{1+\delta^2}$
9/2	13/2	9/2	0.1570	0.03881

cession of the angular momentum vector of the excited nucleus, giving rise to an attenuation of the angular anisotropy of the emitted radiation.[68] In most cases one expects the largest effect to arise from the quadrupole coupling to the electric field. This coupling may be especially strong at the interstitial positions reached by the recoiling nucleus. The conditions are somewhat similar to those encountered in the α decay of the very heavy elements, where the α-γ correlations are observed to be appreciably attenuated[69] even for lifetimes as short as $5\cdot 10^{-10}$ second.[70] The excited states involved in these α-γ correlations are expected to have especially large quadrupole moments (see Sec. V B.2). It may thus be concluded that a lower limit to the lifetimes for which the attenuation effects may be of significance in Coulomb excitations is of the order 10^{-10} second.

d. Higher order interaction effects.—In Secs. II A and B, the probability for exciting the nucleus is treated in first-order perturbation theory. Under most experimental conditions so far studied, the probability for excitation in a single encounter is very small, and such a treatment therefore well justified. The influence of higher order effects giving rise to multiple excitations is considered in Sec. II D. Such effects become of special importance in the case of high projectile charge and large bombarding energies.

II D. Higher Order Excitation Effects

In the theory of Coulomb excitation presented in the preceding sections, the probability for excitation of the nucleus was calculated in lowest order perturbation theory. In most of the experiments which have so far been performed, this approximation is well justified since the excitation probability in a single encounter is very small compared to unity. Thus, for an $E2$ excitation with $\xi=0$ the excitation probability, in a backward scattering, is given by [see (II A.2) and (II A.28) and Table II.8]

$$P=d\sigma_{E2}/d\sigma_R$$

$$=207\frac{A_1}{Z_1^2Z_2^4}E_{\text{Mev}}{}^3B(E2), \qquad \text{(II D.1)}$$

where $B(E2)$ is measured in units of $e^2\cdot(10^{-24}\text{ cm}^2)^2$.

As an example, one finds for 6-Mev α particles on a target with $Z_2=70$ a value for P of about 0.01 for $B(E2)\approx 5$ corresponding to the largest transition probabilities encountered (see Table IV.2).

It is seen, however, that the probability, P, may become appreciable for large bombarding energies, which may especially be employed with highly charged projectiles.[71] Moreover, even when the probability is small, higher order effects may be observed if the direct transition to the final state is weak.

We shall in this section consider the treatment of higher order effects based on the classical description of the projectile orbit, and briefly discuss a few applications.

II D.1. Cross Sections to Second Order

To second order in the charge of the projectile, the amplitude for a transition from the initial nuclear state

[68] H. Frauenfelder, Chapter XIX of *Beta- and Gamma-Ray Spectroscopy* (edited by K. Siegbahn) (North Holland Publishing Company, Amsterdam, 1955).

[69] See e.g., J. O. Newton, *Progress in Nuclear Physics* (edited by O. R. Frisch) (Pergamon Press, London, 1954), Volume 4, p. 256.

[70] While none of these lifetimes has been directly measured, they are expected to be similar to those for the first excited states in Th^{232} and U^{238} inferred from the Coulomb excitation cross sections (Table IV.2).

[71] See G. Breit and J. P. Lazarus, Phys. Rev. **100**, 942 (1955).

i to the final state f is given by

$$b_{if}^{(2)}=b_{if}(\omega)+\sum_z b_{izf}, \qquad \text{(II D.2)}$$

where b_{if} is the first-order amplitude (II A.6) and where[20]

$$b_{izf}=\frac{1}{(i\hbar)^2}\int_{-\infty}^{\infty}dt\langle f|\mathcal{H}(t)|z\rangle \times e^{i\omega_2 t}\int_{\infty}^{t}dt'\langle z|\mathcal{H}(t')|i\rangle e^{i\omega_1 t'}. \qquad \text{(II D.3)}$$

The summation in (2) is to be performed over all intermediate states z, including the initial and final state. The frequencies ω_1, ω_2, and ω, are given by

$$\omega_1=(E_z-E_i)/\hbar,\quad \omega_2=(E_f-E_z)/\hbar,\quad \omega=\omega_1+\omega_2=(E_f-E_i)/\hbar, \qquad \text{(II D.4)}$$

where the energies of the initial, intermediate, and final states are denoted by E_i, E_z, and E_f, respectively.

In the evaluation of the double integral (3), it is convenient to introduce the unit step function

$$\epsilon(t-t')=-\lim_{\delta\to 0^+}\frac{1}{2\pi i}\int_{-\infty}^{\infty}\frac{e^{-i(t-t')q}}{q+i\delta}dq=\begin{cases}1 & t>t'\\ 0 & t<t'.\end{cases} \qquad \text{(II D.5)}$$

We then get

$$b_{izf}=-\lim_{\delta\to 0^+}\frac{1}{2\pi i}\int_{-\infty}^{\infty}\frac{dq}{q+i\delta}b_{iz}(\omega_1+q)b_{zf}(\omega_2-q)=\tfrac{1}{2}b_{iz}(\omega_1)b_{zf}(\omega_2)+\frac{i}{2\pi}\mathcal{P}\int_{-\infty}^{\infty}\frac{b_{iz}(\omega_1+q)b_{zf}(\omega_2-q)}{q}dq, \qquad \text{(II D.6)}$$

where $\mathcal{P}$ stands for the principal part of the integral.

In the focal system the b_{if} are pure imaginary and for electric excitation are given by [see (II A.14), (II A.16), and (II A.24)]

$$b_{if}(\omega)=-i\frac{4\pi Z_1 e}{\hbar v}\sum_{\lambda\mu}\frac{(-1)^{I_i-M_i}}{2\lambda+1}\begin{pmatrix}I_i & \lambda & I_f\\ -M_i & \mu & M_f\end{pmatrix}\langle I_i\|\mathfrak{M}(E\lambda)\|I_f\rangle a^{-\lambda}Y_{\lambda\mu}\left(\frac{\pi}{2},0\right)I_{\lambda\mu}(\vartheta,\xi). \qquad \text{(IID.7)}$$

Thus, the last form of (6) represents a decomposition of b_{izf} into real and imaginary parts.

Inserting (7) into (6) one obtains the following expression for the amplitude to second order:

$$b_{if}^{(2)}=-i\frac{4\pi Z_1 e}{\hbar}\sum_{k\kappa}\frac{(-1)^{I_i-M_i}}{2k+1}\begin{pmatrix}I_i & k & I_f\\ -M_i & \kappa & M_f\end{pmatrix}[\langle I_i\|\mathfrak{M}(Ek)\|I_f\rangle S_{Ek,\kappa}+T_{k\kappa}], \qquad \text{(II D.8)}$$

where

$$T_{k\kappa}=\frac{2\pi Z_1 e}{\hbar v^2}\sum_{z\lambda_1\lambda_2}\frac{(-1)^{\lambda_1+\lambda_2+I_i+I_f+k+\kappa}}{a^{\lambda_1+\lambda_2}}\frac{(2k+1)^2}{(2\lambda_1+1)(2\lambda_2+1)}\langle I_i\|\mathfrak{M}(E\lambda_1)\|I_z\rangle \times\langle I_z\|\mathfrak{M}(E\lambda_2)\|I_f\rangle\begin{Bmatrix}\lambda_1 & \lambda_2 & k\\ I_f & I_i & I_z\end{Bmatrix}[\beta_{k,-\kappa}(\lambda_1\lambda_2\xi_1\xi_2\vartheta)-i\alpha_{k,-\kappa}(\lambda_1\lambda_2\xi_1\xi_2\vartheta)]. \qquad \text{(II D.8a)}$$

We have here employed the relation (II A.6) and have introduced the real functions $\alpha_{k\kappa}$ and $\beta_{k\kappa}$ through the definitions

$$\alpha_{k\kappa}(\lambda_1\lambda_2\xi_1\xi_2,\vartheta)=\sum_{\mu_1\mu_2}\begin{pmatrix}\lambda_1 & \lambda_2 & k\\ \mu_1 & \mu_2 & \kappa\end{pmatrix}Y_{\lambda_1\mu_1}\left(\frac{\pi}{2},0\right)Y_{\lambda_2\mu_2}\left(\frac{\pi}{2},0\right)I_{\lambda_1\mu_1}(\vartheta,\xi_1)I_{\lambda_2\mu_2}(\vartheta,\xi_2), \qquad \text{(II D.9)}$$

and

$$\beta_{k\kappa}(\lambda_1\lambda_2\xi_1\xi_2,\vartheta)=\sum_{\mu_1\mu_2}\begin{pmatrix}\lambda_1 & \lambda_2 & k\\ \mu_1 & \mu_2 & \kappa\end{pmatrix}Y_{\lambda_1\mu_1}\left(\frac{\pi}{2},0\right)Y_{\lambda_2\mu_2}\left(\frac{\pi}{2},0\right)\frac{1}{\pi}\mathcal{P}\int_{-\infty}^{\infty}\frac{d\xi'}{\xi'}I_{\lambda_1\mu_1}(\vartheta,\,\xi_1+\xi')I_{\lambda_2\mu_2}(\vartheta,\,\xi_2-\xi'). \qquad \text{(II D.10)}$$

The two indices refer to the two transitions in the double excitation; thus $\xi_1=\omega_1 a/v$, $\xi_2=\omega_2 a/v$, and $\xi=\xi_1+\xi_2$.

The differential cross section to second order is obtained by inserting (8) into (II A.5 and 4) and may be written in the form

$$d\sigma=d\sigma^{(1)}+d\sigma^{(1,2)}+d\sigma^{(2)}. \qquad \text{(II D.11)}$$

The first term is the first-order excitation cross section (II A.28). The second term represents the interference between first-order and second-order transitions and receives a contribution only from the imaginary part of (6) (the real part of (8a)). Additional terms of the same order of magnitude as the third term in (9) (proportional to Z_1^4) may arise from cross terms between first- and third-order terms in the transition amplitude.

Performing the summation over the magnetic quantum numbers, the two last terms in (11) take the form

$$d\sigma^{(1.2)}=(-1)^{I_i+I_f}16\pi^3\left(\frac{Z_1e}{\hbar v}\right)^3 a^2\sin^{-4}\frac{\vartheta}{2}d\Omega$$

$$\times\sum_{z,\lambda,\lambda_1,\lambda_2}(-1)^{\lambda+\lambda_1+\lambda_2}\begin{Bmatrix}\lambda & \lambda_1 & \lambda_2\\ I_z & I_f & I_i\end{Bmatrix}\frac{\langle I_i\|\mathfrak{M}(E\lambda)\|I_f\rangle\langle I_i\|\mathfrak{M}(E\lambda_1)\|I_z\rangle\langle I_z\|\mathfrak{M}(E\lambda_2)\|I_f\rangle}{a^{\lambda+\lambda_1+\lambda_2}(2\lambda+1)(2\lambda_1+1)(2\lambda_2+1)(2I_i+1)}$$

$$\times\sum_\mu(-1)^\mu Y_{\lambda\mu}\left(\frac{\pi}{2},0\right)I_{\lambda\mu}(\vartheta,\xi_1+\xi_2)\beta_{\lambda,-\mu}(\lambda_1\lambda_2\xi_1\xi_2,\vartheta),\quad\text{(II D.12)}$$

and

$$d\sigma^{(2)}=16\pi^4\left(\frac{Z_1e}{\hbar v}\right)^4 a^2\sin^{-4}\frac{\vartheta}{2}d\Omega$$

$$\cdot\sum_{zz'\lambda_1\lambda_1'\lambda_2\lambda_2'}(-1)^{\lambda_1+\lambda_1'+\lambda_2+\lambda_2'}\frac{\langle I_i\|\mathfrak{M}(E\lambda_1)\|I_z\rangle\langle I_z\|\mathfrak{M}(E\lambda_2)\|I_f\rangle\langle I_i\|\mathfrak{M}(E\lambda_1')\|I_{z'}\rangle\langle I_{z'}\|\mathfrak{M}(E\lambda_2')\|I_f\rangle}{a^{\lambda_1+\lambda_1'+\lambda_2+\lambda_2'}(2\lambda_1+1)(2\lambda_2+1)(2\lambda_1'+1)(2\lambda_2'+1)(2I_i+1)}$$

$$\times\sum_{k\kappa}(2k+1)\begin{Bmatrix}k & \lambda_1 & \lambda_2\\ I_z & I_f & I_i\end{Bmatrix}\begin{Bmatrix}k & \lambda_1' & \lambda_2'\\ I_{z'} & I_f & I_i\end{Bmatrix}[\alpha_{k\kappa}(\lambda_1\lambda_2\xi_1\xi_2,\vartheta)\alpha_{k\kappa}(\lambda_1'\lambda_2'\xi_1'\xi_2',\vartheta)$$

$$+\beta_{k\kappa}(\lambda_1\lambda_2\xi_1\xi_2,\vartheta)\beta_{k\kappa}(\lambda_1'\lambda_2'\xi_1'\xi_2',\vartheta)].\quad\text{(II D.13)}$$

Since the summations over the magnetic quantum numbers have been performed, the summation index z (or z') in (12) and (13) refers only to summation over different energy levels.

Also intermediate levels with energy well above that of the final state may give a significant contribution, since $\beta_{k\kappa}$ for large $\xi_1(\approx-\xi_2)$ behaves approximately as ξ_1^{-1} in contrast to the strong exponential dependence on ξ_1, which is characteristic of $I_{\lambda\mu}(\xi_1)$ and thus also of $\alpha_{k\kappa}$. Even for an intermediate energy transfer (E_z-E_i) comparable with, or larger than, the projectile energy, the above results remain valid provided only that E_f-E_i is small compared to the projectile energy. In fact, as may be seen in more detail from a quantum-mechanical treatment, the effective energy loss of the projectile in the intermediate state is that associated with those values of $\xi_1+\xi'$ which contribute the main part of the integral in (10), and which are of the order $\xi/2$.

In the summation over the multipole orders in (12) and (13) the main contribution will usually arise from the lowest value of λ compatible with the spin and parity selection rules for the nuclear matrix elements.

The coefficients $\alpha_{k\kappa}$ and $\beta_{k\kappa}$ needed for the evaluation of the cross sections (12) and (13) may be calculated from the classical integrals $I_{\lambda\mu}$. For $\lambda=2$ and positive ξ, these are given in Table II.12.[72] The $I_{\lambda\mu}$ for negative arguments are obtained by means of (II E.54).

The angular distribution of the γ quanta following Coulomb excitation will also be subject to second-order corrections.[73] Since the second-order amplitude (8) has the same dependence on the magnetic quantum numbers as the first-order amplitude (7), the angular distribution to second order is obtained from (II A.80–81) by the substitution

$$\langle I_i\|\mathfrak{M}(E\lambda)\|I_f\rangle S_{\lambda\mu}\rightarrow$$
$$\langle I_i\|\mathfrak{M}(E\lambda)\|I_f\rangle S_{\lambda\mu}+T_{\lambda\mu}.\quad\text{(II D.13a)}$$

II D.2. Interference Effects

An interesting case where interference between the first- and the second-order terms in (2) may become significant is that in which the intermediate state z is one of the magnetic substates of the final nuclear level.[71] The transition $z\rightarrow f$ then occurs through the interaction of the projectile with the quadrupole moment of the final state.

From (6) and (7) one obtains for the order of magnitude of b_{izf} [see (II C.7) and (II C.8)] in the case $\lambda=2$

$$\frac{b_{izf}}{b_{if}}\approx5\frac{A_1^{\frac{1}{2}}}{Z_1Z_2^2}E_{\text{Mev}}^{\frac{1}{2}}Q_f,\quad\text{(II D.14)}$$

where the quadrupole moment Q_f, which is measured in units of 10^{-24} cm², is related to the reduced matrix element through [see (II A.18) and (V.32a)]

$$\left(\frac{5}{16\pi}\right)^{\frac{1}{2}}Qe=\langle I\|\mathfrak{M}(E2)\|I\rangle$$

$$\times\left(\frac{I(2I-1)}{(2I+1)(I+1)(2I+3)}\right)^{\frac{1}{2}}.\quad\text{(II D.15)}$$

If the initial state (ground state) also possesses a quadrupole moment $(I_i\geqslant1)$, a corresponding effect

[72] For $\lambda=1$, 3, and 4, the $I_{\lambda\mu}$ are tabulated in reference 88.
[73] Breit, Gluckstern, and Russell, Phys. Rev. (submitted for publication).

arises from intermediate transitions to the substates of this level.

A more detailed calculation has been made for the particular case $I_i=0$, $I_f=2$, and $\xi_1=\xi=0.4$. For the differential excitation cross section at 90° one obtains from (12)

$$d\sigma_{90^\circ}=d\sigma_{90^\circ}^{(1)}[1+1.55Z_1^{-1}Z_2^{-2}A_1^{\frac{1}{2}}E_{\mathrm{Mev}}^{\frac{3}{2}}Q_f]. \qquad \text{(II D.16)}$$

Effects of similar order of magnitude are expected in the angular distribution of the de-excitation γ rays.[71,73]

It is of interest that the interference effects considered are linear in Q and thus provide a means of determining not only the magnitude, but also the sign of the quadrupole moment of the excited state. The present estimates indicate, however, that the effects become of significance only for high bombarding energies and thus especially for heavy ions.

II D.3. Double Excitations

Another important second-order effect is that of a double $E2$ excitation leading to a final state which cannot be reached directly from the ground state by an $E2$ excitation. The order of magnitude of the cross section for such a double excitation may be obtained from (6) and, provided $E_z \lesssim E_f$, is given approximately by [see (II A.2) and (II A.4)]

$$\sigma_{E2,\,E2}\approx\tfrac{1}{4}a^{-2}\sigma_{E2}(I_i\rightarrow I_z)\sigma_{E2}(I_z\rightarrow I_f). \qquad \text{(II D.17)}$$

A direct numerical evaluation of (11) for the specific case $I_i=0$, $I_z=2$, and $I_f=4$, and for $\xi_1=\xi_2=\xi=0$ gives

$$\sigma_{E2,\,E2}=0.0218\left(\frac{Z_1e}{\hbar v}\right)^4 a^{-6} \times B(E2,0\rightarrow2)B(E2,2\rightarrow4), \qquad \text{(II D.18)}$$

which corresponds rather closely to (17).

On account of the large values of $B(E2)$ in collective excitations, the cross section (18) may become appreciable for large bombarding energies (see numerical estimates in Sec. V B.2). Also excitations of higher order than the second may become feasible. The corresponding cross sections can be estimated in analogy to (17).

The excitation of the $I=4$ state may also take place by a direct $E4$ transition with the cross section (II A.30). For $\xi_1=\xi_2=\xi=0$, the ratio of the two cross sections is found to be

$$\frac{\sigma_{E2,\,E2}}{\sigma_{E4}}=0.19\frac{A_1Z_1^2}{E_{\mathrm{Mev}}}\,\frac{B(E2,0\rightarrow2)B(E2,2\rightarrow4)}{e^2B(E4,0\rightarrow4)}. \qquad \text{(II D.19)}$$

If one would employ the single particle estimate (II A.58) for the $B(E\lambda)$ one would obtain a value of about 1/3 for the ratio (19) in the case of 20-Mev α particles. However, if the $E2$ transitions are of collective type, the ratio (19) may be several orders of magnitude larger.

The cross section for the excitation of the $I=4$ state also contains an interference term between the direct $E4$ transition and the double $E2$ transition. As an example the differential cross section at 90° has been evaluated from (12) and (13) for the case $\xi_1=0.2$, $\xi_2=0.4$, with the result

$$d\sigma_{90^\circ}=d\sigma_{90^\circ}^{(1)}\left[1+0.17\left(\frac{Z_1^2A_1}{E_{\mathrm{Mev}}}\,\frac{B(E2,0\rightarrow2)B(E2,2\rightarrow4)}{e^2B(E4,0\rightarrow4)}\right)^{\frac{1}{2}}+0.12\frac{Z_1^2A_1}{E_{\mathrm{Mev}}}\,\frac{B(E2,0\rightarrow2)B(E2,2\rightarrow4)}{e^2B(E4,0\rightarrow4)}\right]. \qquad \text{(II D.20)}$$

The square root of the reduced transition probabilities $B(E\lambda)$ is to be taken with the sign of the reduced matrix elements.

II D.4. Polarization Effects in Elastic Scattering

The virtual excitations also give rise to a modification of the elastic scattering cross section.[74–76] Such polarization effects are especially simple to estimate if the frequencies of the virtual excitations are large compared to the inverse collision time ($\xi_1=-\xi_2\gg1$). Under these conditions, one may for each position of the projectile consider the static polarization of the nucleus and derive the resulting potential which reacts on the projectile.

Expanding the interaction (II A.8) and (II A.9) in multipoles [see (II A.10)] one obtains by a perturbation calculation

$$V_{\mathrm{pol}}(r_p)=4\pi Z_1^2e^2\sum_{\lambda=1}^{\infty}(2\lambda+1)^{-2}r_p^{-2\lambda-2} \times\sum_{z\neq i}\frac{B(E\lambda,i\rightarrow z)}{E_i-E_z} \qquad \text{(II D.21)}$$

for the potential energy of the projectile, arising from the nuclear polarization. In obtaining (21) we have averaged over initial orientations M_i of the target nucleus and summed over M_z; thus, the sum over z only refers to summation over different energy levels.

A similar term in the potential energy may arise from the polarization of the projectile if this is a composite nucleus.

The effect on the elastic scattering cross section may now be obtained by inserting V_{pol} as a perturbing potential into the Schroedinger equation for the scattering process.[76] If the scattering can be treated classically ($\eta\gg1$), the polarizing effect may also be obtained by inserting (21) into the classical equations of motion.

[74] P. Debye and W. Hardmeier, Physik. Z. 27, 196 (1926).
[75] N. F. Ramsey, Phys. Rev. 83, 659 (1951), and Malenka, Kruse, and Ramsey, reference 7.
[76] Breit, Hull, and Gluckstern, Phys. Rev. 87, 74 (1952).

If the target nucleus possesses a spin, additional effects on the elastic scattering cross section may arise from the interaction of the projectile with the static electric moments of the nucleus. For aligned target nuclei this interaction gives rise to changes in the cross section linear in the nuclear moments. This linear term, however, vanishes when averaged over the orientations M_i, but there remains a second-order term which may be comparable with the effect of (21).

In most cases, it is to be expected that the principal polarization effect arises from the quadrupole interaction, on account of the high excitation frequencies associated with the main part of the dipole oscillator strength. For many nuclei the most important quadrupole excitations have rather low frequencies (see Chapter V), and it may then not be possible to consider the nucleus simply in terms of its static polarizability. In such cases $\xi_1=-\xi_2\gtrsim 1$, and it becomes necessary to treat in more detail the coupled motion of projectile and nucleus.

II E. Appendices

II E.1. Emission of Bremsstrahlung

The emission of bremsstrahlung in the collision between the projectile and the nucleus, which constitutes an important background effect in the Coulomb excitation experiments (see Sec. III B.3), may be treated in close analogy to the nuclear excitation process. The first theoretical treatments of the bremsstrahlung process were based on a classical description of the particle orbit.[77] The quantum-mechanical theory of the electric dipole bremsstrahlung was given by Sommerfeld.[78]

The cross section for scattering of the projectile into the solid angle $d\Omega$ with emission of a photon with wave number between q and $q+dq$ is given by [see (II B.2), (II B.9), and (II B.25)]

$$d\sigma=\frac{m_f^2}{4\pi^2\hbar^4}\frac{v_f}{v_i}\frac{R}{\pi}\times\sum_{\lambda\mu\pi}\left|\left\langle 1_{\pi\lambda\mu}\mathbf{k}_f\left|\frac{1}{c}\int\mathbf{j}_p\cdot\mathbf{A}d\tau\right|0\mathbf{k}_i\right\rangle\right|^2 dqd\Omega. \quad \text{(II E.1)}$$

We have here assumed the nucleus to be infinitely heavy and have neglected the spin of the projectile. The matrix element represents a transition from an initial state with no photons present to a final state with one photon of multipole order λ, μ, and parity π. Using the multipole expansion (II B.4), we obtain from (1) by means of (II B.7)

$$d\sigma=\frac{m_f^2c}{\pi^2\hbar^3}\frac{v_f}{v_i}\sum_{\lambda\mu}\frac{(\lambda+1)q^{2\lambda+1}}{\lambda[(2\lambda+1)!!]^2}\times\{|\langle\mathbf{k}_f|\mathfrak{M}_p(E\lambda,\mu)|\mathbf{k}_i\rangle|^2+|\langle\mathbf{k}_f|\mathfrak{M}_p(M\lambda,\mu)|\mathbf{k}_i\rangle|^2\}dqd\Omega, \quad \text{(II E.2)}$$

where the multipole moments $\mathfrak{M}_p(\lambda,\mu)$ are defined by (II B.16) and (II B.17) by replacing $\mathbf{j}_n$ by $\mathbf{j}_p$.

In the nonrelativistic case, the multipole moments are given by (II A.13) and (II A.39). For the electric part of the bremsstrahlung cross section, which is the most important, one thus obtains

$$d\sigma_E=\frac{1}{\pi^2}\left(\frac{m_1c}{\hbar}\right)^2\frac{Z_1^2e^2}{\hbar c}\frac{v_f}{v_i}\sum_\lambda\frac{q^{2\lambda+1}(\lambda+1)}{\lambda[(2\lambda+1)!!]^2}\times|\langle\mathbf{k}_f|r^\lambda Y_{\lambda\mu}(\theta,\phi)|\mathbf{k}_i\rangle|^2 d\Omega dq. \quad \text{(II E.3)}$$

The effect of the nuclear recoil may be taken into account by replacing m_1 with the reduced mass m_0 and the multipole moment in (3), which refers to the center of the nucleus, by the combined moment of projectile and nucleus with respect to the center of mass. The latter replacement simply corresponds to the substitution

$$Z_1\rightarrow(A_1+A_2)^{-\lambda}\cdot[Z_1A_2^\lambda+(-1)^\lambda Z_2A_1^\lambda]. \quad \text{(II E.4)}$$

The reduction of the matrix element in (3) to radial matrix elements can be made in complete analogy to the case of Coulomb excitation [see (II B.45)], and the resultant cross sections can be derived directly from (II B.47) and (II B.48). For the total electric cross section for emission of a photon in the wave-number interval dq, one obtains

$$d\sigma_E=\sum_{\lambda=1}^{\infty}d\sigma_{E\lambda}, \quad \text{(II E.5)}$$

with

$$d\sigma_{E\lambda}=\frac{e^2}{\hbar c}(A_1+A_2)^{-2\lambda}[Z_1A_2^\lambda+(-1)^\lambda Z_2A_1^\lambda]^2\left(\frac{c}{v_i}\right)^2\times a^2(qa)^{2\lambda+2}f_{E\lambda}{}^b(\eta_i,\xi)\frac{dq}{q}, \quad \text{(II E.6)}$$

and

$$f_{E\lambda}{}^b(\eta_i,\xi)=\frac{16(\lambda+1)k_ik_f}{\lambda[(2\lambda-1)!!]^2(2\lambda+1)}a^{-2\lambda-4}\times\sum_{l_il_f}(2l_i+1)(2l_f+1)\begin{pmatrix}l_i & l_f & \lambda\\ 0 & 0 & 0\end{pmatrix}^2|M_{l_il_f}{}^\lambda|^2. \quad \text{(II E.7)}$$

In these expressions, a is the symmetrized distance of closest approach defined by (II A.86).

The radial matrix elements are defined by (II B.46) and can be expressed in terms of hypergeometric func-

[77] H. A. Kramers, Phil. Mag. **46**, 836 (1923); G. Wentzel, Z. Physik **27**, 257 (1924); see also L. Landau and E. Lifshitz, *The Classical Theory of Fields* (Addison-Wesley Press, Cambridge, 1951), pp. 197 ff.

[78] A. Sommerfeld, reference 44, pp. 495 ff.

tions, as is shown in Sec. II B.4. The bremsstrahlung matrix elements, however, are more elementary than the Coulomb excitation matrix elements, since they can all be expressed, from the monopole matrix elements (II B.56), through recursion formulas (see, e.g., II B.68).

We shall here especially consider the electric dipole bremsstrahlung, in which case the matrix elements can be directly related to the $E1$ Coulomb excitation matrix elements.[79] The connection[49] is given through the equation of motion

$$m_0\frac{d^2\mathbf{r}}{dt^2}=\frac{Z_1Z_2e^2}{r^3}\mathbf{r}, \qquad \text{(II E.8)}$$

which leads to

$$\langle\mathbf{k}_f|rY_{1\mu}|\mathbf{k}_i\rangle=-\frac{Z_1Z_2e^2}{m_0\omega^2}\langle\mathbf{k}_f|r^{-2}Y_{1\mu}|\mathbf{k}_i\rangle \qquad \text{(II E.9)}$$

or

$$M_{l,\,l\pm1}{}^{+1}=-\frac{Z_1Z_2e^2}{m_0\omega^2}M_{l,\,l\pm1}{}^{-2}$$

$$=-\frac{4\eta_i\eta_f}{(\eta_f{}^2-\eta_i{}^2)^2}a^3M_{l,\,l\pm1}{}^{-2}. \qquad \text{(II E.10)}$$

By inserting (9) into (3) and comparing the result with the Coulomb excitation f function (II B.34), one obtains[80]

$$f_{B1}{}^b(\eta_i,\xi)=\frac{24}{\pi^2\xi^4}\left[\frac{\eta_i\eta_f}{(\eta_i+\eta_f)^2}\right]^2 f_{E1}(\eta_i,\xi). \qquad \text{(II E.11)}$$

The dipole bremsstrahlung cross section thus takes the form

$$d\sigma_{B1}=\frac{3}{2\pi^2}\frac{e^2}{\hbar c}\left(\frac{Z_1}{A_1}-\frac{Z_2}{A_2}\right)^2$$

$$\times\left(\frac{\hbar}{Mc}\right)^2\eta_i{}^2f_{E1}(\eta_i,\xi)\frac{dq}{q}, \qquad \text{(II E.12)}$$

where M is the proton mass. Introducing numerical values for the constants involved, one obtains [see (II C.8)]

$$d\sigma_{E1}=1.225\cdot10^{-8}Z_1{}^2Z_2{}^2\left(\frac{Z_1}{A_1}-\frac{Z_2}{A_2}\right)^2$$

$$\times A_1E_{\text{Mev}}{}^{-1}f_{E1}(\eta_i,\xi)\frac{dE_x}{E_x}\ \text{barns} \qquad \text{(II E.13)}$$

where E_x is the photon energy. If one inserts for $f_{E1}(\eta_i,\xi)$ in (12) the exact expression (II E.64), one gets the bremsstrahlung formula of Sommerfeld. Numerical values for the f_{E1} function are given in Sec. II C.2.[81]

[79] C. J. Mullin and E. Guth, reference 4.

[80] See also L. C. Biedenharn, Phys. Rev. **102**, 262 (1956); K. Alder and A. Winther, CERN report T/KA-AW-4 (1955).

[81] S. Drell and K. Huang, Phys. Rev. **99**, 686 (1955) have evaluated the Sommerfeld expression for the bremsstrahlung in a specific case by expanding in powers of ξ [see (II E.66)].

The relative intensity of consecutive multipole contributions to the bremsstrahlung is at most of the order $(qa)^2\approx[\xi(v/c)]^2$. The magnetic multipole contributions are reduced with respect to the electric ones by a factor $(v/c)^2$. Thus, in most cases, the electric dipole bremsstrahlung strongly dominates. However, due to the factor $Z_1/A_1-Z_2/A_2$, the $E1$ cross section may vanish for α-particle bombardment on light nuclei. In such instances, the bremsstrahlung is mainly of electric quadrupole and magnetic dipole type.

The angular distribution of the bremsstrahlung γ quanta may also be evaluated in a similar way as the angular distribution of de-excitation γ quanta in Coulomb excitation (see Sec. II A.4 and Sec. II B.5). For pure electric λ-pole bremsstrahlung, one has the following angular distribution function

$$W_{\mathbf{k}_i\mathbf{k}_f}(\Omega_\gamma)$$

$$=\sum_\sigma\Big|\sum_\mu\langle\mathbf{k}_f|r^\lambda Y_{\lambda\mu}(\theta,\phi)|\mathbf{k}_i\rangle D_{\mu\sigma}{}^\lambda(\mathcal{R})\Big|^2, \qquad \text{(II E.14)}$$

where the rotation matrix $D(\mathcal{R})$ represents the transition amplitude for emission of a 2^λ-pole photon in the direction Ω_γ and with polarization σ. By the usual technique of γ-γ correlation, one obtains from (14)

$$W_{\mathbf{k}_i\mathbf{k}_f}(\Omega_\gamma)=\sum_{\mu\mu'k\kappa}\langle\mathbf{k}_f|r^2Y_{\lambda\mu}|\mathbf{k}_i\rangle\langle\mathbf{k}_f|r^\lambda Y_{\lambda\mu'}|\mathbf{k}_i\rangle^*$$

$$\times\begin{pmatrix}\lambda & \lambda & k\\ 1 & -1 & 0\end{pmatrix}\begin{pmatrix}\lambda & \lambda & k\\ -\mu & \mu' & \kappa\end{pmatrix}$$

$$\times(2k+1)^{\frac{1}{2}}Y_{k\kappa}(\Omega_\gamma). \qquad \text{(II E.15)}$$

For $\lambda=1$, the matrix elements are proportional to the dipole Coulomb excitation matrix elements, and in this case one has by comparing with (II A.68) and (II B.82)

$$W_{\mathbf{k}_i\mathbf{k}_f}(\Omega_\gamma)=1+\sum_\kappa\tfrac{1}{2}a_{2\kappa}{}^{E1}(\vartheta,\varphi,\eta_i,\xi)Y_{k\kappa}(\Omega_\gamma), \qquad \text{(II E.16)}$$

where the a coefficients are those occurring in (II A.66). If one integrates over all proton directions, one obtains in the electric dipole case[82,80]

$$W(\vartheta_\gamma)=1+\tfrac{1}{2}a_2{}^{E1}(\eta_i,\xi)P_2(\cos\vartheta_\gamma), \qquad \text{(II E.17)}$$

where ϑ_γ is the direction of the γ quantum with respect to the incoming beam of projectiles. The coefficient $a_2{}^{E1}$ is defined by (II B.84) and (II B.85) and is given numerically in Sec. II C.4.

II E.2. Born Approximation

Under experimental conditions where the Coulomb repulsion is sufficiently strong to prevent the projectile from entering into the nucleus, the parameter η is large (see introduction to Chapter II). Although the Born approximation cannot be applied in such cases, it

[82] The angular distribution in the classical treatment has been considered by G. Wentzel (reference 77).

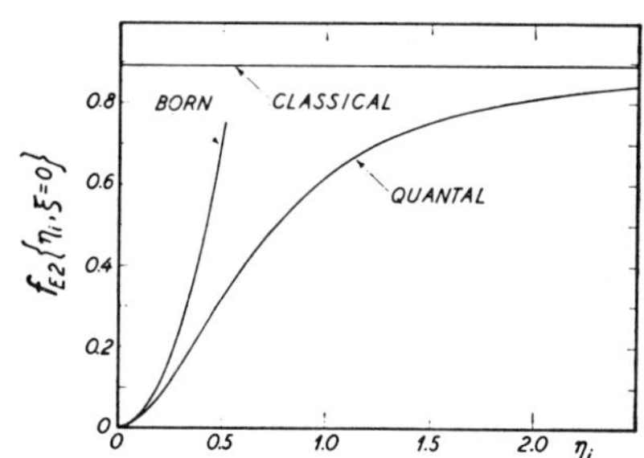

FIG. II.9. Comparison of quantal, classical, and Born approximation for $E2$ Coulomb excitation. The total f functions for $E2$ Coulomb excitation as given by the quantum-mechanical, classical, and Born approximation calculations are plotted as functions of η_i for the case of vanishing energy loss ($\xi=0$).

nevertheless provides an interesting limit of the general theory presented above. Moreover, for high energy projectiles with $\eta\ll 1$, some of the Born approximation results may find application, although under these circumstances the effects of the penetration of the projectile into the nucleus must also be taken into account.

In the Born approximation, the initial and final scattering states are considered as plane waves and the matrix element in (II B.34) thus takes the form

$$\langle \mathbf{k}_f | r^{-\lambda-1} Y_{\lambda\mu}(\theta,\phi) | \mathbf{k}_i \rangle = 4\pi i^\lambda \int_0^\infty j_\lambda(Kr) r^{-\lambda+1} dr \cdot Y_{\lambda\mu}(\mathbf{K}), \quad \text{(II E.18)}$$

where we have used the expansion

$$e^{i(\mathbf{k}_i-\mathbf{k}_f)\cdot \mathbf{r}} = e^{i\mathbf{K}\cdot\mathbf{r}} = \sum_{lm} 4\pi i^l j_l(Kr) Y_{lm}(\mathbf{K}) Y_{lm}^*(\theta,\phi) \quad \text{(II E.19)}$$

and denoted the difference between the wave numbers for the initial and final state by $\mathbf{K}$. The integration over r in (18) leads to

$$\langle \mathbf{k}_f | r^{-\lambda-1} Y_{\lambda\mu}(\theta,\phi) | \mathbf{k}_i \rangle = \frac{4\pi i^\lambda}{(2\lambda-1)!!} K^{\lambda-2} Y_{\lambda\mu}(\mathbf{K}), \quad \text{(II E.20)}$$

and the differential cross section function [see (II B.34)] is thus given by[79,83]

$$df_{E\lambda} = \frac{16\pi}{[(2\lambda+1)!!]^2} a^{2\lambda-2} k_i k_f |\mathbf{k}_i - \mathbf{k}_f|^{2\lambda-4} d\Omega. \quad \text{(II E.21)}$$

The total cross section function is easily obtained by integration over the angles. Since

$$|\mathbf{k}_i - \mathbf{k}_f|^2 = k_i^2 + k_f^2 - 2k_i k_f \cos\vartheta, \quad \text{(II E.22)}$$

one finds, for $\lambda\neq 1$,

$$f_{E\lambda} = \frac{16\pi^2}{(\lambda-1)[(2\lambda+1)!!]^2} \times \{(\eta_i+\eta_f)^{2(\lambda-1)} - (\eta_f-\eta_i)^{2(\lambda-1)}\}, \quad \text{(II E.23)}$$

and for $\lambda=1$

$$f_{E1} = \frac{32\pi^2}{9} \ln\frac{\eta_i+\eta_f}{\eta_f-\eta_i}. \quad \text{(II E.24)}$$

These expressions are expected to coincide with the exact quantum-mechanical expressions for $\eta_i<\eta_f\ll 1$. A comparison for the case $\lambda=2$ and $\xi=0$ is shown in Fig. II.9. It is seen that the Born approximation results deviate appreciably from the exact ones already for rather small values of η.

As regards the differential cross section given by (21), it is of interest that the angular distribution is isotropic in the especially important case of $\lambda=2$.

For large values of η, the Born approximation greatly overestimates the excitation cross section. This is associated with the neglect of the Coulomb repulsion which implies that the small distances give too large a contribution. Thus, the expressions (23) and (24) also do not show the adiabatic behavior for large values of ξ. One may improve the approximation by introducing a cutoff in the radial integral (18) for small distances.[83] If one choses this cutoff at the distance of closest approach, $2a$, one obtains values for the total cross section in rather good agreement with the exact theory for small values of ξ. However, the differential cross section remains essentially incorrect for $\eta>1$.

The angular distribution of the de-excitation γ quanta can also be easily evaluated in the Born approximation. According to (II A.68) and (II B.82), one obtains by means of (20)

$$b_k^{E\lambda} = \begin{pmatrix} \lambda & \lambda & k \\ 1 & -1 & 0 \end{pmatrix}^{-1} \sum_\mu \begin{pmatrix} \lambda & \lambda & k \\ \mu & -\mu & 0 \end{pmatrix} (-1)^\mu \times \int |Y_{\lambda\mu}(\mathbf{K})|^2 K^{2\lambda-4} d\Omega, \quad \text{(II E.25)}$$

where the integration is over all directions of $\mathbf{k}_f$. In the case $\xi=0$, the polar angle for $\mathbf{K}$ is equal to $\pi/2+\vartheta/2$, where ϑ is the deflection angle, and the integral is thus proportional to [see (22)]

$$\int_0^\pi \left| Y_{\lambda\mu}\left(\frac{\pi}{2}+\frac{\vartheta}{2}, 0\right)\right|^2 (1-\cos\vartheta)^{2\lambda-4} \sin\vartheta d\vartheta. \quad \text{(II E.26)}$$

The sum over μ can then be evaluated explicitly with

[83] R. Huby and H. C. Newns, reference 4.

the result[84]

$$a_k^{E\lambda}=\frac{2\lambda(\lambda+1)\Gamma(2\lambda-5/2)\Gamma(2\lambda-2)}{[2\lambda(\lambda+1)-k(k+1)]\Gamma(2\lambda-k/2-5/2)\Gamma(2\lambda+k/2-2)}. \quad \text{(II E.27)}$$

For $\lambda=1$ one obtains

$$a_2^{E1}=1, \quad \text{(II E.28)}$$

while for $\lambda=2$

$$a_2^{E2}=\tfrac{1}{2} \quad \text{and} \quad a_4^{E2}=\tfrac{1}{16}. \quad \text{(II E.29)}$$

It is also of interest to consider the radial matrix elements in the Born approximation. The radial wave functions are in this case spherical Bessel functions and the matrix element takes the form

$$M_{l_i l_f}{}^{-\lambda-1}=\int_0^\infty j_{l_f}(k_f r)r^{-\lambda+1}j_{l_i}(k_i r)dr, \quad \text{(II E.30)}$$

the evaluation of which leads to[85]

$$M_{l_i l_f}{}^{-\lambda-1}=\frac{\pi}{2^{\lambda+1}}k_i^{\lambda-2}\left(\frac{k_f}{k_i}\right)^{l_f}\frac{\Gamma\left(\dfrac{l_i+l_f-\lambda+2}{2}\right)}{\Gamma(l_f+\frac{3}{2})\Gamma\left(\dfrac{l_i-l_f+\lambda+1}{2}\right)}F\left(\frac{l_i+l_f-\lambda+2}{2},\frac{l_f-l_i-\lambda+1}{2},l_f+\tfrac{3}{2};\frac{k_f^2}{k_i^2}\right), \quad \text{(II E.31)}$$

where F is the ordinary hypergeometric function (II E.84) which, in this case, may be expressed by elementary functions. A special discussion of the radial matrix elements in the limit of large l is given in Sec. II E.7.

II E.3. Excitation by Means of Electrons

Although the present article is concerned with electromagnetic excitations produced by heavy projectiles, we shall in this paragraph briefly consider the nuclear excitations by fast electrons.[86] This process has been treated in the Born approximation[87] which is expected to be valid for light target nuclei. We shall give here an equivalent treatment which leads to cross sections in a form analogous to those derived for Coulomb excitation.

The general expressions (II B.8) and (II B.25) for the transition matrix element and the cross section are equally valid for electron excitations. In the Born approximation, the wave functions for the scattering states are plane waves

$$|\mathbf{k}\rangle=|u\rangle e^{i\mathbf{k}\cdot\mathbf{r}}, \quad \text{(II E.32)}$$

where $|u\rangle$ is the spinor and $\mathbf{k}$ the wave number of the electron. It is in this case convenient to perform the integration over the coordinate $\mathbf{r}$ already in the expression (II B.8) and afterwards to perform the integration over the wave number q of the photon. This leads to the following result for the transition matrix element

$$\begin{aligned}\langle f|\mathcal{H}^{(1)}|i\rangle=\frac{16\pi^2}{\lambda(2\lambda+1)!!}K^\lambda e i^{\lambda+1}\sum_{\lambda\mu}(-1)^\mu\Big\{&\langle I_fM_f|\mathfrak{M}(E\lambda,-\mu,K)|I_iM_i\rangle\frac{\langle u_f|\boldsymbol{\alpha}|u_i\rangle}{K^2-\kappa^2}\cdot\frac{\mathbf{K}}{K}\times\mathbf{L}_KY_{\lambda\mu}(\mathbf{K})\\ &-\langle I_fM_f|\mathfrak{M}(M\lambda,-\mu,K)|I_iM_i\rangle\frac{\langle u_f|\boldsymbol{\alpha}|u_i\rangle}{K^2-\kappa^2}\cdot\mathbf{L}_KY_{\lambda\mu}(\mathbf{K})\\ &+i\langle I_fM_f|\mathfrak{M}(C\lambda,-\mu,K|I_iM_i\rangle\lambda\frac{\langle u_f|u_i\rangle}{K^2}Y_{\lambda\mu}(\mathbf{K})\Big\}.\end{aligned} \quad \text{(II E.33)}$$

The wave numbers $\mathbf{K}$ and κ represent the momentum and energy transfer in the collision and are given by

$$\mathbf{K}=\mathbf{k}_i-\mathbf{k}_f, \quad \text{(II E.34)}$$

and

$$\kappa=\frac{\Delta E}{\hbar c}. \quad \text{(II E.35)}$$

The operator $\mathbf{L}_K$ is defined by (II A.35) and operates on $\mathbf{K}$. In the derivation of (33), we have used the identities

$$\int e^{i\mathbf{K}\cdot\mathbf{r}}\mathbf{L}(j_\lambda(qr)Y_{\lambda\mu}(\theta,\phi))d^3\tau =2\pi^2 i^\lambda q^{-2}\delta(q-K)\mathbf{L}_KY_{\lambda\mu}(\mathbf{K}), \quad \text{(II E.36)}$$

[84] A. Erdélyi *et al.*, *Higher Transcendental Functions* (McGraw-Hill Book Company, Inc., New York, 1953), Vol. I, p. 171.

[85] P. 401 of reference 40.

[86] For recent experimental results obtained in high energy electron scattering, see J. H. Fregeau and R. Hofstadter, Phys. Rev. **99**, 1503 (1955).

[87] L. I. Schiff, Phys. Rev. **96**, 765 (1954), which also contains references to earlier work.

and

$$\int e^{i\mathbf{K}\cdot\mathbf{r}}\nabla\times\mathbf{L}(j_\lambda(qr)Y_{\lambda\mu}(\theta,\phi))d^3\tau$$
$$=2\pi^2 i^{\lambda-1}q^{-2}\delta(q-K)\mathbf{K}\times\mathbf{L}_K Y_{\lambda\mu}(\mathbf{K}). \quad \text{(II E.37)}$$

The nuclear transition operators $\mathfrak{M}(E\lambda,\mu,K)$ and $\mathfrak{M}(M\lambda,\mu,K)$ are defined by (II B.16) and (II B.17), where κ is to be replaced by K. Whereas in the Coulomb excitation the wave number dependence of the multipole moments is usually unimportant, since ($\kappa R_0\ll 1$), the K dependence of the nuclear moments in (33) is essential. The last term in (33) arises from the multipole expansion of the instantaneous Coulomb interaction, and the transition operator involved is defined by

$$\mathfrak{M}(C\lambda,\mu,K)=(2\lambda+1)!!K^{-\lambda}$$
$$\times\int\rho_n j_\lambda(Kr)Y_{\lambda\mu}(\theta,\phi)d\tau. \quad \text{(II E.38)}$$

In the limit $K\to 0$, this operator approaches the moment (II A.11).

While the multipole moments $\mathfrak{M}(E\lambda)$ and $\mathfrak{M}(M\lambda)$ vanish for $\lambda=0$, the moment (38) also gives rise to electric monopole transitions.[41]

The differential cross section (II B.25) is now easily obtained by performing the summation over the electron spin indices and the nuclear magnetic quantum numbers. Using the identity

$$\sum_\mu(\mathbf{k}_i\cdot\mathbf{K}\times\mathbf{L}_K Y_{\lambda\mu}(\mathbf{K}))(\mathbf{k}_f\cdot\mathbf{K}\times\mathbf{L}_K Y_{\lambda\mu}(\mathbf{K}))^*$$
$$=\frac{\lambda(\lambda+1)(2\lambda+1)}{8\pi}[K^2\mathbf{k}_i\cdot\mathbf{k}_f-(\mathbf{k}_i\cdot\mathbf{K})(\mathbf{k}_f\cdot\mathbf{K})], \quad \text{(II E.39)}$$

and similar relations, one obtains

$$d\sigma=\sum_{\lambda=0}^{\infty}d\sigma_{E\lambda}+\sum_{\lambda=1}^{\infty}d\sigma_{M\lambda}, \quad \text{(II E.40)}$$

where

$$d\sigma_{E\lambda}=\left(\frac{e}{\hbar c}\right)^2\frac{4\pi(\lambda+1)}{\lambda[(2\lambda+1)!!]^2}\frac{K^{2\lambda}}{k_i^2}\left\{\frac{\lambda}{\lambda+1}B(C\lambda,K)V_L(\vartheta)+B(E\lambda,K)V_T(\vartheta)\right\}d\Omega, \quad \text{(II E.41)}$$

and

$$d\sigma_{M\lambda}=\left(\frac{e}{\hbar c}\right)^2\frac{4\pi(\lambda+1)}{\lambda[(2\lambda+1)!!]^2}\frac{K^{2\lambda}}{k_i^2}\times B(M\lambda,K)V_T(\vartheta)d\Omega. \quad \text{(II E.42)}$$

The reduced nuclear transition probabilities $B(\lambda,K)$ are given by (II B.23) in terms of the multipole matrix elements involved in (33). The dimensionless functions $V_L(\vartheta)$ and $V_T(\vartheta)$ are given by

$$V_L=k_ik_f\frac{2k_i^2+2k_f^2+4m^2c^2/\hbar^2-\kappa^2-K^2}{K^4}, \quad \text{(II E.43)}$$

$$V_T=k_ik_f\frac{(k_i^2+k_f^2-\kappa^2)K^2-2(\mathbf{k}_i\cdot\mathbf{K})(\mathbf{k}_f\cdot\mathbf{K})}{K^2(K^2-\kappa^2)^2}, \quad \text{(II E.44)}$$

and may also be expressed in terms of k_i, k_f, and ϑ as

$$V_L(\vartheta)=\frac{4\frac{m^2c^2}{\hbar^2k_ik_f}+\frac{k_i}{k_f}+\frac{k_f}{k_i}-\frac{\kappa^2}{k_ik_f}+2\cos\vartheta}{\left(\frac{k_i}{k_f}+\frac{k_f}{k_i}-2\cos\vartheta\right)^2}, \quad \text{(II E.45)}$$

$$V_T(\vartheta)=\frac{\left(\frac{k_i}{k_f}\right)^2+\left(\frac{k_f}{k_i}\right)^2+4-\frac{\kappa^2}{k_ik_f}\left(\frac{k_i}{k_f}+\frac{k_f}{k_i}\right)-2\left(2\frac{k_i}{k_f}+2\frac{k_f}{k_i}-\frac{\kappa^2}{k_ik_f}\right)\cos\vartheta+2\cos^2\vartheta}{\left(\frac{k_i}{k_f}+\frac{k_f}{k_i}-2\cos\vartheta\right)\left[\frac{k_i}{k_f}+\frac{k_f}{k_i}-\frac{\kappa^2}{k_ik_f}-2\cos\vartheta\right]^2}. \quad \text{(II E.46)}$$

It is noted that the angular dependence of the cross sections is contained not only in the functions V_L and V_T and in the factor $K^{2\lambda}$, but also in the nuclear transition probabilities.

II E.4. Classical Orbital Integrals

The orbital integrals $I_{\lambda\mu}(\vartheta,\xi)$ [defined by (II A.26)] are the basic functions in the classical theory of Coulomb excitation. In addition they provide an approximation, valid for η or l large compared with unity, for the radial matrix elements in the quantum-mechanical treatment [see (II B.100) and Table II.1].

In this paragraph, we shall discuss some properties of these functions, which are given by the integral

$$I_{\lambda\mu}(\vartheta,\xi)=\int_{-\infty}^{\infty}e^{i\xi(\epsilon\sinh w+w)}\times\frac{[\cosh w+\epsilon+i(\epsilon^2-1)^{\frac{1}{2}}\sinh w]^\mu}{(\epsilon\cosh w+1)^{\lambda+\mu}}dw, \quad \text{(II E.47)}$$

with

$$\epsilon = \frac{1}{\sin(\vartheta/2)}. \quad \text{(II E.48)}$$

Values of $I_{\lambda\mu}$ may be obtained directly by a numerical integration of (47). However, due to the oscillations of the integrand, which are especially pronounced for large ξ, it is convenient to translate the path of integration by an amount of $i(\pi/2)$, whereby one obtains

$$I_{\lambda\mu}(\vartheta,\xi) = e^{-(\pi/2)\xi} \int_{-\infty}^{\infty} e^{-\xi\epsilon \cosh w} e^{i\xi w} \frac{[i \sinh w + \epsilon - (\epsilon^2-1)^{\frac{1}{2}} \cosh w]^{\mu}}{(i\epsilon \sinh w + 1)^{\lambda+\mu}} dw. \quad \text{(II E.49)}$$

Numerical values of $I_{\lambda\mu}(\vartheta,\xi)$ calculated in this way are given in Table II.12 for $\lambda=2$ and for $\lambda=1$, 3, and 4 in reference 88.

A series representation of the classical integrals can be obtained by performing the limiting process $\eta \to \infty$ in the radial matrix elements. For the integrals with $\mu=-\lambda$ one gets directly from (II B.58) by a simple confluence [see (II E.105)]

$$I_{\lambda,-\lambda}(\vartheta,\xi) = 2^{\lambda} \sin^{\lambda}\frac{\vartheta}{2} \exp\left[-\xi\left(\frac{\vartheta}{2} - \frac{\pi}{2} + \cot\frac{\vartheta}{2}\right)\right] \left\{ \frac{|\Gamma(\lambda+i\xi)|^2}{(2\lambda-1)!} \Psi_2(-2\lambda+1, -\lambda+1-i\xi, -\lambda+1+i\xi; z, z^*) \right.$$

$$\left. + 2\, \mathcal{R}e[e^{-\pi\xi}\Gamma(-\lambda-i\xi) z^{\lambda+i\xi} \Psi_2(-\lambda+1+i\xi, \lambda+1+i\xi, -\lambda+1+i\xi; z, z^*)] \right\}, \quad \text{(II E.50)}$$

with

$$z = \frac{\xi}{2}\left(\cot\frac{\vartheta}{2} - i\right) = e^{-i(\vartheta/2)} \frac{\xi}{2 \sin\frac{\vartheta}{2}}. \quad \text{(II E.51)}$$

We have here used the relation (II B.100) between the radial matrix elements and the classical integrals, and the limiting formula (II B.102). The confluent Appell function Ψ_2 is defined in (II E.105). The integral with $\mu=\lambda$ is given by (II B.59) which leads to the relation

$$I_{\lambda,\lambda}(\vartheta,\xi) = (-1)^{\lambda} e^{-\pi\xi} I_{\lambda,-\lambda}(-\vartheta, \xi). \quad \text{(II E.52)}$$

The integrals with $|\mu| \neq \lambda$ can be obtained from the expression (II B.62) by inserting the expansion of the function F_3 in terms of F_2 functions (II E.104) and (II E.97) and then performing the confluence. In this way, one obtains, however, a nonterminating series of Ψ_2 functions (see reference 45). A more convenient form is obtained by means of the recursion formulas for the orbital integrals. These may be derived from the recursion relations for the radial matrix elements (see Sec. II B.4) by performing the limit $\eta \to \infty$. From (II B.72) one thus obtains the following relation

$$\frac{2\xi}{3} I_{20}(\vartheta,\xi) = 2 \tan^2\frac{\vartheta}{2}\left(\frac{\partial I_{22}}{\partial\vartheta} - \frac{\partial I_{2,-2}}{\partial\vartheta}\right) + \xi(I_{22} + I_{2,-2})$$

$$+ \left(\tan^3\frac{\vartheta}{2} - \tan\frac{\vartheta}{2}\right)(I_{22} - I_{2,-2}). \quad \text{(II E.53)}$$

As a function of the parameter ξ the classical integrals possess the symmetry property

$$I_{\lambda\mu}(\vartheta, -\xi) = I_{\lambda,-\mu}(\vartheta,\xi), \quad \text{(II E.54)}$$

which follows directly from (47).

For $\xi=0$, the integrals may be evaluated in terms of elementary functions (see Sec. II E.6).

In the limit of $\xi \gg 1$, the integrals decrease exponentially, reflecting the adiabatic character of the excitation process, and the resultant f functions contain the factor $e^{-2\pi\xi}$.[89] More detailed expressions appropriate to this limit have been obtained by the method of steepest descent.[5]

As a function of ϑ or ϵ the classical integrals have symmetry properties of the type (52). For $\vartheta=\pi$ (or $\epsilon=1$), the orbital integrals are independent of μ. For $\vartheta \ll 1$ ($\epsilon \gg 1$), the $I_{\lambda\mu}$ are simply related to the integrals for straight line orbits, as is discussed in Sec. II E.7.

II E.5. Electric Dipole Excitations

For $\lambda=1$, the classical integrals can be expressed in terms of Hankel functions.[90,5] By a partial integration, one may write (49) in the form

$$I_{1,\pm1}(\vartheta,\xi) = -\frac{\xi e^{-(\pi/2)\xi}}{(\epsilon^2-1)^{\frac{1}{2}}} \int_{-\infty}^{\infty} e^{-\xi\epsilon \cosh w + i\xi w}$$

$$\times (i \sinh w + \epsilon \mp (\epsilon^2-1)^{\frac{1}{2}} \cosh w) dw, \quad \text{(II E.55)}$$

which, by means of the integral representation

$$K_{\nu}(z) = \int_0^{\infty} e^{-z \cosh t + \nu t} dt \quad \text{(II E.56)}$$

for the Hankel function of imaginary argument,[91] can

[88] K. Alder and A. Winther, Kgl. Danske Videnskab. Selskab Mat. fys. Medd. 31, No. 1 (1956).

[89] L. Landau, reference 1.

[90] G. Wentzel, Z. Physik 27, 257 (1924).

[91] See Vol. II, p. 82 of reference 84.

TABLE II.12. The classical orbital integrals for $E2$ Coulomb excitation. The table lists the values of the classical orbital integrals $I_{2\mu}(\vartheta,\xi)$. The first column gives the deflection angle ϑ in degrees. The second column gives the values of λ and μ ($\lambda=2$ for all the integrals listed). The subsequent columns give the values of $I_{2\mu}$ for the ξ value indicated; these entries are given in the form of a number followed by the power of ten by which it should be multiplied. The data are taken from reference 88.

ϑ	$\lambda\cdot\mu$	$\xi=0.0$	$\xi=0.1$	$\xi=0.2$	$\xi=0.3$	$\xi=0.4$	$\xi=0.5$
10°	2.2	5.064 (−3)	8.675 (−4)	1.895 (−4)	4.425 (−5)	1.069 (−5)	2.637 (−6)
	2.0	1.332 (−2)	6.505 (−3)	2.280 (−3)	7.311 (−4)	2.245 (−4)	6.716 (−5)
	2.−2	5.064 (−3)	1.195 (−2)	7.765 (−3)	3.637 (−3)	1.468 (−3)	5.442 (−4)
20°	2.2	2.010 (−2)	7.251 (−3)	2.957 (−3)	1.257 (−3)	5.467 (−4)	2.413 (−4)
	2.0	4.687 (−2)	3.417 (−2)	2.040 (−2)	1.137 (−2)	6.111 (−3)	3.211 (−3)
	2.−2	2.010 (−2)	4.063 (−2)	4.092 (−2)	3.206 (−2)	2.217 (−2)	1.423 (−2)
30°	2.2	4.466 (−2)	2.117 (−2)	1.080 (−2)	5.658 (−3)	3.013 (−3)	1.621 (−3)
	2.0	9.323 (−2)	7.628 (−2)	5.381 (−2)	3.582 (−2)	2.309 (−2)	1.457 (−2)
	2.−2	4.466 (−2)	7.763 (−2)	8.538 (−2)	7.699 (−2)	6.242 (−2)	4.737 (−2)
40°	2.2	7.799 (−2)	4.290 (−2)	2.466 (−2)	1.443 (−2)	8.538 (−3)	5.089 (−3)
	2.0	1.471 (−1)	1.268 (−1)	9.683 (−2)	7.022 (−2)	4.942 (−2)	3.410 (−2)
	2.−2	7.799 (−2)	1.217 (−1)	1.359 (−1)	1.291 (−1)	1.119 (−1)	9.142 (−2)
50°	2.2	1.191 (−1)	7.214 (−2)	4.482 (−2)	2.815 (−2)	1.779 (−2)	1.130 (−2)
	2.0	2.048 (−1)	1.817 (−1)	1.451 (−1)	1.106 (−1)	8.192 (−2)	5.955 (−2)
	2.−2	1.191 (−1)	1.716 (−1)	1.904 (−1)	1.843 (−1)	1.647 (−1)	1.397 (−1)
60°	2.2	1.667 (−1)	1.083 (−1)	7.117 (−2)	4.700 (−2)	3.113 (−2)	2.067 (−2)
	2.0	2.636 (−1)	2.380 (−1)	1.957 (−1)	1.539 (−1)	1.178 (−1)	8.858 (−2)
	2.−2	1.667 (−1)	2.258 (−1)	2.468 (−1)	2.401 (−1)	2.176 (−1)	1.881 (−1)
70°	2.2	2.193 (−1)	1.505 (−1)	1.033 (−1)	7.093 (−2)	4.870 (−2)	3.344 (−2)
	2.0	3.215 (−1)	2.938 (−1)	2.463 (−1)	1.980 (−1)	1.551 (−1)	1.193 (−1)
	2.−2	2.193 (−1)	2.826 (−1)	3.035 (−1)	2.944 (−1)	2.682 (−1)	2.340 (−1)
80°	2.2	2.755 (−1)	1.977 (−1)	1.406 (−1)	9.966 (−2)	7.043 (−2)	4.969 (−2)
	2.0	3.770 (−1)	3.475 (−1)	2.955 (−1)	2.413 (−1)	1.920 (−1)	1.502 (−1)
	2.−2	2.755 (−1)	3.402 (−1)	3.586 (−1)	3.456 (−1)	3.147 (−1)	2.755 (−1)
90°	2.2	3.333 (−1)	2.485 (−1)	1.823 (−1)	1.327 (−1)	9.609 (−2)	6.934 (−2)
	2.0	4.292 (−1)	3.980 (−1)	3.420 (−1)	2.825 (−1)	2.275 (−1)	1.801 (−1)
	2.−2	3.333 (−1)	3.968 (−1)	4.106 (−1)	3.922 (−1)	3.558 (−1)	3.114 (−1)
100°	2.2	3.912 (−1)	3.016 (−1)	2.274 (−1)	1.694 (−1)	1.253 (−1)	9.219 (−2)
	2.0	4.772 (−1)	4.446 (−1)	3.850 (−1)	3.208 (−1)	2.608 (−1)	2.083 (−1)
	2.−2	3.912 (−1)	4.507 (−1)	4.580 (−1)	4.329 (−1)	3.904 (−1)	3.406 (−1)
110°	2.2	4.473 (−1)	3.555 (−1)	2.746 (−1)	2.089 (−1)	1.575 (−1)	1.179 (−1)
	2.0	5.205 (−1)	4.866 (−1)	4.239 (−1)	3.556 (−1)	2.911 (−1)	2.342 (−1)
	2.−2	4.473 (−1)	5.002 (−1)	4.992 (−1)	4.665 (−1)	4.176 (−1)	3.627 (−1)
120°	2.2	5.000 (−1)	4.086 (−1)	3.227 (−1)	2.504 (−1)	1.920 (−1)	1.460 (−1)
	2.0	5.586 (−1)	5.236 (−1)	4.583 (−1)	3.864 (−1)	3.180 (−1)	2.572 (−1)
	2.−2	5.000 (−1)	5.439 (−1)	5.333 (−1)	4.925 (−1)	4.372 (−1)	3.773 (−1)
130°	2.2	5.476 (−1)	4.594 (−1)	3.705 (−1)	2.926 (−1)	2.279 (−1)	1.758 (−1)
	2.0	5.913 (−1)	5.554 (−1)	4.878 (−1)	4.130 (−1)	3.412 (−1)	2.771 (−1)
	2.−2	5.476 (−1)	5.805 (−1)	5.593 (−1)	5.101 (−1)	4.487 (−1)	3.845 (−1)
140°	2.2	5.887 (−1)	5.063 (−1)	4.165 (−1)	3.345 (−1)	2.645 (−1)	2.067 (−1)
	2.0	6.182 (−1)	5.817 (−1)	5.122 (−1)	4.349 (−1)	3.605 (−1)	2.937 (−1)
	2.−2	5.887 (−1)	6.088 (−1)	5.766 (−1)	5.192 (−1)	4.522 (−1)	3.845 (−1)
150°	2.2	6.220 (−1)	5.479 (−1)	4.593 (−1)	3.748 (−1)	3.005 (−1)	2.379 (−1)
	2.0	6.393 (−1)	6.022 (−1)	5.314 (−1)	4.522 (−1)	3.756 (−1)	3.067 (−1)
	2.−2	6.220 (−1)	6.282 (−1)	5.849 (−1)	5.199 (−1)	4.480 (−1)	3.777 (−1)
160°	2.2	6.466 (−1)	5.828 (−1)	4.976 (−1)	4.124 (−1)	3.351 (−1)	2.684 (−1)
	2.0	6.545 (−1)	6.170 (−1)	5.451 (−1)	4.646 (−1)	3.865 (−1)	3.161 (−1)
	2.−2	6.466 (−1)	6.381 (−1)	5.840 (−1)	5.122 (−1)	4.367 (−1)	3.647 (−1)
170°	2.2	6.616 (−1)	6.101 (−1)	5.303 (−1)	4.460 (−1)	3.670 (−1)	2.973 (−1)
	2.0	6.636 (−1)	6.258 (−1)	5.534 (−1)	4.720 (−1)	3.931 (−1)	3.218 (−1)
	2.−2	6.616 (−1)	6.383 (−1)	5.743 (−1)	4.969 (−1)	4.188 (−1)	3.465 (−1)
180°	2.2	6.667 (−1)	6.288 (−1)	5.561 (−1)	4.745 (−1)	3.953 (−1)	3.237 (−1)
	2.0	6.667 (−1)	6.288 (−1)	5.561 (−1)	4.745 (−1)	3.953 (−1)	3.237 (−1)
	2.−2	6.667 (−1)	6.288 (−1)	5.561 (−1)	4.745 (−1)	3.953 (−1)	3.237 (−1)

TABLE II.12.—*Continued.*

ϑ	$\lambda\cdot\mu$	$\xi=0.6$	$\xi=0.7$	$\xi=0.8$	$\xi=0.9$	$\xi=1.0$	$\xi=1.2$
10°	2.2	6.598 (−7)	1.668 (−7)	4.250 (−8)	1.090 (−8)	2.807 (−9)	1.886 (−10)
	2.0	1.975 (−5)	5.739 (−6)	1.652 (−6)	4.724 (−7)	1.343 (−7)	1.071 (−8)
	2.−2	1.908 (−4)	6.438 (−5)	2.110 (−5)	6.765 (−6)	2.131 (−6)	2.031 (−7)
20°	2.2	1.077 (−4)	4.841 (−5)	2.190 (−5)	9.959 (−6)	4.546 (−6)	9.567 (−7)
	2.0	1.661 (−3)	8.496 (−4)	4.309 (−4)	2.171 (−4)	1.088 (−4)	2.696 (−5)
	2.−2	8.685 (−3)	5.117 (−3)	2.936 (−3)	1.650 (−3)	9.119 (−4)	2.681 (−4)
30°	2.2	8.795 (−4)	4.799 (−4)	2.631 (−4)	1.448 (−4)	7.997 (−5)	2.457 (−5)
	2.0	9.064 (−3)	5.578 (−3)	3.404 (−3)	2.064 (−3)	1.245 (−3)	4.473 (−4)
	2.−2	3.438 (−2)	2.415 (−2)	1.655 (−2)	1.113 (−2)	7.368 (−3)	3.114 (−3)
40°	2.2	3.050 (−3)	1.836 (−3)	1.109 (−3)	6.717 (−4)	4.079 (−4)	1.514 (−4)
	2.0	2.321 (−2)	1.563 (−2)	1.044 (−2)	6.932 (−3)	4.577 (−3)	1.971 (−3)
	2.−2	7.174 (−2)	5.464 (−2)	4.068 (−2)	2.975 (−2)	2.144 (−2)	1.076 (−2)
50°	2.2	7.203 (−3)	4.604 (−3)	2.953 (−3)	1.894 (−3)	1.219 (−3)	5.063 (−4)
	2.0	4.270 (−2)	3.031 (−2)	2.134 (−2)	1.493 (−2)	1.039 (−2)	4.972 (−3)
	2.−2	1.142 (−1)	9.080 (−2)	7.069 (−2)	5.412 (−2)	4.088 (−2)	2.257 (−2)
60°	2.2	1.375 (−2)	9.164 (−3)	6.114 (−3)	4.084 (−3)	2.731 (−3)	1.224 (−3)
	2.0	6.570 (−2)	4.824 (−2)	3.514 (−2)	2.543 (−2)	1.831 (−2)	9.368 (−3)
	2.−2	1.572 (−1)	1.281 (−1)	1.024 (−1)	8.054 (−2)	6.254 (−2)	3.658 (−2)
70°	2.2	2.298 (−2)	1.579 (−2)	1.085 (−2)	7.464 (−3)	5.134 (−3)	2.432 (−3)
	2.0	9.053 (−2)	6.803 (−2)	5.072 (−2)	3.757 (−2)	2.768 (−2)	1.483 (−2)
	2.−2	1.980 (−1)	1.636 (−1)	1.328 (−1)	1.061 (−1)	8.384 (−2)	5.083 (−2)
80°	2.2	3.502 (−2)	2.466 (−2)	1.735 (−2)	1.221 (−2)	8.583 (−3)	4.242 (−3)
	2.0	1.159 (−1)	8.853 (−2)	6.710 (−2)	5.053 (−2)	3.785 (−2)	2.096 (−2)
	2.−2	2.345 (−1)	1.952 (−1)	1.598 (−1)	1.290 (−1)	1.030 (−1)	6.383 (−2)
90°	2.2	4.991 (−2)	3.586 (−2)	2.572 (−2)	1.843 (−2)	1.319 (−2)	6.748 (−3)
	2.0	1.407 (−1)	1.088 (−1)	8.350 (−2)	6.364 (−2)	4.825 (−2)	2.737 (−2)
	2.−2	2.654 (−1)	2.217 (−1)	1.823 (−1)	1.480 (−1)	1.188 (−1)	7.464 (−2)
100°	2.2	6.757 (−2)	4.938 (−2)	3.601 (−2)	2.621 (−2)	1.905 (−2)	1.002 (−2)
	2.0	1.643 (−1)	1.282 (−1)	9.928 (−2)	7.637 (−2)	5.842 (−2)	3.373 (−2)
	2.−2	2.901 (−1)	2.424 (−1)	1.996 (−1)	1.623 (−1)	1.307 (−1)	8.270 (−2)
110°	2.2	8.781 (−2)	6.515 (−2)	4.818 (−2)	3.555 (−2)	2.617 (−2)	1.411 (−2)
	2.0	1.859 (−1)	1.462 (−1)	1.140 (−1)	8.827 (−2)	6.799 (−2)	3.979 (−2)
	2.−2	3.079 (−1)	2.569 (−1)	2.114 (−1)	1.719 (−1)	1.385 (−1)	8.786 (−2)
120°	2.2	1.103 (−1)	8.294 (−2)	6.213 (−2)	4.639 (−2)	3.454 (−2)	1.903 (−2)
	2.0	2.054 (−1)	1.623 (−1)	1.272 (−1)	9.905 (−2)	7.669 (−2)	4.534 (−2)
	2.−2	3.189 (−1)	2.652 (−1)	2.176 (−1)	1.767 (−1)	1.422 (−1)	9.014 (−2)
130°	2.2	1.346 (−1)	1.025 (−1)	7.765 (−2)	5.861 (−2)	4.410 (−2)	2.477 (−2)
	2.0	2.222 (−1)	1.763 (−1)	1.388 (−1)	1.085 (−1)	8.431 (−2)	5.024 (−2)
	2.−2	3.232 (−1)	2.675 (−1)	2.188 (−1)	1.771 (−1)	1.422 (−1)	8.980 (−2)
140°	2.2	1.602 (−1)	1.234 (−1)	9.447 (−2)	7.202 (−2)	5.470 (−2)	3.126 (−2)
	2.0	2.362 (−1)	1.880 (−1)	1.484 (−1)	1.164 (−1)	9.071 (−2)	5.437 (−2)
	2.−2	3.210 (−1)	2.643 (−1)	2.151 (−1)	1.735 (−1)	1.388 (−1)	8.717 (−2)
150°	2.2	1.865 (−1)	1.451 (−1)	1.122 (−1)	8.632 (−2)	6.613 (−2)	3.840 (−2)
	2.0	2.472 (−1)	1.972 (−1)	1.560 (−1)	1.226 (−1)	9.578 (−2)	5.765 (−2)
	2.−2	3.131 (−1)	2.561 (−1)	2.074 (−1)	1.664 (−1)	1.326 (−1)	8.264 (−2)
160°	2.2	2.127 (−1)	1.671 (−1)	1.304 (−1)	1.012 (−1)	7.813 (−2)	4.605 (−2)
	2.0	2.552 (−1)	2.039 (−1)	1.615 (−1)	1.271 (−1)	9.945 (−2)	6.002 (−2)
	2.−2	3.000 (−1)	2.437 (−1)	1.961 (−1)	1.565 (−1)	1.240 (−1)	7.665 (−2)
170°	2.2	2.380 (−1)	1.887 (−1)	1.485 (−1)	1.162 (−1)	9.036 (−2)	5.398 (−2)
	2.0	2.600 (−1)	2.079 (−1)	1.649 (−1)	1.298 (−1)	1.017 (−1)	6.146 (−2)
	2.−2	2.825 (−1)	2.278 (−1)	1.821 (−1)	1.444 (−1)	1.138 (−1)	6.962 (−2)
180°	2.2	2.616 (−1)	2.092 (−1)	1.660 (−1)	1.308 (−1)	1.024 (−1)	6.194 (−2)
	2.0	2.616 (−1)	2.092 (−1)	1.660 (−1)	1.308 (−1)	1.024 (−1)	6.194 (−2)
	2.−2	2.616 (−1)	2.092 (−1)	1.660 (−1)	1.308 (−1)	1.024 (−1)	6.194 (−2)

TABLE II.12.—*Continued.*

ϑ	$\lambda \cdot \mu$	$\xi=1.4$	$\xi=1.6$	$\xi=1.8$	$\xi=2.0$	$\xi=4.0$
10°	2.2	1.282 (−11)	8.788 (−13)	6.068 (−14)	4.213 (−15)	1.294 (−26)
	2.0	8.426 (−10)	6.566 (−11)	5.078 (−12)	3.904 (−13)	2.362 (−24)
	2.−2	1.858 (−8)	1.651 (−9)	1.433 (−10)	1.222 (−11)	1.464 (−22)
20°	2.2	2.034 (−7)	4.355 (−8)	9.385 (−9)	2.032 (−9)	5.342 (−16)
	2.0	6.595 (−6)	1.598 (−6)	3.844 (−7)	9.193 (−8)	4.709 (−14)
	2.−2	7.587 (−5)	2.086 (−5)	5.612 (−6)	1.484 (−6)	1.479 (−12)
30°	2.2	7.613 (−6)	2.374 (−6)	7.440 (−7)	2.342 (−7)	2.546 (−12)
	2.0	1.587 (−4)	5.574 (−5)	1.944 (−5)	6.737 (−6)	1.411 (−10)
	2.−2	1.269 (−3)	5.031 (−4)	1.953 (−4)	7.455 (−5)	2.972 (−9)
40°	2.2	5.652 (−5)	2.122 (−5)	7.996 (−6)	3.024 (−6)	2.025 (−10)
	2.0	8.377 (−4)	3.527 (−4)	1.473 (−4)	6.119 (−5)	7.780 (−9)
	2.−2	5.215 (−3)	2.463 (−3)	1.140 (−3)	5.189 (−4)	1.225 (−7)
50°	2.2	2.113 (−4)	8.856 (−5)	3.723 (−5)	1.569 (−5)	3.034 (−9)
	2.0	2.348 (−3)	1.098 (−3)	5.093 (−4)	2.349 (−4)	8.468 (−8)
	2.−2	1.206 (−2)	6.284 (−3)	3.212 (−3)	1.616 (−3)	1.052 (−6)
60°	2.2	5.505 (−4)	2.481 (−4)	1.121 (−4)	5.071 (−5)	1.947 (−8)
	2.0	4.732 (−3)	2.366 (−3)	1.174 (−3)	5.787 (−4)	4.049 (−7)
	2.−2	2.073 (−2)	1.148 (−2)	6.235 (−3)	3.337 (−3)	4.089 (−6)
70°	2.2	1.154 (−3)	5.478 (−4)	2.604 (−4)	1.239 (−4)	7.655 (−8)
	2.0	7.842 (−3)	4.105 (−3)	2.131 (−3)	1.099 (−3)	1.203 (−6)
	2.−2	2.991 (−2)	1.720 (−2)	9.721 (−3)	5.415 (−3)	1.004 (−5)
80°	2.2	2.096 (−3)	1.035 (−3)	5.112 (−4)	2.525 (−4)	2.209 (−7)
	2.0	1.145 (−2)	6.189 (−3)	3.319 (−3)	1.768 (−3)	2.647 (−6)
	2.−2	3.847 (−2)	2.269 (−2)	1.315 (−2)	7.521 (−3)	1.842 (−5)
90°	2.2	3.443 (−3)	1.755 (−3)	8.932 (−4)	4.543 (−4)	5.177 (−7)
	2.0	1.531 (−2)	8.473 (−3)	4.651 (−3)	2.535 (−3)	4.757 (−6)
	2.−2	4.565 (−2)	2.735 (−2)	1.612 (−2)	9.379 (−3)	2.772 (−5)
100°	2.2	5.254 (−3)	2.746 (−3)	1.433 (−3)	7.462 (−4)	1.048 (−6)
	2.0	1.920 (−2)	1.082 (−2)	6.041 (−3)	3.350 (−3)	7.413 (−6)
	2.−2	5.102 (−2)	3.086 (−2)	1.838 (−2)	1.081 (−2)	3.617 (−5)
110°	2.2	7.569 (−3)	4.044 (−3)	2.153 (−3)	1.144 (−3)	1.904 (−6)
	2.0	2.296 (−2)	1.310 (−2)	7.412 (−3)	4.163 (−3)	1.040 (−5)
	2.−2	5.440 (−2)	3.307 (−2)	1.981 (−2)	1.172 (−2)	4.243 (−5)
120°	2.2	1.042 (−2)	5.670 (−3)	3.074 (−3)	1.661 (−3)	3.188 (−6)
	2.0	2.643 (−2)	1.523 (−2)	8.700 (−3)	4.933 (−3)	1.349 (−5)
	2.−2	5.585 (−2)	3.400 (−2)	2.041 (−2)	1.211 (−2)	4.582 (−5)
130°	2.2	1.380 (−2)	7.637 (−3)	4.206 (−3)	2.306 (−3)	5.000 (−6)
	2.0	2.950 (−2)	1.713 (−2)	9.856 (−3)	5.628 (−3)	1.645 (−5)
	2.−2	5.551 (−2)	3.375 (−2)	2.025 (−2)	1.202 (−2)	4.629 (−5)
140°	2.2	1.770 (−2)	9.942 (−3)	5.552 (−3)	3.085 (−3)	7.437 (−6)
	2.0	3.211 (−2)	1.874 (−2)	1.084 (−2)	6.224 (−3)	1.909 (−5)
	2.−2	5.365 (−2)	3.251 (−2)	1.945 (−2)	1.152 (−2)	4.426 (−5)
150°	2.2	2.206 (−2)	1.256 (−2)	7.104 (−3)	3.994 (−3)	1.057 (−5)
	2.0	3.418 (−2)	2.003 (−2)	1.163 (−2)	6.703 (−3)	2.126 (−5)
	2.−2	5.056 (−2)	3.048 (−2)	1.816 (−2)	1.071 (−2)	4.036 (−5)
160°	2.2	2.680 (−2)	1.545 (−2)	8.838 (−3)	5.023 (−3)	1.444 (−5)
	2.0	3.569 (−2)	2.097 (−2)	1.221 (−2)	7.053 (−3)	2.288 (−5)
	2.−2	4.654 (−2)	2.787 (−2)	1.651 (−2)	9.689 (−3)	3.532 (−5)
170°	2.2	3.181 (−2)	1.854 (−2)	1.072 (−2)	6.149 (−3)	1.903 (−5)
	2.0	3.660 (−2)	2.154 (−2)	1.256 (−2)	7.265 (−3)	2.388 (−5)
	2.−2	4.190 (−2)	2.489 (−2)	1.464 (−2)	8.538 (−3)	2.976 (−5)
180°	2.2	3.691 (−2)	2.173 (−2)	1.268 (−2)	7.337 (−3)	2.421 (−5)
	2.0	3.691 (−2)	2.173 (−2)	1.268 (−2)	7.337 (−3)	2.421 (−5)
	2.−2	3.691 (−2)	2.173 (−2)	1.268 (−2)	7.337 (−3)	2.421 (−5)

be expressed by

$$I_{1,\pm1}(\vartheta,\xi)=-2\xi e^{-(\pi/2)\xi}\times\left[K_{i\xi}'(\xi\epsilon)\pm\frac{(\epsilon^2-1)^{\frac{1}{2}}}{\epsilon}K_{i\xi}(\xi\epsilon)\right]. \quad \text{(II E.57)}$$

The K' represents the derivative of the function (56) with respect to the argument.

The integral over the square of $I_{1,\pm1}$, which is needed for the total cross-section function [see (II A.31)], can also be expressed in terms of Hankel functions by means of the Lommel integral formulas.[92] This leads to

$$f_{E1}(\xi)=-\frac{32\pi^2}{9}\cdot e^{-\pi\xi}\xi K_{i\xi}(\xi)K_{i\xi}'(\xi). \quad \text{(II E.58)}$$

In the limit of $\xi\rightarrow0$, this expression diverges with the following asymptotic behavior

$$f_{E1}(\xi)=\frac{32\pi^2}{9}\ln\frac{2}{\gamma\xi}(1-\pi\xi+\cdots). \quad \text{(II E.59)}$$

The number γ is given by

$$\gamma=e^C=1.781\cdots, \quad \text{(II E.60)}$$

where C is the Euler constant.

In the limit $\xi\gg1$, one obtains from (58) the following asymptotic formula

$$f_{E1}(\xi)=\frac{32\pi^3}{9\sqrt{3}}e^{-2\pi\xi}(1+0.218\xi^{-\frac{2}{3}}+\cdots). \quad \text{(II E.61)}$$

Also the quantum-mechanical formulas for the electric dipole excitation cross sections can be expressed in an especially simple form. The matrix element between the scattering states in (II B.34) is equivalent to that involved in the bremsstrahlung cross section (see Sec. II E.1) and may be evaluated by expressing the Coulomb wave functions in parabolic coordinates.[78]

The resulting expression for the differential f function may be written[79,83,5]

$$df_{E1}(\vartheta,\eta_i,\xi)=\frac{32\pi^3\eta_i\eta_f}{9\xi^2}\frac{e^{2\pi\eta_i}}{(e^{2\pi\eta_i}-1)(e^{2\pi\eta_f}-1)}\times\frac{d}{dx}\left\{-x\frac{d}{dx}|F(-i\eta_i,-i\eta_f,1;x)|^2\right\}d\Omega, \quad \text{(II E.62)}$$

where F is the hypergeometric function (II E.84) of the variable

$$x=-\frac{4\eta_i\eta_f}{\xi^2}\sin^2\frac{\vartheta}{2}. \quad \text{(II E.63)}$$

[92] See p. 133 ff of reference 40.

For the total f function (II B.39) one obtains[79,83,5]

$$f_{E1}(\eta_i,\xi)=-\frac{32\pi^4}{9}\frac{e^{2\pi\eta_i}}{(e^{2\pi\eta_i}-1)(e^{2\pi\eta_f}-1)}(-x_0)\times\frac{d}{dx_0}|F(-i\eta_i,-i\eta_f,1;x_0)|^2, \quad \text{(II E.64)}$$

with

$$x_0=-\frac{4\eta_i\eta_f}{\xi^2}. \quad \text{(II E.65)}$$

Since the variable x_0 is always larger than unity, it is necessary for the numerical evaluation of (64) to use the analytic continuation (II E.87) of the hypergeometric function. After differentiation one obtains in this manner

$$f_{E1}(\eta_i,\xi)=-\frac{32\pi^3}{9}\frac{\eta_i\eta_f}{\xi}\frac{1}{e^{2\pi\xi}-1}\times\mathcal{I}m\left\{\frac{1}{\eta_i}F\left(i\eta_i,i\eta_i,1-i\xi;\frac{1}{x_0}\right)\times\left[F\left(1-i\eta_i,-i\eta_i,1+i\xi;\frac{1}{x_0}\right)+e^{i\varphi}F\left(1-i\eta_f,-i\eta_f,1-i\xi;\frac{1}{x_0}\right)\right]+\eta_i\rightleftarrows\eta_f\right\}, \quad \text{(II E.66)}$$

where $\eta_i\rightleftarrows\eta_f$ implies the addition of terms with η_i and η_f interchanged, and where

$$\varphi=2\arg\{\Gamma(i\xi)\Gamma(i\eta_i)/\Gamma(i\eta_f)\}+\xi\ln|x_0|. \quad \text{(II E.66a)}$$

In the limit of $\xi\ll1$, the expression (66) reduces to

$$f_{E1}(\eta_i,\xi)=\frac{32\pi^2}{9}\left(\ln\frac{2\eta}{\xi}+\psi(1)-\mathcal{R}e\{\psi(i\eta)\}\right), \quad \text{(II E.66b)}$$

where ψ is the logarithmic derivative of the Γ function.

The classical limit ($\eta_i\rightarrow\infty$) of $f_{E1}(\eta_i,\xi)$ may be obtained by performing a confluence in the hypergeometric functions in (64). This leads to the expression (58).

The $E1$ Coulomb excitation process is closely related to the problem of the excitation and ionization of atoms by fast charged particles. The atomic stopping power may thus be written in the form

$$\frac{1}{N}\frac{dE}{dx}=\sum_f\sigma_{E1}(i\rightarrow f)(E_f-E_i), \quad \text{(II E.67)}$$

where N is the number of atoms per unit volume and dE/dx the energy loss of the particle per unit path length.

In order to compare (67) with the usual form for the stopping power, it is convenient to introduce the atomic oscillator strength for the transition $i\rightarrow f$ defined by

$$s_{if}=\frac{8\pi}{9}\frac{m}{e^2\hbar^2}B(E1;i\rightarrow f)(E_f-E_i), \quad \text{(II E.67a)}$$

where m is the electronic mass and where s_{if} is normalized in such a manner that the total oscillator strength equals the number of electrons in the atom. The expression (67) may then be written in the form [see (II B.37)]

$$\frac{1}{N}\frac{dE}{dx}=4\pi\frac{Z_1^2e^4}{mv_i^2}\sum_f s_{if}\frac{9}{32\pi^2}f_{E1}(\eta_i,\xi_{if}), \quad \text{(II E.67b)}$$

where η and ξ refer to the collision between the incident particle and an atomic electron; the influence of the nuclear field on the motion of the particle is of minor importance. In the case of collisions with fast particles, the values of ξ_{if} for the important transitions are small compared to unity, so that we may use the asymptotic form of f_{E1} for $\xi\ll1$.

When the collision can be treated by classical mechanics ($\eta\gg1$), it is thus seen from (59) that (67b) gives the classical stopping formula.[93a] In the opposite limit of $\eta\ll1$, where (24) applies, one obtains the stopping formula derived in Born approximation.[93b] The more general quantum-mechanical expression for the stopping power,[93c] valid for all η, is obtained from (67b) by inserting (66b).

II E.6. Limit of $\xi=0$

In the limiting case of $\xi=0$, several expressions from the general theory of Coulomb excitation reduce appreciably. In the classical theory, the exponential factor in the orbital integrals (47) disappears and the resulting integrals can be performed explicitly in terms of elementary functions.[5] Thus, one obtains

$$\begin{aligned} I_{\lambda\mu}(\vartheta,0)&=(\epsilon^2-1)^{-\lambda+\frac{1}{2}}\int_{-\phi_0}^{\phi_0}e^{i\mu\phi}(\epsilon\cos\phi-1)^{\lambda-1}d\phi\\ &=(2\pi)^{\frac{1}{2}}(\lambda-1)!\epsilon^{-\frac{1}{2}}(\epsilon^2-1)^{-(\lambda/2)+1/4}\\ &\quad\times P_{\mu-\frac{1}{2}}^{-\lambda+\frac{1}{2}}\left(\frac{1}{\epsilon}\right). \quad \text{(II E.68)}\end{aligned}$$

We have here introduced the azimuthal angle ϕ of the projectile given by [see (II A.22)]

$$\tan\phi=\frac{(\epsilon^2-1)^{\frac{1}{2}}\sinh w}{\epsilon+\cosh w}, \quad \text{(II E.69)}$$

and the limits ϕ_0 are

$$\phi_0=\tan^{-1}(\epsilon^2-1)^{\frac{1}{2}}=\frac{\pi}{2}-\frac{\vartheta}{2}. \quad \text{(II E.70)}$$

[93] (a) N. Bohr, Phil. Mag. **25**, 10 (1913); see also reference 19. (b) H. A. Bethe, Ann. Physik (5) **5**, 325 (1930). (c) F. Bloch, Ann. Physik (5) **16**, 285 (1933).

In the last expression in (68) the integral has been expressed in terms of Legendre functions of half-integer order.[94]

For the lowest multipole orders, one obtains the following explicit expressions

$$\begin{aligned}
I_{1,\pm1}(\vartheta,0)&=2\sin\frac{\vartheta}{2},\\
I_{2,\pm2}(\vartheta,0)&=\frac{2}{3}\sin^2\frac{\vartheta}{2},\\
I_{20}(\vartheta,0)&=2\tan^2\frac{\vartheta}{2}\left[1-\frac{\pi-\vartheta}{2}\tan\frac{\vartheta}{2}\right],\\
I_{3,\pm3}(\vartheta,0)&=\frac{4}{15}\sin^3\frac{\vartheta}{2},\\
I_{3,\pm1}(\vartheta,0)&=2\frac{\sin^3\frac{\vartheta}{2}}{\cos^4\frac{\vartheta}{2}}\left\{\frac{2+\sin^2\frac{\vartheta}{2}}{3}-\frac{\pi-\vartheta}{2}\tan\frac{\vartheta}{2}\right\}, \quad \text{(II E.71)}\\
I_{4,\pm4}(\vartheta,0)&=\frac{4}{35}\sin^4\frac{\vartheta}{2},\\
I_{4,\pm2}(\vartheta,0)&=2\frac{\sin^4\frac{\vartheta}{2}}{\cos^6\frac{\vartheta}{2}}\left\{\frac{8+9\sin^2\frac{\vartheta}{2}-2\sin^4\frac{\vartheta}{2}}{20}-\frac{3}{4}\frac{\pi-\vartheta}{2}\tan\frac{\vartheta}{2}\right\},\\
I_{40}(\vartheta,0)&=2\tan^6\frac{\vartheta}{2}\left\{\frac{2}{3\sin^2\frac{\vartheta}{2}}+\frac{11}{6}-\frac{\pi-\vartheta}{2}\frac{3+5\tan^2\frac{\vartheta}{2}}{2\tan\frac{\vartheta}{2}}\right\}.
\end{aligned}$$

The differential f functions are given directly in terms of these integrals by means of (II A.29) and (II A.51) and the results are illustrated in Fig. II.7.

The total f function is obtained from df by an integration over the deflection angle ϑ. This integration can also be simply performed, and one obtains

$$\begin{aligned}
f_{E2}(0)&=\frac{8\pi^2}{25}\left(\frac{\pi^2}{16}-\frac{1}{3}\right)=0.8954,\\
f_{E3}(0)&=\frac{8\pi^2}{49}\left(\frac{8}{45}-\frac{\pi^2}{64}\right)=0.03797, \quad \text{(II E.72)}\\
f_{M2}(0)&=\frac{8\pi^2}{25}\left(\frac{\pi^2}{16}-\frac{5}{9}\right)=0.1936.
\end{aligned}$$

For $\lambda=1$, the f functions diverge in the limit $\xi\rightarrow0$.

[94] See p. 159 of reference 84.

The a coefficients in the angular distribution of the de-excitation γ rays (II A.75) also involve integrals over ϑ of the orbital integrals. From (II A.75), (II A.76), and (II A.77) one obtains

$$a_2^{E1}(0)=1,$$

$$a_2^{E2}(0)=\frac{21\pi^2-208}{3\pi^2-16}=-0.05425,$$

$$a_4^{E2}(0)=-\frac{441\pi^2-4352}{48(3\pi^2-16)}=-0.0007587, \quad \text{(II E.73)}$$

$$a_2^{M1}(0)=1.$$

Also in the quantum-mechanical treatment, the case $\xi=0$ is especially simple. For the matrix elements with $l_i-l_f=\pm\lambda$ the last term of (II B.58) vanishes (for $\lambda>1$) and the first F_2 function is unity. For $\xi=0$, one thus obtains for these matrix elements

$$M_{l,\,l+\lambda}{}^{-\lambda-1}(\xi=0)=M_{l+\lambda,\,l}{}^{-\lambda-1}(\xi=0)=(2k)^{\lambda-2}\frac{[(\lambda-1)!]^2}{(2\lambda-1)!}\left|\frac{\Gamma(l+1+i\eta)}{\Gamma(l+\lambda+1+i\eta)}\right|, \quad \text{(II E.74)}$$

where $\eta=\eta_i=\eta_f$. This formula may also be seen to hold for $\lambda=1$. The matrix elements with $|l_i-l_f|<\lambda$ may most easily be obtained by means of the recursion formulas (II B.72) and (II B.68). For the lowest multipole orders, one obtains the following expressions[95]

$$M_{l,\,l+1}{}^{-2}=M_{l+1,\,l}{}^{-2}=\frac{1}{2k}\frac{1}{|l+1+i\eta|},$$

$$M_{l,\,l+2}{}^{-3}=M_{l+2,\,l}{}^{-3}=\frac{1}{6}\frac{1}{|l+1+i\eta|\,|l+2+i\eta|},$$

$$M_{l,\,l}{}^{-3}=\frac{1}{2l(l+1)(2l+1)}[2l+1-\pi\eta+2\eta\,\mathcal{I}m\psi(l+1+i\eta)], \quad \text{(II E.75)}$$

$$M_{l,\,l+3}{}^{-4}=M_{l+3,\,l}{}^{-4}=\frac{k}{15}\frac{1}{|l+1+i\eta|\,|l+2+i\eta|\,|l+3+i\eta|},$$

$$M_{l,\,l+1}{}^{-4}=M_{l+1,\,l}{}^{-4}=\frac{k}{3l(l+1)(l+2)(2l+1)(2l+3)\,|l+1+i\eta|}\times\{3|l+1+i\eta|^2[2l+1-\pi\eta+2\eta\,\mathcal{I}m\psi(l+1+i\eta)]-l(l+1)(2l+1)\}.$$

The imaginary part of the logarithmic derivative ψ of the Γ function can be expressed by elementary functions through the relation

$$\mathcal{I}m\psi(l+1+i\eta)=\pi\coth\pi\eta+\eta^{-1}-2\eta\sum_{n=0}^{l}\frac{1}{n^2+\eta^2}. \quad \text{(II E.76)}$$

In the classical limit ($\eta\rightarrow\infty$), the matrix elements (75) are related to the above calculated orbital integrals (71) by means of (II B.100).

To obtain the differential and the total f function as well as the γ-ray angular distribution functions, a summation over the angular momentum has to be performed [see (II B.48), (II B.50), and (II B.85)]. For large l, the terms in these sums decrease as $l^{-2\lambda+1}$. For $\lambda=1$, the total f function as well as the b coefficients diverge, and one obtains

$$a_2^{E1}(\eta,0)=1. \quad \text{(II E.77)}$$

For $\lambda>1$, the convergence is rather slow and may be improved by employing the Euler sum formula. It should be pointed out that the convergence for $\xi\neq0$ is more rapid due to the adiabatic cutoff for high l's; the sum then becomes a geometric series.

II E.7. Limit of Large Orbital Angular Momenta

An interesting limit of the Coulomb excitation matrix elements is that of $l\gg1$.[96] As pointed out in Sec. II B.6, the radial matrix elements in this limit can be expressed by means of (II B.100). This result can be obtained by employing the WKB approximation or by performing a confluence in the explicit expressions for the radial matrix elements, and holds for $l\gg1$ irrespective of the value of η.

If $l\gg\eta$, the deflection angle of the associated classical orbit is small ($\vartheta\approx2\eta/l$) and the orbits approach straight lines. It is thus of interest to compare the $I_{\lambda\mu}$ with the corresponding integrals for straight line orbits given by

[95] For $\lambda=2$ these results have been given by L. C. Biedenharn and C. M. Class, Phys. Rev. **98**, 691 (1955).

[96] This limit has been studied by Gluckstern, Lazarus, and Breit, reference 18.

[see (II A.24)]

$$I_{\lambda\mu}{}^{st}(\vartheta,\xi)=a^{\lambda}\cdot v\int_{-\infty}^{\infty}e^{i\omega t}\frac{[p+ivt]^{\mu}}{[p^2+(vt)^2]^{\frac{1}{2}(\lambda+\mu+1)}}dt$$
$$=\epsilon^{-\lambda}\int_{-\infty}^{\infty}e^{i\xi\epsilon\sinh w}\frac{[1+i\sinh w]^{\mu}}{(\cosh w)^{\lambda+\mu}}dw, \quad \text{(II E.78)}$$

where p is the impact parameter. We have introduced the parameters $\vartheta=2a/p$ and $\epsilon=p/a$ which, for $p\gg a$, correspond to the deflection angle and eccentricity of the hyperbolic orbit with the impact parameter p.

The integral (78) is the same as the limit of (47) for $\epsilon\gg1$ except for the phase factor $e^{i\xi w}$. The effect of this factor may be seen by transforming (47) according to (49). In the latter form the phase factor $e^{i\xi w}$ can be neglected for large ϵ and we thus have the relation

$$I_{\lambda\mu}(\vartheta,\xi)\approx e^{-(\pi/2)\xi}I_{\lambda\mu}{}^{st}(\vartheta,\xi), \quad \text{(II E.79)}$$

holding for $\vartheta\ll1$.

The large difference between the orbital integrals for straight line and hyperbolic orbits in the case of $\xi\gtrsim1$ is associated with the fact that for such values of ξ the integral is very sensitive to the impact parameter. Thus, an increase of p by the amount a, which represents the order of magnitude of the displacement during the collision, implies a reduction of $I_{\lambda\mu}$ by a factor of the order of that involved in (79).

Since the $I_{\lambda\mu}{}^{st}$ corresponds to the neglect of the Coulomb force on the motion of the projectile, these integrals are for large l related to the Born approximation radial matrix elements by an equation analogous to (II B.100). From (79) we thus obtain

$$M_{l_il_f}{}^{-\lambda-1}\approx e^{-(\pi/2)\xi}M_{l_il_f}{}^{-\lambda-1}\text{ (Born appr.)}, \quad \text{(II E.80)}$$

holding for l_i, $l_f\gg1$. This relation shows that the Coulomb phase in the wave functions, for large l, gives rise to a simple factor, independent of l.[97]

The integral (78) can be expressed by means of the Whittaker function [98] in the form

$$I_{\lambda\mu}{}^{st}(\vartheta,\xi)=(-1)^{(\lambda+\mu)/2}\epsilon^{-1}\left(\frac{\xi}{2\epsilon}\right)^{(\lambda-1)/2}$$
$$\times\Gamma\left(\frac{-\lambda+\mu+1}{2}\right)W_{-\mu/2,-\lambda/2}(2\xi\epsilon). \quad \text{(II E.81)}$$

For large values of ϵ ξ one obtains by employing the asymptotic expansion of the Whittaker function[99,5]

$$I_{\lambda\mu}{}^{st}(\vartheta,\xi)\approx\frac{2\pi}{\Gamma\left(\frac{\lambda+1-\mu}{2}\right)}$$
$$\times e^{-\xi\epsilon}\xi^{(\lambda-\mu-1)/2}(2\epsilon)^{-(\lambda+\mu+1)/2}. \quad \text{(II E.82)}$$

For the radial matrix elements in the limit $l\gg1$, one thus obtains the result

$$M_{l_il_f}{}^{-\lambda-1}\approx\frac{k^{\lambda-2}}{4\eta^{\lambda}}\frac{2\pi}{\Gamma\left(\frac{\lambda+1-\mu}{2}\right)}e^{-(l/\eta+\pi/2)\xi}$$
$$\times\xi^{(\lambda-\mu-1)/2}(2l/\eta)^{-(\lambda+\mu+1)/2}. \quad \text{(II E.83)}$$

A more accurate result may be obtained by an expansion of the Born approximation radial matrix elements (31) employing (86).

II E.8. Some Properties of Hypergeometric Functions

In this paragraph, we shall collect some formulas for hypergeometric functions which are of interest in the theory of Coulomb excitation.[100]

The ordinary hypergeometric function of one variable is defined by the series expansion

$$F(\alpha,\beta,\gamma;z)=\sum_m\frac{\alpha_m\beta_m}{\gamma_m m!}z^m, \quad \text{(II E.84)}$$

where

$$a_m=\frac{\Gamma(a+m)}{\Gamma(a)}=a(a+1)\cdots(a+m-1). \quad \text{(II E.85)}$$

This series is only convergent for $|z|<1$. However, the analytic continuation can again be expressed by hypergeometric functions. We note especially the Kummer transformation

$$F(\alpha,\beta,\gamma;z)=(1-z)^{-\alpha}F\left(\alpha,\gamma-\beta,\gamma;\frac{z}{z-1}\right) \quad \text{(II E.86)}$$

and the relation in terms of the reciprocal argument

$$F(\alpha,\beta,\gamma;z)=\frac{\Gamma(\gamma)\Gamma(\beta-\alpha)}{\Gamma(\beta)\Gamma(\gamma-\alpha)}(-z)^{-\alpha}F(\alpha,1-\gamma+\alpha,1-\beta+\alpha;1/z)$$
$$+\frac{\Gamma(\gamma)\Gamma(\alpha-\beta)}{\Gamma(\alpha)\Gamma(\gamma-\beta)}(-z)^{-\beta}F(\beta,1-\gamma+\beta,1-\alpha+\beta;1/z). \quad \text{(II E.87)}$$

[97] In the applications of the relations (79) and (80) in reference 96 the exponential factor has been omitted.
[98] See p. 274 of reference 84.
[99] See p. 278, reference 84.
[100] Most of these formulas may be found in reference 84 or in the treatise by P. Appell and J. Kampé de Feriet, *Fonctions Hypergéométriques etc.* (Gauthiers Villars, Paris, 1926).

The analytic continuation is also given by the integral representation

$$F(\alpha,\beta,\gamma;\,z)=\frac{\Gamma(\gamma)}{\Gamma(\beta)\Gamma(\gamma-\beta)}\int_0^1 u^{-\beta-1}(1-u)^{\gamma-\beta-1}(1-uz)^{-\alpha}du, \tag{II E.88}$$

valid for $\Re e\beta>0$ and $\Re e(\gamma-\beta)>0$.

When one of the parameters α or β tends to infinity while z becomes small, the function (84) approaches the confluent hypergeometric function

$${}_1F_1(\alpha,\gamma;\,z)=\lim_{\beta\to\infty} F(\alpha,\beta,\gamma;\,z/\beta)=\sum_m \frac{\alpha_m}{\gamma_m m!}z^m, \tag{II E.89}$$

which is convergent for all z.

From the formula (86) one obtains the Kummer transformation for the confluent function

$${}_1F_1(a,\gamma;\,z)=e^z{}_1F_1(\gamma-\alpha,\,\gamma;\,-z). \tag{II E.90}$$

An integral representation of the function ${}_1F_1$ is given by

$${}_1F_1(\alpha,\gamma;\,z)=\frac{\Gamma(\gamma)}{\Gamma(\alpha)\Gamma(\gamma-\alpha)}\int_0^1 e^{zt}t^{\alpha-1}(1-t)^{\gamma-\alpha-1}dt, \tag{II E.91}$$

valid for $\Re e\alpha>0$ and $\Re e(\gamma-\alpha)>0$.

Among the hypergeometric functions of more than one variable, the simplest are the so-called Appell functions. We shall here be concerned with the functions F_1, F_2, and F_3 defined by the series

$$F_1(\alpha,\beta,\beta',\gamma;\,x,y)=\sum_{mn}\frac{\alpha_{m+n}\beta_m\beta_n'}{\gamma_{m+n}m!n!}x^my^n \qquad \begin{matrix}|x|<1\\ |y|<1\end{matrix}, \tag{II E.92}$$

$$F_2(\alpha,\beta,\beta',\gamma,\gamma';\,x,y)=\sum_{mn}\frac{\alpha_{m+n}\beta_m\beta_n'}{\gamma_m\gamma_n'm!n!}x^my^n \qquad |x|+|y|<1, \tag{II E.93}$$

$$F_3(\alpha,\alpha',\beta,\beta',\gamma;\,x,y)=\sum_{mn}\frac{\alpha_m\alpha_n'\beta_m\beta_n'}{\gamma_{m+n}m!n!}x^my^n \qquad \begin{matrix}|x|<1\\ |y|<1\end{matrix}, \tag{II E.94}$$

whose regions of convergence are indicated.

These functions have properties similar to those of the hypergeometric functions of one variable. Thus, for the function F_2 there exist transformations of the Kummer type

$$\begin{aligned}F_2(\alpha,\beta,\beta',\gamma,\gamma';\,x,y)&=(1-y)^{-\alpha}F_2\left(\alpha,\,\beta,\,\gamma'-\beta',\,\gamma,\,\gamma';\,\frac{x}{1-y},\,\frac{y}{y-1}\right)\\ &=(1-x-y)^{-\alpha}F_2\left(\alpha,\,\gamma-\beta,\,\gamma'-\beta',\,\gamma,\,\gamma';\,\frac{x}{x+y-1},\,\frac{y}{x+y-1}\right).\end{aligned} \tag{II E.95}$$

For special values of the parameters, the Appell functions reduce according to the following relations

$$F_2(\alpha,\beta,\beta',\alpha,\alpha;\,x,y)=(1-x)^{-\beta}(1-y)^{-\beta'}F\left(\beta,\,\beta',\,\alpha;\,\frac{xy}{(1-x)(1-y)}\right), \tag{II E.96}$$

$$F_2(\alpha,\beta,\beta',\gamma,\alpha;\,x,y)=(1-y)^{-\beta'}F_1\left(\beta,\,\alpha-\beta',\,\beta',\,\gamma;\,x,\,\frac{x}{1-y}\right), \tag{II E.97}$$

$$F_1(\alpha,\beta,\beta',\gamma;\,x,y)=(1-y)^{-\beta'}F_3\left(\alpha,\,\gamma-\alpha,\,\beta,\,\beta',\,\gamma;\,x,\,-\frac{y}{1-y}\right). \tag{II E.98}$$

The analytic continuation of the function F_3 can be expressed by four F_2 functions of the arguments x^{-1}, y^{-1} as follows:

$$
\begin{aligned}
F_3(\alpha,\alpha',\beta,\beta',\gamma;x,y)
&=\frac{\Gamma(\gamma)\Gamma(\beta-\alpha)\Gamma(\beta'-\alpha')}{\Gamma(\beta)\Gamma(\beta')\Gamma(\gamma-\alpha-\alpha')}(-x)^{-\alpha}(-y)^{-\alpha'}F_2\left(\alpha+\alpha'+1-\gamma,\,\alpha,\,\alpha',\,\alpha+1-\beta,\,\alpha'+1-\beta';\frac{1}{x},\frac{1}{y}\right)\\
&+\frac{\Gamma(\gamma)\Gamma(\beta-\alpha)\Gamma(\alpha'-\beta')}{\Gamma(\beta)\Gamma(\alpha')\Gamma(\gamma-\alpha-\beta')}(-x)^{-\alpha}(-y)^{-\beta'}F_2\left(\alpha+\beta'+1-\gamma,\,\alpha,\,\beta',\,\alpha+1-\beta,\,\beta'+1-\alpha';\frac{1}{x},\frac{1}{y}\right)\\
&+\frac{\Gamma(\gamma)\Gamma(\alpha-\beta)\Gamma(\beta'-\alpha')}{\Gamma(\alpha)\Gamma(\beta')\Gamma(\gamma-\beta-\alpha')}(-x)^{-\beta}(-y)^{-\alpha'}F_2\left(\beta+\alpha'+1-\gamma,\,\beta,\,\alpha',\,\beta+1-\alpha,\,\alpha'+1-\beta';\frac{1}{x},\frac{1}{y}\right)\\
&+\frac{\Gamma(\gamma)\Gamma(\alpha-\beta)\Gamma(\alpha'-\beta')}{\Gamma(\alpha)\Gamma(\alpha')\Gamma(\gamma-\beta-\beta')}(-x)^{-\beta}(-y)^{-\beta'}F_2\left(\beta+\beta'+1-\gamma,\,\beta,\,\beta',\,\beta+1-\alpha,\,\beta'+1-\alpha';\frac{1}{x},\frac{1}{y}\right). \qquad \text{(II E.99)}
\end{aligned}
$$

The analytic continuation of the function F_2 can, in the general case, not be expressed in terms of Appell functions, but may be given by the integral representation

$$
F_2(\alpha,\beta,\beta',\gamma,\gamma';x,y)
=\frac{\Gamma(\gamma)\Gamma(\gamma')}{\Gamma(\beta)\Gamma(\beta')\Gamma(\gamma-\beta)\Gamma(\gamma'-\beta')}\int_0^1\int_0^1 du\,dv\,u^{\beta-1}v^{\beta'-1}(1-u)^{\gamma-\beta-1}(1-v)^{\gamma'-\beta'-1}(1-ux-vy)^{-\alpha}, \qquad \text{(II E.100)}
$$

valid for $\Re e\beta>0$, $\Re e\beta'>0$, $\Re e(\gamma-\beta)>0$, and $\Re e(\gamma'-\beta')>0$. One of the integrations can be performed according to (88) yielding the result

$$
F_2(\alpha,\beta,\beta',\gamma,\gamma';x,y)=\frac{\Gamma(\gamma')}{\Gamma(\beta')\Gamma(\gamma'-\beta')}\int_0^1 dv\,v^{\beta'-1}(1-v)^{\gamma'-\beta'-1}(1-vy)^{-\alpha}F\left(\alpha,\,\beta,\,\gamma;\frac{x}{1-vy}\right). \qquad \text{(II E.101)}
$$

A similar integral representation of F_1 is given by

$$
F_1(\alpha,\beta,\beta',\gamma;x,y)=\frac{\Gamma(\gamma)}{\Gamma(\alpha)\Gamma(\gamma-\alpha)}\int_0^1 du\,u^{\alpha-1}(1-u)^{\gamma-\alpha-1}(1-ux)^{-\beta}(1-uy)^{-\beta'}, \qquad \text{(II E.102)}
$$

valid for $\Re e\alpha>0$ and $\Re e(\gamma-\alpha)>0$.

There exist a large number of relations by which one may expand one Appell function in terms of other hypergeometric functions. An expansion of F_2 is obtained from (101) by transforming the F function according to (87) and (86). By the integral representation (102), this leads to

$$
\begin{aligned}
F_2(\alpha,\beta,\beta',\gamma,\gamma';x,y)&=\frac{\Gamma(\beta)\Gamma(\gamma')\Gamma(\gamma-\beta)\Gamma(\beta'-\alpha)}{\Gamma(\gamma)\Gamma(\beta')\Gamma(\gamma'-\alpha)}(-y)^{-\alpha}\left(\frac{y}{y-1}\right)^{\alpha-\gamma'+1}\\
&\quad\times\sum_m\frac{(1-\beta')_m(1-\gamma'+\alpha)_m}{(1-\beta'+\alpha)_m m!}(y-1)^{-m}F_1\left(\beta,\,-m,\,\alpha-\gamma'+1+m,\,\gamma;x,\frac{x}{1-y}\right)\\
&+\frac{\Gamma(\beta)\Gamma(\gamma')\Gamma(\gamma-\beta)\Gamma(\alpha-\beta')}{\Gamma(\gamma)\Gamma(\alpha)\Gamma(\gamma'-\beta')}(-y)^{-\beta'}\left(\frac{y}{y-1}\right)^{\alpha-\beta'+1}\\
&\quad\times\sum_m\frac{(1-\alpha)_m(1-\gamma'+\beta')_m}{(1+\beta'-\alpha)_m m!}(y-1)^{-m}F_1\left(\beta,\,\alpha-\beta'-m,\,\beta'-\gamma'+1+m,\,\gamma;x,\frac{x}{1-y}\right). \qquad \text{(II E.103)}
\end{aligned}
$$

A similar relation for F_3 is given by

$$
F_3(\alpha,\alpha',\beta,\beta',\gamma;x,y)=(1-x)^{-\alpha}\sum_m\frac{(\gamma-\beta-\beta')_m\alpha_m'}{\gamma_m m!}(-y)^mF_1\left(\gamma-\beta+m,\,\alpha,\,\alpha'+m,\,\gamma+m;\frac{x}{x-1},y\right). \qquad \text{(II E.104)}
$$

For large parameters, the Appell functions reduce to confluent functions. Thus, for the F_2 function with β, β' large, one obtains

$$
\Psi_2(\alpha,\gamma,\gamma';x,y)=\lim_{\beta,\beta'\to\infty}F_2\left(\alpha,\beta,\beta',\gamma,\gamma';\frac{x}{\beta},\frac{y}{\beta'}\right)=\sum_{m,n}\frac{\alpha_{m+n}}{\gamma_m\gamma_n'm!n!}x^my^n, \qquad \text{(II E.105)}
$$

which is convergent for all x and y.

CHAPTER III. EXPERIMENTAL CONDITIONS

In the present chapter, we consider the conditions for the experimental investigations of the Coulomb excitation process. Section III A deals with the requirements on the ion beam, and the following sections treat the problems connected with the observation of the nuclear excitations. These can be detected by observing either the γ rays (III B) or the internal conversion electrons (III C) which are emitted in the decay of the excited states. It is also possible to detect directly the inelastically scattered projectiles (III D).

III A. Beam Requirements

The range of projectile energies which can be employed in the excitation of a given nuclear level is limited on the low-energy side by the condition that the collision time must not be longer than the nuclear period, since otherwise the collision becomes adiabatic and the excitation cross section small. On the other hand, for too high bombarding energies, the projectiles may penetrate into the nucleus, and the interpretation of the observed excitations then becomes more difficult due to the onset of proper nuclear reactions.

For the Coulomb barrier we may write

$$E_B = \frac{Z_1 Z_2 e^2}{R}, \qquad \text{(III.1)}$$

where R is the effective radius of interaction which may be represented by

$$R = r_0 A_2^{\frac{1}{3}} + \rho. \qquad \text{(III.2)}$$

The radius of the projectile is denoted by ρ and is taken to be zero in the case of protons. If one neglects ρ and assumes $r_0 = 1.5 \times 10^{-13}$ cm, one obtains the approximate estimate

$$E_B \simeq Z_1 Z_2 A_2^{-\frac{1}{3}} \text{ Mev}. \qquad \text{(III.3)}$$

Even for bombarding energies somewhat smaller than (1), there may be a significant quantum-mechanical penetration of the barrier. This effect is less important when heavier projectiles or target nuclei are involved (see Fig. III.1). Furthermore, even if the cross section for compound nucleus formation exceeds that for Coulomb excitation, it may still be possible to observe the latter effect, since the compound nucleus usually decays predominantly into other channels than that corresponding to the inelastic scattering (see Sec. IV A.5).

The low-energy limit to the bombarding energy may be expressed by the condition $\xi \lesssim 1$ [see (II A.27)]. According to (II C.13), this condition may also be written

$$E_{\text{Mev}} \gtrsim 0.2 Z_1 (A_1/Z_1)^{\frac{1}{3}} (Z_2 \Delta E_{\text{Mev}})^{\frac{2}{3}}, \qquad \text{(III.4)}$$

where E_{Mev} and ΔE_{Mev} are the bombarding energy and the excitation energy in Mev. In (4) we have neglected the center-of-mass corrections and the relative energy loss $\Delta E/E$.

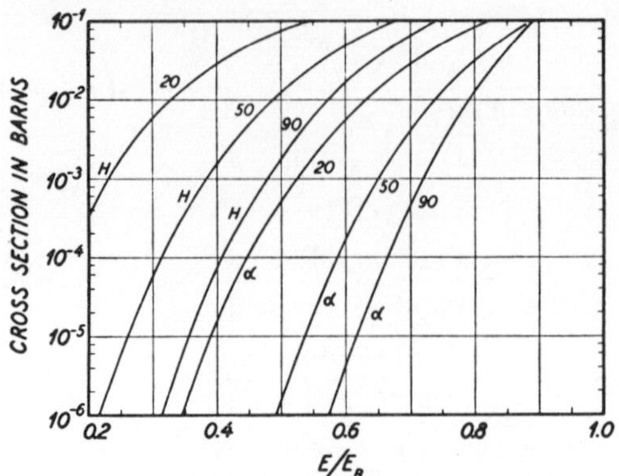

FIG. III.1. Cross sections for compound nucleus formation. The figure gives theoretical estimates of the cross sections as a function of the ratio E/E_B, where E is the kinetic energy in the center-of-mass system, and where E_B is the height of the Coulomb barrier. The curves are labeled H for protons and α for α particles, whereas the numbers indicate the charge number Z_2 of the target element. The cross sections are taken from the tables given by J. M. Blatt and V. F. Weisskopf [*Theoretical Nuclear Physics* (John Wiley and Sons, Inc., New York, 1952)], and correspond to an effective interaction radius given by (III.2) with $r_0 = 1.5 \cdot 10^{-13}$ cm. The value of ρ is taken to be zero for protons, and $1.2 \cdot 10^{-13}$ cm for α particles.

If the conditions (4) and $E < E_B$ are expressed in terms of ξ, one obtains, employing the estimate (3)

$$\left(\frac{A_1}{Z_1}\frac{A_2}{Z_2}\right)^{\frac{1}{2}} \cdot \frac{\Delta E_{\text{Mev}}}{13} \lesssim \xi \lesssim 1 \qquad \text{(III.5)}$$

for the usable range of ξ values.

From (5) it follows that the various types of accelerated ions can be used in approximately the same range of ξ values. Moreover, it is seen that, by employing sufficiently high bombarding energies, it may be possible to excite levels with ΔE as high as 5 Mev. Since, however, ΔE must be small compared to E, it is necessary in the Coulomb excitation of such high-lying levels to employ high energies, and thus rather heavy projectiles, especially in the case of light target nuclei.

For a given value of ξ, the cross section for an excitation of multipole order $E\lambda$ is proportional to $Z_1^2 (A_1/Z_1)^{2\lambda/3}$ [see (II C.13), (II C.15), and (II C.16)]; thus, the largest cross sections are obtained with the heavier projectiles. The advantage of the heavier projectiles is even greater in the case of higher order excitations (see Sec. II D).

In order to obtain the same ξ value for the different projectiles, it is necessary that they be accelerated to energies which are proportional to $Z_1 (A_1/Z_1)^{\frac{1}{3}}$. For a given acceleration voltage this may be approximately achieved, provided the ions can be completely stripped of electrons. However, if this is not the case, the relative magnitude of the excitation cross sections obtainable with different projectiles depends essentially on the available voltage.

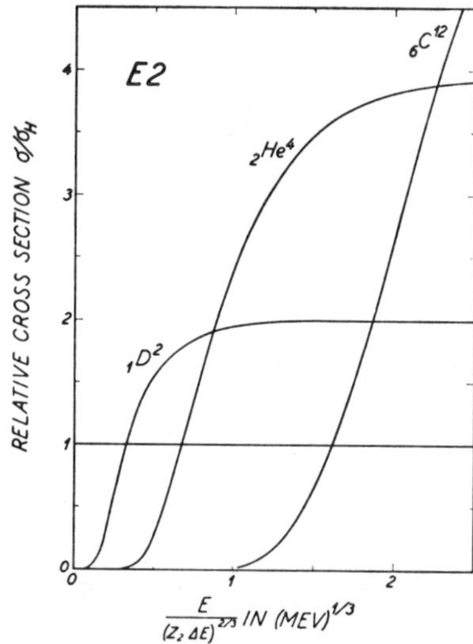

FIG. III.2. Relative excitation cross sections for different projectiles. The curves show the ratio of the theoretical $E2$ cross sections for Coulomb excitation with deuterons, α particles, and carbon ions, to those for protons of the same energy. The ratios also depend on the excitation energy ΔE and the atomic number Z_2 of the target nuclei, but can be expressed approximately as a function of the single parameter $E(Z_2\Delta E)^{-\frac{2}{3}}$.

For projectiles having the same energy, the cross sections for $E2$ excitations are compared in Fig. III.2. The cross sections are proportional to the mass of the projectiles when the excitation energy is so low that the ξ values are much smaller than unity. In such cases, the excitations are best produced by bombardments with the heavier particles, in contrast to the higher lying levels which are more easily excited by means of protons. In the case of thick target experiments the yield for protons is always larger than for heavier projectiles of the same energy, due to the larger range of the protons. On the other hand, the background radiations produced by α-particle bombardments are in general considerably smaller than those produced by protons (see later discussion), and this may, therefore, often be a compensating consideration.

Apart from the question of which type of projectile provides the optimum conditions for Coulomb excitation under given experimental circumstances, it is often a great advantage to be able to compare the yields for different projectiles. Not only does such a comparison constitute a very direct test of the Coulomb excitation character of the process, but it may also yield additional information about the multipole order and the excitation energy (see Sec. IV A.1).

The absolute values of the excitation cross sections depend on the reduced transition probabilities $B(\lambda)$. The largest cross sections are associated with the collective excitations of low energy, which are produced by $E2$ transitions (see Chapter V). As an example, cross sections of the order of millibarns are observed for the excitation of rotational levels in heavy elements with 2-Mev protons. Because of the nonresonant character of the process, such cross sections imply thick target yields of the order of 10^{-7} excitation per proton. Thus, the demands on the current as well as on the energy homogeneity of the beam are often rather modest.

So far most Coulomb excitation experiments have been performed by means of protons, deuterons, and α particles, accelerated in electrostatic generators. It has also been shown that it is feasible to make such experiments with external ion beams from cyclotrons.[101]

III B. Measurements of De-Excitation Gamma Rays

III B.1. Detection Technique

The great sensitivity and simplicity of the scintillation spectrometers employed in γ-ray measurements

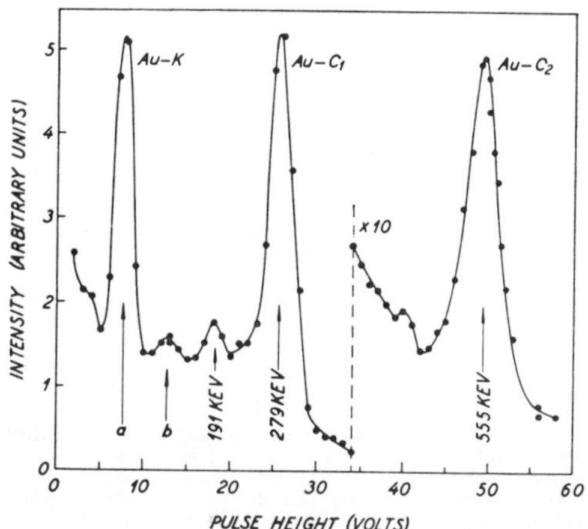

FIG. III.3. Gamma rays from the Coulomb excitation of gold. The figure shows the pulse-height spectrum observed with a crystal scintillation spectrometer [Cook, Class, and Eisinger, Phys. Rev. **96**, 658 (1954)]. The excitation was produced by bombardment of a thick Au target with 3-Mev protons. The peaks C_1 and C_2 correspond to the transitions from the first two strongly excited levels to the ground state, whereas the peak a is due to the characteristic x-rays which follow the ionization of the K shell. The 191-kev peak corresponds to a weaker excitation of a level at 268 kev (see Table IV.2), and b is an escape peak (Compton peak) belonging to the C_1 line.

have rendered these instruments the most widely used detectors in experiments on Coulomb excitation. Typical examples of the pulse-height spectra which have been obtained in this way are given in Figs. III.3, 4, and 5.

The comparatively poor energy resolution of the scintillation spectrometers is often a disadvantage, espe-

[101] Such experiments have been carried out in Zürich with 7-Mev protons (private communication from P. Marmier) and in Leningrad with 15 Mev $(N^{14})^{+++}$ ions (Alkhazov *et al.*, see reference 15).

cially for low-energy radiation, where the background of x-rays is large (see Sec. III B.3). Some improvements can be obtained by using the spectrometer in combination with various absorbers. The absorption coefficients are strongly energy dependent in this region and the effect of the absorbers is therefore dependent on the energy distribution of the radiation. This is illustrated in Fig. III.6, where the background peak is seen to be suppressed relative to the peak due to Coulomb excitation, when the absorber thickness is increased. In cases where a peak is composite, this may be revealed by a change in the shape of the peak when the absorbers are introduced. The measurement of the absorption coefficient can also sometimes be useful in providing an independent energy determination which makes it possible to avoid misinterpretations of the experimental spectra, e.g., due to coincidences or to the so-called escape peaks.

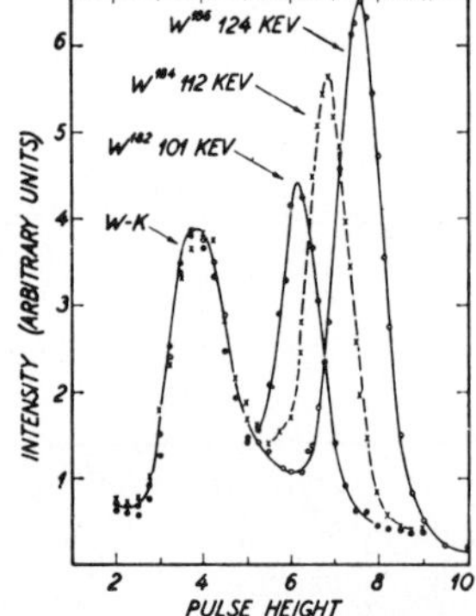

Fig. III.4. Gamma rays from the excitation of separated isotopes of tungsten. The γ rays are observed with a crystal spectrometer and result from bombardments of thick WO_3 target of the separated isotopes with 2.5-Mev protons. The pulse-height spectrum is taken from McClelland, Mark, and Goodman [Phys. Rev. 93, 904 (1953)]. The three γ rays represent the first excited states in the even-A isotopes.

The large yields for the Coulomb excitation processes, which are encountered for low excitation energies, also make feasible the use of proportional counters for the detection. The comparatively high resolution of this type of counter may in such cases be of greater importance than the correspondingly lower efficiency. The Coulomb excitation of some of the heaviest elements has recently been studied in this manner (see Fig. III.7).[101a]

Besides the efficiency and simplicity of the scintillation detector, the observation of the γ radiation has several other intrinsic advantages associated with the relatively small scattering or absorption in the target. This facilitates the measurements of angular distributions, since the correction for the target thickness is usually small. An example of the measured γ-ray angular distributions is shown in Fig. III.8.

[101a] *Note added in proof.*—The very high resolution of the bent crystal spectrometer has recently been employed in an experiment performed with the high current (~100 mA) from a linear accelerator (private communication from H. Mark).

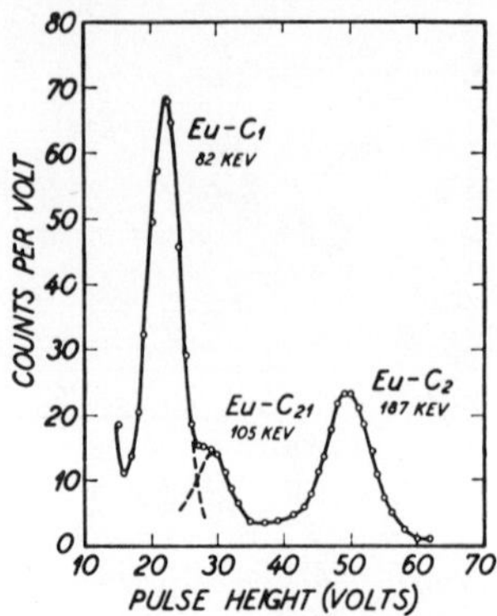

Fig. III.5. Gamma rays from the excitation of europium. The γ rays resulting from a bombardment of a thick Eu_2O_3 target with 6-Mev α particles are observed with a crystal spectrometer. The pulse-height spectrum is taken from N. P. Heydenburg and G. M. Temmer [Phys. Rev. **100**, 150 (1955)]. All three lines are assigned to the isotope Eu^{153}; the C_1 and C_2 lines correspond to the ground-state transitions from the first two rotational excitations, respectively, whereas the C_{21} line represents the cascade transition from the second to the first level.

If one wishes to determine directly the total γ-ray yield, one may either employ a 2π-geometry,[102] or one

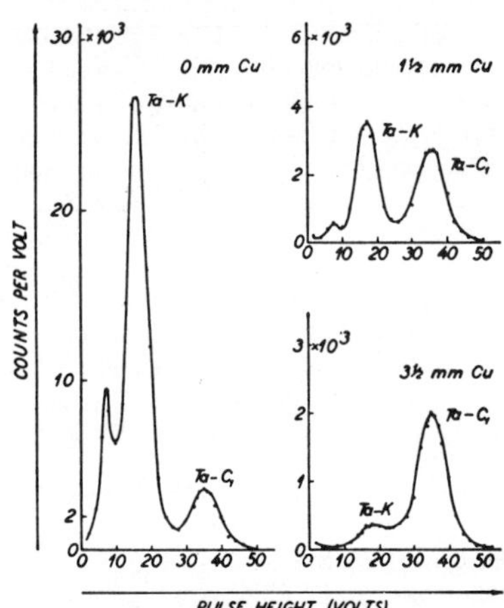

Fig. III.6. Effect of absorbers in γ-ray measurements. The figure illustrates the effect of Cu absorbers inserted between the crystal and the target when Ta is bombarded by 1.75-Mev protons. The pulse-height spectra, obtained with a thick target, are taken from T. Huus and Č. Zupančič [Kgl. Danske Videnskab. Selskab Mat.-fys. Medd. 28, No. 1 (1953)]. With a 3.5-mm copper absorber the characteristic x-rays from the K shell are strongly reduced, while the 137-kev C_1 line from the decay of the first excited nuclear state is much less affected. The spectra also show an escape peak associated with the x-rays.

[102] N. P. Heydenburg and G. M. Temmer, Phys. Rev. **100**, 150 (1955).

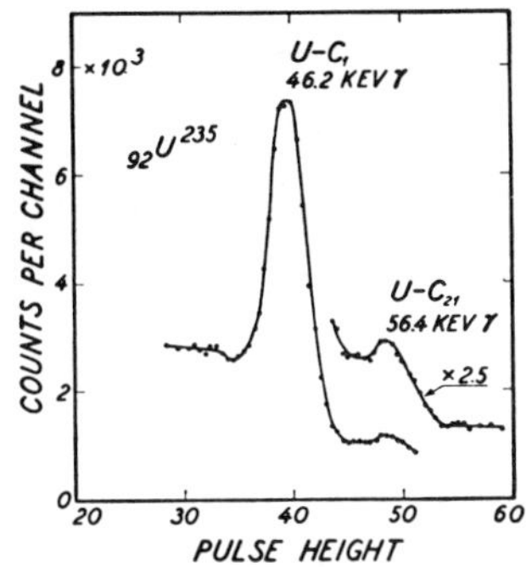

FIG. III.7. Gamma rays from U^{235} observed with a proportional counter. The pulse-height spectrum is obtained with a xenon-filled proportional counter when a target of U^{235} is bombarded by 3-Mev α particles. The figure is reproduced from data communicated to us by J. O. Newton (unpublished). The lines correspond to the ground-state transition from the first rotational state and the cascade transition from the second rotational state.

may make the observations at an angle of 55 or 125 degrees with respect to the beam.[103] For these angles, the P_2 function in (II C.26) and (II C.29) vanishes, and since the coefficient of P_4 in (II C.29) is almost always very small, one observes a yield approximately proportional to the cross section averaged over all angles.

III B.2. Thick Target Yields

The small scattering and absorption of the γ rays in the target make it possible to employ thick targets in the measurements of the excitation cross sections and of the angular distributions. The determination of the cross section from the observed yield then involves either a differentiation of the yield as a function of the bombarding energy, or an integration of the theoretical excitation function along the trajectory of the projectile in the target.

It is convenient to express the result of the latter calculation in terms of an effective target thickness δE_λ which is related to the true thick target yield by

$$Y=\sigma(E_0)\frac{E_0 N}{(dE/ds)_0}\frac{\delta E_\lambda}{E_0}, \tag{III.6}$$

where Y is the fraction of the incoming particles which produce the nuclear excitation and N the density of the investigated atoms in the target. The stopping power of the target material is denoted by dE/ds and is evaluated at the bombarding energy E_0. Thus, the fraction $\delta E_\lambda/E_0$ represents the ratio of the observed yield to that which would result if the excitation cross section σ and the stopping power were independent of the energy of the projectile and had the values corresponding to the energy E_0.

The calculation of δE_λ has been performed assuming $dE/ds \sim E^{-0.55}$. This energy dependence represents rather well the stopping power of protons and α particles in almost the entire range from the lowest energies employed in Coulomb excitation experiments and up to energies equal to the Coulomb barrier.[104]

By means of the theoretical excitation cross section (II C.15) one then obtains

$$\frac{\delta E_\lambda}{E_0}=\frac{1}{u_\lambda(\nu,\zeta_0)}\int_{\zeta_0}^{1} u_\lambda(\nu,\zeta)\frac{d\zeta}{\zeta}, \tag{III.7}$$

where the functions u_λ are defined by

$$u_\lambda(\nu,\zeta)=\zeta^{1.45-2\lambda}(1-\zeta)^{\lambda-1}f_{E\lambda}(\nu,\zeta). \tag{III.8}$$

The relations between the parameters (ν,ζ) and (η_i,ξ) are given by Eqs. (II C.11) and (II C.12), and the subscript zero indicates that the values correspond to the bombarding energy E_0. The values of δE_λ computed from these formulas are given in Fig. III.9 as a function of ξ_0, for the case $\nu \to 0$, which corresponds to the classical limit. The results are rather insensitive to the assumed energy dependence for the stopping power, due to the rapid variation of σ with the energy of the projectile for all but the smallest ξ values. Even in the extreme case of $\xi_0=0$, the value of δE_2 will be changed by only 8% of its magnitude if, instead of $E^{-0.55}$, one employs the rather different energy dependences $E^{-0.3}$

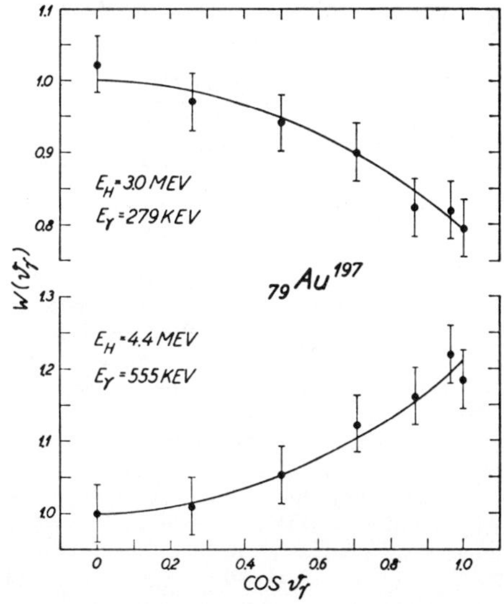

FIG. III.8. Angular distribution of γ rays from gold. The figure shows the angular distribution of the two intense γ rays resulting from the Coulomb excitation of Au^{197} (see Fig. III.3). The data is taken from Cook, Class, and Eisinger [Phys. Rev. **96**, 658 (1954)]. The curves represent a least-square fit to the experimental data. For the 555-kev γ ray the distribution corresponds to the sequence $3/2(E2)7/2(E2)3/2$ of spins and multipolarities (see Table IV.2). For the 279-kev γ ray the angular distribution indicates the sequence $3/2(E2)5/2(M1+E2)3/2$ with an $E2$ intensity of approximately 40% in the decay radiation.

[103] P. H. Stelson and F. K. McGowan, Phys. Rev. **99**, 112 (1955).

[104] Cf. J. Lindhard and M. Scharff, Kgl. Danske Videnskab. Selskab Mat. fys. Medd. **27**, No. 15 (1953).

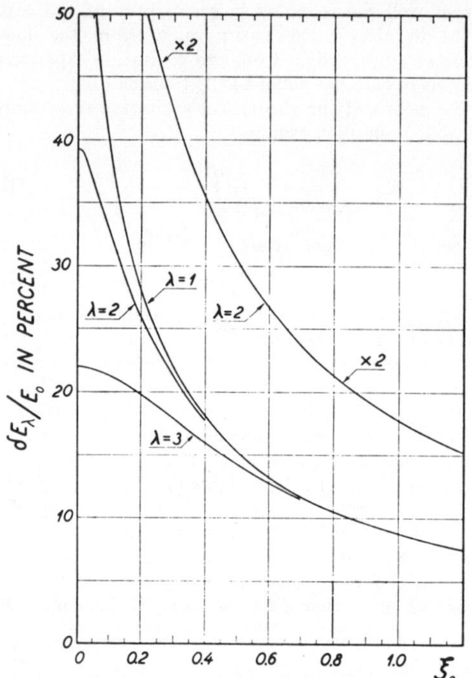

FIG. III.9. Effective target thickness. The curves give the ratio $\delta E_\lambda/E_0$ which enters into the determination of the theoretical thick target yields [see (III.6)]. The ratio has been computed from (III.7) by means of the classical f functions ($\nu=0$) for electric excitation of multipole order $\lambda=1$, 2, and 3; the stopping power has been assumed to depend on the energy of the projectile as $E^{-0.55}$. The abscissa gives the ξ value corresponding to the bombarding energy E_0.

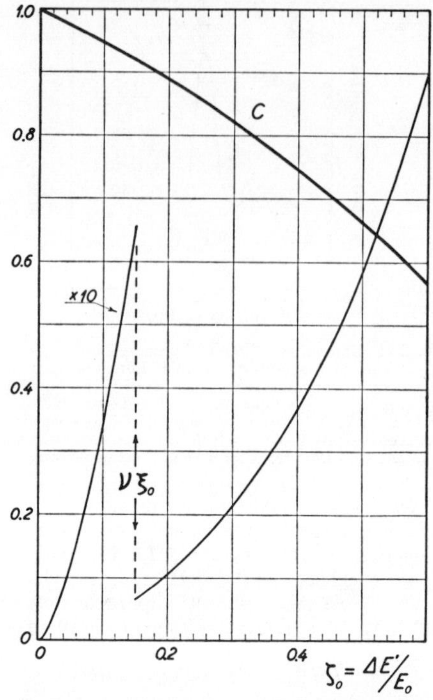

FIG. III.10. Correction factor to the effective target thickness. The quantity C gives the factor by which the value read from Fig. III.9 should be multiplied in order to take into account the finite value of ν [see (II C.10)]. This correction factor may be represented approximately as a function of the single parameter ζ_0 [see (II C.4) and (II C.5)]. The ξ value (II C.12), to be employed in the reading of Fig. III.9, can be obtained from the $\nu\xi_0$ curve, which is shown in the present figure.

or $E^{-0.7}$ for the stopping power. The curves in Fig. III.9 should thus be applicable to all target materials, including compounds, and the uncertainties are expected in most cases to be less than 2%. Also the effect of the energy straggling, which is neglected in (8), is smaller than this amount.

For finite values of the parameter ν, slightly different curves are obtained, but, to an accuracy of better than a few percent, they can be found from the curves for $\nu=0$ by multiplying with a correction factor which is a function of the product $\nu\xi_0$ or ζ_0, only. This correction factor C is given in Fig. III.10 together with a curve for the determination of the ξ value [see (II C.12)].

For the angular distribution coefficients $\bar{a}_k^{E\lambda}$ appropriate to thick target measurements, one obtains in a similar way the expressions

$$\bar{a}_k^{E\lambda}(\nu,\zeta_0)=\frac{\displaystyle\int_{\zeta_0}^1 a_k^{E\lambda}(\nu,\zeta)u_\lambda(\nu,\zeta)\frac{d\zeta}{\zeta}}{\displaystyle\int_{\zeta_0}^1 u_\lambda(\nu,\zeta)\frac{d\zeta}{\zeta}},\qquad\text{(III.9)}$$

where $a_k^{E\lambda}$ are the thin target coefficients [see (II B.83), Fig. II.8]. The values for the thick target coefficients are to a good approximation the same as those for the thin target coefficients, if the latter are evaluated for a bombarding energy which is smaller than the actual one by the factor $(1+\delta E_\lambda/E_0)$. The estimated errors are less than three percent under the condition that the coefficients can be considered to depend linearly on ζ within energy intervals of the order of δE_λ.

The multiple scattering of the projectiles in the target gives rise to an angular spread of at most a few degrees for a target thickness of δE_λ.[105] The effect on the angular distribution of the γ rays is thus of minor importance.

III B.3. Background Radiation

When one studies the radiations following Coulomb excitation, it is of course not only important that the absolute yield is sufficient to give a reasonable counting rate, but also that the yield relative to the existing back-

[105] See, for example, reference 19 and also T. Huus, Kgl. Danske Videnskab. Selskab Mat. fys. Medd. **26**, No. 4 (1951).

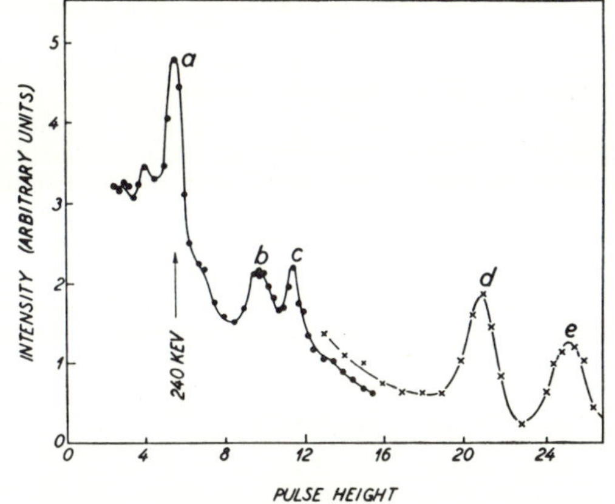

FIG. III.11. Effect of impurities in γ-ray measurements. The pulse-height spectrum, obtained with a crystal, has been observed in the bombardment of a thick Lu_2O_3 target with 2.6-Mev protons [McClelland, Mark, and Goodman, Phys. Rev. **97**, 1191 (1955)]. In addition to the peak a which corresponds to a 240-kev γ ray from the cross-over transition from the second excited state in Lu^{175}, other peaks are observed, arising from the presence of light elements in the target. The peak b at 439 kev is assigned to sodium impurities, and the peak c at 490 kev to the O^{16} (p,γ) process. The peaks d and e at 0.843 Mev and 1.017 Mev, respectively, are ascribed to inelastic scattering in aluminum, contained in the target material as an impurity; nominal purity of the sample was given as 99.9%.

ground radiations is high enough to be detectable in the actual experiments.

The background arises partly from external sources, such as the radiations from the accelerator, or from reactions with impurities in the target and with substances chemically bound to the element under investigation. Thus, oxide targets emit a strong γ radiation in the region of a few hundred kev when bombarded with protons (see Fig. III.11), and a line at 342 kev when bombarded with α particles.[102] In addition to this type of background, there is the background radiation due to processes taking place in the atoms of the investigated element itself. The production of the latter kind of background radiation can of course not be avoided. However, it can be discriminated against, if coincidence measurements can be performed,[106] or if the nuclear decay involves a sufficient delay.[107] When such possibilities do not exist, the best that can be done is to choose the experimental conditions so as to give the smallest possible ratio of background to nuclear radiation. It is therefore important to know how the atomic processes depend on the various parameters of the bombardment.

In the region of low γ-ray energies, the most important background process is the emission of the characteristic x-rays which follow the ionizations produced by the projectiles (see Figs. III.3, III.4, III.6, and III.13). The theoretical cross sections for the ionization of the K shell have been computed in Born approximation for nonrelativistic electron wave functions.[108] The result may be written in the form

$$\sigma_K \simeq Z_1^2(E_{\text{Mev}}/A_1)^4(36/Z_2)^{12}10^{-24}\ \text{cm}^2, \quad \text{(III.10)}$$

provided the K shell binding energy exceeds the maximum energy which a free electron can acquire in a collision with the projectile. Even for bombarding energies close to the Coulomb barrier, this condition is fulfilled for $Z_2>40$.

The experimental cross sections[109] are found to be somewhat larger than given by (10) for protons in the energy range employed in Coulomb excitation; thus, in the case of 4-Mev protons on tantalum, the observed cross sections are about five times larger than the estimate (10). The discrepancy has been ascribed partly to the inadequacy of the Born approximation, partly to relativistic effects in the electron motion.[110] However, the dependence of the cross section on the various parameters is approximately represented by the formula (10). If these x-rays constitute the dominating background, it is of no advantage to employ bombarding energies much higher than those for which the cross section for Coulomb excitation increases approximately as E^4, because then the signal to noise ratio will begin to decrease. For $E2$ excitations, this condition corresponds to $\xi\simeq 0.5$, as can be seen from Fig. III.12. From Eqs. (10) and (II C.13), (II C.15), and (II C.17) it also

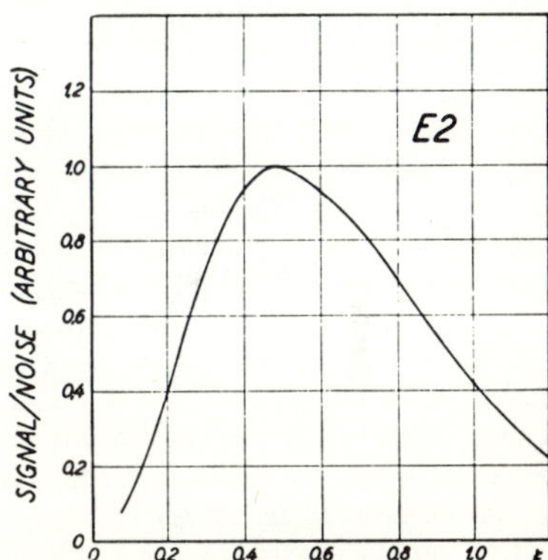

FIG. III.12. Ratio of Coulomb excitation to production of characteristic x-rays. The curve gives the ratio between the theoretical cross sections for $E2$ Coulomb excitation and ionization of the K shell, as a function of ξ. It is seen that an optimum is obtained for a bombarding energy corresponding to $\xi=0.5$. For this ξ value the signal to noise is proportional to $(A_1/Z_1)^4$. The same ξ dependence of the signal to noise ratio applies to the background of δ rays in the electron measurements.

[106] See, for example, reference 103, and G. M. Temmer and N. P. Heydenburg, Bull. Am. Phys. Soc. Ser. II **1**, 43 (1956).

[107] T. Huus and A. Lundén, Phil. Mag. **45**, 966 (1954).

[108] W. Henneberg, Z. Physik **86**, 592 (1933).

[109] Lewis, Simmons, and Merzbacher, Phys. Rev. **91**, 943 (1953); T. Huus and Č. Zupančič, reference 10 (on p. 17 of this reference, read "larger" instead of "smaller").

[110] Lewis *et al.* (see reference 109), and D. Jamnik and Č. Zupančič, Kgl. Danske Videnskab. Selskab Mat. fys. Medd. **31**, No. 2 (1956).

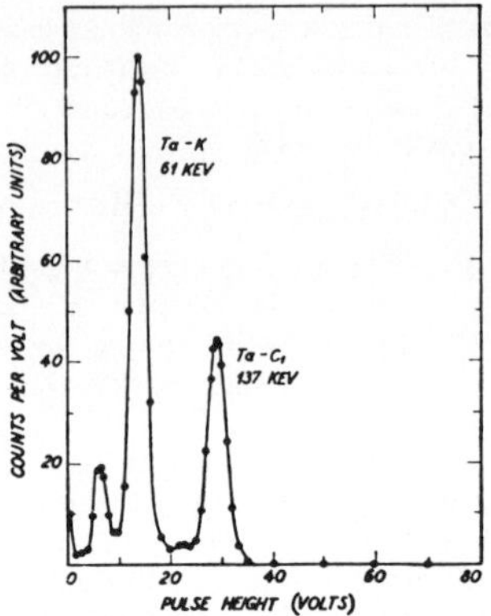

FIG. III.13. X-rays from internal conversion of the nuclear excitation. The pulse-height spectrum, obtained with a crystal, shows the relative strength of the K x-rays and the 137-kev γ ray from tantalum bombarded with 3-Mev α particles [G. M. Temmer and N. P. Heydenburg, Phys. Rev. **93**, 351 (1954)]. No absorbers were employed. The K conversion coefficient for the nuclear radiation is about 1.7, and the major part of the K peak is therefore accounted for by the internal conversion of the γ ray. In the case of proton bombardment the main part of the K x-rays arises from the direct ionization of the K shell (see Fig. III.6).

follows that at the optimum the signal to noise ratio will be proportional to $(A_1/Z_1)^4$, which is 16 times larger for α particles than for protons. Thus, in the α-particle experiments, the observed K x-ray peak is usually small and can sometimes be accounted for nearly exclusively by the effect of the internal conversion of the nuclear radiation (see Fig. III.13).

At γ-ray energies well above the K shell binding energy, one observes in the case of proton bombardment a background radiation which can be ascribed to bremsstrahlung associated with the deflection of the protons in the nuclear field[111] (see Fig. III.14). The cross section for this process is given in Sec. II E.1, where it is shown that the variation with the bombarding energy and the angle of observation is the same as for the $E1$ Coulomb excitation. If from Eq. (II E.13) one computes the corresponding thick target yield by means of Eq. (7) for the effective target thickness (see Fig. III.9), one finds that the total yield for all angles, multiplied by $E\cdot Z_2^{-5/2}$, to a good approximation is a function of the parameter ξ_0 only. This is confirmed by the measured yields[103] which furthermore show a ξ dependence in conformity with the theory. Also the predicted absolute intensity seems to be in agreement with the experimental evidence[112] within the rather large uncertainties of the available data. However, the possibility exists that there may be additional sources of background radiation, such as, e.g., bremsstrahlung associated with the ionization of the inner atomic shells.

The $E1$ excitation cross sections, and thus also the bremsstrahlung, increase with the bombarding energy in very nearly the same way as do the $E2$ excitation cross sections, except for the very high bombarding energies, where the latter become relatively greater, as illustrated by Fig. III.15. In the region of the spectra where the bremsstrahlung is the important background, one thus obtains a nearly constant signal to noise ratio in the case of an $E2$ excitation decaying to the ground state ($E_x=\Delta E$). The signal equals the noise for a partial $B(E2)$ value (see Sec. IV B) given by

$$\epsilon B(E2)\simeq\left(\frac{Z_1}{A_1}\right)^{2/3}\cdot\left(\frac{Z_1}{A_1}-\frac{Z_2}{A_2}\right)^2\left(\frac{Z_2}{85}\right)^{8/3}\times\left(\frac{100}{E_x}\right)^{4/3}\frac{dE_x}{E_x}e^2 10^{-48}\ \mathrm{cm}^4, \qquad \text{(III.11)}$$

where E_x is measured in kev and where dE_x is the resolution of the spectrometer. For a cascade γ ray, the signal-to-noise ratio is usually considerably smaller than for the ground-state decay, and increases with the bombarding energy.

In the case of α-particle bombardment, the bremsstrahlung is very weak due to the fact that the projectiles have nearly the same charge to mass ratio as the target nuclei [see (II E.13)]. The continuous background is indeed also found to be very low in the α-particle measurements, as illustrated by Fig. III.5.

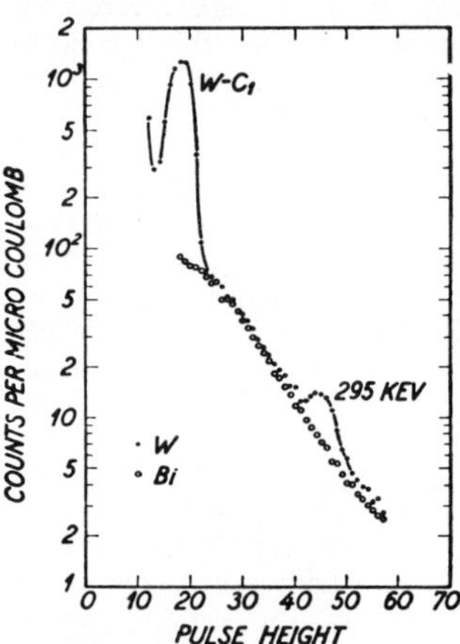

FIG. III.14. Proton bremsstrahlung. The figure shows the pulse-height spectra obtained by bombarding thick targets of natural W and Bi with 4-Mev protons [P. H. Stelson and F. K. McGowan, Phys. Rev. **99**, 112 (1955)]. The C_1 peak is a composite peak corresponding to the first rotational states in the even-A isotopes (see Fig. III.4), whereas the 295-kev peak is assigned to the odd isotope W^{183}. The radiation in the region between the two peaks can be ascribed to proton bremsstrahlung and has practically the same yield for W as for Bi, which give no nuclear radiation.

[111] Č. Zupančič and T. Huus, Phys. Rev. **94**, 205 (1954).

[112] Mark, McClelland, and Goodman, as quoted in reference 81. Also the measurements in reference 111 agree within the experimental error if the correct expression (II E.59) is used rather than the expression (II E.24), which was employed in this reference.

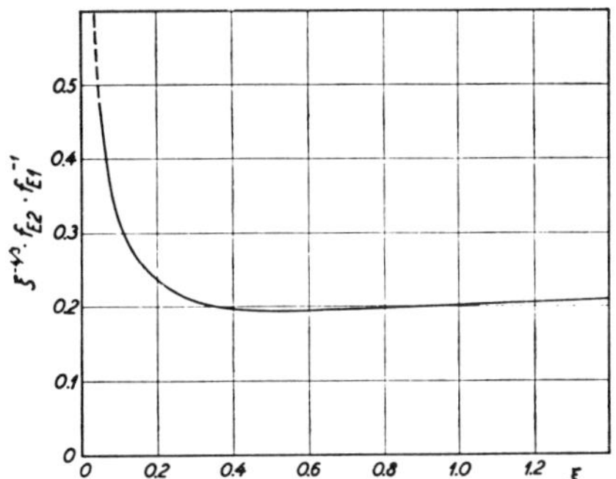

FIG. III.15. Comparison of theoretical cross sections for $E2$ and $E1$ Coulomb excitation. The ordinate is proportional to the ratio of the two cross sections, for the same value of ξ. It is seen that the $E1$ and $E2$ excitation functions are nearly identical over a wide range of ξ values. The excitation function for the dipole bremsstrahlung is the same as for the $E1$ Coulomb excitation (see Sec. II E.1), and the curve therefore also represents the signal-to-noise ratio for $E2$ excitation compared with bremsstrahlung. Thus, high bombarding energies, corresponding to small ξ values, are the most advantageous as far as this type of background radiation is concerned.

III C. Measurements of Conversion Electrons

III C.1. Detection Technique

The study of the internal conversion electrons emitted in the decay of the excited states is to some extent complementary to the γ-ray measurements. For heavy elements and low-energy transitions, an appreciable or even major fraction of the excitations will decay by the emission of such electrons, which may therefore be rather easily detected. Moreover, the derived excitation cross sections may be less sensitive to the value of the conversion coefficients.

Figures III.16–III.20 show some spectra of conversion electrons produced by Coulomb excitation. They have been measured by double-focusing magnetic spectrometers of the wedge-gap type,[113] which are convenient for the purpose. Such spectrometers readily allow the target to be "viewed" from the same side as that turned against the bombarding particles, so that the electrons do not have to penetrate a target support. The comparatively high resolving power is often of particular advantage, because of the great similarity of many of the nuclear spectra (see Figs. III.4 and III.17), and because of the relatively small energy difference between the successive transitions in rotational cascade decays (see Figs. III.5 and III.16).

The fact that conversion electrons from more than one of the atomic shells can be observed makes it possible to obtain additional information by this method. From the measured energy difference between the K and L conversion lines one can unambiguously assign the element in which the excitation has taken place, and from the intensity ratio between the K and L peaks one obtains information about the multipolarity of the radiation (see Figs. III.16 and III.17). With a somewhat higher resolution it should also be possible to determine the multipolarities from a comparison between the lines of the various L subshells. The theoretical angular distribution coefficients have only been partially evaluated. For $M1$ conversion in the K shell, the estimated anisotropies are rather small (see reference 31).

The strong interaction between the electrons and the target atoms implies that in general thin targets have to be employed in the experiments, if one wants to preserve the high resolution. It is important that the target be homogeneous and that the beam remains focused on the same spot, in particular since the conversion lines often appear on top of a strong continuous background. Thick targets can be used when the electrons have a relatively high energy, so that they can penetrate with sufficient ease the layer corresponding to the effective target thickness for the projectiles.

III C.2. Background Effects

If special precautions are not taken, there may be a considerable background due to the large number of scattered beam particles in the spectrometer, but it is rather easy to trap these by means of an appropriate set of stops. Also the background effects resulting from the presence of light atoms in the target are relatively harmless, since these elements have small conversion coefficients and give negligible contributions to the stopping electrons (see later discussion). This is an advantage when, for practical reasons, one employs

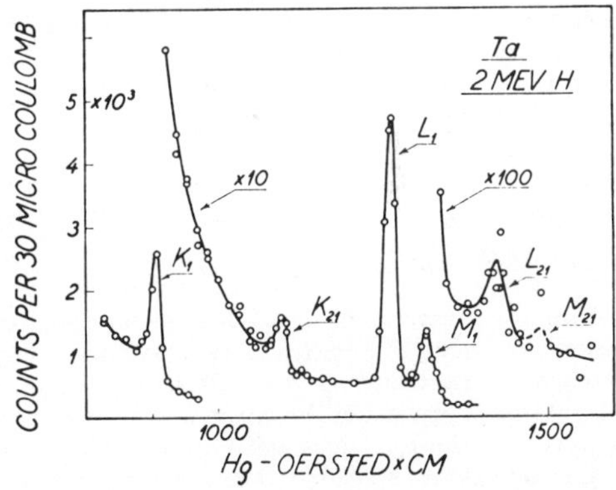

FIG. III.16. Internal conversion electrons from the excitation of tantalum with protons. The figure shows the spectrum of electrons from a 0.3 mg/cm² thin Ta target bombarded with 2-Mev protons [T. Huus and J. H. Bjerregaard, Phys. Rev. **92**, 1579 (1953)]. The measurements are made with a magnetic spectrometer of the wedge-gap type. The K, L, and M conversion lines with the indices 1 and 21 are assigned to the ground-state transition from the first rotational state of Ta^{181}, and to the cascade transition from the second to the first rotational state, respectively. The large K/L ratios indicate predominantly $M1$ transitions.

[113] Kofoed-Hansen, Lindhard, and Nielsen, Kgl. Danske Videnskab. Selskab Mat. fys. Medd. **25**, No. 16 (1950).

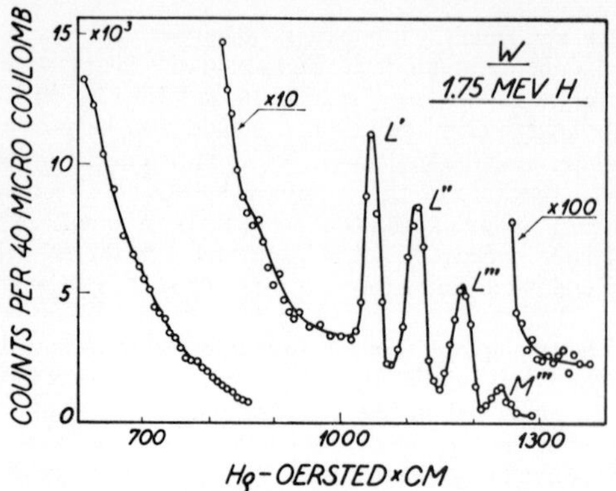

FIG. III.17. Conversion electrons from the excitation of tungsten. The spectrum shows the conversion lines observed in the bombardment of thin targets of natural W by 1.75-Mev protons [T. Huus and J. H. Bjerregaard, Phys. Rev. **92**, 1579 (1953)]. The peaks labeled L', L'', and L''' are predominantly due to the L conversion of the ground-state transitions from the first rotational levels in the even-A isotopes W^{182}, W^{184}, and W^{186}. The strong continuous background at the lower momenta is due to the production of stopping electrons. These conceal the presence of the K lines, but the fact that these lines are not clearly visible implies that the K/L ratios are small, in accordance with the $E2$ character of the transitions.

chemical compounds, such as oxides, for the target preparation. Similarly, the target support gives rise to no difficulties if it is made of light materials.

As in the case of the γ-ray measurements, however, atomic processes in the target element under investigation give rise to background effects which cannot be avoided. The maximum energy which a free electron can acquire in a collision with the projectile is less than 20 kev, even for bombarding energies close to the barrier. Collisions with the outer atomic electrons therefore do not give rise to any significant background. The tightly bound electrons, however, may be ejected with much higher energies, and such δ rays constitute the main background radiation in the electron experiments. The observed yield per energy interval can be represented approximately by the semiempirical expression[111]

$$d\sigma \simeq 2Z_1^2(E_{\mathrm{Mev}}/A_1)^4 Z_2^4 E_\delta^{-7} dE_\delta 10^{-24}\ \mathrm{cm}^2, \qquad \text{(III.12)}$$

where E_δ is the kinetic energy of the ejected electrons measured in kev. The cross section increases very strongly with decreasing E_δ, as is illustrated by Figs. III.16–III.19. The yield of the δ rays increases with Z_2 in contrast to the total ionization cross sections [see (10)]. It is therefore difficult to measure the conversion electrons from the decay of the first excited state of the very heavy elements, and for this reason the method has been applied mostly to the study of somewhat lighter nuclei.

The cross section (12) depends on the bombarding energy and the type of projectile in the same way as the cross section (10), and thus the largest signal to noise ratio for $E2$ excitations is again obtained for $\xi \simeq 0.5$ (see Fig. III.12). For bombarding conditions corresponding to this ξ value, the signal equals the noise, as represented by (12), for a partial $B(E2)$ value (see Sec. IV B) given by

$$\epsilon B(E2) \simeq \left(\frac{Z_1}{A_1}\right)^4 \left(\frac{Z_2}{85}\right)^8 \left(\frac{100}{\Delta E}\right)^4 \times \frac{dE_\delta}{E_\delta}\left(\frac{\Delta E}{E_\delta}\right)^6 e^2 10^{-48}\ \mathrm{cm}^4, \qquad \text{(III.13)}$$

where the excitation energy ΔE is measured in kev, and where E_δ equals the energy of the observed conversion electrons, also measured in kev. It is evident from (13) that the B values corresponding to the noise are the smallest for the heavier projectiles. Consequently, it sometimes proves to be an advantage to use deuterons rather than protons for the excitation of the lowest states (see Fig. III.19), even though the background of penetrating radiation, which is always generated in deuteron bombardments, gives rise to some difficulties. If a sufficiently high acceleration voltage is available, the best results are obtained with α particles (see Figs. III.18 and III.20).

In estimating thick target yields of ejected electrons by means of Eq. (12), it must be taken into account that they come only from a rather thin surface layer. For heavier elements, the effective thickness of this layer is approximately given by[114]

$$t_\infty \simeq (E_\delta/50)^2\ \mathrm{mg/cm}^2, \qquad \text{(III.14)}$$

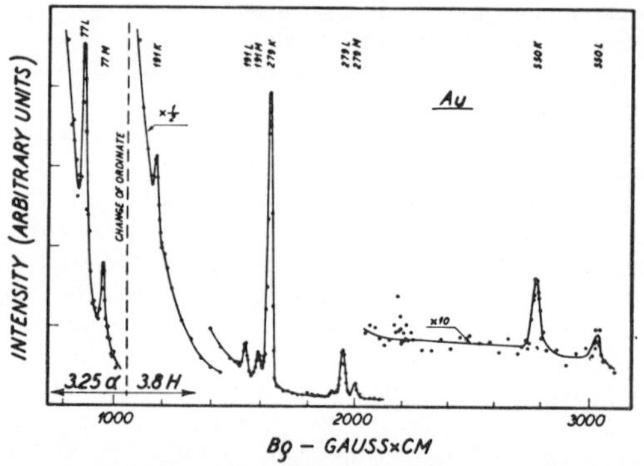

FIG. III.18. Conversion electrons from the excitation of gold. For the assignment of the observed transitions, confer the level scheme in Table IV.2. For the low electron energies the best results are obtained with α particles, and this part of the curve is reproduced from E. M. Bernstein and H. W. Lewis [Phys. Rev. **100**, 1345 (1955)]. For the high energies, the spectrum has been obtained by bombardment with protons. This part of the spectrum represents results obtained by M. S. Moore and C. M. Class (private communication).

[114] See Huus, Bjerregaard, and Elbek, Kgl. Danske Videnskab. Selskab Mat. fys. Medd. **30**, No. 17 (1956).

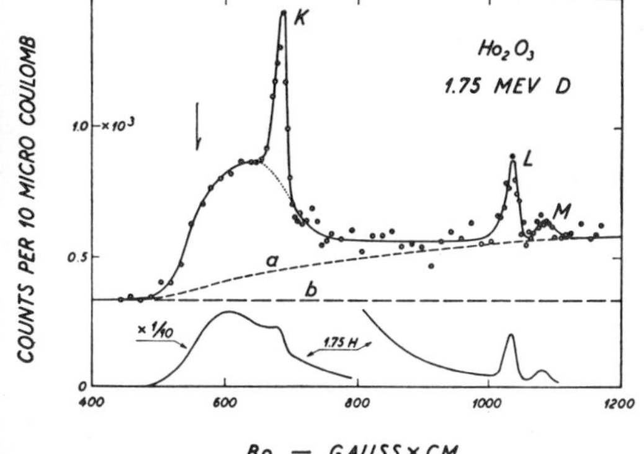

Fig. III.19. Conversion electrons from the excitation of holmium. The observed K, L, and M lines [Huus, Bjerregaard, and Elbek, Kgl. Danske Videnskab. Selskab Mat. fys. Medd. **30**, No. 17 (1956)], are associated with the first rotational state in Ho^{165}. The excitations were produced by the bombardment with 1.75-Mev deuterons which give a relatively small background of stopping electrons (dotted line), as is evident from the comparison with the curve for protons of the same energy, which is also shown in the figure. The arrow marks the cutoff due to the counter window. The dashed line a indicates the background contributions from the generation of β activities and the production of neutrons. The contribution from the latter effect alone is indicated by the dashed line b.

where E_s represents the energy, in kev, with which the electrons emerge from the surface.

In principle, the ejected electrons can be used for calibration of the target thickness, since the rate of their production depends in a smooth way on the atomic number of the target material. However, if the target thickness is not considerably smaller than t_∞, the calibration will be dependent on the homogeneity of the targets. The calibrations may therefore usually be performed more reliably by means of the intensity of the elastically scattered projectiles which are not so easily influenced by the structure of the target.

III D. Measurements of Inelastically Scattered Projectiles

Perhaps the most straightforward method of detection in the Coulomb excitation experiments is to measure directly the inelastically scattered projectiles. This method has the special advantage that each particle group corresponds to the excitation of a definite level, and that the yield is a direct measure of the cross section for the excitation, irrespective of the mode of decay. An example of a spectrum of inelastically scattered protons in a heavy element is shown in Fig. III.21. The measurements have been performed by means of a magnetic spectrometer of high resolving power. Because of the correspondingly small transmission, the particles were detected by means of a photographic plate.

A high resolution can only be obtained with thin targets, and a thickness determination must therefore be included in the measurements. The elastic scattering offers a convenient means for yield calibrations, and in the present case it is even not necessary to know the transmission of the spectrometer, since the solid angle is practically the same for two lines which are close to each other. However, a comparison cannot be made in a single exposure due to the widely different intensities and will, consequently, be dependent on the calibration of a beam integrator. The uncertainties introduced in this way are not of any great significance, in particular if approximately the same currents are employed in the two exposures. One thus directly compares the cross sections for Coulomb excitation with the Rutherford cross section, and the reduced nuclear transition probabilities $B(\lambda)$ derived from such a procedure should therefore be very reliable. At present, the accuracy of the analysis is limited to some extent by the fact that the differential excitation cross sections have only been calculated theoretically in the classical approximation (see Sec. II C.3).

The large cross sections for elastic scattering imply that even extremely small contaminations in the targets give rise to peaks in the spectra (see Fig. III.21), but these lines can be identified by the way in which they move with respect to the main Rutherford line, when the bombarding energy or the angle of observation is changed. Elastic scattering from the target nuclei will, however, give rise to a continuous background if the beam employed for the bombardment is not completely free of energy degraded particles. Even if the beam is passed through a magnetic analyzer before it strikes the target, there may still be a significant background due to scattering from stop edges etc. For this reason, it is in general preferable to observe in the backward directions, where the elastic scattering relative to the Coulomb excitations is the smallest. The ratio of the cross sections for $E2$ Coulomb excitation

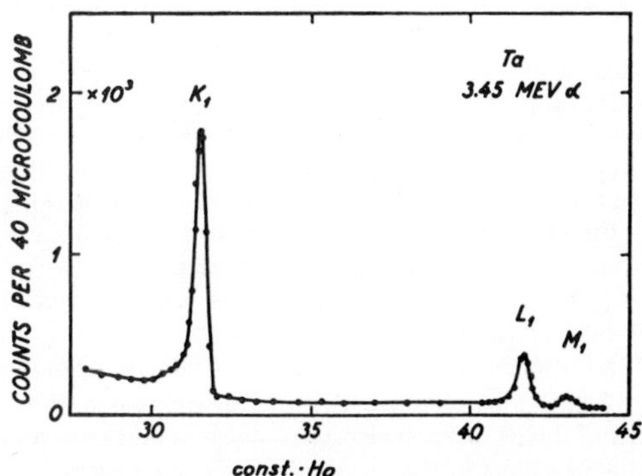

Fig. III.20. Conversion electrons from the excitation of tantalum with α particles. The spectrum shows the conversion lines from the decay of the first rotational excitation of Ta^{181} [E. M. Bernstein and H. W. Lewis, Phys. Rev. **100**, 1345 (1955)]. The background of stopping electrons is seen to be much smaller than in the case of proton bombardments (see Fig. III.16).

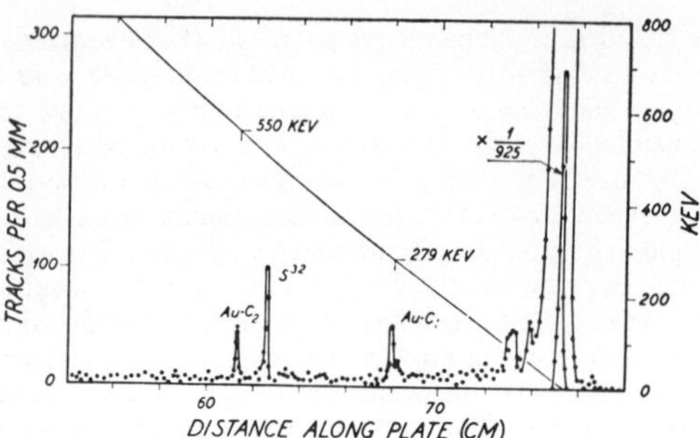

FIG. III.21. Spectrum of protons scattered from gold. The figure shows the energy spectrum of protons scattered from a 0.1 mg/cm^2 thin Au target (B. Elbek, and C. K. Bockelman, to appear in Phys. Rev.). The measurements were made with a magnetic spectrometer of high resolving power, and the particles were detected by means of a photographic plate. The exposure corresponded to approximately 4 millicoulomb. The angle of observation was 130° and the bombarding energy 6 Mev. The energy intervals between the inelastic groups and the strong peak from elastic scattering can be obtained from the calibration curve shown in the figure. The two strongly excited states in Au^{197} correspond to the peaks C_1 and C_2, and there is also an indication of the more weakly excited 268-kev level (see Table IV.2). The peak labeled S^{32} is due to elastic scattering from a contamination of sulfur.

and elastic scattering is proportional to $A_1Z_1^{-2}E^3$, and it is therefore advantageous to employ high bombarding energies. For energies of the order of the Coulomb barrier, the signal to noise increases as A_1Z_1.

The energy region over which the inelastic groups can be observed extends from the elastic peak down to the continuous background from the target support. This free region is related to the recoil energy; the extension increases with the mass and energy of the projectile and with the scattering angle, and decreases with increasing mass of the nuclei in the target support. Light elements in the target support may, on the other hand, give rise to nuclear reactions with the emission of charged particles. Aluminum has been used as a support in experiments with protons, at energies about 6 Mev.[115]

Because of the great strength of the elastic scattering, it is desirable that the spectrometer gives a very sharp image, but even then the elastic peak will always have a significant low-energy tail due to the energy straggling in the target. In the study of the low excitation energies, it is therefore necessary to employ very thin targets, even when the observations are made on particles which have penetrated the target and, thus, on the average have lost the same energy.

In addition to the above-mentioned contributions to the background radiation there will, just as for the γ rays and the conversion electrons, be contributions from atomic processes in the target. Thus, the considerations made earlier with regard to the effect of the bremsstrahlung also apply here, with the supplementary remark that the backward angles of observation favor the $E2$ Coulomb excitations as compared to the $E1$ bremsstrahlung, for which the angular distribution is given by the functions $df_{E1}(\vartheta)$ (see Fig. II.7). The processes leading to the ionization of the inner atomic shells, which give rise to an important background in the measurements on the decay radiations (see foregoing), are of less importance in the detection of the inelastically scattered projectiles, since the angular distribution of the particles responsible for the ionization is expected to be rather strongly peaked in the forward direction.

[115] B. Elbek (to be published).

CHAPTER IV. NUCLEAR DATA OBTAINED FROM COULOMB EXCITATION

In this chapter, we discuss the analysis of the experimental results on Coulomb excitation in terms of the theory given in Chapter II. This analysis confirms the accuracy of the theoretical description of the excitation process and leads to the determination of the nuclear parameters involved in the theory. The chapter also contains a compilation of the experimental results that have been obtained from Coulomb excitation investigations.

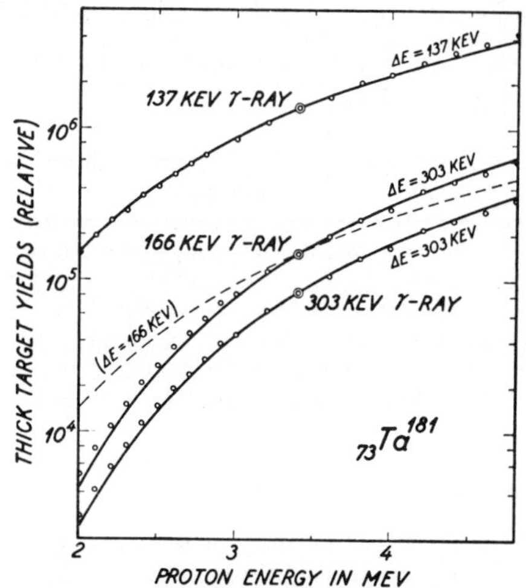

FIG. IV.1. Excitation functions for levels in Ta^{181}. The figure gives the thick target yield of the three γ rays observed in proton bombardment of Ta^{181}. The experimental data are taken from P. H. Stelson and F. K. McGowan, Phys. Rev. **99**, 112 (1955). The full drawn curves give the theoretical energy dependence of the yield, assuming $E2$ Coulomb excitation [see (II C.15) and (III.6) and Figs. II.5 and 6 and III.9 and 10]. The stopping power has been assumed to vary as $E^{-0.55}$, and the curves are normalized to the experimental value at 3.4 Mev, as indicated by the large circles. The theoretical curves are rather sensitive to the excitation energy ΔE. It is seen that the 137 kev and 303 kev γ rays have excitation functions with $\Delta E = E_\gamma$ and thus represent ground-state transitions, while the 166-kev γ ray has an excitation function with $\Delta E = 303$ kev and is thus associated with a cascade decay of a level at this energy (see Fig. V.7). For comparison, the excitation function corresponding to $\Delta E = 166$ kev is drawn with a broken curve. The contribution to the 137-kev radiation resulting from the 303-kev excitation decaying by cascade has not been subtracted from the experimental yield. The correction amounts to about 10% at the highest bombarding energies employed.

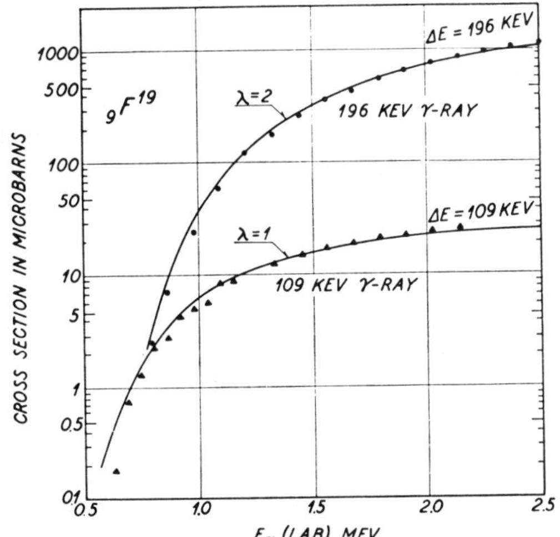

FIG. IV.2. Excitation function for levels in F^{19}. The figure shows the measured excitation cross sections for the 109-kev and 196-kev γ rays observed in α bombardment of a thin target of CaF_2 [Sherr, Li, and Christy, Phys. Rev. **96**, 1258 (1954)]. The theoretical excitation functions given by the full drawn curves are obtained from (II C.15) and Figs. II.4, 5, and 6, by assuming $E1$ excitation with $\Delta E=109$ kev for the 109-kev γ ray, and $E2$ excitation with $\Delta E=196$ kev for the 196-kev γ ray. The excitation functions are not sensitive to the multipole order, but the assumed values of λ are those indicated by other experimental evidence (see the references in Table IV.2). The theoretical curves are normalized to the experimental cross sections at $E_\alpha=1.55$ Mev.

IV A. Analysis of Experimental Data

IV A.1. Excitation Function and Relative Yields

The theoretical expressions for the Coulomb excitation yields as a function of the bombarding energy are independent of the nuclear structure. It is thus possible with considerable certainty to identify an observed radiation as resulting from Coulomb excitation by a measurement of its yield function. Examples of well-measured yield functions are given in Figs. IV.1 and IV.2. It is seen that the theoretical expressions reproduce the observed relative yields over a range in which the cross sections vary by several orders of magnitude.

The yield function depends on the multipole order λ and the excitation energy ΔE, and may thus be used to determine these two quantities. The sensitivity of the yield curve to the excitation energy, ΔE, may often be exploited to decide whether an observed radiation represents a ground-state decay or a cascade radiation from a higher lying state. As an example, Fig. IV.1 clearly shows that the observed 166-kev gamma ray in Ta^{181} originates from an excited state with an energy of about 303 kev. This fact is also directly confirmed by the observation of coincidences between the 166-kev and 137-kev γ rays (see Table IV.2). Similarly, the yield of the 110-kev radiation from Tm^{169} has been shown to indicate that this transition results from the excitation of a 119-kev level decaying to a 9-kev state (see Table IV.2).

The possibility of determining the multipole order of the excitation process on the basis of the yield curve is illustrated in Fig. IV.3. While it would be rather easy to recognize higher multipole orders ($\lambda=3$ or 4), it is usually difficult to distinguish $E2$ from $E1$ on the basis of the excitation function.[115a] These two multipole transitions have very nearly parallel yield functions, except for small ξ values corresponding to high bombarding energies or low excitation energies (see Fig. III.15).

An alternative method for determining the multipole order of the excitation is provided by a comparison of the yield for two different bombarding particles. If, for instance, one choses bombarding energies corresponding to the same value of ξ, the cross section for an excitation of order $E\lambda$ is, to a first approximation, proportional to $Z_1^2(A_1/Z_1)^{2\lambda/3}$ [see (II C.13), (II C.15), and (II C.16)]. Thus, the ratio of the cross sections for proton and alpha-particle bombardments would differ for $E1$ and $E2$ excitation by about a factor of 1.6. This method for determining λ has been used, for example, to establish the $E2$ character of the 446-kev transition in Na^{23} (see Fig. IV.7) and of the 100-kev transition in W^{182} (see Table IV.2).

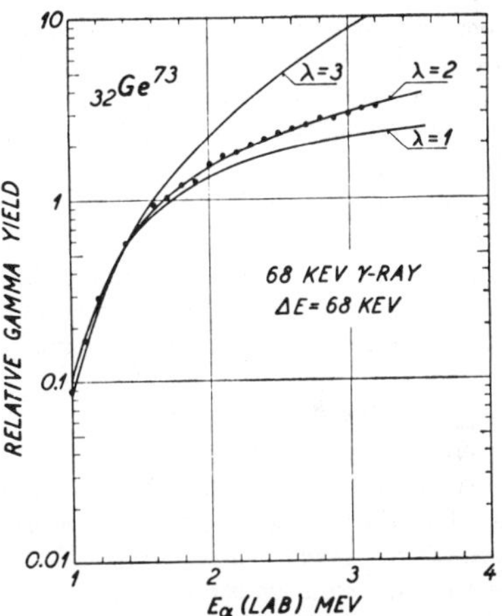

FIG. IV.3. Dependence of excitation function on multipole order. The figure shows the thin target yield of the 68 kev γ ray observed in α bombardment of Ge^{73} [G. M. Temmer and N. P. Heydenburg, Phys. Rev. **96**, 426 (1954)]. The full drawn curves give the theoretical excitation functions for $E1$, $E2$, and $E3$ Coulomb excitation, assuming $\Delta E=68$ kev [see (II C.15) and Fig. II.4]. The curves are normalized to the experimental value at 1.4 Mev. The possibility of distinguishing in the present case between $E1$ and $E2$ excitation on the basis of the yield function is associated with the rather small ξ values for the excitation ($\xi=0.14$ for $E_\alpha=3$ Mev).

[115a] *Note added in proof.*—Recently, an $E3$ excitation process, leading to the 40 kev isomeric level in Rh^{103}, has been identified as of $E3$ type on the basis of the measured excitation function (G. A. Jones and W. R. Phillips, presented at the Amsterdam Conference on Nuclear Reactions, July, 1956).

IV A.2. Angular Distribution of Decay Radiation

The angular distribution of the radiation following Coulomb excitation provides information on the spins and parities of the states involved as well as on the multipole order of the excitation mode and the decay.

Extensive angular distribution measurements have been made[17] of the γ rays from even-even nuclei which should follow the unique $0(E2)2(E2)0$ correlation which is given by [see (II C.29) and Table II.11]

$$W(\vartheta)=1+0.357a_2^{E2}P_2(\cos\vartheta)+1.143a_4^{E2}P_4(\cos\vartheta), \qquad \text{(IV.1)}$$

where the coefficients a_2 and a_4 are characteristic of the Coulomb excitation process. The observed distributions have been analyzed to yield experimental values of these coefficients which are seen in Figs. IV.4 and IV.5 to be in approximate agreement with the theoretical values given in Fig. II.8. It seems that the small differences between the experimental and theoretical values are not outside the experimental uncertainties.

As discussed in Sec. II C.5, there may be in certain cases important effects on the angular distribution resulting from the precession of the nuclear spin in the excited state. In the present cases, however, these effects are expected to be very small due to the short lifetimes of the states involved [$\tau_{\frac{1}{2}}\approx 1\cdot 10^{-11}$ sec for Cd^{114} and $\tau_{\frac{1}{2}}\approx 4\cdot 10^{-11}$ sec for Pt^{194}, as determined from the absolute yield of the Coulomb excitation of these levels (see Sec. IV A.4 and Table IV.2)]. Moreover, for the excited states involved, the static quadrupole moments are expected to be small, even though the transition moments are rather large (see Sec. V C.2). Also the higher order effects in the excitation process are expected to be small in the present circumstances (see Sec. II D.2).[116]

Besides these investigations of the even-even nuclei, a number of measurements of the angular distribution of the γ rays from odd-A nuclei have also been made (see, e.g., Fig. III.8). The analysis of these experiments by means of the theoretical expressions in Sec. II C.4 has yielded the spin determinations and multipole assignments listed in Table IV.2.

In the frequently occurring case of an $E2$ excitation followed by a mixed $M1+E2$ decay, there may often be an ambiguity in the mixing ratio δ as determined from the angular distribution of the γ rays. This ambiguity may be removed by a measurement of the polarization of the γ quantum[117] [see (II A.78a)].

[116] It has been suggested (reference 17) that the multiple scattering of the projectile in the target may give rise to an important correction to the measured angular distribution of the γ rays. However, this effect appears to be very small under most experimental conditions (see the comments in Sec. III B.2).

[117] P. H. Stelson and F. K. McGowan, Bull. Am. Phys. Soc. Ser. II, 1, 164 (1956).

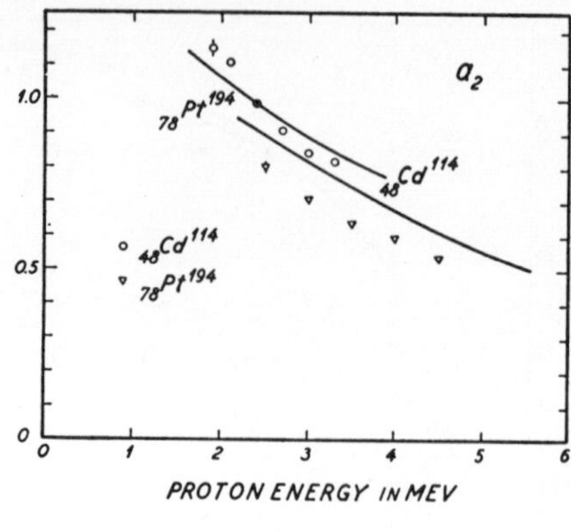

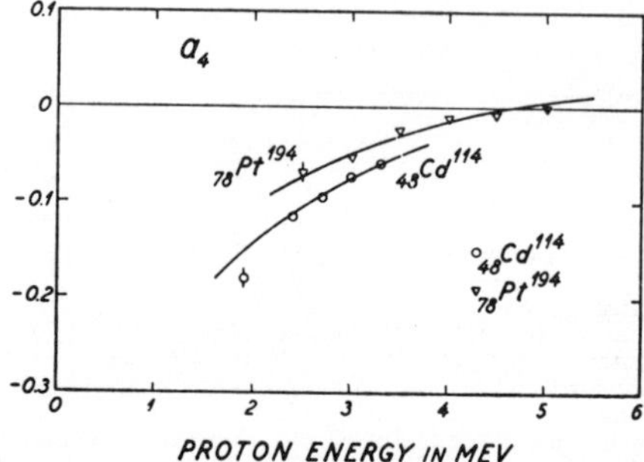

FIGS. IV.4 and IV.5. Angular distribution coefficients a_2 and a_4. The angular distribution of the γ rays following $E2$ Coulomb excitation depends on the excitation process only through the coefficients a_2 and a_4 [see (II C.29)]. The figures plot the experimentally determined a_2 and a_4 coefficients as a function of the proton bombarding energy; the data is taken from the thick target measurements by F. K. McGowan and P. H. Stelson [Phys. Rev. **99**, 127 (1955) and unpublished data, quoted in Goldstein *et al.* (Phys. Rev. **100**, 436 (1955)]. The full drawn curves give the theoretical thick target values for a_2 and a_4 obtained from Fig. II.8 by employing the thick target correction described in Sec. III B.2; the values of δE_2 involved in this correction are taken from Fig. III.9.

IV A.3. Angular Distribution of Inelastically Scattered Particles

The angular distribution of inelastically scattered particles depends only on the multipole order of the excitation, but not on the spins of the states involved. The measured angular distributions of the inelastically scattered protons from Au^{197} are compared in Fig. IV.6 with the theoretical distributions for $\lambda=1$ and 2, obtained in the classical approximation (see Sec. II C.3). The exact quantum-mechanical angular distributions have not so far been evaluated.

IV A.4. Absolute Yields

From the measured absolute cross sections one may derive the reduced transition probability $B(E\lambda)$ by means of the theoretical expressions in Sec. II C.2, and the values obtained in this manner are listed in Table

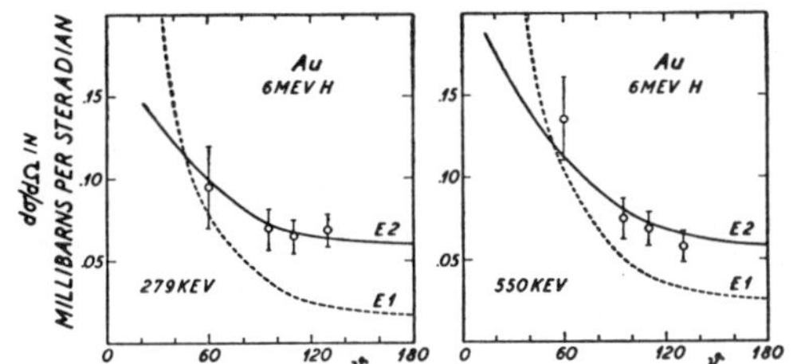

FIG. IV.6. Angular distribution of inelastically scattered protons from Au^{197}. The figure shows the differential cross sections in millibarns per steradian for excitation of the 279-kev and 550-kev levels in Au^{197} with protons of 6 Mev (B. Elbek and C. K. Bockelmann, to appear in Phys. Rev.) The full drawn curves give the theoretical cross sections obtained from (II C.15) and Fig. II.7, assuming $E2$ excitation (see also Sec. II C.3); the $B(E2)$ values are determined so as to give the best fit to the experimental points. The measured angular dependence of the cross sections agrees rather well with the classical theory (the value of η in the present experiment is about 5) and also the absolute values of the cross sections are in approximate agreement with those expected on the basis of the γ-ray yield measurements (see the $B(E2)$ values in Table IV.2). The angular dependence of the cross sections is rather sensitive to the multipole order of the excitation process. This is illustrated by the broken curves which give the theoretical $E1$ differential cross sections, normalized to the same total cross section as the $E2$ curves.

IV.2. The reduced transition probability also determines the lifetime for the inverse radiative transition of order $E\lambda$ [see (II A.56) and (II A.57)]. One thus obtains for the transition probability for γ emission from the excited state I_f to the ground state I_i

$$T_\gamma(E1;I_f\to I_i)=1.59\times10^8(\Delta E)^3 \times B(E1;I_i\to I_f)\frac{2I_i+1}{2I_f+1}\ \text{sec}^{-1}, \quad \text{(IV.2)}$$

and

$$T_\gamma(E2;I_f\to I_i)=1.23\times10^{-2}(\Delta E)^5 \times B(E2;I_i\to I_f)\frac{2I_i+1}{2I_f+1}\ \text{sec}^{-1}, \quad \text{(IV.3)}$$

where ΔE is measured in kev and $B(E\lambda)$ in units of e^2 $(10^{-24}\ \text{cm}^2)^\lambda$.

In a number of cases, measurements are available of both the cross section for Coulomb excitation and the corresponding radiative lifetime, thus providing two independent measurements of $B(E\lambda)$. The comparison of these determinations is contained in Table IV.1; the agreement appears in all cases to be within the experimental error.

In many cases, Coulomb excited states may decay in several modes, either to the ground state with a mixed multipole transition or by a cascade to some other nuclear level. In these cases, measurements of multipole mixtures and branching ratios together with the absolute cross section for Coulomb excitation provide a determination of the absolute transition probabilities for the alternative modes. In this way, a number of $M1$ decay probabilities have been determined (see Table V.5). In a similar manner, it has been possible to determine the transition probabilities for certain $E1$ transitions representing alternative modes of decay for levels in Se and Ag populated by $E2$ Coulomb excitation (see Table IV.2 and the references given there).

IV A.5. Coulomb Excitation at Higher Bombarding Energies

The above analysis refers to experimental arrangements in which the bombarding energy is sufficiently low that penetration into the nucleus can be neglected. The electromagnetic interaction is then the only mechanism for exciting the nucleus.

Already for energies appreciably below the Coulomb barrier, however, the projectile may penetrate to the

TABLE IV.1. Comparison of lifetime determinations with Coulomb excitation yield measurements. The table lists, in columns two and three, the spins of the nuclear ground state, I_0, and of the excited state, I. The measured half-lives, $\tau_{1/2}$, listed in column five, are taken from the review by A. W. Sunyar, Phys. Rev. **98**, 653 (1955) and the additional references listed below. The number in parentheses in column five gives the power of ten for the observed lifetime measured in seconds. The half-lives yield the reduced transition probabilities $B(E2;I_0\to I)$ by means of (IV.3) and the relation $(\tau_{1/2})^{-1}=1.4\,T_\gamma(E2)\,(1+\delta-2)\,(1+\alpha)$ where $T_\gamma(E2)$ is the transition probability per second for $E2$ γ radiation, while δ^2 is the ratio of $E2$ to $M1$ γ-ray intensity, and α is the total conversion coefficient. The values of δ and α are taken from column six of Table IV.2. The $B(E2)$ values obtained from lifetime determinations are given in column six in units of $e^2\times10^{-48}$ cm^4, and are compared with the corresponding quantities obtained from the Coulomb excitation cross sections (see column seven of Table IV.2).

A similar comparison as for the $E2$ transitions in the table can be made for the 110-kev $E1$ transition in $F^{19}(I_0=1/2, I=1/2)$. The measured half-life of 7×10^{-10} sec yields by means of (IV.2) the value $4.8\times10^{-30}\,e^2$ cm^2, while the Coulomb excitation cross section gives $B(E1)=2.3\times10^{-30}\,e^2$ cm^2.

Additional references for $\tau_{1/2}$: Thirion, Barnes, and Lauritsen, Phys. Rev. **94**, 1076 (1954) (F^{19}; 110-kev transition); Fiehrer, Lehmann, Leveque, and Pick, Compt. rend. **241**, 1746 (1955) (F^{19}; 197-kev transition); H. Schopper, Z. Physik **144**, 476 (1956) (V^{51}); F. R. Metzger, Phys. Rev. **101**, 286 (1956) (Ge^{72} and Ge^{74}); F. R. Metzger, Phys. Rev. **98**, 200 (1955) (Hg^{202}); H. deWaard, Phys. Rev. **99**, 1045 (1955), and R. E. Azuma and G. M. Lewis, Phil. Mag. **46**, 1034 (1955) (Tl^{203}); T. R. Gerholm (private communication) (Pb^{207}).

Nucleus	I_0	I	ΔE (kev)	$\tau_{1/2}$ (sec)	Lifetime $B(E2)$	Coulomb excitation $B(E2)$
$_9F^{19}$	1/2	5/2	197	6(−8)	0.01	0.003
$_{23}V^{51}$	7/2	5/2	325	1.0(−10)	0.008	0.006
$_{32}Ge^{72}$	0	2	835	3.2(−12)	0.19	0.26
Ge^{74}	0	2	595	1.3(−11)	0.28	0.30
$_{62}Sm^{152}$	0	2	122	1.4(−9)	3.3	3.1
$_{64}Gd^{154}$	0	2	123	1.2(−9)	3.6	4.5
$_{68}Er^{166}$	0	2	81	1.7(−9)	5.7	6.8[b]
$_{72}Hf^{176}$	0	2	89	1.35(−9)	5.3	6.0
Hf^{180}	0	2	93	1.4(−9)	4.9	5.0
$_{74}W^{182}$	0	2	100	1.27(−9)	4.3	5.6
$_{80}Hg^{198}$	0	2	411	2.1(−11)	1.1	0.8
Hg^{199}	1/2	5/2	159	2.4(−9)	0.35	0.26
Hg^{202}	0	2	439	2.2(−11)	0.8	0.5
$_{81}Tl^{203}$	1/2	3/2	280	1.4(−10)[a]	0.28	0.12
$_{82}Pb^{207}$	1/2	5/2	569	9(−11)	0.031	0.028

[a] This value, obtained from a direct measurement of the delay, differs considerably from the value deduced from the resonance scattering cross section [F. R. Metzger and W. B. Todd, Phys. Rev. **95**, 627(A) (1954)].
[b] The unresolved transitions from all the even erbium isotopes are assumed to have the same $B(E2)$ value.

nuclear surface and thus initiate proper nuclear reactions. Estimates of the expected reaction cross sections are given in Fig. III.1. In light elements, the radiation resulting from such nuclear reactions is characterized by a resonant structure which is superimposed on the more slowly varying yield of the Coulomb excitation. In heavier elements, the level spacing of the compound nucleus is usually below the energy resolution of the incident beam, and in addition the levels may overlap due to the effect of neutron emission.

Even when the average cross section for compound nucleus formation exceeds the Coulomb excitation cross section, it may still be possible to observe the latter, since the compound nucleus will usually decay preferentially through other channels, such as (p,n), (α,n), (α,p), and also the elastic channel (see Fig. IV.8).

Moreover, for light elements where the resonance structure can be resolved, the cross section between resonances may result mainly from Coulomb excitation, and a quantitative determination of the Coulomb excitation yield may then be possible (see Fig. IV.7). In this connection it is significant that the interference between Coulomb excitation and the contribution of a particular resonance is confined to a few angular momenta of the projectile, while the total Coulomb excitation yield results from many angular momenta.

For bombarding energies equal to, or greater than, the Coulomb barrier, the compound nucleus formation takes place with a large probability. Still, the observed inelastic scattering leading to the low-lying states of the target, appears to result from a direct interaction, since the yields greatly exceed those expected from the decay of the compound nucleus.[118] In these reactions one expects, however, besides the Coulomb interaction, an important contribution to the direct excitation from the interaction with the nuclear field. Moreover, the change of the projectile orbit and of the electric multipole fields when the projectile is inside the nucleus imply an essential modification of the calculations given in Chapter II (see, e.g., Sec. II E.3).

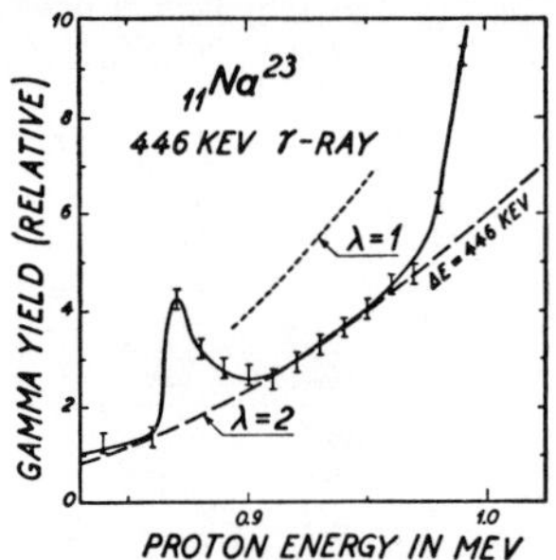

FIG. IV.7. Coulomb excitation of sodium by protons. The figure shows the yield of the 446-kev γ ray from a thin target of NaCl bombarded with protons. [G. M. Temmer and N. P. Heydenburg, Phys. Rev. **98**, 1198(A) (1955) and private communication]. Between the resonances due to compound nucleus formation one observes a smoothly rising background yield which may be ascribed to Coulomb excitation. It is possible to determine the multipole order of the Coulomb excitation by comparing with the yield observed in the excitation with α particles (loc. cit.). The dashed curves correspond to the cross sections expected for $\lambda=1$ and 2 on the basis of the observed cross section for excitation with α particles (see II C.15). The close agreement of the measured cross section with the theoretical curve for $E2$ excitation also confirms that the yield away from resonances is primarily due to Coulomb excitation.

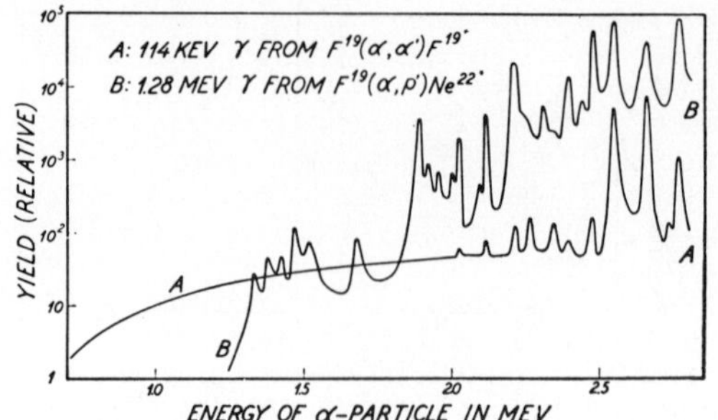

FIG. IV.8. Gamma rays from Coulomb excitation and compound nucleus formation in F^{19} bombarded with α particles. The figure shows the thin target yields of the 114-kev γ ray from the first excited state in F^{19} and the 1.28-Mev γ ray from the first excited state of Ne^{22} formed by an (α,p') process on F^{19} [Sherr, Li, and Christy, Phys. Rev. **96**, 1258 (1954)]. For bombarding energies below 1.2 Mev, the penetration of the α particle through the Coulomb barrier is very small (see Fig. III.1) and the cross section for compound nucleus formation is small compared to that for Coulomb excitation. With increasing bombarding energy, σ_{comp} increases rapidly and soon becomes larger than σ_{coul}. However, even for $E_\alpha \sim 2$ Mev, at which energy the average value of σ_{comp} is an order of magnitude larger than σ_{coul}, the yield of the 114-kev γ ray is only very little affected by the compound nucleus formation, since the probability that the compound nucleus decays by inelastic α-emission is small. Finally, for $E_\alpha \gtrsim 2.5$ Mev, the Coulomb excitation yield of the 114-kev γ ray is overshadowed by the resonance yield from compound nucleus formation.

In some cases, it may be possible to separate the simple Coulomb excitation effect resulting from the particles which have not passed through the nucleus, by observing the inelastically scattered particles in the forward directions. Thus, if $\eta \gg 1$, so that the projectile orbits can be described in the classical approximation, the particles passing outside the nucleus will be scattered into angles less than a critical value ϑ_c. This angle depends on the ratio

$$x = \frac{E}{E_B}, \tag{IV.4}$$

between the projectile energy E and the Coulomb barrier E_B [see (III.1)] and is given by [see (II A.22) and (II A.23)]

$$\vartheta_c = 2\sin^{-1}\left(\frac{1}{2x-1}\right) \approx \frac{115°}{2x-1} \quad (\text{for } x \gtrsim 2). \tag{IV.5}$$

---

[118] See, e.g., P. C. Gugelot, Phys. Rev. **93**, 425 (1954); Schrank, Gugelot, and Dayton, Phys. Rev. **96**, 1156 (1954), and also the review by H. McManus, Brookhaven report On the Statistical Aspects of the Nucleus, 1955.

Besides the particles resulting from Coulomb excitation there may also be a contribution from the nuclear interactions to the inelastic scattering in the forward direction. The relative magnitude of the two contributions will depend on the transition matrix elements as well as on the motion of the projectile inside the nucleus. The conditions for observing the Coulomb excitation appear to be the most favorable if the projectile has only a small chance of traversing the target nucleus, as in the case of α particles incident on heavy nuclei. Moreover, the strong increase of the $E1$ Coulomb excitation cross section in the forward direction may facilitate the detection of such excitations. The Coulomb excitation origin of an observed inelastic scattering may be tested if it is possible to measure the angular distribution or the dependence of the yield on the energy and charge of the projectile.

For $\eta<1$, the quantum mechanical diffraction effects are of more importance than the deflection in the Coulomb field, and the angle ϑ_c loses its significance.

IV B. Compilation of Experimental Results

The results obtained from Coulomb excitation investigations, reported in the literature or available to us by private communication prior to approximately April, 1956, are summarized in Table IV.2. A description of the entries contained in the various columns is given below.

Column I. Nucleus

In cases where the element bombarded consists of more than one isotope, the isotopic assignments of the observed radiation have been made by means of

1. use of separated (or enriched) isotopes, as noted under "comments" (column V),
2. identification of the observed radiation with that found in other reactions where the isotope is known, as indicated under "other processes" (column IX),
3. the general systematics of the excitation energies and cross sections for even-even nuclei, as listed under "comments." Thus, certain observed lines which fit into the established trends (see, e.g., Fig. V.3) may be assigned to an appropriate even-even isotope of the element investigated, while observed lines whose energies differ greatly from those of neighboring even isotopes can be ascribed to odd-A isotopes.

For some of the data, none of these methods of assignment is available, or the radiation is known to be composed of unresolved contributions from several isotopes; in such cases, only the element is listed in column I, and not the mass number.

Column II. Decay Energy

The energies of the observed decay transition following Coulomb excitation are listed in kev. When the method of detection involves the inelastic particle groups, the value listed in this column equals the excitation energy. If the excitation is detected by means of gamma radiation or conversion electrons, the observed decay may represent a cascade from a higher excited level. (See columns V and VIII for the information available on this point.)

A separate entry is made for each experiment, and the references are given in column III.

Column III. Bombarding Conditions

The range of bombarding energies is given in Mev. The projectile used is denoted by, p, proton, d, deuteron, α, alpha particle, N, nitrogen ions. References are given by means of an abbreviation, e.g., (M1) which refers to the bibliography listed at the end of the table. These references apply also to the decay energies and yield measurements listed in the previous and succeeding columns.

Column IV. Partial Reduced Transition Probability $\epsilon B(E2)$

The column lists the information regarding the nuclear transition probability which can be derived directly from the measured yield, assuming only a knowledge of the energy and multipole order of the excitation. Where the excitation energy is uncertain, the listed values are based on the arbitrary assumption that the decay takes place to the ground state. Since all excitations identified so far have been found to be of electric quadrupole type, with the exception of the weakly excited 109-kev level in F^{19}, we have assumed $E2$ character in all except this one case.

The quantity listed in this column is the partial reduced transition probability $\epsilon B(E2)$, where $B(E2)$ [see (II C.15)] is measured in units of $e^2\times10^{-48}$ cm^4, and where ϵ is the fraction of the excitations which decay through the observed mode. The detected radiation indicated in parenthesis is denoted as follows:

(γ) electromagnetic radiation,
(e_K), (e_L), etc. conversion electrons from the K shell, L shell, etc.,
(p') inelastic protons.

Thus, if the detected radiation is a γ ray (γ_j), the decay fraction is given by

$$\epsilon(\gamma_j)=\frac{f_j}{\sum_i (1+\alpha_i)f_i}, \tag{IV.6}$$

where f_i are the relative intensities of the various γ transitions by which the excited state may decay, and α_i are the corresponding total conversion coefficients. In the case of conversion electron detection, the decay fraction is given by (6) multiplied by the conversion coefficient for the conversion line in question. For the inelastic protons, $\epsilon(p')=1$.

If the Coulomb excitation populates several levels in the same nucleus, the radiation from the lower levels may partly result from a cascade decay of a higher level. In the determination of $\epsilon B(E2)$, it is then necessary to establish what part of the measured radiation is due to the direct excitation of the radiating level. In most cases, however, the correction for cascade is small compared with the accuracy of the yield determinations, and we have not attempted to include it except when it has already been taken into account by the original experimenters.

When the isotope responsible for the observed radiation has not been assigned, $\epsilon B(E2)$ has been calculated assuming 100% abundance for the responsible isotope. Where no yield determination is available, only the detected radiation is listed in this column.

Column V. Comments

This column contains a brief summary of other information on the observed levels, which has been obtained from Coulomb excitation. The abbreviations employed are

ex.func. $\Delta E = 127$ The measured yield as a function of energy has been found to be consistent with Coulomb excitation with an excitation energy ΔE equal to the listed value, assuming multipole order $E2$. If the yield function also determines the multipole order of the excitation process, the notation $\lambda = 2$ is added. While the excitation function is rather sensitive to the excitation energy, it often does not distinguish between $E1$ and $E2$ transitions (see Sec. IV A.1).

sep.iso. The isotope assignment has been established by Coulomb excitation experiments, employing enriched isotopes.

$\gamma(\theta)$ The angular distribution of the γ radiation has been measured. The spins and multipolarity of the transitions which have been deduced from the observed angular distribution are indicated by $I_i(E\lambda)I_f(L)I_{ff}$, where I_i, I_f, and I_{ff} are the spins of the initial state, Coulomb excited state, and final state, respectively, while $E\lambda$ and L are the multipolarity of the Coulomb excitation process and of the subsequent γ radiation, respectively.

δ The ratio of the amplitudes of $E2$ and $M1$ γ radiation in a mixed transition. For the definition of the sign of δ, see Sec. II A.4 and Table II.11.

$\sigma(p):\sigma(\alpha)$ gives $\lambda = 2$. The measured ratio of the Coulomb excitation cross sections for protons and α particles implies $E2$ excitation.

α_K, K/L The K conversion coefficient α_K and K/L ratio measured in Coulomb excitation experiments.

$\gamma(100)$–$\gamma(200)$coinc. The two indicated γ rays are found to be in coincidence.

$\gamma(100):\gamma(200)$ The value listed is the measured intensity ratio of two γ rays assumed to originate from the same excited level.

Column VI. Multipole Order and Conversion Coefficients of the Decay Radiation

For even-even nuclei, the excited states are assumed to be of (2+) character and to decay by pure $E2$ radiation. For odd-A nuclei, the decays are often of mixed $M1$ and $E2$ type, and the column lists the percentage of the γ decays which are of $E2$ type; the information is obtained from the angular correlation measurements or K/L ratios. In a number of cases, the transitions can be classified as $\Delta I = 2$, and thus as pure $E2$ decays, on the basis of the rotational interpretation of the levels involved (see column VIII). For some of the rotational transitions with $\Delta I = 1$, where no other evidence is available, the multipole mixture has been calculated from observed branching ratios [(V.10 and V.17); see also Fig. V.7]. In these cases the $E2$ percentage is listed in parenthesis.

In addition, the column gives the conversion coefficients employed in the derivation of the $B(E2)$ values in column VII. The listed values for the K shell conversion coefficients α_K for $Z<50$ are taken from the calculations appropriate to a point nucleus.[119] The theoretical values for $Z>50$ include the effect of the finite nuclear size.[120] The L-shell conversion coefficients α_L have been obtained from the α_K values by assuming a K/L ratio equal to that for a point nucleus.[119] The total conversion coefficients, α, are obtained by assuming $\alpha = \alpha_K + 1.3\alpha_L$, in order to approximately take into account the conversion in the higher shells.

Column VII. Reduced Nuclear Transition Probability $B(E2)$

In cases where sufficient data are available, a total $B(E2)$ for the excitation process may be computed from the $\epsilon B(E2)$ values listed in column IV and the assumed conversion coefficients (column VI). The $B(E2)$ value is listed opposite the radiation which represents the ground-state decay mode of the level in question. The value given is a weighted average of the various experimental yield determinations.[121]

Column VIII. Level Scheme

The suggested level schemes are based on the Coulomb excitation measurements as well as the evidence from other sources indicated in column IX. Only levels which have been observed in Coulomb excitation experiments are included. The spin assignments listed in parenthesis are based on the assumed rotational char-

[119] Rose, Goertzel, Spinrad, Harr, and Strong, Phys. Rev. **83**, 79 (1951) and Rose, Goertzel, and Swift, privately circulated tables.

[120] L. A. Sliv, privately circulated tables; see also L. A. Sliv, J. Exptl. Theoret. Phys. U.S.S.R. **21**, 770 (1951) and L. A. Sliv and M. A. Listengarten, *ibid.* **22**, 29 (1952).

[121] There appears to be an unresolved discrepancy between the γ-ray yield measurements in the series of experiments reported in (M1, M3, M4, and M6) and those of other experimenters (see column IV). Rather than attempt to average such conflicting determinations, we have arbitrarily omitted the former values from the averages listed in column VII.

TABLE IV.2. Results obtained from Coulomb excitation. The description of this table is given in (IV B) of the text.

I Nucleus	II Decay energy	III Bombarding conditions	IV $\epsilon B(E2)$	V Comments	VI Multipole order and conversion coefficients	VII $B(E2)$	VIII Level schemes	IX Other processes
$_9F^{19}$	109 112 113 124	0.6–2.8α (S1) 0.7–1.2α (J1) 1–3α (T2; H9) 15N (A1)	(γ) (γ) (γ) (γ)	ex. func. $\Delta E=110$, $\lambda=1$ or 2 (S1; J1; H9) $\gamma(\theta)$: $1/2(E1)1/2(E1)1/2$(S1) $B(E1) 2.3\times10^{-30}e^2x$ cm^2(S1) $=1\times10^{-29}e^2x$ cm^2(J1)			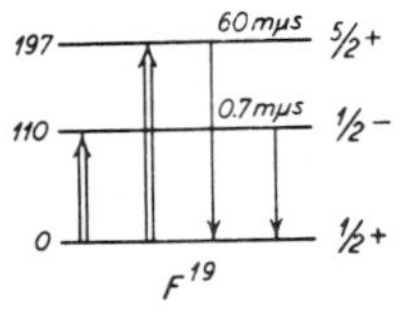 197 — 60 mμs — 5/2+ 110 — 0.7 mμs — 1/2− 0 — 1/2+ F^{19}	110 $Ne^{21}(d\alpha)$; $F(pp')$; $F(nn'\gamma)$; $O^{19}(\beta^-\gamma)$; and others
	196 195 196 205	0.8–2.8α (S1) 0.7–1.2α (J1) 1–3α (H9; T2) 15N (A1)	0.003 (γ) 0.002 (γ) (γ) (γ)	ex. func. $\Delta E=196$ (S1; H9) $\sigma(p):\sigma(\alpha)$ gives $\lambda=2$(T1; B9) $\gamma(\theta)$: $1/2(E2)5/2(E2)1/2$(S1) no cascade transition via 110-kev level observed		0.003		197 $Ne^{21}(d\alpha)$; $F(pp')$; $F(nn'\gamma)$; $O^{19}(\beta^-\gamma)$; and others
$_{11}Na^{23}$	446 435	1.5–3.6α (T2) 15N (A1)	0.013 (γ) (γ)	ex. func. $\Delta E=446$(T2) $\sigma(p):\sigma(\alpha)$ gives $\lambda=2$(T1)		0.013		439 $Na(nn'\gamma)$; $Na(pp'\gamma)$; $Ne^{23}(\beta^-\gamma)$; and others
$_{22}Ti^{46}$	890	6α (H2)	0.15 (γ)			0.15		890 $Sc^{46}(\beta^-\gamma)$
Ti^{47}	160	1.2–3.4α (T2)	0.040 (γ)	ex. func. $\Delta E=160$(T2)		0.040		160 $Sc^{47}(\beta^-\gamma)$
Ti^{48}	990	6α (H2)	0.083 (γ)	γ(750) also observed from 6α, but assigned to $Ti^{48}(\alpha n\gamma)Cr^{51}$(H2)		0.083		990 $Sc^{48}(\beta^-\gamma)$ $V^{48}(\beta^+\gamma)$
$_{23}V^{50}$	225	6.5α (H10)	0.011 (γ)	sep. iso. (H10)		0.011		
V^{51}	320 325 320	1.6–3.4α (T2) 1.5p (M4) 15N (A1)	0.0056 (γ) (γ) (γ)	ex. func. $\Delta E=320$(T2)	7%E2	0.0056		323 $Ti^{51}(\beta^-\gamma)$ 320 $Cr^{51}(\epsilon\gamma)$ 321 $V(pp')$; $V(nn'\gamma)$
$_{24}Cr^{53}$	155(?)	1.3p (M4)	0.015 (γ)	sep. iso. (M4)				
$_{25}Mn^{55}$	128 131 ~125 128 127	1–3.4α (T2) 0.5–2.5p (M4) 1.75p (H3) 3α (B5) 15N (A1)	0.075 (γ) 0.087 (γ) 0.0009 (e_K) (e_K,γ) (γ)	ex.func. $\Delta E=128$(T2; M4) $\sigma(p):\sigma(\alpha)$ gives $\lambda=2$(T2) $\alpha_K=0.0144$; $\gamma(\theta)$: $5/2(E2)$ $7/2(M1)5/2$(B5)	$E2<2\%$	0.07		130 $Mn(pp')$ 126 $Mn(nn'\gamma)$
	590	15N (A1)	(γ)					

TABLE IV.2.—*Continued.*

I	II	III	IV	V	VI	VII	VIII	IX
Nucleus	Decay energy	Bombarding conditions	$\epsilon B(E2)$	Comments	Multipole order and conversion coefficients	$B(E2)$	Level schemes	Other processes
$_{26}Fe^{56}$	854	6α (H2)	0.10 (γ)			0.10	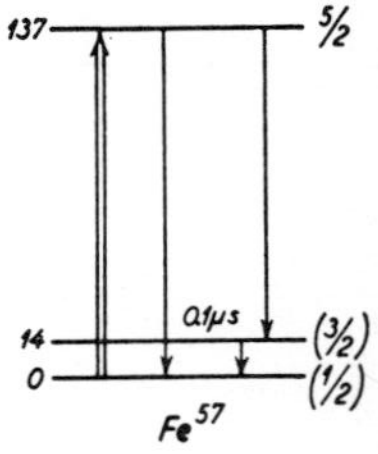	845 $Co^{56}(\beta^+\gamma)$ 845 $Mn^{56}(\beta^-\gamma)$ 850 Fe($nn'\gamma$) 845 Fe(pp')
Fe^{57}	14	1–3α (T2)	(γ)	sep.iso. (T2); fed by 122-kev transition from 137-kev level			137 — 5/2; 14 — (3/2); 0 — (1/2); 0.1μs; Fe^{57}	14 $Co^{57}(\beta^+\gamma)$
	123 122 123	1–3α (T2) 1.75*p* (H3) 0.6*p* (L1)	0.051 (γ) 0.0007 (e_K) ~0.015 (γ)	sep.iso. (T2; H3) ex.func. $\Delta E = 137$ (T2)	$\alpha_K = 0.017$			123 $Co^{57}(\beta^+\gamma)$ 117 $Mn^{57}(\beta^-\gamma)$
	137 137	1–3α (T2) 1.75*p* (H3)	(γ) 0.0004 (e_K)	sep.iso. (T2; H3); γ(137) weak compared to the cascade γ(122)	$\alpha_K = 0.19$	0.05		134 $Co^{57}(\beta^+\gamma)$ 131 $Mn^{57}(\beta^-\gamma)$
$_{28}Ni^{58}$	1450	3–4.5*p* (S2)	≲0.1 (γ)	nonresonant background observed between strong resonances (S2)		~0.1		1.47 $Ni^{58}(nn'\gamma)$ 1.453 $Ni^{58}(pp')$
$_{29}Cu^{63}$	990	7α (T7)	0.029 (γ)			0.029	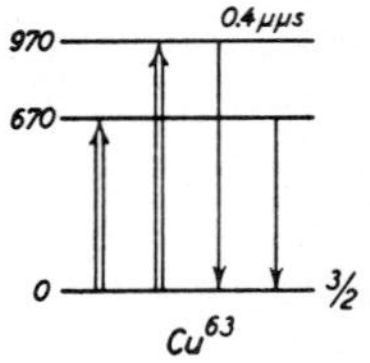 970; 670; 0 — 3/2; 0.4μμs; Cu^{63}	960 $Zn^{63}(\beta^+\gamma)$ 960 Cu($nn'\gamma$) 968 $Cu^{63}(pp')$
	690	7α (T7)	0.010 (γ)			0.010		669 $Cu^{63}(pp')$ 660 Cu($nn'\gamma$)
Cu^{65}	1150	7α (T7)	0.027 (γ)			0.027		1120 $Ni^{65}(\beta^-\gamma)$ 1112 $Zn^{65}(\epsilon\gamma)$ 1130 Cu($nn'\gamma$)
	815	7α (T7)	0.0087 (γ)			0.0087		
$_{30}Zn^{64}$	1000	7α (T7)	0.11 (γ)			0.11		970 $Ga^{64}(\beta^+\gamma)$
Zn^{66}	1040	7α (T7)	0.087 (γ)			0.087	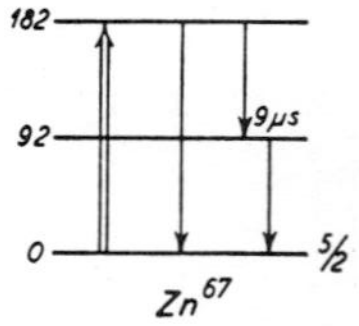 182; 92; 0 — 5/2; 9μs; Zn^{67}	1044 $Cu^{66}(\beta^-\gamma)$ 1050 $Ga^{66}(\epsilon\gamma)$
Zn^{67}	90	3 and 6α (T2)		sep.iso. (T2); ex.func. $\Delta E = 182$ (T2)				92 $Cu^{67}(\beta^-\gamma)$ 90; 92 $Ga^{67}(\epsilon\gamma)$
	182	3 and 6α (T2)	0.032 (γ)	sep.iso. (T2)		0.04		182 $Cu^{67}(\beta^-\gamma)$ 182 $Ga^{67}(\epsilon\gamma)$
Zn^{68}	1170	7α (T7)						1100 $Ga^{68}(\beta^+\gamma)$
$_{32}Ge^{70}$	1020	6α (H2)	0.077 (γ)			0.077		1036 $Ga^{70}(\beta^-\gamma)$ 1070 $As^{70}(\beta^+\gamma)$ 1050 Ge($nn'\gamma$)
Ge^{72}	830	6α (H2)	0.26 (γ)			0.26		834 $Ga^{72}(\beta^-\gamma)$ 835 $As^{72}(\beta^+\gamma)$ 820 Ge($nn'\gamma$)

TABLE IV.2.—*Continued.*

I	II	III	IV	V	VI	VII	VIII	IX
Nucleus	Decay energy	Bombarding conditions	$\epsilon B(E2)$	Comments	Multipole order and conversion coefficients	$B(E2)$	Level schemes	Other processes
Ge^{73}	68.7 72	1–3α (T2) 15N (A1)	0.085 (γ) (γ)	sep.iso.; ex.func. $\Delta E=69$ and $\lambda=2$(T2)				
Ge^{74}	593	6α (H2)	0.30 (γ)			0.30		596 $As^{74}(\beta^+\gamma)$ 610 $Ge(nn'\gamma)$
Ge^{76}	556	6α (H2)	0.33 (γ)	isotope assigned on basis of systematics		0.33		
$_{33}As^{75}$	68	3α (T3)	(γ)				As^{75}: 814; 574; 280 (5/2); 200 (1/2); 0 3/2	67 $Se^{75}(\epsilon\gamma)$ 66 $Ge^{75}(\beta^-\gamma)$
	199 200	3 and 7α(T3; T7) 2p (P1)	0.025 (γ) 0.016 (γ)			0.022		200 $Se^{75}(\epsilon\gamma)$ 199 $Ge^{75}(\beta^-\gamma)$
	283 280	3 and 7α(T3; T7) 2p (P1)	0.071 (γ) 0.05 (γ)			0.06		281 $Se^{75}(\epsilon\gamma)$
	574	7α (T7)	0.072 (γ)			0.072		
	814	7α (T7)	0.066 (γ)			0.066		
$_{34}Se^{74}$	635	7α (T7)	0.21 (γ)	sep.iso.(T7)		0.21		635 $As^{74}(\beta^-\gamma)$
Se^{76}	567	7α (T7)	0.43 (γ)	sep.iso.(T7)		0.43		560 $As^{76}(\beta^-\gamma)$ 560 $Br^{76}(\beta^+\gamma)$
Se^{77}	86	7α (T7)	(γ)	sep.iso.(T7)			Se^{77}: 457 (3/2−); 244 (5/2−); 160 7/2+ (17s); 0 1/2−	87 $As^{77}(\beta^-\gamma)$
	160	7α (T7)	(γ)	sep.iso.; isomeric trans. $T_{1/2}=17$ sec. fed by 86 kev trans. from 244-kev level; $\gamma(160):\gamma(244)=0.02$(T7)				160 $As^{77}(\beta^-\gamma)$ 160 $Br^{77}(\epsilon\gamma)$
	210	7α (T7)	(γ)	sep.iso.; cascade trans. from 457-kev level; $\gamma(210)$–$\gamma(244)$ coinc. $\gamma(210):\gamma(457)=0.28$(T7)				
	244 238	3 and 7α(T3; T7) 15N (A1)	0.12 (γ) (γ)	sep.iso.(T7)		0.12		246 $As^{77}(\beta^-\gamma)$ 237 $Br^{77}(\epsilon\gamma)$ $Se(n\gamma)$
	457 452	3 and 7α(T3; T7) 15N (A1)	0.68 (γ) (γ)	sep.iso.(T7)		0.9		
Se^{78}	615	7α (T7)	0.36 (γ)	sep.iso.(T7)		0.36		~600 $Se(n\gamma)$
Se^{80}	654	7α (T7)	0.23 (γ)	sep.iso.(T7)		0.23		
Se^{82}	880	7α (T7)	0.056 (γ)	sep.iso.(T7)		0.056		

TABLE IV.2.—*Continued.*

I	II	III	IV	V	VI	VII	VIII	IX
Nucleus	Decay energy	Bombarding conditions	$\epsilon B(E2)$	Comments	Multipole order and conversion coefficients	$B(E2)$	Level schemes	Other processes
$_{35}Br^{79}$	44	3α (H1)	(γ)				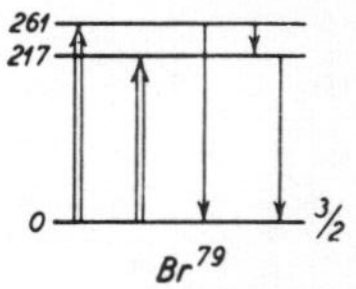 261, 217, 0; 3/2; Br^{79}	44.5 $Kr^{79}(\beta^+\gamma)$
	219	p, α (W1)	0.025 (γ)	sep.iso. (W1)				217.3 $Kr^{79}(\beta^+\gamma)$
	213	3α (H1)	(γ)					
	266	3α (H1)	(γ)					261.3 $Kr^{79}(\beta^+\gamma)$
Br^{81}	278	p, α (W1)	0.031 (γ)	sep.iso. (W1)				
$_{37}Rb^{85}$	150	3α (T3)						150 $Kr^{85m}(\beta^-\gamma)$ 150 $Sr^{85m}(\epsilon\gamma)$
$_{41}Nb^{93}$	850	(T8)	(γ)	ex.func. $\Delta E=710$ (P2)				
	710	2.75–4p (P2)	(γ)					
$_{42}Mo^{94}$	871	6α (H2)	0.29 (γ)			0.29		874 $Tc^{94}(\beta^+\gamma)$ ~900 $Nb^{94m}(\beta^-\gamma)$
Mo^{95}	204	3 and 6α (T3; H2)	0.070 (γ)	ex.func. $\Delta E=200$ (S4)		0.055		201 $Tc^{95}(\epsilon\gamma)$
	212	2.1–2.9p (S4)	0.040 (γ)					
	199	2.5p (M4)	0.057 (γ)					
	210	15N (A1)	(γ)					
Mo^{96}	778	6α (H2)	0.31 (γ)			0.4		771 $Tc^{96}(\epsilon\gamma)$ 770 $Nb^{96}(\beta^-\gamma)$
Mo^{98}	786	6α (H2)	0.27 (γ)	sep.iso. (H2)		0.4		
$Mo^{96,98}$	770	2.4–3p (S4)	0.59 (γ)	ex.func. $\Delta E=770$ (S4)				
Mo^{100}	528	6α (H2)	0.66 (γ)	sep.iso. (T7; H2) ex.func. $\Delta E=540$ (S4)		0.64		
	525	2.5p (M4)						
	540	2.1–3p (S4)	0.61 (γ)					
	540	15N (A1)	(γ)					
$_{44}Ru^{99}$	89.5	3 and 7α (H1; T6)	0.054 (γ)					89 $Rh^{99}(\beta^+\gamma)$
Ru^{100}	540	7α (T6)	0.30 (γ)			0.30	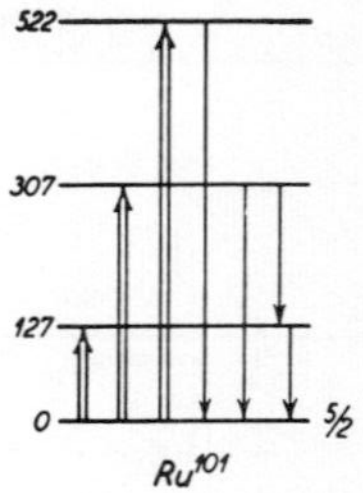 522, 307, 127, 0; 5/2; Ru^{101}	535 $Rh^{100}(\beta^+\gamma)$ 550 $Tc^{100}(\beta^-\gamma)$
Ru^{101}	127	3 and 7α (H1; T7)	0.061 (γ)					
	307	7α (T7)	0.036 (γ)					307 $Tc^{101}(\beta^-\gamma)$ 307 $Rh^{101}(\epsilon\gamma)$
	522	7α (T7)	0.041 (γ)					
	180	7α (T7)		cascade trans. from 307-kev level; $\gamma(180)-\gamma(127)$ coinc. (T7)				
Ru^{102}	473	7α (T6)	0.63 (γ)			0.63		475 $Rh^{102}(\beta^+\gamma)$
Ru^{104}	362	7α (T6)	1.04 (γ)			1.0		

Table IV.2.—*Continued.*

I	II	III	IV	V	VI	VII	VIII	IX
Nucleus	Decay energy	Bombarding conditions	$_eB(E2)$	Comments	Multipole order and conversion coefficients	$B(E2)$	Level schemes	Other processes
$_{45}$Rh^{103}	295 305 295	6α (H7) 2.1–2.9p (S4) 15N (A1)	0.18 (γ) 0.23 (γ) (γ)	ex.func. $\Delta E = 305$ (S4); $\gamma(\theta): 1/2(E2)3/2(M1+E2)1/2$; $\delta = -0.18$ or -1.17 (M7)		0.21	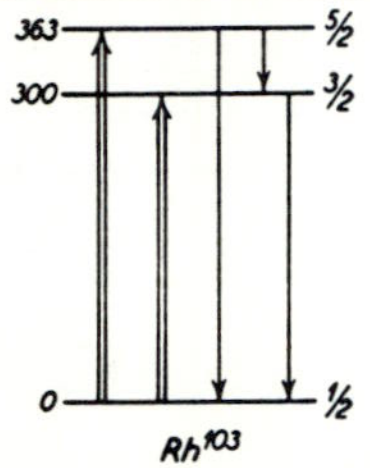 363 — 5/2 300 — 3/2 0 — 1/2 Rh^{103}	300 Pd^{103}($e\gamma$)
	357 365 358	6α (H7) 2.1–2.9p (S4) 15N (A1)	0.32 (γ) 0.32 (γ) (γ)	ex.func. $\Delta E = 365$ (S4); $\gamma(\theta): 1/2(E2)5/2(E2)1/2$ (M7)		0.36		367 Pd^{103}($e\gamma$)
	65	2.1–2.9p (S4)	0.020 (γ)	cascade trans. from 360-kev level (S4)	$\alpha = 1$			65 Pd^{103}($e\gamma$)
$_{46}$Pd^{104}	550 575	6α (T5) 2.1–2.9p (S4)	0.46 (γ) 0.54 (γ)	sep.iso. (T5) ex.func. $\Delta E = 575$ (S4)		0.50		556 Ag^{104}($\beta^+\gamma$) 556 Rh^{104}($\beta^-\gamma$)
Pd^{105}	266 270	6α (T5) 2.5p (M4)	0.04 (γ) 0.0084 (γ)	sep.iso. (T5; M4)		0.04		
	433 430	6α (T5) 2.5p (M4)	0.17 (γ) 0.057 (γ)	sep.iso. (T5; M4)		0.17		
Pd^{106}	510 500 520	6α (T5) 2.5p (M4) 2.1–2.9p (S4)	0.59 (γ) 0.18 (γ) 0.72 (γ)	sep.iso. (T5) $\gamma(\theta): 0(E2)2(E2)0$ (M7) ex.func. $\Delta E = 520$ (S4)		0.66		512 Ag^{106}($\beta^+\gamma$) 513 Rh^{106}($\beta^-\gamma$)
Pd^{108}	424 425 445	6α (T5) 2.5p (M4) 2.1–2.9p (S4)	0.78 (γ) 0.21 (γ) 1.0 (γ)	sep.iso. (T5) $\gamma(\theta): 0(E2)2(E2)0$ (M7) ex.func. $\Delta E = 445$ (S4)		0.89		430 Ag^{108}($e\gamma$)
Pd^{110}	370 365 380	6α (T5) 2.5p (M4) 2.1–2.9p (S4)	1.04 (γ) 0.26 (γ) 0.95 (γ)	sep.iso. (T5) ex.func. $\Delta E = 380$ (S4) $\gamma(\theta): 0(E2)2(E2)0$ (M7)		1.0		
$_{47}$Ag^{107}	318 319 323	6α (H7) 4α (F1) 1.75p (H4; H3)	0.12 (γ) 0.16 (γ) 0.0022 (e_K)	sep.iso. (H7; F1) ex.func. $\Delta E = 318$ (H7)	$\alpha_K = 0.014$	0.16	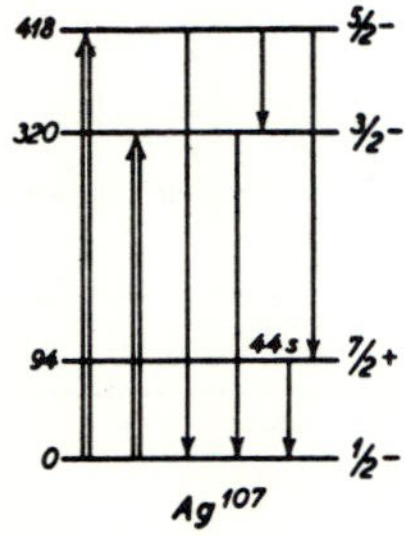 418 — 5/2− 320 — 3/2− 94 — 7/2+ (44 s) 0 — 1/2− Ag^{107}	
	413 419	6α (H7) 3p (F1)	0.21 (γ) 0.23 (γ)	sep.iso. (H7; F1)		0.28		
Ag^{109}	305 306 308	6α (H7) 4α (F1) 1.75p (H4; H3)	0.13 (γ) 0.18 (γ) 0.0026 (e_K)	sep.iso. (H7; F1)	$\alpha_K = 0.019$	0.17		
	400 412	6α (H7) 3p (F1)	0.23 (γ) 0.31 (γ)	sep.iso. (H7; F1)		0.31		
Ag	90	1.35–2p (H4)	(e)	isomeric level fed by trans. from levels with $\Delta E \sim 400$ kev; branching ratio $e(90):\gamma(400) = 0.005:1$ (H4)				94 Ag^{107m}(γ) 88 Ag^{109m}(γ)

TABLE IV.2.—*Continued.*

I Nucleus	II Decay energy	III Bombarding conditions	IV $\epsilon B(E2)$	V Comments	VI Multipole order and conversion coefficients	VII $B(E2)$	VIII Level schemes	IX Other processes
Ag	320	1.75p (H4)	0.1 (γ)	composite line from both isotopes;			412 — 5/2−; 306 — 3/2−; 88 — 7/2+ (39s); 0 — 1/2−; Ag^{109}	
	315	2.8p (M4)	0.079 (γ)	ex.func. $\Delta E=325$(S4)				
	325	2.1–2.9p(S4)	0.22 (γ)	$\gamma(\theta)$:1/2($E2$)3/2($M1+E2$)1/2				
	320	p,α (B8)	(e)	$\delta=-0.19$ or -1.14(M7)				
	310	15N (A1)	(γ)					
	418	2.8p (M4)	0.086 (γ)	composite line from both isotopes;				
	427	2.1–2.9p(S4)	0.34 (γ)	ex.func. $\Delta E=427$(S4)				
	420	p,α (B8)	(e)	$\gamma(\theta)$:1/2($E2$)5/2($E2$)1/2(M7)				
	409	15N (A1)	(γ)					
	104	2.1–2.9p(S4)	0.016 (γ)	cascade trans. from 400-kev to 300-kev levels in both isotopes(S4)	$\alpha=0.3$			
$_{48}Cd^{110}$	654	6α (T5)	0.41 (γ)	sep.iso.(T5)		0.41		656 $Ag^{110}(\beta^-\gamma)$; 654 $In^{110}(\beta^-\gamma)$
Cd^{111}	340	6α (T5)	0.19 (γ)	sep.iso.(T5; M4)		0.19		340 $Ag^{111}(\beta^-\gamma)$; 330 $In^{111}(e\gamma)$
	330	2.8p (M4)	0.027 (γ)					
Cd^{112}	620	6α (T5)	0.46 (γ)	sep.iso.(T5; M4)		0.46		620 $Ag^{112}(\beta^-\gamma)$
	610	2.8p (M4)	0.13 (γ)					
Cd^{113}	290	6α (T5)	0.11 (γ)	sep.iso.(T5; M4)		0.11		
	290	2.8p (M4)	0.058 (γ)					
	297	15N (A1)	(γ)					
	550	6α (T5)	<0.2 (γ)					
Cd^{114}	550	6α (T5)	0.55 (γ)	sep.iso.(T5; M4)		0.55		556 $In^{114}(e\gamma)$; 550 Cd($n\gamma$)
	545	2.8p (M4)	0.14 (γ)					
	543	15N (A1)	(γ)					
Cd^{116}	508	6α (T5)	0.62 (γ)	sep.iso.(T5)		0.62		
$_{49}In^{115}$	500	3.8p (T3)	(γ)	ex.func. indicates compound nucleus formation (D1)				500 $Cd^{115}(\beta^-\gamma)$
	500	3p (M4)	0.058 (γ)					
	562	15N (A1)	(γ)					
$_{51}Sb$	~160	3α (T3)		isotope uncertain				

TABLE IV.2.—*Continued.*

I Nucleus	II Decay energy	III Bombarding conditions	IV $\epsilon B(E2)$	V Comments	VI Multipole order and conversion coefficients	VII $B(E2)$	VIII Level schemes	IX Other processes
$_{52}Te^{120}$	560	6.5α (H10)	0.55 (γ)	sep.iso. (H10)		0.55		
Te^{122}	570	6.5α (H10)	0.47 (γ)	sep.iso. (H10)		0.47	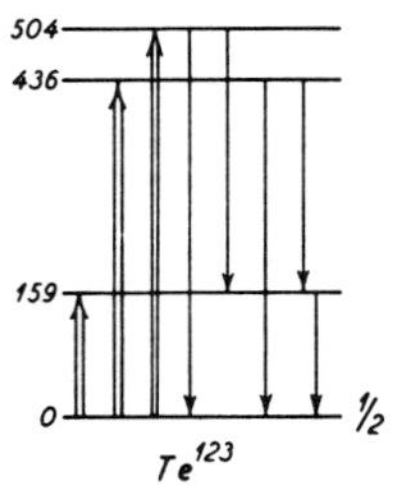	565 $Sb^{122}(\beta^-\gamma)$
Te^{123}	159	4α, p (F1)	0.014 (γ)	sep.iso. ex.func. $\Delta E = 160$ (F1)	$\alpha = 0.2$	0.017	504, 436, 159, 0, ½, Te^{123}	159 $I^{123}(\epsilon\gamma)$ $Te^{123m}(\gamma\gamma)$
	274	3.7p (F1)	(γ)	sep.iso., cascade trans. from 436-kev level; $\gamma(274)-\gamma(159)$ coinc. (F1)				
	342	3.7p (F1)	(γ)	sep.iso., cascade trans. from 504-kev level; $\gamma(342)-\gamma(159)$ coinc. (F1)				
	436	3.7p (F1)	(γ)	sep.iso. (F1)				
	504	3.7p (F1)	(γ)	sep.iso. (F1)				
Te^{124}	608	6.5α (H10)	0.39 (γ)	sep.iso. (H10)		0.39		605 $Sb^{124}(\beta^-\gamma)$
Te^{125}	435	3p (F1)	0.44 (γ)	sep.iso. (F1)		0.44		425 $Sb^{125}(\beta^-\gamma)$
	633	3.5p (F1)	0.26 (γ)	sep.iso. (F1)		0.26		637 $Sb^{125}(\beta^-\gamma)$
Te^{126}	662	6.5α (H10)	0.32 (γ)	sep.iso. (H10)		0.32		650 $I^{126}(\epsilon\gamma)$
Te^{128}	750	6.5α (H10)	0.28 (γ)	sep.iso. (H10)		0.28		750 $I^{128}(\epsilon\gamma)$
Te^{130}	850	6.5α (H10)	0.26 (γ)	sep.iso. (H10)		0.26		
$_{53}I^{127}$	57	3α (H1)	(γ)					57 $Xe^{127}(\epsilon\gamma)$
	60	α (D1)	(γ)					62 $I(nn'\gamma)$
	60	15N (A1)	(γ)					
	205	3α (H1)	0.039 (γ)			0.039		200 $Xe^{127}(\epsilon\gamma)$
	212	2p (M4)	(γ)					208 $I(nn'\gamma)$
	203	1.5–3.5p,α (D1)	0.039 (γ)					
	205	15N (A1)	(γ)					
	392	2.0–3.5p (D1)	0.019 (γ)			0.019		390 $I(nn'\gamma)$
	438	2–3.5p (D1)	0.0060 (γ)			0.006		435 $I(nn'\gamma)$
	631	2–3.5p (D1)	0.10 (γ)			0.10		632 $I(nn'\gamma)$
	751	2.5–3.5p (D1)	0.08 (γ)			0.08		750 $I(nn'\gamma)$
	941	2.5–3.5p (D1)	0.36 (γ)			0.36		950 $I(nn'\gamma)$

TABLE IV.2.—*Continued.*

I	II	III	IV		V	VI	VII	VIII	IX
Nucleus	Decay energy	Bombarding conditions	$\epsilon B(E2)$		Comments	Multipole order and conversion coefficients	$B(E2)$	Level schemes	Other processes
$_{55}Cs^{133}$	85	3α (T3)		(γ)					81 $Xe^{133}(\beta^{-}\gamma)$ 82 $Ba^{133}(\epsilon\gamma)$
$_{56}Ba$	60	3α (H1)		(γ)	isotope uncertain				
	118	3α (H1)		(γ)	isotope uncertain				
$_{57}La^{139}$					$\gamma(166)$ not observed with 6α $B(E2)<10^{-2}$(H8)				
$_{59}Pr^{141}$					$\gamma(145)$ not observed with 6α $B(E2)<10^{-2}$(H8)				
$_{60}Nd^{145}$	70	3 and 6α(H1; H8)	~0.06	(γ)	sep.iso. (H8)	$\alpha=4$	~0.3		
	71	$2.2p$ (S3)		(γ)					
Nd^{146}	455	6α (H8)	0.25	(γ)	sep.iso. (H8)		0.25		455 $Pr^{146}(\beta^{-}\gamma)$
Nd^{148}	300	3 and 6α(T6; H8)	0.68	(γ)	sep.iso. (H8)	$\alpha=0.04$	0.69		
	300	2.2–3.3p(S3)	0.64	(γ)	ex.func. $\Delta E=300$(S3)				
Nd^{150}	128	3 and 6α(H1; H8)	1.32	(γ)	sep.iso. (H8)	$\alpha=0.90$	2.3		
	131	1.5–3.3p(S3)	1.20	(γ)	ex.func. $\Delta E=131$ and $\lambda=2$;	$\alpha_L=0.25$			
	132	$1.75p$ (H3)	0.31	(e_L)	$\gamma(\theta):0(E2)2(E2)0$(S3)				
$_{62}Sm^{148}$	562	6α (H8)	0.50	(γ)	sep.iso. (H8; M6; S5)		0.50		
	550	$2.9p$ (M6)	0.74	(γ)	$\gamma(\theta):0(E2)2(E2)0$(S5)				
	562	p (S5)		(γ)					
Sm^{150}	337	6α (H8)	0.94	(γ)	sep.iso. (H8; M6; S5)	$\alpha=0.04$	0.98		337 $Sm^{149}(n\gamma)$
	335	$2.9p$ (M6)	0.51	(γ)	$\gamma(\theta):0(E2)2(E2)0$				340 $Pm^{150}(\beta^{-}\gamma)$
	337	$2.6p$ (S5)		(γ)	ex.func. $\Delta E=337$(S5)				
Sm^{152}	122	3 and 6α(H1; H8)	1.52	(γ)	sep.iso. (H8; M6; S5)	$\alpha_L=0.36$	3.1		121.8 $Eu^{152}(\epsilon\gamma)$
	125	$2.9p$ (M6)	0.43	(γ)	$\gamma(\theta):0(E2)2(E2)0$	$\alpha=1.2$			
	123	$1.75p$ (H3)	0.56	(e_L)	ex.func. $\Delta E=124$ and $\lambda=2$(S5)				
	124	$2.3p$ (S5)		(γ)					
	122	$6p$ (E2)	2.66	(p')					
Sm^{154}	82	3 and 6α(H1; H8)	0.80	(γ)	sep.iso. (H8; M6; S5)	$\alpha_L=2.0$	4.5		
	84	$2.9p$ (M6)	0.27	(γ)	$\gamma(\theta):0(E2)2(E2)0$	$\alpha=4.4$			
	83	$1.75p$ (H3)	2.55	(e_L)	ex.func. $\Delta E=84$ and $\lambda=2$(S5)				
	85	$2.1p$ (S5)		(γ)					
	84	$6p$ (E2)	3.60	(p')					
$_{63}Eu^{151}$	195	6.5α (H10)	0.067	(γ)	sep.iso. (H10; C2)	$\alpha=0.3$	0.09		
	193	$3.7p$ (C2)		(e)					
	110	6.5α (H10)	0.024	(γ)	sep.iso. (H10) cascade trans. from 304-kev level	$\alpha=1.3$			

TABLE IV.2.—*Continued.*

I	II	III		IV		V	VI	VII	VIII	IX
Nucleus	Decay energy	Bombarding conditions		$_eB(E2)$		Comments	Multipole order and conversion coefficients	$B(E2)$	Level schemes	Other processes
Eu^{151}	304	6.5α	(H8; H10)	0.22	(γ)	sep.iso. (H10; C2)	α=0.06	0.29		
	300	2.9*p*	(M6)		(γ)					
	304	3.7*p*	(C2)		(*e*)					
	284	3.7*p*	(C2)		(*e*)	sep.iso. (C2) probably cascade trans. from 304-kev level				
Eu^{153}	82	3 and 6α	(T6; H8)	0.60	(γ)	sep.iso. (H10; C2)			192 — (9/2); 83 — (7/2); 0 — 5/2; Eu^{153}	83.4 $Sm^{153}(\beta^-\gamma)$
	85	2.9*p*	(M6)	0.20	(γ)	K/L∼1 (H3)	∼50%*E*2	2.8		
	84	1.75*p*, *d*	(H3)	0.65	(e_L)		α_L=1.3			
	81	6*p*	(E2)	1.9	(*p*′)		α=4.0			
	82.5	3.7α	(C2)		(*e*)					
	105	6α	(T6; H8)	0.18	(γ)	sep.iso. (H10; C2)	(∼50%*E*2)			
	115	2.9*p*	(M6)			cascade trans. from 195-kev level;	α_L=0.4			
	∼111	1.75*p*	(H3)	∼0.05	(e_L)		α=1.5			
	109	3.7α	(C2)		(*e*)	γ(108)−γ(84) coinc. (H8) γ(110):γ(195)=0.38 (M6)				
	187	6α	(T6; H8)	0.29	(γ)	sep.iso. (H10; C2)	100%*E*2			
	200	2.9*p*	(M6)	0.14	(γ)		α_K=0.17			
	195	1.75*p*	(H3)	∼0.051	(e_K)		α=0.25	0.78		
	192	3.7α	(C2)		(*e*)					
$_{64}Gd^{154}$	123	3 and 6α	(H1; H8)	2.32	(γ)	sep.iso. (M6)	α_L=0.43	4.5		123.4 $Eu^{154}(\beta^-\gamma)$
	123	1.9*p*	(M6)	1.0	(γ)		α=1.2			
	124	1.75*p*	(H3)	∼0.73	(e_L)					
Gd^{155}	60	1.75*p* and α	(B7)	0.48	(e_L)	sep.iso. (B7)	5%*E*2 α_L=1.6 α=10	3.3	145 — (7/2); 60 — (5/2); 0 — 3/2; Gd^{155}	59.8 $Eu^{155}(\beta^-\gamma)$
	145	6α	(H8)	0.105	(γ)	sep.iso. (H8; B7)	100%*E*2			
	140	2.9*p*	(M6)		(γ)					
	146	1.75*p*	(B7)	0.059	(e_K)					
Gd^{156}	89	6α	(H8)	1.90	(γ)	sep.iso. (M6)	α_L=1.8	7.7		88.8 $Gd^{155}(n\gamma)$
	89	1.9*p*	(M6)	0.74	(γ)		α=3.9			$Eu^{156}(\beta^-\gamma)$
	90	1.75*p*	(H3)	2.8	(e_L)					
	91	6*p*	(E2)	5.0	(*p*′)					
Gd^{157}	55	1.75*p* and α	(B7)	0.54	(e_L)	sep.iso. (B7)	4%*E*2 α_L=2.0 α=13	3.5	131 — (7/2); 55 — (5/2); 0 — 3/2; Gd^{157}	
	131	6α	(H8)	0.086	(γ)	sep.iso. (H8; B7)	100%*E*2	1.3		
	127	2.9*p*	(M6)		(γ)					
	131	6*p*	(E2)	1.2	(*p*′)					
	132	1.75*p*	(B7)	0.032	(e_L)					
Gd^{158}	79	6α	(H8)	1.75	(γ)	sep.iso. (M6)	α_L=3.1	10		79.1 $Gd^{157}(n\gamma)$
	80	1.9*p*	(M6)	0.63	(γ)		α=6.0			

Table IV.2.—*Continued.*

I Nucleus	II Decay energy	III Bombarding conditions	IV $\epsilon B(E2)$	V Comments	VI Multipole order and conversion coefficients	VII $B(E2)$	VIII Level schemes	IX Other processes
Gd^{158}	79 80	6p (E2) 1.75p (H3)	~4 (p') 3.4 (e_L)					
Gd^{160}	76 76 76 75	6α (H8) 1.9p (M6) 1.75p (H3) 6p (E2)	1.79 (γ) 0.73 (γ) 3.4 (e_L) ~4 (p')	sep.iso. (M6)	$\alpha_L=3.6$ $\alpha=7.0$	11		
$_{65}Tb^{159}$	58 58	1.75d (H3) 6p (E2)	0.45 (e_L) 1.88 (p')		(1.3%$E2$) $\alpha_L=1.5$ $\alpha=11$	2.4	138 (7/2) 58 (5/2) 0 3/2 Tb^{159}	57.5 $Gd^{159}(\beta^-\gamma)$
	79 77 81	6α (H8) 2.9p (M6) 1.75d (H3)	0.33 (γ) (γ) (e_L)	ex.func. $\Delta E=136$ $\gamma(79)-\gamma(57)$ coinc. (T4)	(1.3%$E2$) $\alpha=4.2$			
	136 135	6α (H8) 6p (E2)	0.041 (γ) 1.14 (p')		100%$E2$ $\alpha=0.8$	1.4		
	167(?)	2.9p (M6)	(γ)					
$_{66}Dy^{162}$	82 82	1.75p (H3) 6p (E2)	3.2 (e_L) ~6 (p')	isotope assignment on basis of systematics	$\alpha_L=3.2$ $\alpha=5.8$	6.7		
Dy^{164}	75 72	1.75p (H3) 6p (E2)	4.8 (e_L) ~6 (p')		$\alpha_L=4.7$ $\alpha=8.1$	8.9		72.8 $Ho^{164}(\epsilon\gamma)$
$Dy^{161,163}$	166	6α (T6; H8)	0.24 (γ)	isotope assignment on basis of systematics				170 $Ho^{161}(\epsilon\gamma)$
Dy	76	3 and 6α (T6; H8)	0.41 (γ)	a composite line				
$_{67}Ho^{165}$	94 96 95	3 and 6α (T6; H8) 1.75p, d (H3) 6p (E2)	0.78 (γ) 0.32 (e_L) 1.7 (p')	K/L~5 (H3)	4%$E2$ $\alpha_L=0.45$ $\alpha=3.1$	2.6	211 (11/2) 95 (9/2) 0 7/2 Ho^{165}	94 $Dy^{165}(\beta^-\gamma)$
	116 112	1.75p (H3) 6α (T4)	0.07 (e_L) 0.15 (γ)	cascade from 211-kev level $\gamma(116)-\gamma(94)$ coinc. (T4)	(4%$E2$) $\alpha_L=0.26$ $\alpha=1.9$			
	206 212	6α (T7; H12) 1.9p (H6)	0.024 (γ) (e_K)		100%$E2$ $\alpha=0.2$	0.63		

TABLE IV.2.—*Continued.*

I Nucleus	II Decay energy	III Bombarding conditions	IV $\epsilon B(E2)$	V Comments	VI Multipole order and conversion coefficients	VII $B(E2)$	VIII Level schemes	IX Other processes
$_{68}Er^{167}$	172	6α (T6; H8)	0.064 (γ)	isotope assignment on basis of systematics				
Er	79	3 and 6α (T6; H8)	0.83 (γ)	a composite line due to all isotopes	$\alpha_L = 4.2$	5.7		80 $Ho^{166}(\beta^-\gamma)$
	81	1.75*p* (H3)	2.8 (e_L)		$\alpha = 7.1$			80 $Ho^{166m}(\beta^-\gamma)$
	79	6*p* (E2)	5.1 (p')					80 $Tm^{166}(\epsilon\gamma)$
$_{69}Tm^{169}$	109	6α (H8)	1.40 (γ)	cascade trans. from 119-kev level established by ex.func.; $K/L = 7$ (H3)	(2%$E2$)		118 — 5/2; 8 — 3/2; 0 — 1/2; Tm^{169}	109.9 $Yb^{169}(\epsilon\gamma)$
	111	1.75*p* (H3)	0.32 (e_L)		$\alpha_L = 0.32$			
					$\alpha = 2.4$			
	119	1.75*p* (H3)	0.11 (e_L)		100%$E2$	3.8		118.3 $Yb^{169}(\epsilon\gamma)$
	119	6*p* (E2)	3.47 (p')		$\alpha_L = 0.77$			
					$\alpha = 1.7$			
$_{70}Yb^{173}$	180	6α (T6; H8)	0.065 (γ)	isotope assigned on basis of systematics				
Yb	78	3 and 6α (T3; H8)	0.42 (γ)	composite line from all isotopes	$\alpha_L = 5.8$	4.8		
	77	1.75*p* (H3)	2.7 (e_L)		$\alpha = 9$			
	76	6*p* (E2)	5.7 (p')					
	66(?)	1.75α (H3)	(e_L)					
	110	6α (H8)	0.037 (γ)					
$_{71}Lu^{175}$	114	3 and 6α (T6; H8)	0.86 (γ)	$K/L \sim 5.5$ (H3)	12%$E2$	2.9	251 — (11/2); 113 — 9/2; 0 — 7/2; Lu^{175}	113.4 $Hf^{175}(\epsilon\gamma)$
	112	2.6*p* (M1)	0.35 (γ)		$\alpha_L = 0.44$			114 $Yb^{175}(\beta^-\gamma)$
	114	1.75*p* (H3)	0.36 (e_L)		$\alpha = 2.6$			
	116	6*p* (E2)	2.54 (p')					
	136	6α (T4)	0.22 (γ)	γ(136)—γ(113) coinc. (T4)	(12%$E2$)			137.6 $Yb^{175}(\beta^-\gamma)$
					$\alpha = 1.4$			
	250	3 and 6α (T6; H8)	0.11 (γ)		100%$E2$	0.67		251.0 $Yb^{175}(\beta^-\gamma)$
	240	2.6*p* (M1)	(γ)		$\alpha = 0.14$			
Lu	180	3 and 6α (H1; T6; H8)	0.022 (γ)	isotope uncertain				

TABLE IV.2.—*Continued.*

I	II	III	IV	V	VI	VII	VIII	IX
Nucleus	Decay energy	Bombarding conditions	$_{e}B(E2)$	Comments	Multipole order and conversion coefficients	$B(E2)$	Level schemes	Other processes
$_{72}Hf^{176}$	87	6α (H8)	0.88 (γ)	sep.iso. (M1; H8)	$\alpha=5.8$	6.0	Hf^{177}: 250 (11/2); 113 (9/2); 0 7/2	89 $Lu^{176}(\beta^-\gamma)$
	87	1.5p (M1)	1.7 (γ)					89 $Lu^{176m}(\beta^-\gamma)$
	90	1.75α (H3)	(e_L)					
Hf^{177}	112	6α (H8)	0.93 (γ)	sep.iso. (M1; H8)	(60% $E2$)	2.4		112.97 $Lu^{177}(\beta^-\gamma)$
	112	p (M1)	1.1 (γ)		$\alpha_L=0.87$			
	114	1.75p (H3)	0.39 (e_L)		$\alpha=2.5$			
	136	6α (T4)	0.11 (γ)	γ(138)–γ(113) coinc. (T4)	(60% $E2$) $\alpha=1.3$			
	250	6α (H8)	0.30 (γ)	sep.iso. (M1; H8)	100% $E2$	0.63		249.69 $Lu^{177}(\beta^-\gamma)$
	235	2.6p (M1)	(γ)		$\alpha=0.15$			
Hf^{178}	90	6α (H8)	1.28 (γ)	sep.iso. (M1; H8)	$\alpha=5.4$	6.6		
	91	1.5p (M1)	1.3 (γ)					
Hf^{179}	119	6α (H8)	0.77 (γ)	sep.iso. (M1; H8)	(14% $E2$)	2.9	Hf^{179}: 262 (13/2); 121 (11/2); 0 9/2	
	122	p (M1)	0.49 (γ)		$\alpha_L=0.40$			
	123	1.75p (H3)	0.19 (e_L)		$\alpha=2.3$			
	141	6α (T4)	0.066 (γ)	γ(141)–γ(121) coinc. (T4)	(14% $E2$) $\alpha=1.5$			
	260	6α (H8)	0.029 (γ)		100% $E2$	0.20		
	250	2.6p (M1)	(γ)		$\alpha=0.13$			
$Hf^{177,179}$	112	4p (S4)	0.59 (γ)					
	248	4p (S4)	(γ)					
Hf^{180}	93	6α (H8)	1.14 (γ)	sep.iso. (M1; H8)	$\alpha=4.8$	5.0		93.3 $Ta^{180}(e\gamma)$
	92	1.5p (M1)	1.3 (γ)					$Hf^{180m}(\gamma)$
$Hf^{178,180}$	95	1.75p (H3)	2.3 (e_L)		$\alpha_L=3.1$			
Hf	90	4p (S4)	0.88 (γ)	all even isotopes				
$_{73}Ta^{181}$	138	p (M1; M2)	0.58 (γ)	ex.func. $\Delta E=137$ (H5; T3;	16% $E2$	1.9	Ta^{181}: 303 11/2; 136 9/2; 0 7/2	136.25 $Hf^{181}(\beta^-\gamma)$
	137	1–2.2p (H5)	0.7 (γ)	G1; B1; S8; D1)	$\alpha_L=0.29$			136.5 $W^{181}(e\gamma)$
	136	2p (H3; H6)	0.21 (e_L)	$\gamma(\theta)$: 7/2 ($E2$) 9/2 ($E2+M1$) 7/2	$\alpha=1.7$			136 $Ta^{181}(nn'\gamma)$
	136	6α (H8)	0.70 (γ)	(G1; E1)				
	137	3α (T3)	(γ)	$K/L=6.5$ (H3)				
	137	2–4p (G1)	(γ)	$=6.3$ (B4)				
	137	1.2–4.5p (B1)	(γ)					
	139	3p (E1)	(γ)					
	137	1.4–5p, 4α (S4)	0.83 (γ)					
	137	2–3p (D1)	0.60 (γ)					
	137	3.5α (B4)	(e)					
	138	15N (A1)	(γ)					

TABLE IV.2.—*Continued.*

I	II	III	IV	V	VI	VII	VIII	IX
Nucleus	Decay energy	Bombarding conditions	$_eB(E2)$	Comments	Multipole order and conversion coefficients	$B(E2)$	Level schemes	Other processes
Ta^{181}	165	2p (H3; H6)	0.22 (e_K)	cascade trans. from 303-kev	20% $E2$			
	166	p (G1)	(γ)	level; γ(165)–γ(137)	$\alpha_K=0.78$			
	167	6α (H8)	0.13 (γ)	coinc. (G1; H8)	$\alpha=1.0$			
	167	3p (E1)	(γ)	$K/L\sim7$ (H3)				
	166	4p (B1)	(γ)	$\gamma(\theta)$: 7/2($E2$)11/2($E2+M1$)9/2				
	167	1.8–5p (S4)	0.22 (γ)	$\delta=0.51$ (M7)				
	166	2–4p (B8)	(e)	ex.func. $\Delta E=303$ (S4)				
	165	3p (D1)	0.18 (γ)					
	303	1.8–2.2p (H5)	0.1 (γ)	ex.func. $\Delta E=300$ (B1; S4; D1)	100% $E2$	0.57		
	303	2–4p (G1)	(γ)	$\gamma(\theta)$: 7/2($E2$)11/2($E2$)7/2	$\alpha=0.08$			
	309	3p (E1)	(γ)	(G1; $E1$; M7)				
	303	6α (H8)	0.067 (γ)	$\gamma(165):\gamma(303)=1.7$ (G1)				
	303	2.2–4.5p (B1)	(γ)	1.25 (E1)				
	300	p (M1)	0.07 (γ)	1.74 (S4)				
	303	1.4–5p (S4)	0.13 (γ)	1.7 (M5)				
	302	2–3p (D1)	0.12 (γ)	1.52 (D1)				
	301	15N (A1)	(γ)					
$_{74}W^{182}$	100	1.75p (H3; H6)	2.7 (e_L)	sep.iso. (M3)	$\alpha_L=2.4$	5.6		100.09 $Ta^{182}(\beta^-\gamma)$
	101	2.5p (M1; M3)	0.24 (γ)	$\sigma(p):\sigma(d):\sigma(\alpha)$ gives $\lambda=2$ (B2)	$\alpha=4.0$			
	102	p,α (B8)	(e)					
W^{183}	46.5	1.45α (H3)	0.5 (e_M)				292 — 5/2; 99 — 5/2; 46 — 3/2; 0 — 1/2; W^{183}	46.48 $Ta^{183}(\beta^-\gamma)$
	103	2.5p (M1; M3)	0.06 (γ)	sep.iso. (M3)				99.07 $Ta^{183}(\beta^-\gamma)$
	295	4p (S4)	0.055 (γ)	see decay scheme established in $Ta^{183}(\beta^-\gamma)$		0.27		291.7 $Ta^{183}(\beta^-\gamma)$
W^{184}	111	1.75p (H3; H6)	1.7 (e_L)	sep.iso. (M3)	$\alpha_L=1.4$	4.2		110 $Ta^{184}(\beta^-\gamma)$
	112	2.5p (M1; M3)	0.32 (γ)		$\alpha=2.6$			
	112	p,α (B8)	(e)					
W^{186}	123	1.75p (H3; H6)	1.4 (e_L)	sep.iso. (M3)	$\alpha_L=0.9$	4.2		123 $Re^{186}(e\gamma)$
	124	2.5p (M1; M3)	0.39 (γ)		$\alpha=1.7$			125 $Ta^{186}(\beta^-\gamma)$
	124	p,α (B8)	(e)					
W	115	2p (H5)	1.4 (γ)	composite line				
	114	2–4p (G1)	(γ)	ex.func. $\Delta E=114$				
	~120	3α (T3)	(γ)	$\gamma(\theta)$: 0($E2$)2($E2$)0 (G1)				
	112	4p (S4)	1.3 (γ)					
	112	15N (A1)	(γ)					

TABLE IV.2.—*Continued.*

I	II	III	IV	V	VI	VII	VIII	IX
Nucleus	Decay energy	Bombarding conditions	$\epsilon B(E2)$	Comments	Multipole order and conversion coefficients	$B(E2)$	Level schemes	Other processes
$_{75}Re^{185}$	130	p (M1)	0.20 (γ)	sep.iso. (M1; W1; D1)	(3%$E2$)		286 ($9/2$); 126 ($7/2$); 0 $5/2$; Re^{185}	125 $Os^{185}(\epsilon\gamma)$
	125	1.75p (H3)	0.30 (e_L)		$\alpha_L=0.40$	1.4		130 $W^{185}(\beta^-\gamma)$
	126	p,α (W1)	0.30 (γ)		$\alpha=2.8$			
	125	3.2p (D1)	0.22 (γ)					
	160	p,α (W1)	(γ)	sep.iso. (W1; D1)	(3%$E2$)			162 $Os^{185}(\epsilon\gamma)$
	158	3.2p (D1)	0.16 (γ)	$\gamma(160)$–$\gamma(126)$ coinc. (W1)	$\alpha=1.5$			165 $W^{185}(\beta^-\gamma)$
	290	p (M1)	(γ)	sep.iso. (M1; W1; D1)	100%$E2$	0.37		
	286	p,α (W1)	(γ)	$\gamma(286):\gamma(160)=0.25$ (W1)	$\alpha=0.11$			
	280	3.2p (D1)	0.020 (γ)					
Re^{187}	139	p (M1)	0.19 (γ)	sep.iso. (M1; W1; D1)	(3%$E2$)	1.2	303 ($9/2$); 135 ($7/2$); 0 $5/2$; Re^{187}	134.25 $W^{187}(\beta^-\gamma)$
	135	1.75p (H3)	0.18 (e_L)		$\alpha_L=0.35$			
	135	p,α (W1)	0.32 (γ)		$\alpha=2.3$			
	135	3.2p (D1)	0.39 (γ)					
	168	p,α (W1)	(γ)	sep.iso. (W1; D1)	(3%$E2$)			
	163	3.2p (D1)	0.20 (γ)	$\gamma(168)$–$\gamma(135)$ coinc. (W1)	$\alpha=1.2$			
	320	p (M1)	(γ)	sep.iso. (M1; W1; D1)	100%$E2$	0.46		
	303	p,α (W1)	(γ)	$\gamma(303):\gamma(168)=0.21$ (M5)	$\alpha=0.1$			
	300	3.2p (D1)	0.038 (γ)	$=0.21$ (W1)				
Re	130	3α (H1)	(γ)					
	305	6α (T8)	(γ)					
$_{76}Os^{188}$	158	3α (H1)	(γ)					155 $Re^{188}(\beta^-\gamma)$
Os	180	3α (H1)	(γ)					
	188	3α (H1)	(γ)					
	202	3α (H1)	(γ)					
$_{77}Ir^{191}$	115	3.2p (D1)	(γ)	sep.iso. (D1)			355 ($7/2$); 129 $5/2$; 0 $3/2$; Ir^{191}	
	129	1.75p (H3)	0.12 (e_L)	sep.iso. (D1)	~30%$E2$	0.63		129.4 $Pt^{191}(\epsilon\gamma)$
	133	3.2p (D1)	0.15 (γ)		$\alpha=2.5$			129 $Os^{191}(\beta^-\gamma)$
					$\alpha_L=0.6$			
	216	3.2p (D1)	0.33 (γ)	sep.iso. (D1)	$\alpha=0.5$			
	356	3.2p (D1)	0.28 (γ)	sep.iso. (D1)		0.8		

TABLE IV.2.—*Continued.*

I Nucleus	II Decay energy	III Bombarding conditions	IV $\epsilon B(E2)$	V Comments	VI Multipole order and conversion coefficients	VII $B(E2)$	VIII Level schemes	IX Other processes
Ir^{193}	139 143	1.75p (H3) 3.2p (D1)	0.06 (e_L) 0.18 (γ)	sep.iso.(D1)	~30%$E2$ $\alpha=2.0$ $\alpha_L=0.4$	0.47	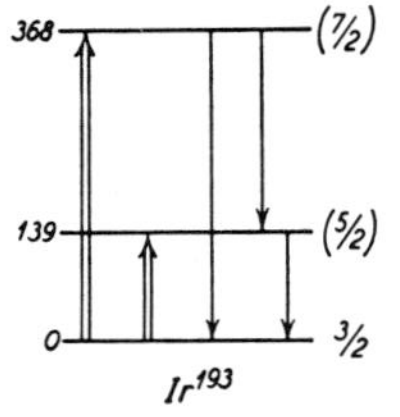	139 $Os^{193}(\beta^-\gamma)$
	230	3.2p (D1)	0.07 (γ)	sep.iso.(D1)	$\alpha=0.4$			
	368	3.2p (D1)	0.21 (γ)	sep.iso.(D1)		0.3		
Ir	133	3.4α (T9)	(γ)	composite line				
	219	3.4α (T9)	(γ)	composite line				
	360	3.4α (T9)	(γ)	composite line				
$_{78}Pt^{194}$	328 330 330	3α and p(H1) 3p (M1) 2.5–5p (S4)	(γ) 0.51 (γ) 1.62 (γ)	sep.iso.(M1) $\gamma(\theta):0(E2)2(E2)0$(M7) ex.func. $\Delta E=330$(S4)	$\alpha=0.07$	1.7		326 $Ir^{194}(\beta^-\gamma)$ 328 $Au^{194}(\epsilon\gamma)$
Pt^{195}	29 29	3α (H1) α (B4)	(γ) (γ)	ex.func. $\Delta E=29$(B4)				31 $Au^{195}(\epsilon\gamma)$
	98 100 97	3α (H1) 4α (S4) 3–4α (B4)	(γ) (γ) (e_L)	ex.func. $\Delta E=126$(B4)			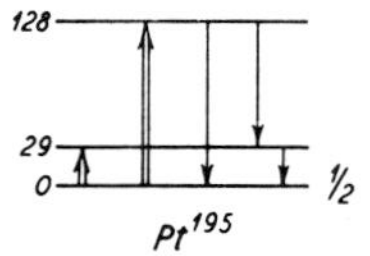	99 $Au^{195}(\epsilon\gamma)$
	128 130	3α (H1) 4α (S4)	(γ) (γ)					129 $Au^{195}(\epsilon\gamma)$
	212 210 210 210 210	3α and p(H1) 3p (M1) 1.75p (H3) 5p (S4) 4p (B4)	(γ) 0.23 (γ) 0.15 (e_K) 0.31 (γ) (e_K)	sep.iso.(M1) $K/L\sim4$(H3) $=6$(B4)	$\alpha\sim0.7$ $\alpha_K\sim0.55$	0.48		
	240	3–5p (S4)	0.15 (γ)		$\alpha\sim0.3$	0.2		
Pt^{196}	360 358	3p (M1) 2.5–5p (S4)	0.26 (γ) 1.2 (γ)	sep.iso.(M5) $\gamma(\theta):0(E2)2(E2)0$(M7) ex.func. $\Delta E=358$(S4)	$\alpha=0.06$	1.3		354 $Au^{196}(\epsilon\gamma)$
Pt^{198}	425 403	3p (M1) 2.5–5p (S4)	0.20 (γ) 1.3 (γ)	ex.func. $\Delta E=403$(S4)	$\alpha=0.04$	1.4		

Level schemes (column VIII): Ir^{193}: levels 0 (3/2), 139 (5/2), 368 (7/2). Pt^{195}: levels 0 (1/2), 29, 128.

TABLE IV.2.—*Continued.*

I	II	III	IV	V	VI	VII	VIII	IX
Nucleus	Decay energy	Bombarding conditions	$\epsilon B(E2)$	Comments	Multipole order and conversion coefficients	$B(E2)$	Level schemes	Other processes
$_{79}Au^{197}$	77	3α and p (H1)	(γ)	ex.func. $\Delta E=77$ (B4)			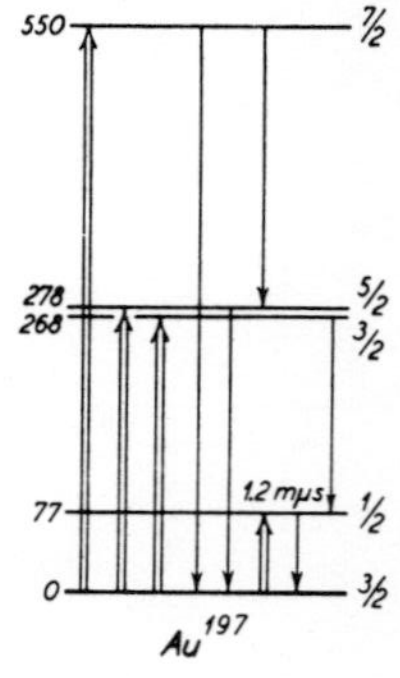 550 — 7/2; 278 — 5/2; 268 — 3/2; 77 — 1/2 (1.2 mμs); 0 — 3/2; Au 197	77 $Pt^{197}(\beta^-\gamma)$ $Hg^{197}(\epsilon\gamma)$
	77	3.25α (B4)	(e)					
	190	3α and p (H1)	(γ)	$\sigma(\alpha):\sigma(p)$ indicates a cascade	α∼1			191 $Pt^{197}(\beta^-\gamma)$ $Hg^{197}(\epsilon\gamma)$
	191	3p (C1)	(γ)	transition from 268-kev				
	195	3.9p (G1)	(γ)	level (S4)				
	191	4p, α (S4)	0.042 (γ)	$K/L=4.2$ (M8)				
	191	3.5p (B4)	(e_L)					
	269	3.8p (M8)	(e)	$K/L=6$ (M8)		0.08		
	268	6p (E3)	(p′)					
	273	4p (S4)	0.02 (γ)	cascade trans. from 550-kev level				
				$\gamma(273)-\gamma(277)$ coinc. (S4)				
	277	3α and p (H1)	(γ)	ex.func. $\Delta E=277$ (C1; G1; S4)	30% E2	0.33		279 $Hg^{197}(\epsilon\gamma)$
	279	2–5p (C1)	0.14 (γ)	$\gamma(\theta):3/2(E2)5/2(E2+M1)3/2$	α=0.32			
	277	2–4p (G1)	0.3 (γ)	$\delta^2=0.6$ (C1)	$\alpha_K=0.24$			
	277	6α (T8)	0.25 (γ)	$\delta^2=0.07$ (G1)				
	281	2p (H3)	0.072 (e_K)	$\delta=-0.75$ (M7)				
	277	1.6–5p (S4)	0.23 (γ)	$K/L=5.5$ (B4)				
	279	3.5p (B4)	(e_K)	$=6.3$ (M8)				
	279	6p (E3)	0.36 (p′)					
	286	15N (A1)	(γ)					
	555	3–5p (C1)	0.26 (γ)	ex.func. $\Delta E=550$ (C1; G1; S4)	100% E2	0.43		
	545	3–4p (G1)	(γ)	$\gamma(\theta):3/2(E2)7/2(E2)3/2$				
	545	6α (T8)	0.56 (γ)	(C1; G1; M7)				
	550	2.2–5p (S4)	0.42 (γ)	cascade trans. via 277-kev level				
	550	6p (E3)	0.42 (p′)	$<5\%$ (C1)				
	580	15N (A1)	(γ)	$<10\%$ (G1)				
				$=4\%$ (S4)				
				$K/L=3.6$ (M8)				
$_{80}Hg^{198}$	411	4p (B6)	1.09 (γ)	sep.iso. (B6)	α=0.04	0.8		411 $Au^{198}(\beta^-\gamma)$
	411	3.2p (D1)	0.4 (γ)					
Hg^{199}	163	3α (H1)	(γ)	sep.iso. (D1; B6)	100% E2	0.26	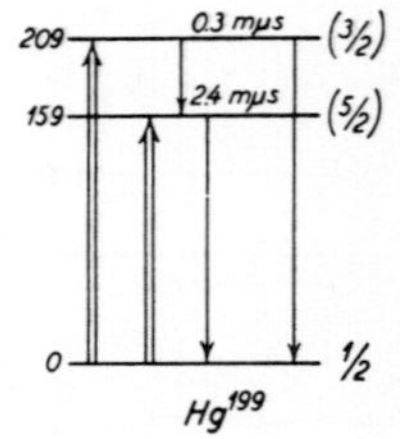 209 — (3/2) (0.3 mμs); 159 — (5/2) (2.4 mμs); 0 — 1/2; Hg 199	159 $Au^{199}(\beta^-\gamma)$
	159	3.2p (D1)	0.084 (γ)		α=0.9			158 $Tl^{199}(\epsilon\gamma)$
	159	3p (B6)	0.19 (γ)					
	209	3.2p (D1)	0.038 (γ)	sep.iso. (D1; B6)	α=0.9	0.10		209 $Au^{199}(\beta^-\gamma)$
	209	3p (B6)	0.071 (γ)					208 $Tl^{199}(\epsilon\gamma)$
Hg^{200}	375	3.2p (D1)	0.47 (γ)	sep.iso. (D1; B6)	α=0.06	0.7		368 $Tl^{200}(\epsilon\gamma)$
	368	4p (B6)	0.80 (γ)					370 $Hg(n\gamma)$
Hg^{202}	439	3.2p (D1)	0.4 (γ)	sep.iso. (D1; B6)	α=0.04	0.5		439 $Tl^{202}(\epsilon\gamma)$
	439	4p (B6)	0.57 (γ)					

TABLE IV.2.—*Continued.*

I	II	III	IV	V	VI	VII	VIII	IX
Nucleus	Decay energy	Bombarding conditions	$\epsilon B(E2)$	Comments	Multipole order and conversion coefficients	$B(E2)$	Level schemes	Other processes
${}_{81}Tl^{203}$	279 279	3–4.5p (B3) 4p, α (S4)	0.09 (γ) 0.11 (γ)	ex.func. $\Delta E=279$(B3)	73%$E2$ $\alpha=0.20$	0.12	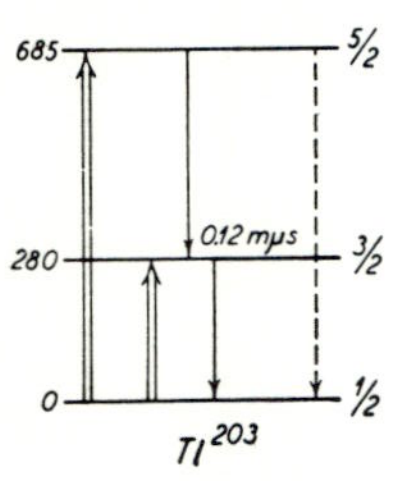 Tl^{203}: 685 — 5/2; 280 — 3/2; 0 — 1/2; 0.12 mμs	280 $Pb^{203}(\epsilon\gamma)$ 280 $Hg^{203}(\beta^{-}\gamma)$
Tl^{205}	205 205	3–4.5p (B3) 4p, α (S4)	0.11 (γ) 0.072 (γ)	ex.func. $\Delta E=205$(B3)	$\alpha\sim0.5$	0.14	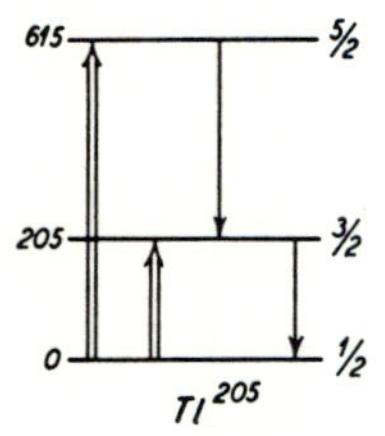 Tl^{205}: 615 — 5/2; 205 — 3/2; 0 — 1/2	203 $Hg^{205}(\beta^{-}\gamma)$
Tl	410 410	4–4.5p (B3) 4p (S4)	(γ) 0.12 (γ)	probably from both isotopes; in coinc. with both $\gamma(205)$ and $\gamma(280)$ (S4)	$\alpha\sim0.1$			405 $Pb^{203}(\epsilon\gamma)$
${}_{82}Pb^{206}$	810	4.5p (S4)	0.14 (γ)	sep.iso. (S4)		0.14		803.3 $Bi^{206}(\epsilon\gamma)$
Pb^{207}	570 580	4.5p (S4) 15N (A1)	0.028 (γ) (γ)			0.028		569 $Bi^{207}(\epsilon\gamma)$
${}_{90}Th^{232}$	50 53 50	3 and 6α(T3; T8) 4α (S4) 2.5–3.3α (D1)	0.057 (γ) 0.0096(γ)		$\alpha=290$	9		$\sim$50 $U^{236}(\alpha\gamma)$
	760	5p (S4)	0.13 (γ)					
	719	4.8p (M8)	(e_K)	$K/L=3.5$(M8)				

TABLE IV.2.—*Continued.*

I Nucleus	II Decay energy	III Bombarding conditions	IV $\epsilon B(E2)$	V Comments	VI Multipole order and conversion coefficients	VII $B(E2)$	VIII Level schemes	IX Other processes
$_{92}U^{233}$	40.4	2.8α (N1)	0.015 (γ)		(80% $E2$) $\alpha=870$	13	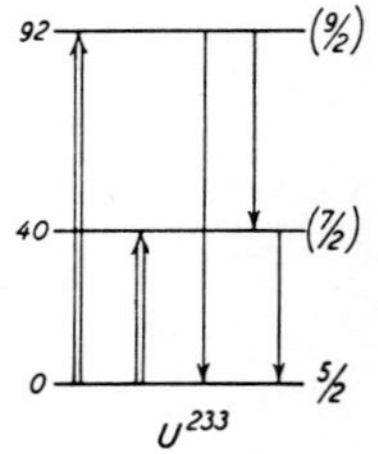 U^{233}: 92 (9/2); 40 (7/2); 0 5/2	40.5 $Pa^{233}(\beta^-\gamma)$
	51.5	2.8α (N1)	0.0070 (γ)	cascade trans. from 92-kev level	(80% $E2$) $\alpha=240$			
	92.3	2.8α (N1)	0.032 (γ)		100% $E2$ $\alpha=21$	2.4		
U^{235}	46.2	2.8α (N1)	0.057 (γ)		(6% $E2$) $\alpha=66$	3.8	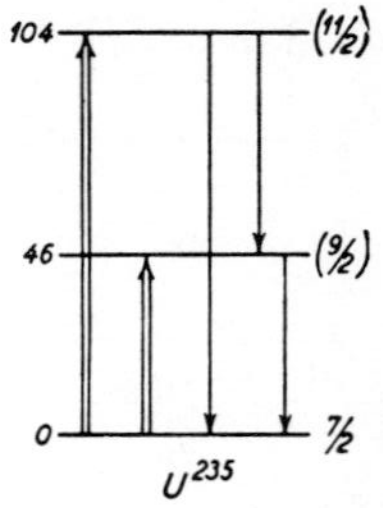 U^{235}: 104 (11/2); 46 (9/2); 0 7/2	
	56.7	2.8α (N1)	0.017 (γ)	cascade trans. from 104-kev level	(6% $E2$) $\alpha=31$			
	103.2	2.8α (N1)	0.0041 (γ)		100% $E2$ $\alpha=15$	0.60		
U^{238}	44 45 44.7	3α (H1; T8) 3.3α (D1) 2.8α (N1)	0.034 (γ) 0.0070 (γ) 0.015 (γ)		$\alpha=680$	13		44 $Pu^{242}(\alpha)$
$_{93}Np^{237}$	33.2	2.8α (N1)	0.026 (γ)		(2% $E2$) $\alpha=150$	4.0	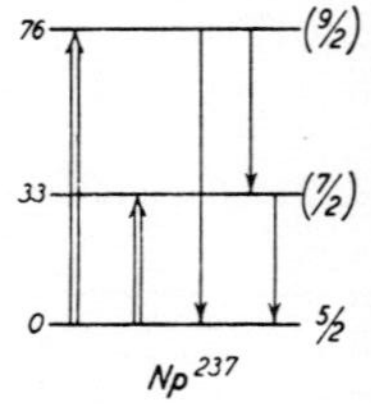 Np^{237}: 76 (9/2); 33 (7/2); 0 5/2	33.20 $Am^{241}(\alpha\gamma)$ 33.2 $U^{237}(\beta^-\gamma)$
	42.6	2.8α (N1)	0.023 (γ)	cascade trans. from 76-kev level	(2% $E2$) $\alpha=68$			
	76	2.8α (N1)	0.0026 (γ)		100% $E2$ $\alpha=60$	1.8		

TABLE IV.2.—*Continued.*

I	II	III	IV	V	VI	VII	VIII	IX
Nucleus	Decay energy	Bombarding conditions	$\epsilon B(E2)$	Comments	Multipole order and conversion coefficients	$B(E2)$	Level schemes	Other processes
$_{94}Pu^{239}$	49.6	2.8α (N1)	0.0073(γ)	probably cascade trans. from 57-kev level	(22%$E2$) $\alpha=150$		57 — 5/2; 8 — 3/2; 0 — 1/2; Pu^{239}	49.40 $Np^{239}(\beta^-\gamma)$
	57.5	2.8α (N1)	0.012 (γ)		100%$E2$ $\alpha=260$	4.2		57.25 $Np^{239}(\beta^-\gamma)$

A 1 Alkhazov, Andreyev, Greenberg, and Lemberg, Nuclear Phys. **2**, 65 (1956).
B 1 R. Barloutaud and T. Grjebine, Compt. rend. **239**, 491 (1954).
2 J. H. Bjerregaard and T. Huus, Phys. Rev. **94**, 204 (1954).
3 Barloutaud, Grjebine, and Riou, Compt. rend. **240**, 1207 (1955).
4 E. M. Bernstein and H. W. Lewis, Phys. Rev. **100**, 1345 (1955).
5 E. M. Bernstein and H. W. Lewis, Phys. Rev. **100**, 1367 (1955).
6 Barloutaud, Grjebine, and Riou, Compt. rend. **242**, 1284 (1956).
7 J. H. Bjerregaard and U. Meyer-Berkhout, Z. Naturforsch. **11a**, 273 (1956).
8 E. M. Bernstein and H. W. Lewis, Phys. Rev. **99**, 617(A) (1955).
9 C. A. Barnes, Phys. Rev. **97**, 1226 (1955).
C 1 Cook, Class, and Eisinger, Phys. Rev. **96**, 658 (1954).
2 C. M. Class and U. Meyer-Berkhout (to be published).
D 1 Davis, Divatia, Lind, and Moffat, Phys. Rev. **103**, 1801 (1956).
E 1 Eisinger, Cook, and Class, Phys. Rev. **94**, 735 (1954).
2 B. Elbek (to be published).
3 B. Elbek and C. K. Bockelman, Phys. Rev. (to be published).
F 1 Fagg, Wolicki, Bondelid, Dunning, and Snyder, Phys. Rev. **100**, 1299 (1955).
G 1 W. I. Goldburg and R. M. Williamson, Phys. Rev. **95**, 767 (1954).
H 1 N. P. Heydenburg and G. M. Temmer, Phys. Rev. **93**, 906 (1954).
2 N. P. Heydenburg and G. M. Temmer, Phys. Rev. **99**, 617(A) (1955) and private communication.
3 Huus, Bjerregaard, and Elbek, Kgl. Danske Vidensk. Selskab Mat. fys. Medd. **30**, No. 17 (1956).
4 T. Huus and A. Lundén, Phil. Mag. **45**, 966 (1954).
5 T. Huus and Č. Zupančič, Kgl. Danske Vidensk. Selskab Mat. fys. Medd. **28**, No. 1 (1953).
6 T. Huus and J. Bjerregaard, Phys. Rev. **92**, 1579 (1953).
7 N. P. Heydenburg and G. M. Temmer, Phys. Rev. **95**, 861 (1954).
8 N. P. Heydenburg and G. M. Temmer, Phys. Rev. **100**, 150 (1955) and privately circulated revisions of yield determinations.
9 N. P. Heydenburg and G. M. Temmer, Phys. Rev. **94**, 1252 (1954).
10 N. P. Heydenburg and G. M. Temmer, Bull. Am. Phys. Soc. Ser. II, **1**, 164 (1956) and private communication.
J 1 G. A. Jones and D. H. Wilkinson, Phil. Mag. **45**, 230 (1954).
L 1 Lemmer, Segaert, and Grace, Proc. Phys. Soc. (London) **68A**, 701 (1955).
M 1 McClelland, Mark, and Goodman, Phys. Rev. **97**, 1191 (1955).
2 C. L. McClelland and C. Goodman, Phys. Rev. **91**, 760 (1953).
3 McClelland, Mark, and Goodman, Phys. Rev. **93**, 904 (1954).
4 Mark, McClelland, and Goodman, Phys. Rev. **98**, 1245 (1955).
5 H. Mark and G. Paulissen, Phys. Rev. **99**, 1654(A) (1955).
6 H. Mark and G. T. Paulissen, Phys. Rev. **100**, 813 (1955).
7 F. K. McGowan and P. H. Stelson, Phys. Rev. **99**, 127 (1955).
8 Moore, Class, Prossler, and Schiffer, Bull. Am. Phys. Soc. Ser. II, **1**, 88 (1956).
N 1 J. O. Newton (private communication).
P 1 E. B. Paul and H. E. Gove (private communication, November, 1953).
2 van Patter, Rothman, Mandeville, and Swann, J. Franklin Inst. **259**, 261 (1955).
S 1 Sherr, Li, and Christy, Phys. Rev. **94**, 1076 (1954) and **96**, 1258 (1955).
2 Schiffer, Moore, and Class, Phys. Rev. (to be published).
3 Simmons, van Patter, Famularo, and Stuart, Phys. Rev. **97**, 89 (1955).
4 P. H. Stelson and F. K. McGowan, Phys. Rev. **99**, 112 (1955).
5 Simmons, Famularo, and Freier, Phys. Rev. **100**, 1265(A) (1955).
T 1 G. Temmer and N. P. Heydenburg, Phys. Rev. **98**, 1198(A) (1955).
2 G. M. Temmer and N. P. Heydenburg, Phys. Rev. **96**, 426 (1954) and privately circulated revisions of yield determinations.
3 G. M. Temmer and N. P. Heydenburg, Phys. Rev. **93**, 351 (1954).
4 G. M. Temmer and N. P. Heydenburg, Bull. Am. Phys. Soc. Ser. II, **1**, 43 (1956) and privately circulated revisions of yield determinations.
5 G. M. Temmer and N. P. Heydenburg, Phys. Rev. **98**, 1308 (1955) and privately circulated revisions of yield determinations.
6 G. M. Temmer and N. P. Heydenburg, Phys. Rev. **99**, 617(A) (1955) and private communication.
7 G. M. Temmer and N. P. Heydenburg, Phys. Rev. **100**, 961(A) (1955) and private communication.
8 G. M. Temmer and N. P. Heydenburg (private communication).
9 G. M. Temmer and N. P. Heydenburg, Phys. Rev. **94**, 1399 (1954).
W 1 Wolicki, Fagg, and Geer, Phys. Rev. **100**, 1265(A) (1955) and private communication.

acter of the excited levels. Where directly measured lifetimes are available, the half-lives are listed, employing the following abbreviations: *s* (seconds), *μs* (10^{-6} sec), *mμs* (10^{-9} sec).

The level populated in Coulomb excitation of an even-even nucleus appears in all cases to be the first excited, 2+, state and no decay scheme is drawn.

Column IX. Other Processes

This column lists other reactions in which levels are observed that may tentatively be identified with those found in Coulomb excitation. The observed energies are given in kev, together with the reaction involved. References to the experimental work may be found in Hollander, Perlman, and Seaborg, Revs. Modern Phys. **25**, 469 (1953), and Nuclear Data Cards, edited by K. Way *et al.*, National Research Council, Washington D. C.

CHAPTER V. COLLECTIVE NUCLEAR EXCITATIONS

An outstanding feature of the nuclear spectra revealed by the Coulomb excitation studies is the systematic occurrence throughout the periodic system of low-energy electric quadrupole transitions of a strength greatly exceeding that which would be associated with the excitation of a single nucleon. The estimate (II A.58) of the reduced transition probability for a single proton transition of $E2$ type gives[122]

$$B(E2)_{sp}=3\cdot 10^{-5}A^{4/3}e^2 10^{-48}\ \text{cm}^4. \qquad \text{(V.1)}$$

Thus, from a comparison with the observed $B(E2)$ values in column VII of Table IV.2, it is seen that most elements exhibit $E2$ transitions of a strength more than 10 times the single particle unit, and that in certain regions transitions occur with a probability exceeding this unit by a factor of more than 100.

These enhanced transitions are clearly due to the cooperative effects of a large number of nucleons, and indeed most of the observed levels can be interpreted in terms of simple collective excitations of rotational or vibrational type. Where this interpretation can be made, the Coulomb excitation experiments yield valuable information on such collective nuclear properties as the equilibrium shape, the deformability, and the inertial parameters associated with the collective motion. In the present chapter, we shall outline the theory of collective nuclear excitations and discuss the evidence obtained from the Coulomb excitation experiments.

V A. Qualitative Considerations

In the analysis of nuclear excitation spectra it is possible to distinguish between two different modes of excitation, the first associated with the motion of individual nucleons and the second with collective types of nuclear motion.[123,124] One may think of the former degrees of freedom as representing the motion of the nucleons in a fixed nuclear potential (the intrinsic nuclear motion), while the latter are associated with variations in the shape and orientation of the nuclear field.

Such a separation of the motion becomes possible when the frequencies of the collective excitations are small compared with those characterizing the intrinsic nucleonic motion and is in many respects analogous to the separation between electronic and nuclear motion in molecules.

When this adiabatic condition is fulfilled, one may treat the equations of motion for the nucleus in two steps. First one considers the nucleonic motion for fixed values of the collective parameters α, specifying the nuclear field; the energy eigenvalues for this motion are denoted by $E_i(\alpha)$. The collective motion superposed on the intrinsic motion is then given by a Hamiltonian of the approximate form

$$H_{\text{coll}}=E_i(\alpha)+\tfrac{1}{2}B_i(\alpha)\dot{\alpha}^2. \qquad \text{(V.2)}$$

The functions $E_i(\alpha)$ are referred to as the potential energy surfaces of the nucleus and play a similar role as in the treatment of molecular vibrations and rotations. In the present discussion we are especially interested in the behavior of the potential energy surfaces near the equilibrium shape.[125]

The second term in (2) gives the kinetic energy of the collective motion, which may be obtained by considering the nucleonic motion for slowly varying α. This kinetic energy can be written as a quadratic expression in the $\dot{\alpha}$, provided all the frequencies of the intrinsic motion are large compared to those of the collective motion, so that the intrinsic motion adjusts adiabatically to the variation in α.

If the intrinsic motion possesses degenerate or close lying energy levels, the adiabatic approximation may partially break down. The nucleus must then be described in terms of a coupled system of collective oscillations and the low energy intrinsic degrees of freedom in question.[124]

---

[122] As already noted in Chapter II, the statistical factor appearing in (II A.58) is somewhat arbitrary; it is the factor appropriate to a two proton excitation of the type $(j^2)_{J=0}\rightarrow(j^2)_{J=2}$ in the limit of large j.

[123] For a recent review of the nuclear independent particle model, see M. G. Mayer and J. H. D. Jensen, *Elementary Theory of Nuclear Shell Structure* (John Wiley and Sons, Inc., New York, 1955).

[124] Collective nuclear oscillations were first considered by N. Bohr and F. Kalckar, Kgl. Danske Videnskab. Selskab Mat. fys. Medd. **14**, No. 10 (1937). The interplay between collective and independent particle motion has been discussed by J. Rainwater, Phys. Rev. **79**, 432 (1950); A. Bohr, Kgl. Danske Videnskab. Selskab Mat. fys. Medd. **26**, No. 14 (1952); D. L. Hill and J. A. Wheeler, Phys. Rev. **89**, 1102 (1953); A. Bohr and B. R. Mottelson, Kgl. Danske Videnskab. Selskab Mat. fys. Medd. **27**, No. 16 (1953).

[125] The behavior of these surfaces for larger deformations has been discussed in connection with the nuclear fission process [N. Bohr and J. A. Wheeler, Phys. Rev. **56**, 426 (1939); D. L. Hill and J. A. Wheeler, reference 124; A. Bohr, *Proceedings of the International Conference on the Peaceful Uses of Atomic Energy* (Columbia University Press, New York, 1956), Vol. 2, p. 151 (Geneva, 1956)].

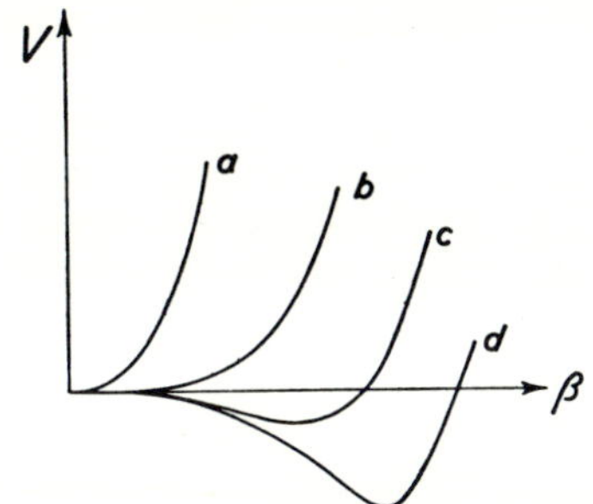

FIG. V.1. Potential energy surfaces for even-even nuclei. The nuclear potential energy V is plotted as a function of the deformation parameter β, which may, for instance, represent the quadrupole eccentricity of the nucleus [see (V.6)]. The various curves are intended to illustrate schematically the behavior of the potential energy surfaces for even-even nuclei as one moves away from closed shells.

The curve a represents a configuration with only relatively few particles outside of closed shells. As particles are added the restoring force decreases though the spherical shape ($\beta=0$) remains stable (curve b). Still further from the closed shell the spherical shape may become unstable (curve c) and the nucleus acquires a nonspherical equilibrium shape. With the addition of still more nucleons the equilibrium eccentricity increases and the minimum in the potential energy surface becomes sharper (curve d).

The curves all refer to the lowest intrinsic state. Additional sets of potential energy surfaces are associated with each excited intrinsic state.

Although the details of the figure have no quantitative significance, the qualitative trends are suggested by simple considerations (see the discussion in the text).

In the earliest treatments of collective nuclear oscillations, one attempted to estimate the potential and kinetic energy in (2) by comparing the nucleus with a liquid drop. It is found, however, that the shell structure in the nucleonic motion has a profound effect on the collective properties of the individual nuclei. Thus, the potential energy surfaces depend essentially on the nucleonic configuration and also the inertial parameters $B_i(\alpha)$ deviate from the hydrodynamical estimates.

The main features of the nuclear potential energy surfaces are determined by the competition between the particles in closed shells, which strongly prefer a spherical nuclear shape, and the particles in unfilled shells which tend to polarize the nucleus and bring about a nonspherical equilibrium shape.[126] The latter tendency is, however, counteracted by the residual interactions between the nucleons, which must be added to the interactions already included in the average field. The residual interactions imply correlations in the nucleonic motion which reduce the net polarizing effect. This reduction is a result of the attractive short-range character of the nuclear forces which favor states of maximum spherical symmetry.[123]

The influence of the residual interactions is the greatest for small deformations of the nuclear shape, as a consequence of the degeneracy of the particle motion in a spherical field. For large nuclear eccentricities, the strong coupling of the individual particles to the nuclear deformation removes the degeneracies, and the residual interactions are then of less importance.

The dependence of the potential energy surfaces on the number of nucleons in unfilled shells is illustrated schematically in Fig. V.1. The figure refers to even-even nuclei, for which the lowest intrinsic state for a spherical shape possesses zero total angular momentum. The spherical density distribution of such a state implies that the average polarizing effect of the particles vanishes. For an even-even nucleus, the spherical shape thus always represents an equilibrium, which may, however, be either stable or unstable. For configurations with only relatively few particles outside of closed shells, the deformation which would result in the absence of residual interactions is small; the coupling between the nucleons is then mainly determined by these interactions and the spherical nuclear shape remains a stable equilibrium. For sufficiently many particles in unfilled shells, however, the deformation caused by the nucleonic motion is large and thus only little affected by the residual interactions; the strongly deformed shape then gives the minimum in the potential energy surface, and the spherical shape is unstable.

On the basis of these qualitative considerations we consider briefly the general features of the collective nuclear excitation spectra for the different configurations.

For a closed shell nucleus, the special stability of the spherical equilibrium shape[127] implies that oscillations in shape would have high frequencies. Since these frequencies may be of the order of those involved in the single particle motion, there may be no sharp distinction between collective and single particle excitations of a closed shell nucleus.[128]

If one or a few nucleons are added to (or subtracted from) a closed shell configuration, the low-energy nuclear states may be approximately described in terms of the motion of these added particles. There exists, however, a weak coupling between this nucleonic motion and the oscillations of the closed shell core, which implies a significant enhancement of the electric multipole transitions between the low-lying levels.

As more nucleons are added to the closed shell configuration, the description of the excitations in terms of the motion of the individual nucleons becomes highly complex, especially due to the effect of configuration mixing. Moreover, the coupling to the closed shell core increases.

Already for nuclei containing relatively few particles in unfilled shells, however, one observes states in the low-energy nuclear spectrum which can be approximately described in terms of simple collective oscillations. The collective behavior of the nucleons may be understood from the fact that the potential energy of

[126] J. Rainwater, reference 124.

[127] S. Gallone and C. Salvetti, Nuovo cimento (9) **10**, 145 (1953). See also the references in footnote 124.

[128] D. Inglis, Phys. Rev. **97**, 701 (1955).

deformation decreases as one moves away from closed shell configurations (see Fig. V.1); therefore, the frequency of collective oscillation soon becomes smaller than the main frequencies of the intrinsic motion.

In the vicinity of the closed shells, where the spherical shape represents a stable equilibrium, the collective excitations correspond to vibrations about this shape. The frequencies of these vibrations are expected to decrease fairly regularly with the addition of particles, corresponding to the decreasing restoring force. Eventually, this tendency may lead to instability of the spherical shape and a resulting nonspherical equilibrium shape (see Fig. V.1).

For such deformed nuclei the collective spectrum separates into excitations of vibrational and rotational type. The first corresponds to oscillations about the equilibrium shape for fixed orientation of the nucleus, while the second represents a collective motion which rotates the nuclear orientation while preserving the shape. Such a separation becomes possible since the nuclear deformation implies that a large mass transport is associated with the rotational motion. This motion can thus take place with small frequency and therefore without affecting the shape (or intrinsic structure) of the nucleus.

The simple character of the rotational motion gives rise to many regularities in the rotational excitation levels, which make them easily identifiable. The rotational states are also especially strongly excited in Coulomb excitation experiments, as a consequence of their low energy and large electric quadrupole transition probabilities. We therefore begin, in Sec. B, with a more detailed discussion of this special type of collective excitation.

The properties of the vibrational modes of excitation, in spherical and deformed nuclei, are at present less well established, but the Coulomb excitation process constitutes one of the most promising methods for a further exploration of these states. In Secs. C and D, we discuss the general characteristics expected for vibrational spectra and summarize the available evidence on these excitations. Finally, in Sec. E, we consider briefly some of the special features of the excitation spectra for nuclei in the closed shell regions.

V B. Rotational Excitations

The occurrence of rotational spectra is a general characteristic of nuclei possessing a nonspherical equilibrium shape. For such nuclei it is possible to separate between a collective rotational motion and the nucleonic motion for fixed nuclear orientation. This latter motion may again separate into vibrational and individual particle components, but will in the present section often be referred to simply as the intrinsic nuclear motion, since the main regularities in the rotational spectra are independent of the details of this intrinsic structure.

V B.1. Energy Spectrum

The rotational spectrum becomes especially simple if the nuclear shape possesses axial symmetry, as appears generally to be the case for the very strongly deformed nuclei.[129,130] The angular momentum coupling scheme is then similar to that of a linear molecule[131] and can be characterized by the three constants of the motion: the total angular momentum I, its projection M on a space-fixed axis, and its projection K on the nuclear symmetry axis (see Fig. V.2).

Since there can be no collective rotations about a symmetry axis (see footnote 147 later), the quantum number K is a constant for each rotational band and represents an intrinsic angular momentum. The rotational spectrum for the nucleus has the same general form as for a molecule and may be written[132]

$$E_I = E_0 + \frac{\hbar^2}{2\mathfrak{J}}\{I(I+1) + a(-1)^{I+\frac{1}{2}}(I+\tfrac{1}{2})\delta_{K,\frac{1}{2}}\}, \quad \text{(V.3)}$$

where E_0 is a constant depending only on the intrinsic structure, while $\mathfrak{J}$ represents the effective moment of inertia about an axis perpendicular to the nuclear symmetry axis. The last term in the brackets, occurring only for states with $K=\frac{1}{2}$, is associated with a decoupling of the spin angular momentum from the rotational motion. The decoupling parameter a can be expressed as an expectation value for the intrinsic motion.[132,133] A similar decoupling effect is well known from molecular spectra (see, e.g., the uncoupling of the electronic spin from the

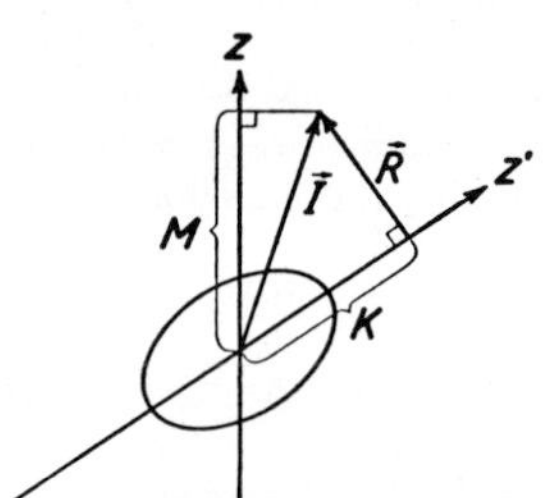

FIG. V.2. Coupling scheme for deformed nuclei. For strongly deformed nuclei possessing axial symmetry, the angular momentum properties may be characterized by the three constants of the motion I, M, and K. While I and M represent the total angular momentum and its component along the fixed z axis, the component of I along the nuclear symmetry axis, z', is denoted by K. The collective rotational angular momentum $\mathbf{R}$ is perpendicular to the z' axis; thus K represents an intrinsic angular momentum.

[129] The principal empirical evidence for the axial symmetry is the observed $I(I+1)$ type of rotational spectra (see, e.g., Fig. V.4). The preference for axial symmetry is also consistent with theoretical estimates of the equilibrium shape for the nuclear shell structure.

[130] Rotational spectra for nuclei without axial symmetry have been considered by C. Marty [Nuclear Phys. 1, 85 (1956)].

[131] A. Bohr, Phys. Rev. 81, 134 (1951).

[132] A. Bohr and B. R. Mottelson, reference 124.

[133] S. G. Nilsson, Kgl. Danske Videnskab. Selskab Mat. fys. Medd. 29, No. 16 (1955).

rotational motion for $^2\Sigma$ states, which leads to rotational spectra with $a=+1$ and $a=-1$).[134]

The value of K for the nuclear ground state may be obtained from a consideration of the individual particle motion in the deformed nucleus. In such a nonspherical field, the angular momenta l_k and j_k of a nucleon are in general not constants of the motion, but for axially symmetric nuclei the nucleon orbitals may be labeled by the constant of the motion Ω_k, which represents the projection on the symmetry axis of the total angular momentum of the nucleon. States which differ only in the sign of Ω_k are degenerate, since they are the same except for the sense of the particle motion around the symmetry axis. This gives rise to an especially simple type of shell structure, in which the particles are filled pairwise in states of opposite Ω_k with no net contribution to K. Thus, for the lowest state of an even-even nucleus, all the particles are in paired orbits, and we have $K=0$. In an odd-A nucleus, the last nucleon occupies an unpaired orbit, and K equals the Ω_k of this orbit.[135]

This coupling scheme, which would apply for independent particle motion, is somewhat modified by the residual interactions between the nucleons. However, there is evidence that these interactions can be considered as acting principally between paired nucleons, so that the above classification remains valid for the ground state of even-even nuclei and the intrinsic states of odd-A nuclei corresponding to the different orbits of the last odd nucleon.[136] On the other hand, the degrees of freedom associated with the excitation of paired nucleons may partly manifest themselves in collective vibrational motion (see Sec. V D).

For nuclear shapes possessing reflection symmetry with respect to a plane perpendicular to the nuclear axis, the possible rotational quantum states are governed by symmetry requirements similar to those applying to homonuclear diatomic molecules.[137,138] Thus, for an intrinsic state with $K=0$, only even or odd values of I are allowed, according to the symmetry of the intrinsic state with respect to a rotation of 180° about an axis perpendicular to the symmetry axis. In particular, for the ground state of an even-even nucleus, only the even values

$$I=0, 2, 4, 6, \cdots \text{ (even parity)} \qquad \text{(V.4)}$$

occur in the rotational spectrum.[139]

For intrinsic states with $K\neq 0$, the allowed values of the nuclear spin are

$$I=K, K+1, K+2, \cdots. \qquad \text{(V.5)}$$

The members of the band all have the same parity which equals the parity of the intrinsic motion. Thus, for an odd-A nucleus with a single unpaired nucleon, the orbit of this last particle determines the parity as well as the K-value of the rotational band.

Even-even nuclei.—The Coulomb excitation experiments have provided one of the most important sources of information on the rotational excitations. These are strongly populated by transitions of electric quadrupole type, and in an even-even nucleus, one should thus excite the first state ($I=2+$) of the lowest rotational band. It has also been found that the Coulomb excitation induces just one strong transition in each even-even nucleus far from closed shells. The excitation energy E_2 of this state is a rather smooth function of the atomic number and decreases as one moves away from closed shells. The energy systematics of the first excited states of even-even nuclei is shown in Fig. V.3.

The regions of large nuclear deformations are characterized by especially small values of the excitation energies E_2, and, as will be discussed below, rotational spectra are only expected in nuclei for which E_2 is less than the critical value indicated by the dotted curve in Fig. V.3. Such small excitation energies are found in the light elements with $A\sim 8$ and 24, and in the heavier elements with $150<A<190$ and $A>222$.

These regions include just the nuclei for which the number of particles in unfilled shells relative to those in closed shells is especially large. In the mass region $40<A<150$ the conditions for the occurrence of large deformations are less favorable, partly due to the effect of the spin orbit coupling which breaks the major shells, and partly due to the neutron excess which implies that the closings of neutron and proton shells occur for different nuclei.

The rotational interpretation of the states populated by Coulomb excitation in the mass regions $150<A<190$ and $A>222$ is confirmed by the observation of higher excited states in the rotational band. These states which are populated in radioactive decay processes are found in even-even nuclei to have the spin sequence (4) with energies corresponding to (3). (See Fig. V.4)[140].

The Coulomb excitation of these higher states would require either a transition of multipole order greater

[134] See G. Herzberg, *Spectra of Diatomic Molecules* (D. van Nostrand Company, Inc., New York, 1950), p. 222.

[135] For calculations of the single particle states in deformed axially symmetric potentials, see S. Moszkowski, Phys. Rev. **99**, 803 (1955); S. G. Nilsson, reference 133; K. Gottfried, Phys. Rev. **103**, 1017 (1956).

[136] Compare the classification of the spins and parities of the ground states and low-lying intrinsic excitations of odd-A nuclei with nonspherical shape in terms of the binding states of the last odd nucleon in an ellipsoidal potential (B. R. Mottelson and S. G. Nilsson, Phys. Rev. **99**, 1615 (1955); K. Gottfried, reference 135). Additional evidence is provided by the systematic occurrence of unhindered α decay in odd-A nuclei [Bohr, Fröman, and Mottelson, Kgl. Danske Videnskab. Selskab Mat. fys. Medd. **29**, No. 10 (1955)].

[137] A. Bohr, reference 124.

[138] K. W. Ford, Phys. Rev. **90**, 29 (1953).

[139] For the spin-parity values of rotational bands associated with vibrational excitations, see Sec. V D.1.

[140] Evidence for rotational bands in the nuclei with $A=24$ and 25 is discussed by Litherland, Paul, Bartholomew, and Gove, Phys. Rev. **102**, 208 (1956). For the nuclei around $A=8$, the consequences of the present description are similar to those which follow from the α-particle model [see the review by D. R. Inglis, Revs. Modern Phys. **25**, 390 (1953)].

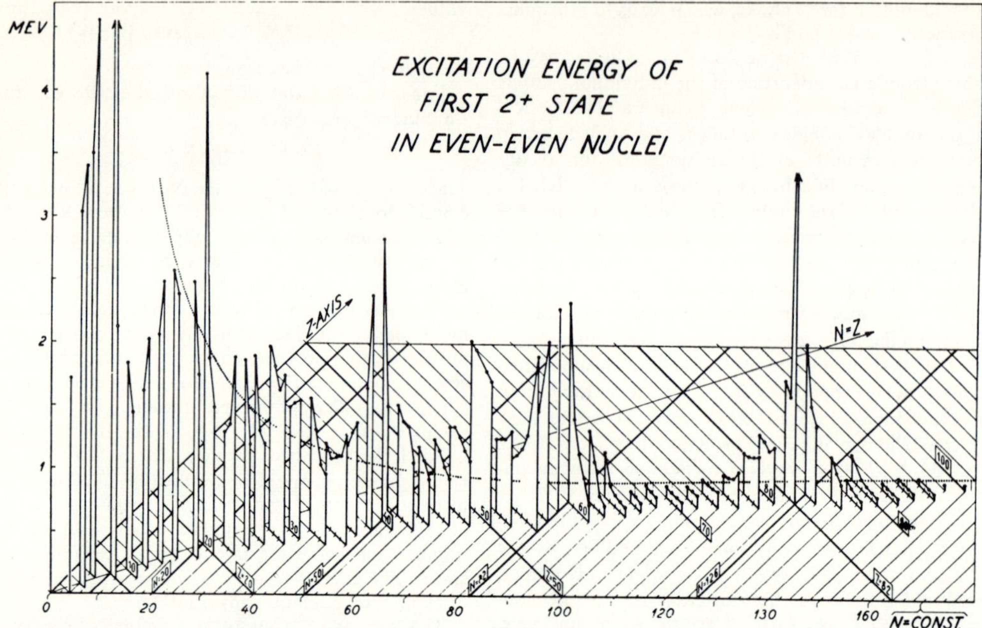

Fig. V.3. Energy systematics of first excited 2+ states in even-even nuclei. The energies of the first excited 2+ states of the even-even nuclei are plotted as a function of neutron number N and proton number Z. The black circles indicate levels that have been observed in Coulomb excitation, while the open circles represent levels which have so far only been observed in radioactivity or nuclear reaction studies.

The rotational spectra occur in the regions farthest from closed shells, where the excitation energies are lowest; in other regions, the excitations have the character of collective quadrupole vibrations (see Sec. V C). The separation between these two regions is approximately given by the criterion (9) which is illustrated by the dotted curve following the stable mass region. Thus, the rotational spectra are found in the regions where the observed first excited states have energies less than this separation line. For the value of $\mathcal{I}_{rig}$ in (9), we have used the relation (7) with $R_0=1.2A^{1/3}\times10^{-13}$ cm, and have estimated the higher order β-dependent corrections by assuming β to have the critical value $0.6v$ with the interaction parameter $v=1.8A^{-1/3}$ (see reference 145).

The figure is a representation of the systematics first discussed by G. Scharff-Goldhaber, Phys. Rev. **90**, 587 (1953) and by P. Preiswerk and P. Stähelin, Nuovo cimento **10**, 1219 (1953). The experimental energies are taken from Table IV.2, and from the following compilations: F. Ajzenberg and T. Lauritsen, reference 187; P. M. Endt and J. C. Kluyver, Revs. Modern Phys. **26**, 95 (1954); K. Way *et al.*, Nuclear Level Schemes, $40\leq A\leq92$, Washington (1955). Additional data are obtained from: Perlman, Bernstein, and Schwartz, Phys. Rev. **92**, 1236 (1953) Pd^{108} and Cd^{108}; L. Grodzins and H. Motz, Phys. Rev. **100**, 1236(A) (1955) Sn^{114}; C. L. McGinnis, Phys. Rev. **98**, 1172(A) (1955) Sn^{120}; Farrelly, Koerts, van Lieshout, Benczer, and Wu, Phys. Rev. **98**, 1172(A) (1955) Sn^{122}; M. J. Glaubman, Phys. Rev. **98**, 645, 1172(A) (1955) Sn^{122}; Benczer, Farrelly, Koerts, and Wu, Phys. Rev. **100**, 955(A) (1955) Te^{128} and Xe^{128}; R. S. Caird and A. C. G. Mitchell, Phys. Rev. **94**, 412 (1954) Xe^{130}; H. N. Brown and R. A. Becker, Phys. Rev. **96**, 1372 (1954) Er^{164}; A. H. W. Aten, Jr., and G. D. de Feyfer, Physica **21**, 543 (1955) Os^{190}; Aten, de Feyfer, Sterk, and Wapstra, Physica **21**, 740 (1955) Os^{190}; M. W. Johns and S. V. Nablo, Phys. Rev. **96**, 1599 (1954) Os^{192} and Pt^{192}; V. E. Krohn and S. Raboy, Phys. Rev. **95**, 1354 (1954) Pb^{204}; I. Bergström and A. H. Wapstra, Phil. Mag. **46**, 61 (1955) Pb^{204}; Mihelich, Schardt, and Segrè, Phys. Rev. **95**, 1508 (1954) Po^{210}; I. Perlman (private communication of work by Asaro, Harvey, Hollander, Perlman, Smith, and Stephens) Em^{218}, Ra^{222}, Th^{222}, Th^{226}, Pu^{240}, Pu^{242}, Fm^{256}; T. O. Passell, UCRL-2528 (1954), U^{236}, Pu^{236}; O. P. Hok and G. J. Sizoo, Physica **20**, 77 (1954) U^{232}; Asaro, Stephens, Harvey, and Perlman, Phys. Rev. **100**, 137 (1955) Cm^{246}, Cm^{248}; Asaro, Stephens, Thompson, and Perlman, Phys. Rev. **98**, 19 (1955) Cf^{250}.

than $E2$, or a multiple $E2$ transition, and has not yet been observed. For an estimate of the cross sections for these processes, see Sec. V B.2.

The moments of inertia derived from the observed rotational spectra of even-even nuclei in the region $150\leq A\leq188$ are plotted in Fig. V.5 as a function of the nuclear quadrupole deformation parameter β. If the nucleus is assumed to have spheroidal shape, β is given by

$$\beta=\frac{4}{3}\left(\frac{\pi}{5}\right)^{\frac{1}{2}}\frac{\Delta R}{R_0}=1.06\frac{\Delta R}{R_0}, \qquad (V.6)$$

where R_0 is the mean nuclear radius and ΔR the difference between the major and minor semiaxis of the spheroid. The values of β employed in Fig. V.5 are obtained from the observed $E2$ transition probabilities, which determine the quadrupole moment of the nuclear shape [see (10) and (12) and Table V.2]. The moments of inertia are plotted in units of the moment

$$\mathfrak{J}_{rig}=\tfrac{2}{5}AMR_0^2(1+0.31\beta+\cdots), \qquad (V.7)$$

associated with a rigid rotation of a spheroid of mass AM about an axis perpendicular to the symmetry axis.

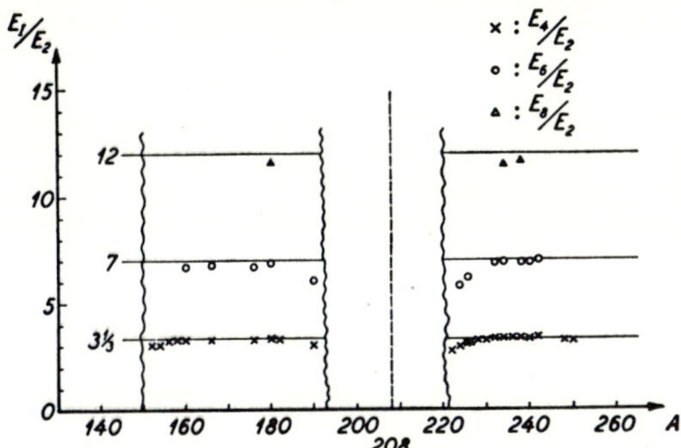

FIG. V.4. Energy ratios of rotational excitations in even-even nuclei. The figure shows the measured ratios of the energies of the higher rotational excitations to the energy of the first excited (2+) state in the regions $150<A<190$ and $A>222$, where rotational spectra are expected (compare Fig. V.3). The horizontal lines are the limiting theoretical ratios obtained from (3), assuming the higher states to have the spins 4+, 6+, 8+; while these spin values are experimentally established in only a few cases, they are in all cases consistent with available data on the decay scheme. The small systematic deviations from the limiting expression (3), which increase with the approach to closed shells and with I, can be interpreted in terms of the perturbation of the intrinsic structure produced by the rotational motion (compare Sec. V B.4).

The experimental data for the figure is taken from the compilation in Chapter XVII in *Beta- and Gamma-Spectroscopy*, edited by K. Siegbahn (North Holland Publishing Company, Amsterdam, 1955), and from: A. H. W. Aten, Jr., and G. D. de Feyfer, Physica **21**, 543 (1955) Os^{190}; Aten, de Feyfer, Sterk, and Wapstra, Physica **21**, 740 (1955) Os^{190}; I. Perlman (private communication of work by Asaro, Harvey, Hollander, Perlman, Smith, and Stephens) Ra^{222}, Th^{226}, U^{234}, Pu^{238}, Pu^{240}, Cm^{242}, Cm^{248}; Asaro, Stephens, and Perlman (submitted for publication) Ra^{224}, Ra^{226}; Goldhaber, der Mateosian, Harbottle, and McKeown, Phys. Rev. **99**, 180 (1955) Th^{228}; F. Asaro and I. Perlman, Phys. Rev. **99**, 37 (1955) Th^{228}; O. P. Hok, Phys. Rev. **99**, 1613 (1955) Th^{228}, Th^{230}, U^{232}, U^{234}; Asaro, Stephens, Thompson, and Perlman, Phys. Rev. **98**, 19 (1955) Cf^{250}.

The empirical moments of inertia are seen from Fig. V.5 to be appreciably smaller than $\mathfrak{J}_{rig}$ and to increase strongly with increasing β. A simple classical model of a rotational motion with these properties is provided by a wave traveling on the surface of a liquid drop. Assuming irrotational flow, this model yields the moment of inertia[141]

$$\mathfrak{J}_{irrot}=\tfrac{2}{5}AMR_0^2\beta^2(0.89+O(\beta^2)) \qquad \text{(V.8)}$$

for a nucleus of spheroidal shape. While the nuclear moments have some of the qualitative features of this irrotational flow model, it is seen from Fig. V.5 that the observed moments are considerably larger than $\mathfrak{J}_{irrot}$.[142]

The nuclear moments of inertia can be interpreted in more detail in terms of the response of the nucleonic motion to the slowly rotating nuclear field.[143,144] It is found[145,146] that, for independent particle motion, the effective moment of inertia would be approximately that corresponding to rigid rotation, but that the residual interactions between the nucleons reduce the moment, which then exhibits a dependence on β of the type observed.[147] Residual interactions so strong as to break down the shell structure would reduce the moment to values approaching $\mathfrak{J}_{irrot}$. The observed moments indicate interactions about three times smaller than this limit. The full drawn curve in Fig. V.5 corresponds to a rough estimate[148] of the moments of inertia for interactions of such a magnitude.

As one approaches the closed shell configurations, the value of β decreases and, eventually, as a consequence of the residual interactions, the nuclear deformation collapses and the equilibrium shape becomes spherical (see Fig. V.1). The nucleus then no longer possesses a rotational spectrum and the collective excitations correspond to vibrations about the spherical equilibrium (see Sec. V C.1).

A criterion for the transition from vibrational to rotational spectra may be obtained by noting that this transition is associated with a change of the nucleonic coupling scheme (see Sec. V A). For the nuclei with spherical equilibrium shape, the coupling of the particles in unfilled shells is determined mainly by the residual interactions, while the development of a stable equilibrium shape is associated with a tendency of the individual nucleonic orbits to align themselves in the deformed nuclear field. Since, for completely independent particle motion, the moment of inertia would have the value (7) corresponding to rigid rotation, the smallest moment compatible with the occurrence of rotational spectra is expected to be a certain fraction of $\mathfrak{J}_{rig}$. This fraction has been estimated on the basis of

[141] See, e.g., H. Lamb, *Hydrodynamics* (Cambridge University Press, New York, 1916), p. 82 ff (see also reference 137).

[142] See A. Bohr and B. R. Mottelson, Phys. Rev. **89**, 316 (1953); reference 132; K. W. Ford, reference 138; **95**, 1250 (1954). For a more detailed comparison with the potential flow model, including the effect of higher multipoles in the nuclear shape, see T. Gustafson, Kgl. Danske Videnskab. Selskab Mat. fys. Medd. **30**, 5 (1955).

[143] D. R. Inglis, Phys. Rev. **96**, 1059 (1954).

[144] The nuclear moment of inertia is also related to the dependence of the collective orientation angles on the nucleonic coordinates. The introduction of orientation angles associated with an irrotational collective flow has been considered by A. Bohr, *Rotational States in Atomic Nuclei* (Ejnar Munksgaard, Copenhagen, 1954); G. Süssmann, Z. Physik **139**, 543 (1954); H. A. Tolhoek, Physica **XXI**, 1 (1955); S. Tomonaga, Progr. Theoret. Phys. **13**, 467 (1955); F. Coester, Phys. Rev. **99**, 170 (1955); R. Nataf, Compt. rend. **240**, 2510 and **241**, 31 (1955); Marumori, Yukawa, and Tanaka, Progr. Theoret. Phys. **13**, 442 (1955); T. Marumori and E. Yamada, *ibid.* **13**, 557 (1955); T. Marumori, *ibid.* **14**, 608 (1955); Lipkin, de Shalit, and Talmi, Nuovo cimento (10)2, 773 (1955); T. Miyazima and T. Tamura, Progr. Theoret. Phys. (to be published); T. Tamura, Nuovo cimento (to be published; F. Villars (privately circulated manuscript); see also the discussion of this approach in reference 145).

[145] A. Bohr and B. R. Mottelson, Kgl. Danske Videnskab. Selskab Mat. fys. Medd. **30**, No. 1 (1955).

[146] S. Moszkowski, Phys. Rev. **103**, 1328 (1956).

[147] In the special case of a rotation about a symmetry axis, the moment of inertia vanishes, since a rotation of the field then has no effect on the nucleonic motion.

[148] This estimate (compare reference 145) is based on a "two nucleon model" in which the nucleons outside of closed shells are represented by two interacting nucleons in p states. Despite the schematic character of this model, it may provide a qualitative description of the competition between the residual interactions and the nuclear shell structure.

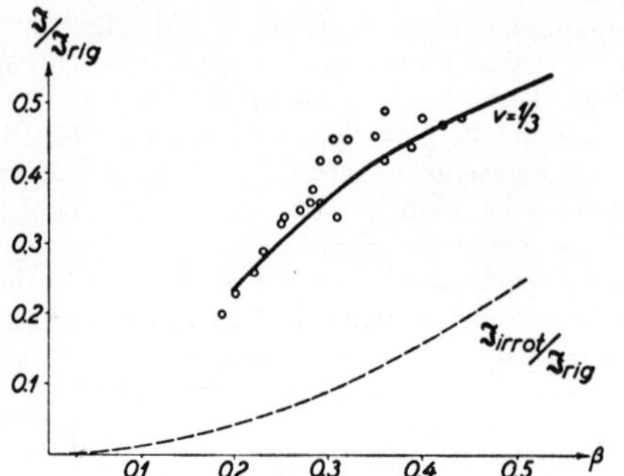

FIG. V.5. Dependence of moments of inertia on the nuclear deformation. The empirical moments of inertia of even-even nuclei in the region $150 < A < 188$ are plotted as a function of the nuclear deformation parameter β. The figure is taken from reference 145 which employed empirical data on $\mathcal{I}$ and Q_0 which is substantially the same as that contained in Table V.2. The moments of inertia are plotted in units of the moment $\mathcal{I}_{\text{rig}}$ associated with a rigid rotation [see (V.7)]. The full drawn curve represents a theoretical estimate based on a simplified model (reference 148). The parameter, v, appearing in this estimate is a measure of the strength of the residual interactions and the value chosen has been adjusted to fit the experimental data.

For comparison the moment of inertia corresponding to irrotational flow (V.8) is shown by the dotted curve.

the simplified model[148] mentioned above, which yields a value of about one quarter of $\mathfrak{J}_{\text{rig}}$. This would imply that the transition from vibrational to rotational spectra could be characterized approximately by a critical value

$$(E_2)_{\text{crit}} \approx 13\hbar^2/\mathfrak{J}_{\text{rig}} \qquad \text{(V.9)}$$

for the energy of the first excited state of an even-even nucleus. Stable equilibrium deformations and rotational spectra occur in this model only for even-even nuclei with E_2 values smaller than (9). However, the coefficient in (9) may have an A dependence which would lie outside the scope of this model. (See Fig. V.3.)

Odd-A nuclei.—For odd-A nuclei, the $E2$ excitation process can populate both the first and second rotational excitations of the ground state [see (5)]. It is indeed found that, in the regions where rotational spectra are found in the even-even nuclei, the Coulomb excitation of odd-A nuclei strongly populates just two states. The energies of the states, identified in this manner as rotational excitations, are listed in Table V.1. It is seen that the ratios of the energies agree well with those calculated from (3). While the assumed spin sequence (5) appears in all cases to be consistent with Coulomb excitation data and with the evidence from the observed radioactive decay schemes, unambiguous spin determinations have been made in only a few cases (see Ta and W in Table IV.2).

For the nuclei with ground-state spin $I_0 = \frac{1}{2}$, the irregular sequence of the observed states reveals the effect of the second term in (3). The value of a obtained from the observed levels is listed in column seven of Table V.1. From the values of $\mathfrak{J}$ and a, the position of the higher members of the band can be calculated. Although these levels are not populated by $E2$ Coulomb excitation, they have been observed in a number of cases in radioactive decays with energies rather accurately given by (3).[149] The values of a can be approximately accounted for on the basis of the wave function for the last odd nucleon.[133,150]

The rotational energy constants $3\hbar^2/\mathfrak{J}$ determined from the odd-A spectra are listed in column five of Table V.1. The corresponding quantity for the related even-even nucleus, obtained by removing the last odd nucleon, is listed in column six of the table and is seen

TABLE V.1. Rotational states in odd-A nuclei populated by Coulomb excitation. The table lists the odd-A nuclei in the regions $152 < A < 190$ and $A > 222$ which have been studied by Coulomb excitation. The only nuclei omitted are the odd isotopes of Dy, Er, and Yb for which it appears that the radiation from the lowest excitations has not been resolved from that of the even isotopes.

The ground state spins I_0, listed in column two, are taken from the compilation of Hollander, Perlman, and Seaborg, Revs. Modern Phys. **25**, 469 (1953) and the additional references listed below. The energies of the first excited state, $E^{(1)}$, and of the second excited state, $E^{(2)}$, are listed in columns three and four. The $E^{(2)}$ value in parenthesis is that calculated from (V.3), assuming the spins I_0+1, I_0+2 for the two excited states. The moment of inertia parameter employed is obtained from the experimental value of $E^{(1)}$ and is listed in column five. For comparison, the corresponding parameter for the neighboring even-even nucleus, obtained by removing the last odd nucleon, is listed in column six.

For nuclei with $I_0 = 1/2$, the rotational spectra involve an additional parameter, "a." For these nuclei, the moment of inertia parameter and the value of "a" determined by means of (V.3) from the measured values of $E^{(1)}$ and $E^{(2)}$ are listed in column five.

Additional references for I_0: D. R. Speck, Phys. Rev. **101**, 1725 (1956) ($Gd^{155,157}$); D. R. Speck and F. A. Jenkins, Phys. Rev. **101**, 1831 (1956) ($Hf^{177,179}$); K. L. van der Sluis and J. R. McNally, Jr., J. Opt. Soc. Am. **44**, 87 (1954) (U^{233}); Hutchinson *et al.*, Phys. Rev. **102**, 292 (1956) (U^{235}); van den Berg *et al.*, Physica **20**, 37 (1954), and Bleaney *et al.*, Phil. Mag. **45**, 991 (1954) (Pu^{239}).

Nucleus	I_0	$E^{(1)}$ (kev)	$E^{(2)}$ (kev)		$3\hbar^2/\mathfrak{J}$ (kev)	$(3\hbar^2/\mathfrak{J})_{e\text{-}e}$ (kev)
$_{63}Eu^{153}$	5/2	83	192	(190)	71	122
$_{64}Gd^{155}$	3/2	60	145	(144)	72	123
Gd^{157}	3/2	55	131	(132)	66	89
$_{65}Tb^{159}$	3/2	58	138	(139)	70	79
$_{67}Ho^{165}$	7/2	95	211	(211)	63	73
$_{69}Tm^{169}$	1/2	8.4	118.3		74 ($a=-0.77$)	80
$_{71}Lu^{175}$	7/2	113.8	251.0	(252.9)	76	78
$_{72}Hf^{177}$	7/2	113	250	(251)	75	89
Hf^{179}	9/2	121	262	(264)	66	90
$_{73}Ta^{181}$	7/2	136	303	(302)	91	93
$_{74}W^{183}$	1/2	46.5	99.1		78 ($a=0.19$)	100
$_{75}Re^{185}$	5/2	126	286	(288)	108	112
Re^{187}	5/2	135	303	(309)	116	123
$_{92}U^{233}$	5/2	40.4	92.1	(92.4)	35	45
U^{235}	7/2	46.2	103.0	(102.7)	31	45
$_{93}Np^{237}$	5/2	33.2	75.8	(75.9)	28	44
$_{94}Pu^{239}$	1/2	7.8	57.2		37 ($a=-0.58$)	43

[149] Tm^{169}; S. E. Johannson, Phys. Rev. **100**, 835 (1955); J. M. Cork *et al.*, Phys. Rev. **101**, 1042 (1956); E. N. Hatch *et al.*, Bull. Am. Phys. Soc. Ser. II **1**, 170 (1956). W^{183}; see Fig. V.8. Pu^{239}; Hollander, Smith, and Mihelich, Phys. Rev. **102**, 740 (1956).

[150] B. R. Mottelson and S. G. Nilsson, Kgl. Danske Videnskab. Selskab Mat. fys. Medd. (to be published); see also the analysis of the Tm^{169} spectrum in Z. Physik **141**, 217 (1955).

to be systematically somewhat greater than the odd-A value. This difference appears to be associated, at least partly, with the difference in the intrinsic excitation spectra for even-even and odd-A nuclei. While the first intrinsic excitation of an even-even nucleus usually has an energy of the order of a Mev in heavy nuclei, the odd-A nuclei exhibit excitations with an average spacing of about two hundred kev, associated with the change of orbit of the last odd particle.[151] The occurrence of low-lying intrinsic excitations in the odd-A nuclei implies that the intrinsic motion is less able to follow the rotational motion in an adiabatic manner, with a resultant increase in the effective moment of inertia (see Sec. V B.4).

V B.2. Excitation Cross Sections

An especially valuable feature of the Coulomb excitation process is the possibility of determining the absolute transition probability for the excitation by a measurement of the cross section. Since the rotational transitions leave the intrinsic structure unaltered, the transition matrix element can be expressed as an expectation value for the intrinsic structure, multiplied by a vector addition coefficient. Thus, one obtains[132] for an $E2$ transition from a state I_i, K to another state I_f, K of the same rotational band, the reduced transition probability [see (II A.18)][152]

$$B(E2;I_i\rightarrow I_f)=\frac{5}{16\pi}e^2Q_0^2\langle I_i2K0|I_i2I_fK\rangle^2, \quad (\text{V}.10)$$

where Q_0 is the electric quadrupole moment of the nuclear shape, defined by

$$eQ_0=\left\langle K\left|\int\rho r^2(3\cos^2\theta'-1)d\tau'\right|K\right\rangle. \quad (\text{V}.11)$$

In (11), ρ is the nuclear charge density and the angle θ' is measured from the intrinsic nuclear axis (z' in Fig. V.2). The wave function for the intrinsic nuclear state is labeled by K.

For a uniformly charged nucleus of spheroidal shape, Q_0 can be expressed in terms of β, given by (6), and one obtains

$$Q_0=\frac{3}{(5\pi)^{\frac{1}{2}}}ZR_0^2\beta(1+0.16\beta+\cdots), \quad (\text{V}.12)$$

where Z is the nuclear charge number. Corresponding to the fact that, for the strongly deformed nuclei, the quadrupole moments are an order of magnitude larger than those associated with a single proton, the transition probabilities (10) are observed to be appreciably larger than the single particle unit (1), in some cases by more than a factor of a hundred.

The intrinsic quadrupole moments Q_0 deduced by means of (10) from the observed cross sections for Coulomb excitation of rotational levels are listed in Tables V.2 and V.3.

For odd-A nuclei, it is possible to compare the intrinsic quadrupole moments deduced from transition probabilities with the expectation values for the quadrupole moment Q in the nuclear ground state, as obtained from atomic hyperfine structure separations. The latter quantity is defined by

$$eQ=\left\langle I, M=I\left|\int\rho r^2(3\cos^2\theta-1)d\tau\right|I, M=I\right\rangle, \quad (\text{V}.13)$$

where the angle θ is measured from the fixed z axis. For a state of a rotational band, the moment Q is related to Q_0 by

$$Q=Q_0\frac{3K^2-I(I+1)}{(I+1)(2I+3)}, \quad (\text{V}.14)$$

TABLE V.2. Moments of inertia and quadrupole moments of even-even nuclei. The table lists the even-even nuclei which exhibit rotational spectra and for which there exists evidence on the transition probabilities. Column two gives the moment of inertia parameter obtained from the energy of the first excited state (2+). Column three gives the intrinsic quadrupole moments obtained from the measured transition probabilities by means of (V.10). The data come partly from Coulomb excitations, see Table IV.1, and the additional data given by A. W. Sunyar, Phys. Rev. **98**, 653 (1955). The deformation parameter, β, in the last column is obtained from the Q_0 values by means of (V.12), assuming $R_0=1.2A^{\frac{1}{3}}\,10^{-13}$ cm.

Nucleus	$\frac{3\hbar^2}{\mathfrak{J}}$ (kev)	$\lvert Q_0\rvert$ (10^{-24} cm²)	β
$_{60}Nd^{150}$	130	4.8	0.25
$_{62}Sm^{152}$	122	5.7	0.28
Sm^{154}	83	6.7	0.33
$_{64}Gd^{154}$	123	6.3	0.30
Gd^{156}	89	8.8	0.41
Gd^{158}	79	10	0.46
Gd^{160}	76	10	0.47
$_{66}Dy^{160}$	86	7.8	0.35
Dy^{162}	82	8.2	0.36
Dy^{164}	73	9.5	0.41
$_{68}Er^{164}$	90	7.8	0.33
Er^{166}, Er^{168}, Er^{170}	80	7.8	0.33
$_{70}Yb^{170}$	84	7.5	0.30
Yb^{172}, Yb^{174}, Yb^{176}	78	7.8	0.31
$_{72}Hf^{176}$	89	7.5	0.29
Hf^{178}	91	8.1	0.31
Hf^{180}	93	7.1	0.27
$_{74}W^{182}$	100	7.1	0.26
W^{184}	112	6.5	0.24
W^{186}	124	6.5	0.24
$_{76}Os^{186}$	137	5.5	0.20
Os^{188}	155	5.1	0.18
$_{90}Th^{232}$	52	10	0.25
$_{92}U^{238}$	44	11	0.28

[151] See, e.g., the difference between the spectra of W^{182} and W^{183} (Murray *et al.*, Phys. Rev. **97**, 1007 (1955)) or between Pu^{238} (reference 176) and Pu^{239} (reference 149).

[152] The relationship of the vector addition coefficients used in this chapter to the equivalent $3j$ symbols employed in Chapter II is given by (II A.17).

and, in the special case of the ground state ($I_0=K$), one obtains

$$Q=Q_0\frac{I_0(2I_0-1)}{(I_0+1)(2I_0+3)}. \qquad \text{(V.15)}$$

The smaller value of Q, as compared with Q_0, reflects the fact that, even for the state $M=I$, the intrinsic nuclear axis is not completely aligned along the fixed z axis.

The intrinsic quadrupole moments obtained from the spectroscopic Q values by means of (15) are compared in Table V.3 with those derived from the Coulomb excitation cross sections by means of (10). The two determinations seem to be consistent, considering the rather large uncertainties involved, especially in the estimate of the nuclear quadrupole moments from the measured hyperfine structure intervals.

As is seen from expression (10), the ratio of the cross sections for the excitation of the first and the second rotational state in an odd-A nucleus is independent of Q_0.[153] Thus, a measurement of this ratio provides a direct test of the nuclear coupling scheme. The available evidence is collected in Table V.4 and compared with the ratios calculated from (10).

The Q_0 values obtained from (10) are plotted in Fig. V.6, and show similar general trends as previously observed in the spectroscopic quadrupole moments.[154] Thus, the nuclear deformations increase strongly as one moves away from closed shell regions, reflecting the polarizing effect of particles outside of closed shells (see Sec. V A). A theoretical estimate of Q_0 may be obtained by calculating the binding energies of the individual nucleons as a function of the nuclear deformation and minimizing the total nuclear energy.[155] The deformations calculated in this manner are shown by the full drawn curve in Fig. V.6.

The excitation of higher members of the ground-state rotational band ($\Delta I\geqslant 3$) may be achieved by multiple $E2$ processes. (See Sec. II D.3.) The cross sections for such processes may become quite large for high bombarding energies, as may be employed especially with heavy ions as projectiles. Thus, for 20-Mev α particles incident on a target with $Z_2=90$, the cross section for a

TABLE V.3. Intrinsic quadrupole moments of odd-A nuclei. The table lists the odd-A nuclei in the regions $152<A<190$ and $A>222$ for which there is evidence on the nuclear deformation from Coulomb excitation cross sections or spectroscopic hyperfine structure measurements. The ground-state spins, I_0, are taken from the references in Table V.1, and the intrinsic quadrupole moments (in column three) are obtained from the experimental transition probabilities in Table IV.2 by means of (V.10). The spectroscopic quadrupole moments, Q, in column four are taken from the compilation of N. F. Ramsey, *Nuclear Moments* (John Wiley and Sons, Inc., New York, 1953), and the additional references given below. From these moments, the intrinsic quadrupole moments in column five are obtained by means of (V.15).

Additional references for Q: D. R. Speck Phys. Rev. **101** 1725 (1956) ($Gd^{155,157}$); J. M. Baker and B. Bleaney Proc. Phys. Soc. (London) **68A** 1090 (1955) (Ho^{165}); Bogle *et al.*, Proc. Phys. Soc. (London) **65A**, 760 (1952) (Er^{167}); T. Kamei, Phys. Rev. **99**, 789 (1955) (Lu^{175}; Ta^{181}); Fred *et al.*, Phys. Rev. **98**, 1514 (1955) (Ac^{227}); K. L. van der Sluis and J. R. McNally, Jr., J. Opt. Soc. Am. **44**, 87 (1954) (U^{235}); Korostyleva *et al.*, J. Exptl. Theoret. Phys. U.S.S.R. **28**, 471 (1955) and N. I. Kaliteevskij and M. P. Chaika, private communication (ratio between moments of U^{233} and U^{235}).

Nucleus	I_0	$\lvert Q_0\rvert$ (Coul. exc.)	Q	Q_0 (hfs)
$_{63}Eu^{153}$	5/2	7.7	2.5	7.0
$_{64}Gd^{155}$	3/2	8.0	1.1	5.5
Gd^{157}	3/2	7.7	1.0	5.0
$_{65}Tb^{159}$	3/2	6.9		
$_{67}Ho^{165}$	7/2	7.8	~2	~4
$_{68}Er^{167}$	7/2		~10	~20
$_{69}Tm^{169}$	1/2	8.0	0	
$_{70}Yb^{173}$	5/2		3.9	11
$_{71}Lu^{175}$	7/2	8.2	5.7	12
$_{72}Hf^{177}$	7/2	7.5		
Hf^{179}	9/2	~7		
$_{73}Ta^{181}$	7/2	6.8	4.3	9.2
$_{75}Re^{185}$	5/2	5.4	2.8	7.8
Re^{187}	5/2	5.0	2.6	7.3
$_{89}Ac^{227}$	3/2		−1.7	−8
$_{92}U^{233}$	5/2	14	~6	~17
U^{235}	7/2	9	~8	~17
$_{93}Np^{237}$	5/2	9		
$_{94}Pu^{239}$	1/2	8.3	0	

TABLE V.4. Relative intensities of rotational excitations. The table lists the odd-A nuclei for which the cross sections for the Coulomb excitation of the first two rotational states have been measured. The ratio of the experimental transition probabilities (compare Table IV.2) is listed in column three together with the theoretical value (in parenthesis) obtained from (V.10). The ground-state spins, I_0, are taken from the references given in Table V.1.

Note added in proof.—A recent systematic study of the ratio of $B(E2)$ values obtained in Coulomb excitation of odd-A nuclei (G. Goldring and G. T. Paulissen, Phys. Rev. **103**, 1314 (1956)) has yielded values in approximate agreement with those listed in Table V.4. The one essential difference occurs for Hf^{179}, for which the new measurements give a ratio of 0.22.

Nucleus	I_0	$\frac{B(E2;\,I_0\to I_0+2)}{B(E2;\,I_0\to I_0+1)}$
$_{63}Eu^{153}$	5/2	0.28 (0.35)
$_{64}Gd^{157}$	3/2	0.38 (0.56)
$_{65}Tb^{159}$	3/2	0.56 (0.56)
$_{67}Ho^{165}$	7/2	0.24 (0.26)
$_{71}Lu^{175}$	7/2	0.23 (0.26)
$_{72}Hf^{177}$	7/2	0.26 (0.26)
Hf^{179}	9/2	0.07 (0.20)
$_{73}Ta^{181}$	7/2	0.29 (0.26)
$_{75}Re^{185}$	5/2	0.27 (0.35)
Re^{187}	5/2	0.39 (0.35)
$_{92}U^{233}$	5/2	0.18 (0.35)
U^{235}	7/2	0.16 (0.26)
$_{93}Np^{237}$	5/2	0.44 (0.35)

[153] Such intensity relations apply quite generally to transitions leading to different members of a rotational band. For applications to β and γ transitions, see Alaga, Alder, Bohr, and Mottelson, Kgl. Danske Videnskab. Selskab Mat. fys. Medd. **29**, No. 9 (1955); for α transitions see Bohr, Fröman, and Mottelson, reference 136; for deuteron stripping reactions, see G. R. Satchler, Phys. Rev. **97**, 1416 (1955).

[154] W. Gordy, Phys. Rev. **76**, 139 (1949). Townes, Foley, and Low, Phys. Rev. **76**, 1415 (1949).

[155] D. Pfirsch, Z. Physik **132**, 409 (1952). S. Moszkowski, reference 135; B. R. Mottelson and S. G. Nilsson, reference 136; K. Gottfried, reference 135.

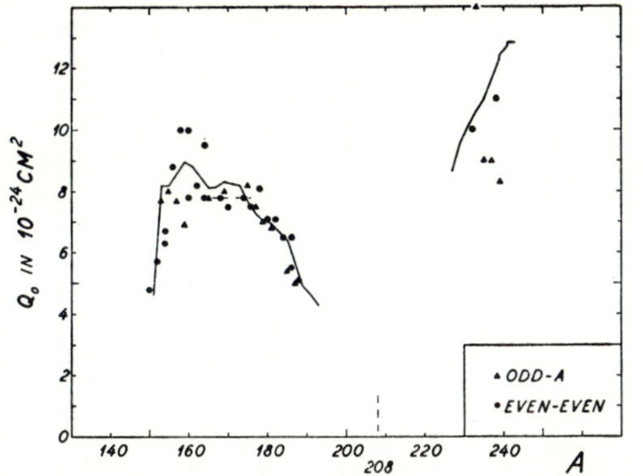

FIG. V.6. Intrinsic quadrupole moments of deformed nuclei. The experimental Q_0 values, determined from $E2$ transition probabilities (compare Tables V.2 and V.3), are plotted as a function of the nuclear mass number. The experimental uncertainties are usually of the order of 10–20%, but may be somewhat greater in the very heavy elements region. The full drawn curve gives a theoretical estimate of Q_0 based upon an independent particle description of the intrinsic structure (B. R. Mottelson and S. G. Nilsson, reference 136 and Kgl. Danske Vidensk. Selskab Mat. fys. Medd. (to be published).

second order $E2$ excitation of the 4+ state of an even-even nucleus is found from (II D.18) and (10) to be about 50 millibarns, assuming $Q_0=10$ barns.

While the cross sections for single or multiple $E2$ excitations of rotational states depend on the nuclear quadrupole moment, the possible occurrence of higher multipole moments in the nuclear shape can in principle be studied by means of the Coulomb excitation of corresponding multipole order populating higher members of the ground-state rotational band. Thus, an $E\lambda$ transition from the ground-state I_iK to the rotational states I_fK would be characterized by the reduced transition probabilities

$$B(E\lambda; I_i\rightarrow I_f)=\langle K|\mathfrak{M}'(E\lambda,0)|K\rangle^2 \times\langle I_i\lambda K0|I_i\lambda I_fK\rangle^2, \quad (V.16)$$

where $\mathfrak{M}_I(E\lambda,0)$ is the intrinsic $E\lambda$ moment defined by (II A.11) with the coordinates referring to the intrinsic nuclear system. The cross sections for higher multipole excitation will usually be appreciably smaller than those for multiple $E2$ transitions. Thus, if one assumes[156] a value for $B(E4;0\rightarrow4)$ of $0.2e^2(10^{-24}\text{ cm}^2)^4$ in a nucleus with $Z_2=90$, one obtains from (II C.15) a cross section of about 0.2 millibarn for $E4$ excitation of a 4+ state with 20-Mev α particles.

V B.3. Magnetic Dipole Decay of Rotational Excitations

Since the successive rotational states in an odd-A nucleus have $\Delta I=1$ [see (5)], the γ radiation emitted in the decay of these states will in general be a mixture of $M1$ and $E2$, although the excitation is of almost pure $E2$ type (see Sec. II A.3). The absolute $E2$ transition probability can be determined from the cross section for Coulomb excitation. Thus, a determination of the relative strength of the $M1$ as compared with the $E2$ radiation in the decay of the first excited state will also yield the absolute $M1$ transition probability. This information can be obtained from angular distributions or internal conversion measurements on the emitted radiation (or from the lifetime of the excited state). The $M1$ transition probability in the cascade transition $(I_0+2\rightarrow I_0+1)$ can be determined from the relative strength of $M1$ and $E2$ in this transition together with the branching ratio between the mixed $M1+E2$ cascade radiation and the pure $E2$ cross-over $(I_0+2\rightarrow I_0)$ decay of the second excited state. If only one of these data is available one may employ the rotational formula (10) to provide the additional relation necessary for the estimate of the absolute $M1$ transition probability in the cascade radiation.

The magnitude of the $M1$ transition probabilities between rotational states can be related to the gyromagnetic ratios, g_K and g_R, of the intrinsic and collective motion. The reduced $M1$ transition probability from a state I_iK to another state I_fK of the same rotational band (with $K\neq\frac{1}{2}$) is given by [132]

$$B(M1; I_i\rightarrow I_f)=\frac{3}{4\pi}\left(\frac{e\hbar}{2Mc}\right)^2(g_K-g_R)^2K^2 \times\langle I_i1K0|I_i1I_fK\rangle^2. \quad (V.17)$$

The relative sign of $M1$ and $E2$ transition amplitudes may also be determined from angular distribution measurements (see Sec. II C.4). This phase is related to the sign of Q_0 and of g_K-g_R and is given by

$$\text{sign}\delta=\text{sign}\frac{g_K-g_R}{Q_0}, \quad (V.18)$$

where δ is the ratio between the reduced $E2$ and $M1$ matrix elements for the transition (see Sec. II A).

The static magnetic moment μ of a state in the rotational band may also be expressed in terms of the gyromagnetic ratios g_K and g_R. Thus, for $K\neq\frac{1}{2}$, one obtains

$$\mu=\frac{K^2}{I+1}(g_K-g_R)+Ig_R. \quad (V.19)$$

For a band with $K=\frac{1}{2}$, the magnetic properties involve an additional parameter b_0 similar to the decoupling parameter, a, in the energy spectrum.[133] For this case, the $M1$ transition probability and the magnetic moment may be written in the form

$$B(M1; I+1\rightarrow I)=\frac{3}{64\pi}\left(\frac{e\hbar}{2Mc}\right)^2\frac{2I+1}{I+1}(g_K-g_R)^2 \times(1+(-1)^{I-\frac{1}{2}}b_0)^2, \quad (V.20)$$

[156] A recent analysis of the fine structure intensities in the α decay of the heavy nuclei [P. O. Fröman, Kgl. Danske Videnskab. Selskab Mat. fys. Medd. (to be published)] indicates $E4$ moments in the nuclear shape corresponding in some cases to values as large as $B(E4;0\rightarrow4)=0.2e^2(10^{-24}\text{ cm}^2)^4$; the single particle unit (II A.58) for $Z_2=90$ corresponds to $B(E4)=0.01e^2(10^{-24}\text{ cm}^2)^4$.

with

$$\text{sign}\delta=\text{sign}\frac{g_K-g_R}{Q_0}(1+(-1)^{I-\frac{1}{2}}b_0), \qquad \text{(V.21)}$$

and

$$\mu=\frac{1}{4(I+1)}(g_K-g_R)$$
$$\times[1-(2I+1)(-1)^{I-\frac{1}{2}}b_0]+Ig_R. \quad \text{(V.22)}$$

A measurement of both the $M1$ transition probability and the ground state magnetic moment can thus yield g_K and g_R separately (and the value of b_0 for $K=\frac{1}{2}$ bands if an additional transition probability or moment is measured). The quantities g_K and b_0 can be related to the intrinsic nucleonic configuration, while g_R provides information on the rotational motion, and may be compared with the value

$$(g_R)_u=\frac{Z}{A}, \qquad \text{(V.23)}$$

corresponding to a flow of uniformly charged nuclear matter.

The g_R values obtained from Coulomb excitation studies and ground state moments of nuclei with $I_0\neq\frac{1}{2}$ are listed in Table V.5. The uncertainties in these values are considerable, but the data may indicate deviations from (23). In fact, one expects such deviations in odd-A nuclei as a consequence of the especially large contribution of the last odd nucleon to the rotational moment of inertia, and thus also to the rotational angular momentum.

The analysis of rotational spectra in odd-A elements

TABLE V.5. Magnetic properties of rotational levels in odd-A nuclei. The table lists the odd-A nuclei with $I_0\neq 1/2$ for which there is evidence on both the ground state magnetic moment, μ, and the $M1$ rotational transition probability. The ground-state spins and magnetic moments in columns two and three are taken from the references given in Tables V.1 and V.3 and the additional references given below. The reduced $M1$ transition probabilities $B(M1)$ are obtained from the measured $B(E2)$ values and the $M1/E2$ ratios given in Table IV.2 by means of the relations (II A.52) and (II A.53). From the values of μ and $B(M1)$ the gyromagnetic ratios, g_R and g_K, are determined from (17) and (19). Since (17) is a quadratic expression, there exist two sets of possible g_K and g_R values, except in the case of Ta^{181} where angular correlation measurements have established the sign of (g_K-g_R) by comparison with the sign of the quadrupole moment [compare (18)]. The g factors in columns four and five are listed in such a manner that the first value of g_R belongs together with the first value of g_K.

The table does not include nuclei with $I_0=1/2$, since the magnetic properties of these nuclei involve an additional parameter b_0 [compare (20), (21), and (22), and the analysis of the W^{183} spectrum in Fig. V.8].

Additional references for μ: J. M. Baker and B. Bleaney, Proc. Phys. Soc. (London) **68A**, 257 (1955) (Tb^{159}); D. R. Speck, Bull. Am. Phys. Soc. Ser. II, **1**, 282 (1956) (Hf^{177} and Hf^{179}).

Nucleus	I_0	μ	g_R	g_K
$_{63}Eu^{153}$	5/2	1.5	0.5, 0.7	0.6, 0.5
$_{64}Gd^{155}$	3/2	−0.30	0.3, −0.7	−0.5, 0.1
Gd^{157}	3/2	−0.37	0.3, −0.7	−0.6, 0.1
$_{65}Tb^{159}$	3/2	1.5	0.1, 1.9	1.6, 0.4
$_{67}Ho^{165}$	7/2	3.3	0.3, 1.6	1.1, 0.8
$_{71}Lu^{175}$	7/2	2.6	0.3, 1.2	0.9, 0.6
$_{72}Hf^{177}$	7/2	0.61	0.3, 0.1	0.1, 0.2
Hf^{179}	9/2	−0 47	0 2, −0.4	−0.2, 0.0
$_{73}Ta^{181}$	7/2	2.1	0.2	0.7
$_{75}Re^{185}$	5/2	3.14	0.5 2.1	1.6 0.9
Re^{187}	5/2	3.18	0.5, 2.0	1.6, 1.0
$_{92}U^{233}$	5/2	0.8	0.4, 0.3	0.3, 0.3
U^{235}	7/2	−0.5	0.1, --0.4	−0.2, −0.1

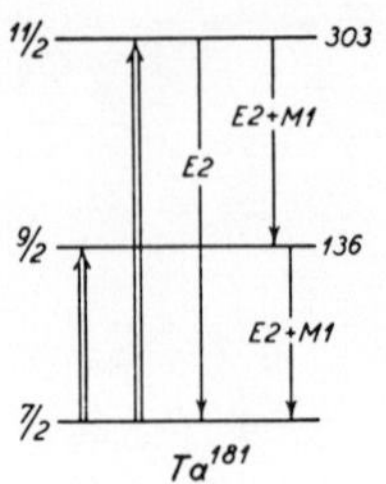

FIG. V.7. Rotational spectrum for Ta^{181}. The Coulomb excitation experiments have determined the energies and spins (from the angular distribution of the γ rays and measured conversion coefficients) of the first two rotational excitations in Ta^{181}. In addition the absolute magnitudes of all the $M1$ and $E2$ transition probabilities have been obtained by combining the Coulomb excitation cross sections with the measured branching ratio between the 167- and 303-kev γ rays, and the $M1$ to $E2$ ratios determined from the γ-ray angular distribution and the conversion coefficients (compare the data in Table IV.2). In addition, the relative phases of the $M1$ and $E2$ radiation have been determined for the 167-kev transition from the angular distribution in Coulomb excitation and for the 136-kev transition from the γ-γ angular correlation following the β decay of Hf^{181} [F. K. McGowan, Phys. Rev. **93**, 471 (1954); Heer *et al.* Helv. Phys. Acta **28**, 336(A) (1955); F. Boehm and P. Marmier, Phys. Rev. **103**, 342 (1956)].

These results test the rotational interpretation of the observed levels in the following ways.

(a) The spin sequence is that given by (V.5) in which we assume $K=7/2$ as determined from the ground state spin, and the parity is the same for all the states.

(b) The observed energy ratio of the $I=11/2$ and $I=9/2$ states is 2.23 ± 0.02 which agrees with the ratio 20/9 obtained from (V.3).

(c) The reduced $E2$ transition probabilities $B(E2; 7/2\rightarrow 9/2)$, $B(E2; 7/2\rightarrow 11/2)$, and $B(E2; 11/2\rightarrow 9/2)$ are found to have the relative values $1:0.29:\sim 1.3$ which may be compared with relative values 1:0.26:0.81 obtained from (V.10).

(d) The reduced $M1$ transition probabilities $B(M1; 11/2\rightarrow 9/2)$ and $B(M1; 9/2\rightarrow 7/2)$ have a ratio of ~ 2 while the theoretical value obtained from (V.17) is 1.53. Moreover the relative phases of M1 and $E2$ radiation are the same in the 136- and 167-kev transition as expected from theory (see V.18).

From the experimental data one determines a number of nuclear parameters characterizing the ground state configuration in Ta^{181}.

(a) The moment of inertia obtained from the rotational energies is given by $3\hbar^2/\mathcal{I}=91$ kev. For a discussion of the interpretation of this value compare Fig. V.5, Table V.1, and the comments in the text.

(b) The intrinsic quadrupole moment determined from the $E2$ transition probabilities by means of (V.10) has the value $Q_0=6.8$. This quantity determines the quadrupole eccentricity parameter which is found from (V.12) to be $\beta=0.25$; see also the discussion of Q_0 in Fig. V.6 and Table V.3.

(c) The $M1$ transition probability and its phase relative to that of the $E2$ transition [together with the assumption of a positive Q_0 as indicated by spectroscopic evidence, (compare Table V.3)] yields by means of (V.17) and (V.18) the value $g_K-g_R=0.5$ for the difference between the intrinsic and collective gyromagnetic ratios. This value, when combined with the measured magnetic moment, yields by means of (V.19) the separate g factors listed in Table V.5.

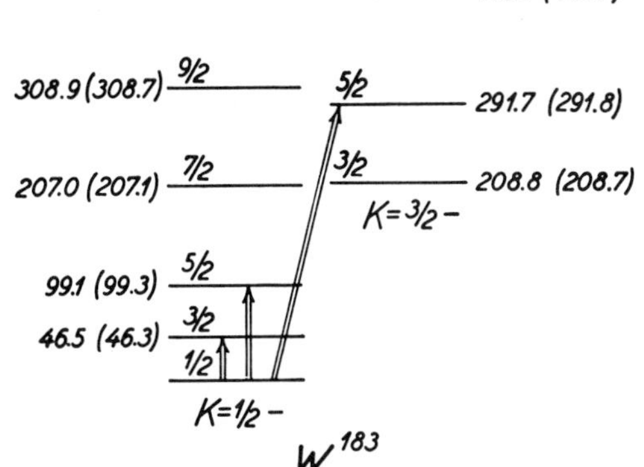

FIG. V.8. Low-lying energy levels in W^{183}. The figure shows the energies, spins, and parities of the low-lying states in W^{183}, populated in the β decay of Ta^{183} [Murray, Boehm, Marmier, and DuMond, Phys. Rev. **97**, 1007 (1955)].

The Coulomb excitation strongly populates the first two excited states of the ground state rotational band (compare Table IV.2). The energies in this band reveal the effect of the decoupling term in (V.3) characteristic of configurations with $K=1/2$. From the energies of the $(I=3/2)$ and $(I=5/2)$ states in this band, one can calculate the parameters $\mathcal{J}$ and a. Using these parameters, the position of the $(I=7/2)$ level calculated from (V.3) agrees well with the observed level, while the calculated energy for the $I=9/2$ state is a few percent below the observed value. Similarly, the relative energies in the $(K=3/2)$ band are found to deviate somewhat from those obtained from (V.3). These perturbations have been interpreted in terms of a coupling between the two bands resulting from the Coriolis effect of the nuclear rotation. By suitably adjusting the matrix element associated with this coupling, the observed energies have been accounted for with great precision (A. K. Kerman, reference 157). The calculated energies are shown in parenthesis.

To a first approximation, the electric quadrupole transition probabilities for the rotational excitations can be discussed in a similar manner as for Ta^{181} (see Fig. V.7), although for W^{183} the data is less complete. However, the coupling to the $(K=3/2)$ band implies small deviations from (V.10), and leads especially to enhanced $E2$ transition probabilities for the excitation of the states in this higher band. The magnitude of the $B(E2)$ values for these transitions provide a direct measure of the admixed amplitudes and can be calculated from the parameters involved in the analysis of the observed energies. The observed $B(E2)$ for the population of the 292-kev level (compare Table IV.2) is in good agreement with the analysis in reference 157, which, in addition, predicts $B(E2)\approx 0.1$ for the excitation of the 209-kev level.

The magnetic parameters, g_K, g_R, and b_0 for the ground-state rotational band may be obtained from the observed relative intensities in the decay of the 207-kev and 99-kev levels together with the estimated Q_0 value of 6.5 and the measured ground state magnetic moment $\mu=0.115$ [P. P. Sogo and C. D. Jeffries, Phys. Rev. **98**, 1316 (1955)]. By means of (V.20) and (V.22) one finds $b_0=0.28$, $g_K=0.30$, and $g_R=0.49$, by a suitable choice of the phases (A. K. Kerman, private communication). The coupling between the bands has an appreciable effect on the magnetic transition probabilities, which have been analyzed in detail in reference 157.

As discussed in Sec. V B.4, the moment of inertia and the magnetic parameters determined directly from the ground state rotational band contain contributions from the coupling between this band and the $K=3/2$ band. In the present case where this coupling has been determined from the analysis of the energy perturbations, it is possible to derive the various parameters which would characterize the ground state band in the absence of this coupling. These "unrenormalized" values are found to be $(3\hbar^2/\mathcal{J})_0=95$ kev, $(a)_0=0.17$, $(b_0)_0=0.46$, $(g_K)_0=0.16$, and $(g_R)_0=0.28$ (A. K. Kerman, reference 157 and private communication). It is of interest that the moment of inertia $\mathcal{J}_0$ is much closer to the moments of neighboring even-even nuclei than the effective moment (compare Table V.1). The coupling between the bands is seen in the present case to increase the effective value of g_R; this is opposite to the general trend for odd neutron nuclei suggested in reference 146.

is summarized in the captions to Figs. V.7 and V.8, which discuss the spectra of Ta^{181} and W^{183}.

V B.4. Coupling Between Rotational and Intrinsic Motion

Rotational spectra of the simple form (3) are obtained when the rotation is so slow that the intrinsic motion can adjust adiabatically to the changing orientation ot the nuclear field. The finite rotational frequency gives rise to small nonadiabatic excitations resulting from the Coriolis and centrifugal forces acting on the nucleons in the body fixed coordinate system. To lowest order in the rotational frequency, these virtual excitations imply an increase in the energy of the nucleus proportional to $I(I+1)$ and thus provide the moment of inertia associated with the rotational motion.[143,145] To higher order, the nonadiabatic effects give rise to a coupling between the rotational and intrinsic motion, which implies deviations from the rotational spectrum (3) and the geometrical relations for the nuclear moments, e.g., (10), (14), (17), and (19).

In odd-A nuclei the nonadiabatic excitations of the last odd nucleon will play a special role, since these do not involve the breaking of any pairs. As noted above, in connection with the discussion of Table V.1, the inertial effect arising from these excitations may be largely responsible for the observed differences in the moments of inertia of even-even and odd-A nuclei. Small deviations from the energy ratios (3) may arise primarily from the higher order effects of the near lying bands associated with the lowest states of the last odd particle. Such effects have been detected (see, e.g., Table V.1 and reference 149) and a detailed analysis[157] has been made for the spectrum of W^{183} (see Fig. V.8). Even when the deviations from (3) are small, the electromagnetic transition probabilities may be appreciably affected. In particular the $E2$ transition probabilities between the coupled bands may be strongly enhanced. In the Coulomb excitation of W^{183}, an enhancement of this type has been observed and has been related quantitatively to the observed energy perturbations[157] (see Fig. V.8). Further studies of this phenomenon would provide valuable information on the relationship between the rotational motion and that of the individual nucleons.

In addition to the specific effects of the coupling between close lying bands, rotational perturbations also arise from the nonadiabatic excitation of higher lying intrinsic states. An important part of such perturbations may be described as rotation-vibration interac-

[157] A. K. Kerman, Kgl. Danske Videnskab. Selskab Mat. fys. Medd. **30**, No. 15 (1956).

tions of a similar kind as in molecules. These interactions imply an energy depression which in first approximation is proportional to $I^2(I+1)^2$.[132] In even-even nuclei, the small deviations of the observed rotational energies from the $I(I+1)$ ratios (see Fig. V.4) are indeed always negative and increase systematically with increasing rotational frequency; there is also some evidence that these perturbations to first order are proportional to $I^2(I+1)^2$.[158]

V C. Vibrations of Spherical Nuclei

V C.1. Classification of Vibrations

The shape oscillations of a spherical nucleus[159] may be classified according to their multipole order λ. The excitation quanta, called phonons, have total angular momentum λ, parity $(-1)^\lambda$, and may be further characterized by their component of angular momentum μ along a space fixed axis.

The vibrational motion is associated with an oscillating electric multipole moment, and it is convenient to normalize the vibrational amplitudes $\alpha_{\lambda\mu}$ with respect to the multipole moments (II A.11) by the relation[160]

$$\mathfrak{M}(E\lambda,\mu)=\frac{3}{4\pi}ZeR_0{}^\lambda\alpha^*{}_{\lambda\mu}. \tag{V.24}$$

This normalization is chosen in such a manner that, in the idealized case of a nucleus with constant density and a sharp surface, the coordinates $\alpha_{\lambda\mu}$ would define the nuclear surface by

$$R(\theta,\varphi)=R_0(1+\sum_{\lambda\mu}\alpha_{\lambda\mu}Y_{\lambda\mu}(\theta,\varphi)). \tag{V.25}$$

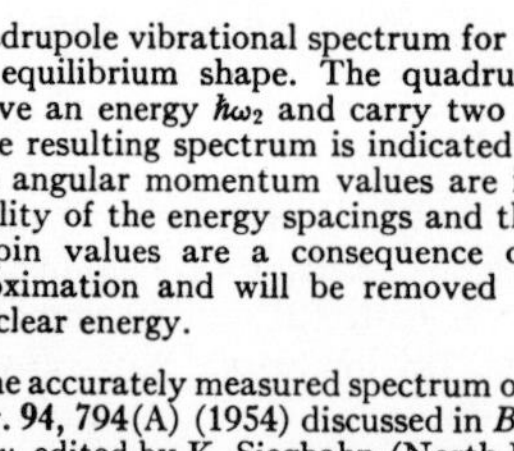

$3\hbar\omega_2$ ———— 0,2,3,4,6 +

$2\hbar\omega_2$ ———— 0,2,4 +

$\hbar\omega_2$ ———— 2 +

———— 0+

FIG. V.9. Quadrupole vibrational spectrum for even-even nuclei with spherical equilibrium shape. The quadrupole vibrational quanta each have an energy $\hbar\omega_2$ and carry two units of angular momentum. The resulting spectrum is indicated in the figure, in which the total angular momentum values are indicated on the right. The equality of the energy spacings and the degeneracy of the different spin values are a consequence of the harmonic oscillator approximation and will be removed by higher order terms in the nuclear energy.

For small amplitudes of oscillation, the energy may be expanded in powers of $\alpha_{\lambda\mu}$ and of the time derivatives $\dot{\alpha}_{\lambda\mu}$, and one obtains to a first approximation

$$H_{\mathrm{coll}}=\sum_{\lambda\mu}(\tfrac{1}{2}B_\lambda|\dot{\alpha}_{\lambda\mu}|^2+\tfrac{1}{2}C_\lambda|\alpha_{\lambda\mu}|^2), \tag{V.26}$$

corresponding to a set of independent harmonic oscillators, with energy quanta

$$\hbar\omega_\lambda=\hbar\left(\frac{C_\lambda}{B_\lambda}\right)^{\frac{1}{2}}. \tag{V.27}$$

While the classification of the nuclear vibrations in multipole orders and the expression (26) are general consequences of the spherical equilibrium shape and small amplitudes of oscillation, the parameters B_λ and C_λ depend on the more detailed structure of the nucleus. The former represents the mass transport associated with the vibration, and it is of interest to compare the observed B_λ in nuclear vibrations with the value

$$(B_\lambda)_{\mathrm{irrot}}=\frac{1}{\lambda}\frac{3}{4\pi}AMR_0{}^2, \tag{V.28}$$

corresponding to the surface oscillations of an irrotational and incompressible liquid drop. The parameter C_λ represents an effective surface tension which, within the scope of the liquid drop model, may be obtained from the surface energy appearing in the semiempirical mass formula (see, e.g., reference 124).

The lowest frequencies of collective vibration are in most cases expected to be of quadrupole type ($\lambda=2$), since a surface deformation with $\lambda=1$ simply represents a center-of-mass displacement.[161]

V C.2. Quadrupole Vibrations of Even-Even Nuclei

The vibrational excitation spectra have the simplest character for the ground-state configuration of even-even nuclei, for which the intrinsic structure does not contribute to the nuclear angular momentum. The quadrupole vibrational spectrum for an even-even nucleus is illustrated schematically in Fig. V.9. The exact equality of the energy separations as well as the degeneracies which result from the harmonic oscillator approximation (26) will be modified by higher order terms in the nuclear energy (see, e.g., references 163 and 166).

The vibrational excitations are characterized by enhanced electric transition probabilities. These can be expressed directly in terms of the parameters B_λ and C_λ, since the vibrational amplitudes have been normalized with respect to the electric multipole moments [see

[158] See, e.g., the accurately measured spectrum of Hf^{180} (Mihelich *et al.*, Phys. Rev. **94**, 794(A) (1954) discussed in *Beta- and Gamma-Ray Spectroscopy*, edited by K. Siegbahn (North Holland Publishing Company, Amsterdam, 1955), Chapter XVII, p. 487.

[159] See, e.g., reference 137.

[160] Equation (24) is meant to apply only to the matrix elements between the vibrational levels, and thus merely to give the normalization of the vibrational amplitude. Taken as a definition for $\alpha_{\lambda\mu}$ in terms of the nucleonic coordinates [see (II A.13)], the relation (24) would imply the assumption of irrotational flow for the collective motion, an assumption which is not in general expected to be valid (see reference 145 and Table V.7).

[161] Collective dipole oscillations ($\lambda=1$) of the neutrons with respect to the protons have been discussed in connection with the nuclear photoeffect. Such oscillations have been estimated to have energies of the order of 10–20 Mev [M. Goldhaber and E. Teller, Phys. Rev. **74**, 1046 (1948); H. Steinwedel and J. H. D. Jensen, Z. Naturforsch. **5a**, 413 (1950)].

TABLE V.6. Vibrational spectra in even-even nuclei. The table lists the even-even nuclei, for which the empirical data on the second excited state is compatible with a 2+ assignment. The energies of the first and second excited states are listed in columns two and three, and their ratio in column four. The $E2$ transition probability from the ground state to the first excited state (compare Table V.7) is given in column five, in units of the single particle estimate (V.1). The cascade transition from the second to the first excited state may proceed by $M1$ or $E2$ radiation, and the ratio of the components, as determined by angular correlations, or conversion coefficients, is shown in column six. Finally, column seven gives the ratio of the reduced $E2$ transition probability of the cross-over and cascade radiation from the second excited state. In the cases where the $M1$ admixture in the cascade radiation is unknown, the value for the ratio in column seven corresponds to the assumption of pure $E2$ radiation for the cascade transition, and is then given in parenthesis. The table is based on the empirical data in Table V.7 and on the references listed below.

Nucleus	E_2 (Mev)	E_2' (Mev)	E_2'/E_2	$\frac{B(E2;0\to 2)}{B_{sp}(E2)}$	$(M1/E2)_{2'\to 2}$	$\frac{B(E2;2'\to 0)}{B(E2;2'\to 2)}$	Reference
${}_{26}Fe^{58}$	0.81	1.62	2.00		0.2	0.01	s
${}_{28}Ni^{60}$	1.33	2.18	1.64	17		$(3\cdot 10^{-3})$	a
${}_{30}Zn^{64}$	1.00	2.27	2.27	15		(0.1)	a
Zn^{66}	1.05	2.40	2.29	11		(0.04)	a
${}_{34}Se^{76}$	0.55	1.19	2.17	45	~ 1	0.1	a
${}_{36}Kr^{82}$	0.77	1.45	1.88			(0.01)	a
Kr^{84}	0.9	1.9	2.1			>0.1	a
${}_{40}Zr^{92}$	0.93	1.83	1.97			(0.06)	a
${}_{44}Ru^{100}$	0.54	1.36	2.52	22		(0.02)	b
Ru^{102}	0.47	1.10	2.34	45		(0.15)	c
${}_{52}Te^{122}$	0.57	1.26	2.21	26	0.1	0.01	d, e, f
Te^{126}	0.65	1.40	2.16	17		(0.004)	g, h
${}_{54}Xe^{126}$	0.39	0.86	2.20			(0.01)	g, h
Xe^{128}	0.46	0.98	2.13			(0.01)	i
${}_{78}Pt^{192}$	0.32	0.61	1.90		0.025	0.008	j, k
Pt^{194}	0.33	0.62	1.88	50	small	0.01	j, l
Pt^{196}	0.35	0.69	1.97	38	0.05	$<4\cdot 10^{-4}$	d, m
${}_{80}Hg^{198}$	0.41	1.09	2.66	29	0.7	0.03	d, n, o–q
${}_{84}Po^{214}$	0.61	1.38	2.26	15	$\gtrsim 1$		r

[a] Nuclear Level Schemes ($40\leq A\leq 92$), edited by Way, King, McGinnis, and van Lieshout, U. S. Atomic Energy Commission, Washington, 1955.
[b] L. Marquez, Phys. Rev. **92**, 1511 (1953).
[c] P. Avignon, Compt. rend. **240**, 176 (1955).
[d] R. M. Steffen, *Proceedings of the 1954 Glasgow Conference* (Pergamon Press, London and New York, 1955), p. 206.
[e] M. J. Glaubman, Phys. Rev. **98**, 645 (1955).
[f] B. Farelly *et al.*, Phys. Rev. **99**, 1440 (1955).
[g] M. L. Perlman and J. P. Welker, Phys. Rev. **95**, 133 (1954).
[h] L. Koerts *et al.*, Phys. Rev. **98**, 1230 (1955).
[i] N. Benczer *et al.*, Phys. Rev. **100**, 955(A) (1955).
[j] M. W. Johns and S. V. Nablo, Phys. Rev. **96**, 1599 (1954).
[k] H. W. Taylor and R. W. Pringle, Phys. Rev. **99**, 1345 (1955).
[l] Mandeville, Varma, and Saraf, Phys. Rev. **98**, 94 (1955).
[m] M. T. Thieme and E. Bleuler, Phys. Rev. **99**, 1646(A) (1955).
[n] D. Schiff and F. R. Metzger, Phys. Rev. **90**, 849 (1953).
[o] C. D. Schrader, Phys. Rev. **92**, 928 (1953).
[p] L. G. Elliot *et al.*, Can. J. Phys. **32**, 153 (1954).
[q] D. Maeder *et al.*, Helv. Phys. Acta **27**, 3 (1954).
[r] O. B. Nielsen, private communication.
[s] Frauenfelder, Levine, Rossi, and Singer, Phys. Rev. **103**, 352 (1956).

(24)]. Thus, for the one phonon excitation, one obtains [see (II A.18)]

$$B(E\lambda;I=0\to I=\lambda)$$
$$=(2\lambda+1)\left(\frac{3}{4\pi}ZeR_0^{\lambda}\right)^2\frac{\hbar}{2(B_\lambda C_\lambda)^{\frac{1}{2}}}. \quad (V.29)$$

In the harmonic oscillator approximation, the transitions to higher states in the vibrational spectrum are forbidden.

The static electric moments of the vibrational excitations vanish to lowest order and are thus expected to be small in spite of the large transition moments. The smallness of the static $E2$ moments for the excited states of an even-even nucleus is a characteristic feature of vibrational as distinct from rotational excitations.

In the decay of vibrational states, $M1$ radiation is forbidden even when ΔI equals 0 or 1, such as in the transition from the second to the first 2+ state in the quadrupole vibrational spectrum. In fact, the magnetic moment associated with the collective motion is proportional to the angular momentum and is thus a constant of the motion which possesses no matrix elements between different energy levels.

The static magnetic moment of a vibrational excitation is given by

$$\mu=g_R\cdot I, \quad (V.30)$$

where the collective g factor is expected to be similar to that associated with the rotational motion of the deformed nuclei (see Sec. V B.3).

The transitions between vibrational states with $\Delta I=0$ may also take place by emission of conversion electrons corresponding to a nuclear electric monopole transition. The matrix elements for these $E0$ transitions may be related to the vibrational parameters and the nuclear compressibility.[162]

V C.3. *Discussion of Empirical Data*

Collective excitation spectra corresponding to vibrations about a spherical equilibrium shape are expected

[162] E. L. Church and J. Weneser, Phys. Rev. **100**, 943 (1955).

in the extensive intermediate regions between the closed shell nuclei and the nuclei with especially many particles in unfilled shells and a resulting nonspherical equilibrium shape (see Sec. V A). In these intermediate regions, the low-lying excited states of even-even nuclei are indeed found to exhibit a pattern which resembles that of quadrupole vibrations about a spherical equilibrium.[163]

The first excited states are always of 2+ type and the energies vary regularly with neutron and proton number, decreasing as one moves away from closed-shell regions (see Fig. V.3). Moreover, the cross sections for Coulomb excitation of these states are found to be an order of magnitude larger than for single particle transitions, exhibiting the collective character of the transitions [see (1) and Table IV.2].

In many cases, a second vibrational excitation with spin 0+, 2+, or 4+ has been observed in radioactive decay processes. The energy of this second state is found to be in almost all cases between 2 and 2.5 times that of the first excited state, and the vibrational character is especially indicated by the observed γ decay. Thus, when the second excited state is of 2+ type, it is found to decay to the first excited 2+ state mainly by $E2$ radiation, although $M1$ radiation would be strongly preferred, assuming single particle matrix elements. In addition, the reduced $E2$ transition probability to the ground state, which would be forbidden for harmonic vibrations, is found to be in most cases several orders of magnitude smaller than that to the first excited state.[164] These features are exhibited in Table V.6 which lists the evidence on the vibrational spectra in those cases where the second excited state has the 2+ character.

The observed deviations of the energy ratios from those obtained in the harmonic oscillator approximation may possibly be accounted for by higher order terms in the collective nuclear Hamiltonian.[165] However, the striking selection rule observed in the $E2$ decay of the second 2+ state suggests that the higher order terms in the potential energy depend only on the parameter $\beta^2=\sum_\mu |\alpha_{2\mu}|^2$, since otherwise this selection rule would no longer hold.[166] The inclusion of such higher order terms in β would leave the two phonon states with spins 2+ and 4+ degenerate, while in most cases the 0+ state would lie higher.

From the energy and $B(E2)$ value for the excitation of the first 2+ state, one can determine the parameters B_2 and C_2 by means of (27) and (29). The values obtained, which are listed in Table V.7, exhibit the im-

[163] G. Scharff-Goldhaber and J. Weneser, Phys. Rev. **98**, 212 (1955).

[164] J. J. Kraushaar and M. Goldhaber, Phys. Rev. **89**, 1081 (1953); R. M. Steffen, *Proceedings of the 1954 Glasgow Conference* (Pergamon Press, London and New York, 1955), p. 206.

[165] The deviations from the harmonic oscillator spectrum have also been discussed in terms of the coupling of the collective vibrations to high-frequency nucleonic excitations (reference 163).

[166] M. Jean and L. Wilets, Compt. rend. **241**, 1108 (1955) and Phys. Rev. **102**, 788 (1956).

TABLE V.7. Vibrational parameters for even-even nuclei. The table lists the even-even nuclei in the regions $A\leq 150$ and $190<A<220$ for which there exists evidence on the transition probability to the first excited (2+) state. The energies, E_2, of these states and the reduced transition probabilities, $B(E2; 0\rightarrow 2)$, are listed in columns two and three. The data are obtained from the Coulomb excitation results in Table IV.2, from the lifetime data in Table IV.1 and from the additional references given below. From the values of $E_2=\hbar\omega$ and of $B(E2)$, the effective surface tension C_2 and mass parameter B_2 for quadrupole vibrations are obtained from (V.27) and (V.29). The B_2 values are given in units of $(B_2)_{\text{irrot}}$ [compare (V.28)]. In the immediate neighborhood of closed shells, the observed excitations may be described in more detail in terms of the excitations of the few particles outside of closed shells (see Sec. V.E).

Additional references: Devons *et al.*, Proc. Phys. Soc. (London) **69A**, 173 (1955) (C^{12} and Ne^{20}); F. R. Metzger, Bull. Am. Phys. Soc. Ser. II, **1**, 40 (1956) (Ni^{60}). The lifetime determinations for Po^{212} and Po^{214} from the $\alpha-\gamma$ branching are discussed in reference 132.

Nucleus	E_2 (kev)	$B(E2;0\rightarrow 2)$ $(e^2\cdot 10^{-48}\ \text{cm}^4)$	C_2 (Mev)	$B_2/(B_2)_{\text{irrot}}$
${}_6C^{12}$	4400	0.009	14	2.9
${}_{10}Ne^{20}$	1630	0.045	5.8	3.6
${}_{22}Ti^{46}$	890	0.15	14	7.3
Ti^{48}	990	0.083	29	12
${}_{26}Fe^{56}$	854	0.10	36	15
${}_{28}Ni^{58}$	1450	$\sim$0.1	77	10
Ni^{60}	1330	0.12	59	9.0
${}_{30}Zn^{64}$	1000	0.11	61	15
Zn^{66}	1040	0.087	86	18
${}_{32}Ge^{70}$	1020	0.077	120	23
Ge^{72}	830	0.23	32	9.2
Ge^{74}	593	0.29	19	10
Ge^{76}	556	0.33	17	9.3
${}_{34}Se^{74}$	635	0.21	32	15
Se^{76}	567	0.43	14	8.1
Se^{78}	615	0 36	20	8.7
Se^{80}	654	0.23	34	13
Se^{82}	880	0.056	190	39
${}_{42}Mo^{94}$	871	0.29	67	11
Mo^{96}	778	0.4	44	8.9
Mo^{98}	786	0.4	47	8.7
Mo^{100}	528	0.64	20	8.0
${}_{44}Ru^{100}$	540	0.30	48	18
Ru^{102}	473	0.63	20	10
Ru^{104}	362	1.0	10	8.2
${}_{46}Pd^{104}$	556	0.50	34	12
Pd^{106}	512	0.66	24	9.5
Pd^{108}	430	0.89	15	8.3
Pd^{110}	375	1.0	12	8.5
${}_{48}Cd^{110}$	654	0.41	58	13
Cd^{112}	620	0.46	49	12
Cd^{114}	550	0.55	37	11
Cd^{116}	508	0.62	32	11
${}_{52}Te^{120}$	560	0.55	48	13
Te^{122}	570	0.47	58	14
Te^{124}	608	0.39	77	16
Te^{126}	662	0.32	105	18
Te^{128}	750	0.28	140	18
Te^{130}	850	0.26	170	18
${}_{60}Nd^{146}$	455	0.25	150	44
Nd^{148}	300	0.69	36	23
${}_{62}Sm^{148}$	562	0.50	100	18
Sm^{150}	337	0.98	31	16
${}_{78}Pt^{194}$	330	1.7	39	13
Pt^{196}	358	1.3	55	17
Pt^{198}	403	1.4	60	13
${}_{80}Hg^{198}$	411	1.0	88	19
Hg^{200}	370	0.7	120	30
Hg^{202}	439	0.6	170	29
${}_{82}Pb^{206}$	803	0.14	1400	72
${}_{84}Po^{212}$	719	0.3	640	39
Po^{214}	606	0.6	270	23

portant influence of the shell structure on the nuclear vibrational properties. The effective surface tension C_2 varies strongly with respect to the liquid drop value, which is of the order of 50 Mev throughout the mass region considered, and the inertial parameter B_2 deviates considerably from the value (28).

While no quantitative analysis of the nuclear vibrational parameters has been given, some of the qualitative trends of the data in Table V.7 can be understood as consequences of the nuclear shell structure and the residual interactions (see Sec. V A). Thus, the effective nuclear surface tension decreases as one moves away from the closed shell regions, as a result of the polarizing effect of the nucleons in unfilled shells. While in the regions of closed shell configurations the C_2 values considerably exceed the liquid drop estimate, the C_2 values are appreciably smaller than this estimate for nuclei with many particles in unfilled shells, corresponding to the approach to instability of the spherical shape (see Fig. V.1). The inertial parameters B_2 systematically exceed those corresponding to irrotational flow. In regions away from closed shells, the increase of the inertia over that for irrotational flow is comparable to, although somewhat larger than, that observed in the rotational motion (see Sec. V B.2, and especially Fig. V.5).

In the more detailed interpretation of the data in Table V.7, it is necessary to take into account that the nucleus does not oscillate as a homogeneous system because of the different behavior of the particles in the unfilled shells from those in the closed shell core. This distinction is of special significance when the number of particles outside of closed shells is small. For such nuclei the vibrational energy is mainly associated with the motion of these few particles; on the other hand, an important and sometimes dominating part of the electric quadrupole moment of the vibration arises from the polarization of the closed shell core by the outside particles. This accounts for the large values of the vibrational parameters in Table V.7 for nuclei in the vicinity of the closed shell regions; in fact, the amplitudes α, normalized by (24), measure essentially the small core deformation, and the kinetic and potential energies per unit of α thus become very large.

With the approach to closed shell configurations the collective description of the low-lying excited states becomes less appropriate, and a more detailed analysis may be given by considering the motion of the individual particles outside of closed shells under the influence of their mutual interactions (see Sec. V E).

V C.4. Octupole Vibrations of Even-Even Nuclei

The odd parity collective vibrations of lowest energy are expected to be of octupole character ($\lambda=3$). In even-even nuclei the one phonon excitation of this type has $I=3$ and negative parity, while the two phonon octupole excitations give rise to states with $I=0+$, $2+$, $4+$, and $6+$. The levels with one quadrupole and one octupole phonon have $I=1-$, $2-$, $\cdots 5-$.

A rather low-lying (3−) state has been observed[167] in the spectrum of Gd^{152}, and may possibly represent a $\lambda=3$ vibrational excitation. Such an interpretation would imply a transition probability for $E3$ Coulomb excitation appreciably in excess of the single particle estimate (II A.58). It is of interest that odd parity states of similar energy have also been observed in neighboring deformed nuclei; in these, the coupling between the quadrupole deformation and the octupole mode may give rise to a lowest odd-parity excitation of $I=1-$ (see Sec. V D.4).

V C.5. Spectra of Odd-A Nuclei

In the regions where the even-even nuclei exhibit collective excitations corresponding to vibrations about a spherical equilibrium, the spectra of the odd-A nuclei are somewhat more complex and in most cases less well understood.

For the description of the low-energy spectrum of an odd-A nucleus, one must consider intrinsic degrees of freedom in addition to the collective motion. The intrinsic motion possesses a nonvanishing angular momentum which must be combined with the collective angular momentum; moreover, low-lying intrinsic excitations may arise from the change of orbital of the last odd nucleon, such as in the case of the nuclear isomers.[168]

The intrinsic nucleonic degrees of freedom are coupled to the collective oscillations, since the latter involve variations in the nuclear field. The effect of the coupling to the quadrupole vibrations depends essentially on the parameter[169]

$$q=\left(\frac{5}{16\pi}\right)^{\frac{1}{2}}\frac{k}{(\hbar\omega_2 C_2)^{\frac{1}{2}}}, \tag{V.31}$$

where the coupling constant k is of the order of magnitude of the average potential energy of a nucleon. Thus, as one moves away from closed shell regions, the coupling is expected to increase as a consequence of the decrease of the effective surface tension and of the vibrational frequencies.[170]

If the coupling could be neglected, one would obtain, for each intrinsic state, a collective spectrum corresponding to the phonon excitations. The states most strongly excited in the Coulomb excitation are the one

[167] H. Kendall and L. Grodzins, Bull. Am. Phys. Soc. Ser. II, 1, 164 (1956) and O. Nathan and M. Waggoner (private communication).

[168] See, e.g., M. Goldhaber and R. D. Hill, Revs. Modern Phys. 24, 179 (1952).

[169] See reference 132. In this reference the strength of the coupling has been discussed in terms of the parameter $x=qj^{-\frac{1}{2}}$.

[170] In reference 132 the increase of the coupling as one moves away from closed shells has been described in terms of the coherence of the individual particles in the unfilled shells in polarizing the nuclear shape. In the present discussion the paired particles are included in the collective degrees of freedom and their polarizing effect is thus reflected in the variation of the vibrational parameters.

phonon quadrupole excitations of the ground state configuration. If the ground state has spin I_0, these excited states have spins $I_0+2, I_0+1, \cdots |I_0-2|$ and energies of the same order of magnitude as the first excited state in the neighboring even-even nuclei. The reduced transition probability for the excitation of these states is given by

$$B(E2; I_0\rightarrow I_f)=\frac{1}{5}\frac{2I_f+1}{2I_0+1}B(E2)_{\text{ph}}, \qquad (\text{V}.32)$$

where $B(E2)_{\text{ph}}$ is the quadrupole phonon excitation probability (29).

When the coupling between the intrinsic and collective motion is weak ($q<1$), the effect on the energy spectrum and transition probabilities may be obtained from a perturbation treatment.[171]

If one estimates the coupling strength q employing the empirical vibrational parameters in Table V.7, one finds for most nuclei $1<q<3$ indicating an intermediate coupling situation. With such values of the coupling the stationary states represent a rather complicated superposition of individual particle and collective motions.[172]

In a few regions ($A\sim 20$, 75, 105, 150, and 192), one obtains $q\gtrsim 4$. For such strong couplings, the last odd particle may appreciably polarize the nuclear shape with a resulting approach to the coupling scheme characteristic of the deformed nuclei[137,173] (see Sec. V B). The strongest transitions observed in Coulomb excitation may then be approximately described as rotational excitations.[174]

In the limits of very weak or very strong coupling, the dominant quadrupole transitions are associated with a few simple excitations. For an intermediate coupling situation the pattern is more complex, but there exists an approximate sum rule which may be of use in the analysis of such spectra. Thus, the summed quadrupole strength $\sum_f B(E2; I_0\rightarrow I_f)$ obtained by adding all the low-energy transitions (i.e., with energies less than a few times $\hbar\omega_2$) is proportional to the average value of β^2 in the nuclear ground state, aside from a small contribution due to the quadrupole transitions of the last odd particle. The sum is, therefore, expected to be approximately the same in odd-A nuclei as in neighboring even-even nuclei, except insofar as the last particle polarizes the nucleus and thereby increases the sum. If the nucleus possesses a ground-state spin $I_0>\frac{1}{2}$, the sum over the final states I_f should include also the ground state quadrupole strength $B(E2; I_0\rightarrow I_0)$. The latter quantity is related to the static electric quadrupole moment Q of the ground state by [see (13) and (II A.18)]

$$B(E2; I_0\rightarrow I_0)=\frac{5}{16\pi}e^2Q^2\frac{(I_0+1)(2I_0+3)}{I_0(2I_0-1)}, \qquad (\text{V}.32\text{a})$$

a relation which is independent of the structure of the nuclear state.

When the total strength observed in an odd-A nucleus is appreciably smaller than $B(E2; 0\rightarrow 2)$ for the transition to the first excited state in the neighboring even-even nucleus, one may conclude that there exist strong quadrupole transitions as yet undetected. A total strength greatly exceeding this value implies an appreciable polarization produced by the last particle and may thus indicate that the coupling scheme is approaching that of a deformed nucleus. In this limit the ground-state strength contributes the fraction $\langle I_0 2 I_0 0 | I_0 2 I_0 I_0\rangle^2 = I_0(2I_0-1)(I_0+1)^{-1}(2I_0+3)^{-1}$ of the total quadrupole strength [see (10)]. For weak or intermediate coupling, the relative contribution of the ground-state strength to the total sum is expected to be smaller than this fraction.

V D. Vibrations of Spheroidal Nuclei

V D.1. Classification of Vibrations

While the lowest collective excitations of the strongly deformed nuclei correspond to rotations with preservation of shape, one may also expect these nuclei to exhibit collective excitations which correspond to vibrations about the equilibrium shape. Only scanty evidence is at present available on such vibrational excitations, but since the Coulomb excitation is well suited for the exploration of these levels we shall briefly outline the expected patterns.

For a nonspherical nucleus, the angular momentum of a vibrational quantum is not a constant of the motion due to the coupling to the nuclear rotation. Still, one may characterize the symmetry of the vibrations by a quantum number λ which represents the number of nodal surfaces and which, in the limit of small nuclear eccentricities, corresponds to the multiple order. The parity of the vibrations is $(-1)^\lambda$. For axially symmetric nuclei, the vibrations may in addition be characterized by the quantum number ν, representing the component of vibrational angular momentum about the symmetry axis. For given λ, the component ν may take the values $0, \pm 1 \cdots \pm\lambda$ but, in contrast to the vibrations of spherical nuclei, the vibrational parameters B and C [see (26)], and thus also the frequency, depend on $|\nu|$

[171] L. L. Foldy and F. J. Milford, Phys. Rev. **80**, 751 (1950); A. Reifman, Z. Naturforsch. **8a**, 505 (1953); M. Trocheris, J. phys. radium **14**, 635 (1953); reference 132; A. K. Kerman, Phys. Rev. **92**, 1176 (1953); F. J. Milford, Phys. Rev. **93**, 1297 (1953); K. W. Ford and C. Levinson, Phys. Rev. **100**, 1 (1955); B. J. Raz, thesis, University of Rochester, 1955; W. W. True, Phys. Rev. **101**, 1342 (1956).

[172] Intermediate coupling calculations have been given by K. Ford, reference 138; reference 132; D. C. Choudhury, Kgl. Danske Videnskab. Selskab Mat. fys. Medd. **28**, No. 4 (1954); reference 163; B. J. Raz, reference 171.

[173] The near instability of the spherical shape for the even-even nuclei in these regions is also indicated by the comparison of the energies of the first excited states with the critical value drawn in Fig. V.3.

[174] See, e.g., the discussion of the spectra of $Ag^{107,109}$ (F1, H4, H7), $Ir^{191,193}$ (*D*1), and Au^{197} (*C*1, *G*1, *S*4); the reference code refers to Table IV.2.

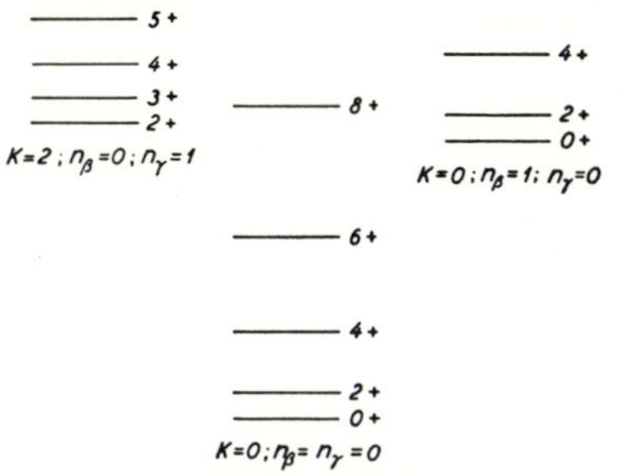

5+
4+
3+
2+
$K=2;\ n_\beta=0;\ n_\gamma=1$

8+

6+

4+

2+
0+
$K=0;\ n_\beta=n_\gamma=0$

4+

2+
0+
$K=0;\ n_\beta=1;\ n_\gamma=0$

FIG. V.10. Quadrupole vibrations of an even-even nucleus with spheroidal shape. The quadrupole vibrations of the nucleus about a spheroidal equilibrium shape separate into two modes, of which the one has $\nu=0$ (β vibration) and the other $|\nu|=2$ (γ vibration). The figure illustrates the ground-state rotational band as well as the rotational bands associated with the first vibrational excitation of these two modes ($n_\beta=1$ and $n_\gamma=1$, respectively). The figure is meant for illustrative purposes only and no significance should be attached to the relative magnitude of the two vibrational frequencies.
It is expected that Coulomb excitation will strongly excite the two states ($n_\beta=1$; $I=2+$) and ($n_\gamma=1$; $I=2+$). These states decay by means of $E2$ radiation to the $I=0$, 2 and 4 members of the ground-state band with relative reduced transition probabilities 1:10/7:18/7 for the β vibration and 1:10/7:1/14 for the γ vibration [compare (V.33)]. $M1$ radiation is expected to be weak in these transitions, even when $\Delta I=0$ or 1.
In an odd-A nucleus, the $n_\beta=1$ vibrational excitation gives rise to a rotational band with $K=K_0$, where K_0 is the ground-state K value. The $n_\gamma=1$ excitation gives rise to two bands with $K=|K_0\pm 2|$.

as well as on λ. In the special case of $\nu=0$, the vibrations preserve the axial symmetry of the nuclear shape.

The rotational band associated with a one-phonon excitation has $K=|K_0+\nu|$, where K_0 is the intrinsic angular momentum of the ground state. The band contains the members $I=K, K+1, K+2, \cdots$ except for a $\nu=0$ vibration in a nucleus with a paired nucleonic configuration (ground-state configuration of even-even nuclei), in which case the band contains the states $I=0+, 2+, 4+, \cdots$ (for λ even) or $I=1-, 3-, 5-, \cdots$ (for λ odd).

V D.2. Transition Probabilities

The reduced transition probability of multipole order EL for the excitation of a vibration with angular momentum component ν may be written[175]

$$B(EL;I_iK_i\rightarrow I_fK_f)=\langle K_f|\mathfrak{M}'(EL,\nu)|K_i\rangle^2 \times\langle I_iLK_iK_f-K_i|I_iLI_fK_f\rangle^2, \quad \text{(V.33)}$$

where the first factor contains the vibrational matrix element of the electric multipole moment (II A.11) expressed in the intrinsic coordinate system. The second factor in (33) gives the relative probabilities for the excitation of the different members of the rotational band associated with the vibration. The formula (33) also applies to the radiative decay of a vibrational excitation, and thus gives the branching ratios to the members of the ground-state band.

[175] In the special case of $K_i=0$ and $K_f\neq 0$ (excitation of $\nu\neq 0$ modes in even-even nuclei), the value of $B(EL)$ is twice that given by (33).

If the electric multipole order, L, of the excitation equals that of the vibration, λ, the transition probability (33) can be expressed in terms of the vibrational parameters $B_{\lambda\nu}$ and $C_{\lambda\nu}$ by means of the relation [compare (29)]

$$\langle K_f|\mathfrak{M}'(E\lambda,\nu)|K_i\rangle^2=\left(\frac{3}{4\pi}ZeR_0^\lambda\right)^2\frac{\hbar}{2(B_{\lambda\nu}C_{\lambda\nu})^{\frac{1}{2}}}. \quad \text{(V.34)}$$

We here assume a normalization of the vibrational amplitudes in analogy to (24).

V D.3. Quadrupole Vibrations

The lowest order shape vibrations have $\lambda=2$ and are of approximately quadrupole type. A deformation of order $\lambda=2$ and $\nu=\pm 1$ is equivalent to a rotation and the only occurring quadrupole vibrations thus have $\nu=0$ (β vibrations) or $\nu=\pm 2$ (γ vibrations). The quadrupole vibrational pattern expected for an even-even nucleus is illustrated in Fig. V.10.

While theoretical estimates of the vibrational frequencies are rather uncertain, due to the influence of the shell structure, the empirical data on the spectra of the even-even nuclei appears to indicate that the quadrupole vibrational energies for the very strongly deformed nuclei may be of the order of a Mev in the heavy nuclei. States with some of the expected properties have been found in a number of even-even nuclei.[176,176a] The determination of the $E2$ transition matrix elements would be of great value for the classification of these levels, since vibrational excitations should be appreciably stronger than single particle transitions. As in the case of spherical nuclei, $M1$ radiation is forbidden in the decay of vibrational excitations, even when $\Delta I=0$ or 1. With decreasing deformation and gradual approach to the transition region, where the nuclear shape becomes spherical, the (2+) rotational excitation increases in frequency and goes over into the one phonon quadrupole vibration. The lowest vibrational excitation goes over into a two phonon state and

[176] See, e.g., Er^{166} (1.46 Mev) J. S. Fraser and J. C. D. Milton, Phys. Rev. **98**, 1173(A) (1955). W^{182} (1.22 Mev) Murray *et al.*, reference 151; Alaga *et al.*, Kgl. Danske Videnskab. Selskab Mat. fys. Medd. **29**, No. 9 (1955). Pu^{238} (1.03 Mev) Rasmussen *et al.*, Phys. Rev. **99**, 42 and 47 (1955).

[176a] *Note added in proof.*—Recent studies of the levels in the one-Mev region in heavy nuclei of the even-even type have revealed the systematic occurrence of states having many of the expected properties associated with both γ- and β-vibrational excitations (private communication from Asaro, Hollander, Perlman, Rasmussen, and Stephens; see also the review article on α radioactivity by I. Perlman and J. Rasmussen, to appear in *Handbook of Physics*, edited by S. Flügge). In this connection, it seems likely that the 760-kev level excited in Th^{232} (see Table IV.2) is a 2+ level corresponding to a β vibration. Its excitation cross section would then indicate a $B(E2)$ value several times greater than the single particle unit (1).

its frequency may thus decrease somewhat with the approach to the transition region.[177]

V D.4. Octupole Vibrations

The lowest odd parity modes ($\lambda=3$) should resemble octupole vibrations, and have $\nu=0, \pm1, \pm2$, and ±3.

Recently, relatively low lying 1− states, and in some cases also 3− and 5− states, have been observed in a number of even-even nuclei in the heavy elements region[178,179] (see Fig. V.11 and Table V.8). Their systematic occurrence suggests an interpretation in terms of odd parity vibrations,[180] and the branching ratios in the $E1$ decays to the ground-state rotational band [see (33)] characterize the states as $K=0$ (and thus $\nu=0$) in all cases (see Table V.8). The lower energy of the $\nu=0$ mode as compared with the vibrations having $|\nu|=1$, 2, and 3 may be associated with a prolate nuclear shape.

The frequencies of these vibrations estimated on the basis of the liquid drop model would be a few Mev in heavy nuclei. However, as in the case of quadrupole vibrations, the shell structure is expected to have an important influence. Thus, the restoring force for odd parity vibrations will be strongly reduced by the occurrence of near lying single particle levels of opposite parity, which might even lead to stable odd parity deformations.

In an even-even nucleus, the odd parity vibrational levels of 1− type can be excited by an $E1$ transition from the ground state; there are two such levels having $|\nu|=0$ and 1, respectively. The transition probability (33) depends on the displacement of the center of charge

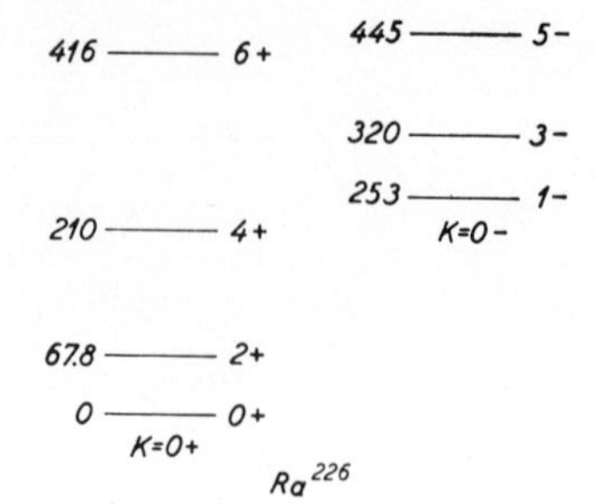

FIG. V.11. Level spectrum of Ra^{226}. The figure shows the levels in Ra^{226} populated in the α decay of Th^{230} [Stephens, Asaro, and Perlman, Phys. Rev. (to be published)]. The observed levels appear to form two rotational bands, of which the first is the usual ground-state band of even-even nuclei with $K=0+$ (compare Fig. V.4). The negative parity levels form a rotational band with $K=0$ as determined by means of (V.33) from the observed relative intensities in the γ decay of these states (compare Table V.8).

As discussed in the text, the observed odd parity states may be associated with a collective vibration of approximately octupole type and with component $\nu=0$. The appreciably larger moment of inertia observed in this band as compared with that in the ground-state band may result from the coupling to the expected octupole vibrational mode with $|\nu|=1$ ($K=1-$).

TABLE V.8. States of 1− type in even-even nuclei. The excitation energies of the observed 1− states in the heavy elements are listed in column three, while the energies of the first excited 2+ states are given in column two. The 1− states decay by $E1$ radiation to the 0+ and 2+ members of the ground state rotational band, and the ratios of the reduced transition probabilities are shown in column four. These may be compared with the theoretical ratios obtained from (V.33) for the assignments $K=0$ and $K=1$ for the 1− states. The table is taken from Stephens, Asaro, and Perlman, Phys. Rev. **100**, 1543 (1955).

Nucleus	$E(2+)$ (kev)	$E(1-)$ (kev)	$B(E1;1-\rightarrow0+)/B(E1;1-\rightarrow2+)$ exp.	theor. ($K=0$)	theor. ($K=1$)
$_{88}Ra^{222}$	112	242	0.48±0.15	0.50	2.00
Ra^{224}	84	217	0.36±0.15		
Ra^{226}	68	253	0.49±0.08		
$_{90}Th^{226}$	73	232	0.51±0.15		
Th^{228}	58	326	0.43±0.08		
$_{94}Pu^{238}$	43	605	0.60±0.15		

with respect to the center of mass. Since this vanishes for a uniformly charged nucleus, the cross section is expected to be relatively small. A contribution to the nuclear dipole moment arises from the polarization of the nuclear charge resulting from the Coulomb forces. Estimates indicate that this effect would lead to $B(E1)$ values several orders of magnitude smaller than the single particle unit (II A.58).

The excitation of the 3− states would involve the $E3$ moment [see (34)], which is expected to be appreciably larger than that of a single particle. The determination of the cross section for $E3$ excitation would thus provide crucial information on the interpretation of these states. If $B(E3)$ for such an excitation were 10 times the value (II A.58) for a nucleus with $Z=90$ [i.e., assuming $B(E3;0\rightarrow3)=0.2e^2(10^{-24}\text{ cm}^2)^3$], the excitation cross section for 20-Mev α particles would be about one millibarn.

V E. Regions of Closed Shells

For nuclei with only a few particles outside of closed shells, a rather detailed analysis of the low-energy excitations may be obtained by considering these particles as moving, under the influence of their mutual interactions, in a fixed central binding field produced by the closed shell core.[181] The weak coupling between the par-

[177] Thus the 633-Mev level in Os^{188} with $I=2+$ may represent a γ vibration in the transition region (Johns *et al.*, Can. J. Phys. **34**, 69 (1956) and Potnis *et al.*, Phys. Rev. **102**, 459 (1956)).

[178] See especially Stephens, Asaro, and Perlman, Phys. Rev. **96**, 1568 (1954); **100**, 1543 (1955), and to be published.

[179] See also the 960-kev level in Sm^{152} which appears to have $I=1-$ and $K=0$ [O. Nathan and M. Waggoner, Nuclear Phys. (to be published).

[180] R. F. Christy (private communication).

[181] Detailed analyses of this type have been discussed, e.g., by D. R. Inglis, reference 140 ($A\leq16$); D. Kurath, Phys. Rev. **101**, 216 (1956) ($A<16$); A. M. Lane, Proc. Phys. Soc. (London) **66A**, 977 (1954) ($A=13$); J. P. Elliot and B. H. Flowers, Proc. Roy. Soc. (London) **A229**, 536 (1955) ($A=18$, 19); M. G. Redlich, Phys. Rev. **99**, 1427 (1955) ($A=18$, 19); S. Goldstein and I. Talmi, Phys. Rev. **102**, 589 (1956) ($A=38$, 40); C. Levinson and K. W. Ford, Phys. Rev. **100**, 13 (1955) ($A=42$, 43); W. W. True, Phys. Rev. **101**, 1342 (1956) ($A=204$); D. E. Alburger and M. H. L. Pryce, Phys. Rev. **95**, 1482 (1954) ($A=206$).

ticles and the excitations of the core may be added as a perturbation and principally contributes to the electric moments associated with the particle motion. The dynamical effects of the core also imply a coupling between the outside particles[182] which may contribute an appreciable part of the apparent interaction between the nucleons.

To the extent that the dynamics of the core can be described in terms of collective oscillations, an approximate expression for the resultant nuclear transition moment may be simply obtained from (II A.13) by including in the sum over k only the extra nucleons, and replacing their charges e_k by the effective radiating charges e_k' given by[132]

$$e_k' = e_k + \frac{3Z'e}{4\pi}\frac{k}{C_\lambda'}\left(\frac{R_0}{r_k}\right)^\lambda_{\text{Av}}, \qquad \text{(V.35)}$$

where Z' and C_λ' are the charge number and effective surface tension of the closed shell core. The coupling constant k in (35) is the same as that employed in (31).

[182] Reference 132, p. 27.

Estimates of C_λ' for a closed shell indicate that the induced charge is of the order of one unit.

Especially clear-cut examples of this effect are provided by the $E2$ Coulomb excitation cross section for the first excited states in Pb^{206} and Pb^{207} (see Table IV.2) and the measured $E2$ decay rate[183] of the first excited state in O^{17}. In these configurations consisting entirely of neutrons outside of closed shells, the observed transition rates imply an effective polarization charge of about one unit.[184,185] A similar enhancement of the electric transition probability is observed for the 196-kev $E2$ excitation of F^{19},[186] and in the $E3$ decay of the 3− states in O^{16} [187] and Pb^{208}.[188]

[183] J. Thirion and V. L. Telegdi, Phys. Rev. **92**, 1253 (1953).

[184] W. W. True, reference 171; J. Raz, reference 171.

[185] The polarization charge (35) contributes not only to the transition rate, but also to the static electric moments. Thus, the very small electric quadrupole moment reported for O^{17} [G. R. Bird and C. H. Townes, Phys. Rev. **94**, 1203 (1954)] seems surprising (see the detailed discussion by J. Raz, reference 171).

[186] J. Elliot and B. H. Flowers, reference 181.

[187] See F. Ajzenberg and T. Lauritsen, Revs. Modern Phys. **27**, 77 (1955).

[188] Elliot *et al.*, Phys. Rev. **93**, 356 (1954).

PHYSICAL REVIEW VOLUME 103, NUMBER 3 AUGUST 1, 1956

Reorientation Effect in Coulomb Excitation*

G. BREIT, R. L. GLUCKSTERN, AND J. E. RUSSELL
Yale University, New Haven, Connecticut

(Received April 23, 1956)

Reprinted with corrections by the authors.

The semiclassical treatment of the effect of the reorientation of the spin of the target nucleus during a collision resulting in Coulomb excitation is presented. It is found that the effect vanishes for zero excitation, but that it may be appreciable for finite excitation. The effect on the angular distribution of the photons is similarly found to vanish for head-on collisions. Three different types of experiments involving the measurement of the angular distribution of the photons are discussed. Typical numerical results for comparison with experiment are presented and the possibilities offered by bombardment with heavy ions are noted.

I. INTRODUCTION

THE fact that finite-amplitude effects can be appreciable in Coulomb excitation has been pointed out by Breit and Lazarus.[1] In the present note, one of the effects is calculated in the semiclassical approximation for a $0\rightarrow 2$ transition. The effect under consideration consists in the reorientation of the nuclear axis caused by the electric field of the bombarding particle after it has excited the nucleus. The change in the nuclear spin direction affects the angular distribution of the γ rays. The latter distribution is affected by other finite-amplitude effects entering in the same order of the calculation such as the excitation $0\rightarrow 4$ followed by de-excitation $4\rightarrow 2$. Since this effect depends on higher (2^4) multipole action, it appears reasonable to neglect it in a preliminary survey of possibilities. The transition chain $0\rightarrow 2\rightarrow 2'$ will also affect the γ angular distribution. In the usual case of Coulomb excitation from ground states of even-even nuclei, the order of the rotational levels is 0, 2, and 4 while levels with $I=2$ occur at higher energies and transitions to them involve changes in quantum numbers additional to the rotational one. For both reasons, their effects may be expected to be less serious although they should be eventually taken into account.

The principal interest in the reorientation effect lies in the possibility which it offers of ascertaining static nuclear quadrupole moments in excited states. The other available method[2] employing intermolecular field involves the theory of solids and complexities regarding the motion of a recoil nucleus through a solid. The reorientation effect in Coulomb excitation appears to be relatively free of such complications, the orbit of the projectile being well defined. It should be pointed out that screening corrections for the influence of atomic electrons are minor for the Coulomb excitation reorientation effect while they are present in the method of intermolecular fields. Since values of quadrupole moments of ground states are usually affected by screening corrections[3] and since the latter are hard to estimate reliably,[4] there may be a special value in the Coulomb excitation reorientation effect for the measurement of these moments.

The calculations presented below are made by means of the semiclassical treatment (SCT) which employs classical mechanics for the relative motion of the target and projectile. Since one expects the reorientation effect to be more serious for the heavier projectiles, this approximation is likely to be satisfactory for a preliminary survey. The calculations are presented with reference to three possible experiments. In the first the angular distribution of the γ rays is measured in the usual manner, the only reference line being the incident charged particle beam. In the second the inelastically scattered charged particles are counted in coincidence with the γ rays, no attempt being made to define the charged particle orbit. In the third the incident beam and the inelastically scattered particle directions are used to define the orbit in coincidence with γ counting. In the third type of experiment there is a maximum possibility of obtaining checks on the parameters entering the interpretation. It is probably the one most seriously affected by the inexactness of the SCT.

The present paper is confined to consideration of the first nonvanishing order in the finite amplitude effects. It is thus concerned with the calculation of cross-term effects arising from the second order probability amplitude effects. Terms quadratic in the second-order probability amplitude corrections are omitted since their inclusion would require the consideration of third order effects in the probability amplitudes. With these approximations, it is found that some special circumstances cause the reorientation effect on the γ-angular distribution to vanish for zero excitation energy. These considerations do not exclude the possibility of detection of finite-amplitude effects for small excitation energies by studies of inelastic scattering. For nonvanishing excitation energies, there remains an effect which, while somewhat affected by the special circumstances which

* This research was supported by the U. S. Atomic Energy Commission and by the Office of Ordnance Research, U. S. Army.

[1] G. Breit and J. P. Lazarus, Phys. Rev. **100**, 942 (1955).

[2] H. Frauenfelder, in *Annual Reviews of Nuclear Science* (Annual Reviews, Inc., Stanford, 1953), Vol. 2, p. 129; see also A. Abragam and R. V. Pound, Phys. Rev. **92**, 943 (1953).

[3] R. Sternheimer, Phys. Rev. **84**, 244 (1951); Foley, Sternheimer, and Tycko, Phys. Rev. **93**, 734 (1954); R. M. Sternheimer and H. M. Foley, Phys. Rev. **102**, 731 (1956).

[4] C. Schwartz, Phys. Rev. **97**, 380 (1955).

make it vanish for zero excitation energy, appears to be large enough for observation.

An additional special circumstance arises for the case of finite excitation by head-on collisions. The total cross section shows an appreciable effect, but the angular distribution remains unaffected in the higher order calculation. Since the large probabilities of excitation occur for head-on collisions, this circumstance makes the reorientation effect smaller than it might have been otherwise. Nevertheless the second-order effects rise rapidly as the impact parameter increases from its vanishing value for head-on collisions.

List of Notation

The following is a list of the more frequently occurring symbols and their definitions in the approximate order of their appearance.

Z_1, Z_2 = nuclear charge of the incident, target particles respectively.

v = velocity of the incident particle.

$\mathbf{r}$, $\mathbf{r}_i$ = displacement of the incident particle from the target nucleus in the frame of reference in which the x axis bisects the orbit hyperbola. The plane of the orbit is the xy plane and the direction of the orbit is such that the incident particle moves from the fourth to the first quadrant of the xy plane in its trajectory.

Φ = azimuth of $\mathbf{r}$ with respect to the x axis.

$\mathbf{r}_p$; r_p, θ_p, φ_p = displacement and components in spherical coordinates of the equivalent nuclear proton in the same frame of reference as that used for $\mathbf{r}$.

θ_{pt} = angle between $\mathbf{r}$ and $\mathbf{r}_p$.

Y_{lm} = spherical harmonic of order l and magnetic quantum number m.

$R_i(r_p)$, $R_f(r_p)$ = radial wave function of the equivalent nuclear proton in the ground and excited states of the target nucleus.

$\mu = 2, 1, 0, -1, -2$ = magnetic sublevels of an $I=2$ state of the excited target nucleus.

$H'(t)$ = time-dependent quadrupole interaction term between the incident particle and the proton in the target nucleus.

$E_{fi} = \hbar\omega_{fi} = \hbar\omega$ = excitation energy of the excited state of the target nucleus.

$H'_{\mu i}(t)$ = matrix element of $H'(t)$ between the target nucleus ground state ($I=0$) and the excited state ($I=2$) sublevel with magnetic quantum number μ.

$H'_{\mu\mu'}(t)$ = matrix element of $H'(t)$ between sublevels μ and μ' of the excited ($I=2$) state of the target nucleus.

$c_0(t)$, $c_\mu(t)$ = time-dependent amplitudes of the nuclear wave function corresponding to the ground-state ($I=0$) and excited-state ($I=2$) sublevels given by μ.

$c = c^{(0)} + c^{(1)} + c^{(2)}$ = separation of the time-dependent amplitudes into terms corresponding to the power of the interaction H', as given in the superscript.

$\langle r^2\rangle_{ff}$, $\langle r^2\rangle_{fi}$ = radial matrix element of r_p^2 between the states R_f, R_f and R_f, R_i respectively.

$2a'$ = closest distance of approach of the incident particle to the target nucleus for head-on collision.

ϵ = eccentricity of the hyperbolic orbit.

$\xi = \eta_f - \eta_i$, $\eta_i = Z_1Z_2e^2/\hbar v_i$, $\eta_f = Z_1Z_2e^2/\hbar v_f$.

$S_\mu^{(2)}$ = amplitude for direct quadrupole Coulomb excitation to the sublevel μ, apart from ϵ- and ξ-independent factors.

S_μ = quantity which replaces $S_\mu^{(2)}$ after inclusion of the reorientation effect.

(θ,φ) = polar angles of the direction of the photon emission in the same coordinate system as that used for $\mathbf{r}$.

$\mathbf{c}$, $\mathbf{s}$ = $\cos\theta$, $\sin\theta$ respectively.

$\delta = \tan^{-1}(\epsilon^2-1)^{\frac{1}{2}}$. The angle $\pi - 2\delta$ is the scattering angle of the incident particle.

$\varphi' = \varphi + \delta$ = azimuthal angle of the photon emission direction for the x-axis directed along the negative of the initial particle velocity and the z axis the same as for φ.

$\mathbf{1}_s$ = unit vector in the direction of polarization of the photon.

$\mathbf{k}$ = propagation vector of the photon.

$F_2(kr) = (\pi/2kr)^{\frac{1}{2}}J_1(kr)$.

θ' = angle of $\mathbf{k}$ with respect to the negative of the incident particle direction.

J = angular distribution, apart from constant factors, of the emitted photons.

$\langle J\rangle$ = average of J with respect to a rotation of the orbit plane about the incident beam direction.

$a_{2\epsilon}(\xi)$, $a_{4\epsilon}(\xi)$ = angular distribution coefficients of the Legendre polynomial of order 2 and 4 for a given orbit eccentricity, ϵ.

$a_2(\xi)$, $a_4(\xi)$ = averages over orbit eccentricity of $a_{2\epsilon}(\xi)$, $a_{4\epsilon}(\xi)$.

$T_0 = \frac{1}{2}S_0^{(2)}$.

$T_c = -(S_2^{(2)} + S_{-2}^{(2)})/2\sqrt{6}$.

$T_s = (S_{-2}^{(2)} - S_2^{(2)})/2\sqrt{6}$.

$\mathcal{T}_0 = \frac{1}{2}S_0$.

$\mathcal{T}_c = -(S_2 + S_{-2})/2\sqrt{6}$.

$\mathcal{T}_s = (S_{-2} - S_2)/2\sqrt{6}$.

$\lambda = Z_1e^2\langle r^2\rangle_{ff}/[7\hbar a'^2 v]$, parameter relating magnitude of the reorientation effect to direct Coulomb excitation.

$D + \delta D$ = separation of the θ'-independent term in $(1/4)\langle J\rangle$ into the result of neglecting reorientation and the change due to reorientation.

$N_2 + \delta N_2$ = separation of the coefficient of $P_2(\theta')$ in $(56/5)\langle J\rangle$ into the result of neglecting reorientation and the change due to reorientation.

$N_4 + \delta N_4$ = separation of the coefficient of $P_4(\theta')$ in $(7/2)\langle J\rangle$ into the result of neglecting reorientation and the change due to reorientation.

$C_1, \cdots C_7$ = coefficients of the angle-dependent terms in a measurement of the correlation between the directions of the scattered particle and the photon. The C's are defined in Eqs. (15), (14.2), (14.3), (14.4), (14.5).

II. PROBABILITY AMPLITUDES

In the interests of simplicity, the calculation will be presented as though there were only one nuclear proton present. As is well known, the generality of application is not affected by doing so provided the results are expressed in terms of appropriate transition and static quadrupole moments. The interaction Hamiltonian is taken as

$$H'(t)=Z_1e^2(r_p^2/r_t^3)P_2(\cos\theta_{pt}), \quad (1)$$

the nuclear proton quantities being distinguished by subscript p, the trajectory quantities by subscript t, and θ_{pt} being the angle between the proton and projectile directions as viewed from the origin. For simplicity the discussion will be carried on as though the bombarded nucleus were infinitely heavy. The nuclear wave function is represented as

$$w_\mu(\mathbf{r}_p)=R_f(r_p)Y_{2\mu}(\theta_p,\varphi_p) \quad (1.1)$$

in the state with $I=2$ and as

$$v_0(\mathbf{r}_p)=R_i(r_p)Y_{00}(\theta_p,\varphi_p) \quad (1.2)$$

in the state with $I=0$. The radial functions are normalized by

$$\int_0^\infty R^2(r_p)r_p^2dr_p=1; \quad (1.3)$$

the latter equation being meant to apply to both radial functions. The wave function is expanded as

$$\psi=c_0v_0+\Sigma_\mu c_\mu w_\mu; \quad (1.4)$$

all other nuclear states being omitted in view of the simplifying assumptions mentioned in the introduction. Employing the zero-order approximation

$$c_0^{(0)}=\exp(-iE_0t/\hbar) \quad (1.5)$$

and the equation

$$\frac{\hbar}{i}\frac{d}{dt}[c_\mu\exp(iE_\mu t/\hbar)]+\{H'_{\mu i}(t)c_0+\Sigma_{\mu'}H'_{\mu\mu'}(t)c_{\mu'}\}\times\exp(iE_\mu t/\hbar)=0, \quad (1.6)$$

one obtains

$$c_\mu=c_\mu^{(1)}+c_\mu^{(2)} \quad (2)$$

with

$$c_\mu^{(1)}=-(i/\hbar)\exp(-iE_\mu t/\hbar)\int_{-\infty}^t H'_{\mu i}(t)\times\exp(iE_{\mu i}t/\hbar)dt, \quad (2.1)$$

$$c_\mu^{(2)}=-(i/\hbar)\exp(-iE_\mu t/\hbar)\times\int_{-\infty}^t \exp(iE_\mu t/\hbar)[H'_{\mu i}(t)c_0^{(1)}+\Sigma_{\mu'}H'_{\mu\mu'}(t)c_{\mu'}^{(1)}(t)]dt, \quad (2.2)$$

where $c_0^{(1)}$ is the lowest-order correction to c_0, the designation of the order being in terms of the power of H' involved. One obtains also

$$c_0^{(1)}=-(i/\hbar)\exp(-iE_0t/\hbar)\times\Sigma_\mu\int_{-\infty}^t H'_{i\mu}(t)\exp(iE_0t/\hbar)c_\mu^{(1)}(t)dt. \quad (2.3)$$

Since the c_μ are being calculated by an iteration procedure which gives perfect normalization when completely carried out, a correction for normalization need not be made and Eq. (2) should be therefore good enough for obtaining $|c_\mu|^2$ correct up to and including terms in H'^3. It may be of interest to note that the normalization sum resulting from the inclusion of $c_\mu^{(2)}$ and without taking $c_0^{(1)}$ into account is already good enough for calculating H'^3 effects on $|c_\mu|^2$ without the consideration of what happens in higher iterations as may be seen as follows. In a given order, the iteration procedure gives only an approximately normalized solution. Calculation of

$$|c_0|^2+\Sigma_\mu|c_\mu|^2$$

shows the presence of a correction factor of the form

$$1+\mathcal{O}(H'^2)$$

to all probability amplitudes. Since the intention is only that of obtaining c_μ to within order H'^2, one may omit this factor and similarly, since $c_0^{(1)}=\mathcal{O}(H'^2)$, the first term in brackets in Eq. (2.2) is $\mathcal{O}(H'^3)$ so that it may be dropped. There results

$$c_\mu(t)=c_\mu^{(1)}(t)-(i/\hbar)\exp(-iE_\mu t/\hbar)\int_{-\infty}^t\exp(iE_\mu t/\hbar)\times\Sigma_{\mu'}H'_{\mu\mu'}(t)c_{\mu'}^{(1)}(t)dt+\mathcal{O}(H'^3). \quad (2.4)$$

From now on, the correction $\mathcal{O}(H'^3)$ will be omitted. The quantity c_μ being needed only for $t=\infty$, it suffices to consider

$$c_\mu(\infty)=c_\mu^{(1)}(\infty)-\hbar^{-2}\exp(-i\omega_ft)\int_{-\infty}^{+\infty}dt''\times\int_{t''}^\infty dt'\Sigma_{\mu'}H'_{\mu\mu'}(t')H'_{\mu'i}(t'')\exp(i\omega_{fi}t'') \quad (3)$$

with the notation

$$\omega_f=E_f/\hbar,\quad \omega_i=E_i/\hbar,\quad \omega_{fi}=E_{fi}/\hbar. \quad (3.1)$$

Employing standard manipulations, one finds

$$\Sigma_{\mu'}H'_{\mu\mu'}(t)H'_{\mu'i}(t')=\frac{5}{(4\pi)^{\frac{3}{2}}}AI_\mu/[r^3(t)r^3(t')] \quad (3.2)$$

with

$$A=Z_1^2e^4\left\{\int_0^\infty r_p^4R_f^2(r_p)dr_p\right\}\times\left\{\int_0^\infty r_p^4R_f(r_p)R_i(r_p)dr_p\right\} \quad (3.3)$$

and

$$I_\mu=\int Y_{2\mu}^*(p)P_2(pp')P_2(pt)P_2(p't')d\Omega_p d\Omega_{p'}$$

$$=(4\pi/5)\int Y_{2\mu}^*(p)P_2(pt)P_2(pt')d\Omega_p. \quad (3.4)$$

Here p stands for θ_p and φ_p, pt for θ_{pt}, etc. The quantities I_μ can be conveniently evaluated in a coordinate system with z-axis perpendicular to the plane tt'. It is found that

$$I_\mu=(4\pi/5)^{\frac{1}{2}}i_\mu/7, \quad (3.5)$$

$$i_\mu=\tfrac{1}{2}(\tfrac{3}{2})^{\frac{1}{2}}(\tau^{-2}+\tau'^{-2}),\ 0,\ 2-3\cos^2(\Phi-\Phi'),\ 0,$$

and

$$\tfrac{1}{2}(\tfrac{3}{2})^{\frac{1}{2}}(\tau^2+\tau'^2) \quad (3.6)$$

for $\mu=2, 1, 0, -1$, and -2, respectively, with

$$\tau=\exp(i\Phi), \quad \tau'=\exp(i\Phi') \quad (3.7)$$

and Φ, Φ' standing for azimuthal angles of $\mathbf{r}_t$ and $\mathbf{r}_{t'}$ respectively.

If one denotes the integrals in (3.3) as $\langle r^2\rangle_{ff}$ and $\langle r^2\rangle_{fi}$, respectively, Eq. (3) becomes

$$c_\mu^{(2)}(\infty)=-\frac{Z_1^2e^4}{7\sqrt{5}}\langle r^2\rangle_{ff}\langle r^2\rangle_{fi}\frac{\exp(i\omega_f t)}{\hbar^2}$$

$$\times\int\int_{t>t'}\frac{i_\mu(t,t')\exp(i\omega_{fi}t')}{r^3r'^3}dtdt' \quad (4)$$

with r and r' standing for values of r at times t and t', respectively. Strictly speaking, the quantity in this equation is not exactly $c_\mu^{(2)}$ in the sense of Eq. (2.2) but is obtained by omitting the first term in brackets in (2.2), causing an error of order H'^3. Introducing the parametric representation[5]

$$x=a'(\epsilon+\cosh w), \quad y=a'(\epsilon^2-1)^{\frac{1}{2}}\sinh w,$$
$$r=a'(1+\epsilon\cosh w), \quad t=(a'/v)(w+\epsilon\sinh w), \quad (4.1)$$

where

$$a'=(Z_1Z_2e^2/Mv^2), \quad (4.1')$$

the quantity M being the reduced mass, one finds that

$$c_\mu^{(1)}(\infty)=-\frac{(4\pi)^{\frac{1}{2}}i}{5\hbar}Z_1e^2\langle r^2\rangle_{fi}\exp(-i\omega_f t)$$

$$\times\int_{-\infty}^{+\infty}r^{-3}Y_{2\mu}^*(t)\exp(i\omega_{fi}t)dt \quad (4.2)$$

has the value

$$c_\mu^{(1)}(\infty)=\frac{iZ_1e^2}{2(\sqrt{5})\hbar a'^2v}\langle r^2\rangle_{fi}\exp(-i\omega_f t)S_\mu^{(2)} \quad (4.3)$$

with

$$S_{\pm2}^{(2)}=-\left(\frac{3}{2}\right)^{\frac{1}{2}}\int_{-\infty}^{+\infty}e^{i\xi(w+\epsilon\sinh w)}$$

$$\times\frac{[\epsilon+\cosh w\mp i(\epsilon^2-1)^{\frac{1}{2}}\sinh w]^2}{(1+\epsilon\cosh w)^4}dw, \quad (4.4)$$

$$S_0^{(2)}=\int_{-\infty}^{+\infty}\frac{e^{i\xi(w+\epsilon\sinh w)}}{(1+\epsilon\cosh w)^2}dw. \quad (4.5)$$

The notation $S_\mu^{(2)}$ is that used by Alder and Winther.[6] The quantity ξ is taken to be

$$\xi=\eta_f-\eta_i \quad (4.6)$$

rather than $\omega_{fi}a'/v$ in order to obtain better agreement with the quantum-mechanical treatment of first-order Coulomb excitation. Here

$$\eta_i=Z_1Z_2e^2/\hbar v_i, \quad \eta_f=Z_1Z_2e^2/\hbar v_f. \quad (4.7)$$

For $E_{fi}=0$, the integrals in Eq. (4) are simplified. Thus one finds in this case

$$\frac{1}{2}\left(\frac{3}{2}\right)^{\frac{1}{2}}\int\int_{>t'}\frac{\tau^{-2}+\tau'^{-2}}{r^3r'^3}dtdt'$$

$$=-\tfrac{1}{2}S_2^{(2)}S_0^{(2)}/(a'^4v^2), \quad (E_{fi}=0), \quad (5)$$

where the vanishing of contributions from $\sin2\Phi+\sin2\Phi'$ can be inferred by a consideration of the transformation $(t,t')\to(-t',-t)$. Since the integral is symmetric, it is possible to remove the condition $t>t'$ and to express the result in terms of an integral over the whole t, t' plane. Substitution in terms of the $S_\mu^{(2)}$ through a comparison with (4.2) and (4.3) yields then the right side of (5). Similarly,

$$\int\int_{t>t'}[2-3\cos^2(\Phi-\Phi')]/(r^3r'^3)dtdt'$$

$$=[\tfrac{1}{4}(S_0^{(2)})^2-\tfrac{1}{2}(S_2^{(2)})^2]/(a'^4v^2), \quad (E_{fi}=0). \quad (5.1)$$

Substitution in Eq. (4) and comparison with Eq. (4.3) gives

$$c_\mu^{(2)}(\infty)/c_\mu^{(1)}(\infty)=\frac{iZ_1e^2}{7\hbar a'^2v}\langle r^2\rangle_{ff}$$

$$\times(-S_0^{(2)},0,\tfrac{1}{2}S_0^{(2)}-(S_2^{(2)})^2/S_0^{(2)},0,-S_0^{(2)}),$$

$$(E_{fi}=0) \quad (5.2)$$

the order being again for $\mu=2, 1, 0, -1$, and -2, respectively. The ratio of second-order to first-order effects is seen to be pure imaginary since the $S_\mu^{(2)}$ are real. The gamma-ray intensities depend on the probabilities of emission from the states μ, which in turn are proportional to $|c_\mu|^2$. According to Eq. (5.2), the

[5] K. A. Ter-Martirosyan, J. Exptl. Theoret. Phys. (U.S.S.R.) 22, 284 (1952).

[6] K. Alder and A. Winther, Phys. Rev. 91, 1578 (1953).

population of the excited levels is affected only to the second order of $c_\mu^{(2)}(\infty)/c_\mu^{(1)}(\infty)$ on account of the purely imaginary value of the ratio, and the effect on the angular distribution disappears in the order worked with here.

The consideration just given is incomplete on account of the presence of interference effects between γ rays emitted from different sublevels μ. A consideration of their effect shows that in the lowest order they appear in the combination

$$\left|-\mathbf{cs}S_0^{(2)}+\frac{1}{\sqrt{6}}\mathbf{cs}(S_2^{(2)}e^{2i\varphi}+S_{-2}^{(2)}e^{-2i\varphi})\right|^2 \quad (5.3)$$

with

$$\mathbf{c}=\cos\theta, \quad \mathbf{s}=\sin\theta, \quad (5.4)$$

where (θ,φ) are the polar angles of the direction of photon emission. From Eq. (4.4), it follows that

$$S_2^{(2)}=S_{-2}^{(2)}, \quad (E_{fi}=0), \quad (5.5)$$

and hence, according to Eq. (5.2) the effect of the second-order correction, may be represented by

$$S_2^{(2)}\rightarrow(1+i\alpha_2)S_2^{(2)}, \quad S_{-2}^{(2)}\rightarrow(1+i\alpha_2)S_{-2}^{(2)},$$
$$S_0^{(2)}\rightarrow(1+i\alpha_0)S_0^{(2)}, \quad (5.6)$$

and the $S_\mu^{(2)}$ as well as α_0, α_2 are real. The cross-product terms arising from (5.2) with the $S_\mu^{(2)}$ corrected for second order effects are seen to be

$$-(4/6^{\frac{1}{2}})\mathbf{c}^2\mathbf{s}^2$$
$$\times\mathrm{Re}\{S_2^{(2)*}S_0^{(2)}(1-i\alpha_2)(1+i\alpha_0)\cos2\varphi\}. \quad (5.7)$$

The α_0 and α_2 survive only in the combination $\alpha_0\alpha_2$ and the angular distribution effect is therefore of a higher order than the effects considered here.

For head-on collisions, which correspond to $\epsilon=1$, one has

$$S_2^{(2)}=-(3/2)^{\frac{1}{2}}S_0^{(2)}, \quad (\epsilon=1, E_{fi}=0), \quad (6)$$

so that

$$\tfrac{1}{2}S_0^{(2)}-(S_2^{(2)})^2/S_0^{(2)}=-S_0^{(2)}, \quad (\epsilon=1, E_{fi}=0). \quad (6.1)$$

According to Eq. (5.2), the $c_\mu^{(2)}/c_\mu^{(1)}$ are all changed in the same ratio and there can be no change in the γ-ray angular distributions emitted for these orbits even apart from the fact that the changes in the c_μ are 90° out of phase with the first order effects.

III. GAMMA ANGULAR DISTRIBUTIONS

The angular distribution of γ rays for quadrupole radiation can be obtained by employing the interaction energy with the transverse electromagnetic field in the $-\int\mathbf{j}\cdot\mathbf{A}d\mathbf{r}$ form, which makes the transition matrix element appear as

$$H''=C\int w_\mu^*\left(\frac{\mathbf{r}}{r}\cdot\mathbf{1}_e\right)e^{i\mathbf{k}\cdot\mathbf{r}}(dv_0/dr)d\mathbf{r}, \quad (7)$$

with C standing for a constant the value of which is immaterial for immediate purposes. The term $\mathbf{1}_e\cdot(\mathbf{r}/r)$ arises from the particle current, which contains $(\hbar/i)\nabla v_0$. Here $\mathbf{k}$ is the propagation vector of the photon while $\mathbf{1}_e$ is its polarization vector. In Eq. (7), effects of all multipoles are included. For the transition in question, there can be no electric or magnetic dipole effects on account of the ΔL selection rule. The electric quadrupole effect arises from the $3iP_1(\cos\theta)F_1(kr)/(kr)$ part of the expansion of $e^{i\mathbf{k}\cdot\mathbf{r}}$ and gives rise therefore to

$$\int Y_{2\mu}^*(\theta_p,\varphi_p)\left[\left(\frac{\mathbf{r}}{r}\right)_p\cdot\mathbf{1}_e\right]P_1\left(\frac{\mathbf{k}}{k}\cdot\left[\frac{\mathbf{r}}{r}\right]_p\right)d\Omega_p \quad (7.1)$$

as the angle-dependent factors in the matrix elements. Employment of relations between rates of change of matrices and their values replaces the dv_0/dr of Eq. (7) by v_0 giving rise to standard forms of quadrupole matrix elements with the same angular factors. The term in $P_3((\mathbf{k}/k)\cdot(\mathbf{r}/r))$ in the expansion of $e^{i\mathbf{k}\mathbf{r}}$ also gives a nonvanishing contribution, but since it is multiplied by $F_3(kr)/kr$ it corresponds to a 2^4 pole and will be omitted. It is convenient to introduce the polarization vectors with direction cosines as follows:

$$\mathbf{1}_a=(\mathbf{c}\cos\varphi, \mathbf{c}\sin\varphi, -\mathbf{s}),$$
$$\mathbf{1}_b=(-\sin\varphi, \cos\varphi, 0), \quad (7.2)$$

the first of which is in the plane through the z axis and the photon direction while the second is perpendicular to that plane. The direction cosines of the nuclear proton are

$$(\mathbf{r}/r)_p=(\sin\theta_p\cos\varphi_p, \sin\theta_p\sin\varphi_p, \cos\theta_p). \quad (7.3)$$

Substitution in Eq. (7.1) gives

$$\int Y_{2,\mu}\cdot\left(\left(\frac{\mathbf{r}}{r}\right)_p\cdot\mathbf{1}_a\right)P_1\left(\frac{\mathbf{k}}{k}\cdot\left[\frac{\mathbf{r}}{r}\right]_p\right)d\Omega_p$$
$$=\left(\frac{4\pi}{5}\right)^{\frac{1}{2}}\frac{1}{\sqrt{6}}\mathbf{cs}\exp(\mu i\varphi), \quad (\mu=2,-2) \quad (7.4)$$

$$\int Y_{2,0}\cdot\left(\left(\frac{\mathbf{r}}{r}\right)_p\cdot\mathbf{1}_a\right)P_1\left(\frac{\mathbf{k}}{k}\cdot\left[\frac{\mathbf{r}}{r}\right]_p\right)d\Omega_p$$
$$=-\left(\frac{4\pi}{5}\right)^{\frac{1}{2}}\mathbf{cs}, \quad (7.5)$$

$$\int Y_{2,\mu}\cdot\left(\left(\frac{\mathbf{r}}{r}\right)_p\cdot\mathbf{1}_b\right)P_1\left(\frac{\mathbf{k}}{k}\cdot\left[\frac{\mathbf{r}}{r}\right]_p\right)d\Omega_p$$
$$=i^{\mu/2}\left(\frac{4\pi}{5}\right)^{\frac{1}{2}}\frac{\mathbf{s}}{\sqrt{6}}\exp(\mu i\varphi), \quad (\mu=2,-2) \quad (7.6)$$

while the other matrix elements vanish. Since the emission of a γ ray leaves the nucleus always in the same state v_0, the γ rays emitted from the sublevels $\mu=2$, 0,

-2 are coherent[7] and one should add the amplitudes obtainable from Eqs. (7.4), (7.5), and (7.6) before taking the square of absolute values. Since the $S_\mu^{(2)}$ are real, it does not matter in the calculation of lowest order effects (i.e., on assumption of infinitesimal amplitude of excitation) whether one uses the quantities in (7.4) and (7.6) or their complex conjugates in the calculation of γ-emission amplitudes. For finite-amplitude effects, however, the $S_\mu^{(2)}$ are replaced in the c_μ by complex numbers and care must be taken therefore regarding having the gamma emission matrix element multiplied by the correct excitation amplitude. The whole process takes place in the order of transitions $i0\rightarrow f\mu\rightarrow i0$, the first arrow occurring by Coulomb excitation and the second by gamma emission. Since, according to (4.2), the $c_\mu^{(1)}$ contains $Y_{2\mu}^*$ under the integral, the element in (7.1) corresponds to absorption of a γ ray and the quantities listed in (7.4), (7.5), and (7.6) should be used multiplied by corresponding $S_\mu^{(2)}$ for the same μ. The relative amplitudes for emission, applicable also in the case of complex quantity replacements for the $S_\mu^{(2)}$, are

$$\begin{aligned} &\mathbf{c}^2\mathbf{s}^2|-S_0+(1/\sqrt{6})(S_2e^{2i\varphi}+S_{-2}e^{-2i\varphi})|^2 \\ &\quad+\tfrac{1}{6}\mathbf{s}^2|S_2e^{2i\varphi}-S_{-2}e^{-2i\varphi}|^2 \\ &=\mathbf{c}^2\mathbf{s}^2|S_0|^2+\tfrac{1}{6}\mathbf{s}^2(\mathbf{c}^2+1)(|S_2|^2+|S_{-2}|^2) \\ &\quad-(2/\sqrt{6})\mathbf{c}^2\mathbf{s}^2\,\mathrm{Re}S_0^*(S_2e^{2i\varphi}+S_{-2}e^{-2i\varphi}) \\ &\quad-\tfrac{1}{3}\mathbf{s}^4\,\mathrm{Re}(S_2S_{-2}^*e^{4i\varphi})\equiv(2/15)J, \end{aligned} \tag{7.7}$$

where the omission of the superscript (2) on $S_\mu^{(2)}$ means that it is replaced by its corrected value in such a way that substitution of S_μ for $S_\mu^{(2)}$ on the right side of Eq. (4.3) causes a change of $c_\mu^{(1)}$ on the left to $c_\mu^{(1)}+c_\mu^{(2)}$. The quantity J in Eq. (7.7) contains the angular correlation of the γ rays with the direction of the incident particle; the factor 2/15 turns out to be convenient in later expressions, but has no other significance.

In the calculation carried out above, the parametric representation of Eq. (4.1) has been used. This corresponds to the x axis directed along the major axis of the hyperbolic orbit. It is more convenient to have the results referred to axes such that the initial particle velocity is along the negative direction of the x axis. This is accomplished by the transformation

$$x'+iy'=(x+iy)e^{i\delta},\quad \delta=\tan^{-1}(\epsilon^2-1)^{\frac{1}{2}}, \tag{7.8}$$

where (x',y',z) are the new axes. The azimuthal angle of the γ ray with respect to the (x',y',z) axes will be called φ'. It is related to φ by

$$\varphi'=\varphi+\delta. \tag{8}$$

In the γ-ray intensity formulas one should substitute therefore $\varphi'-\delta$ for φ.

The relative intensities will first be obtained by averaging over rotations around x'. Such intensities are needed for an experiment in which no coincidences with inelastically scattered particles are used. The intensities so obtained can also be used for an experiment in which the inelastically scattered particles are counted in a cone corresponding to a given scattering angle $\pi-2\delta$ of the inelastically scattered particle. The latter type of experiment yields more information than the former since it gives the angular distribution corresponding to a given orbit eccentricity ϵ. In the evaluation of averages of Eq. (7.7) one needs the following averages over rotations of orbit planes around Ox'.

$$\begin{aligned} \langle(\mathbf{c}^2+1)\mathbf{s}^2\rangle&=(4/5)+(2/7)P_2-(3/35)P_4, \\ \langle\mathbf{c}^2\mathbf{s}^2\rangle&=(2/15)-(1/21)P_2-(3/35)P_4, \\ \langle\mathbf{c}^2\mathbf{s}^2\exp(2i\varphi')\rangle&=(1/7)(P_2-P_4), \\ \langle\mathbf{s}^4\exp(4i\varphi')\rangle&=P_4. \end{aligned} \tag{8.1}$$

The argument of the Legendre function P_L is $\cos\theta'$, with θ' standing for the angle between the photon propagation vector and the negative of the incident beam direction. The averaging is readily performed by introducing the coordinate system of the orbit plane, transforming to a coordinate system fixed with respect to the laboratory, and working in Cartesian coordinates until one is ready to introduce the angle of rotation of the orbit plane by a simple transformation for z and y'. Substitution of these averages in (7.7) gives

$$\begin{aligned} \langle J\rangle=\mathrm{Re}\{&|S_2|^2+|S_0|^2+|S_{-2}|^2 \\ &+(5/14)[|S_2|^2-|S_0|^2+|S_{-2}|^2 \\ &-(\sqrt{6})S_0^*(S_2e^{-2i\delta}+S_{-2}e^{2i\delta})]P_2 \\ &-(3/28)[|S_2|^2+6|S_0|^2+|S_{-2}|^2 \\ &-(10/3)(\sqrt{6})S_0^*(S_2e^{-2i\delta}+S_{-2}e^{2i\delta}) \\ &+(70/3)S_2S_{-2}^*e^{-4i\delta}]P_4\}. \end{aligned} \tag{8.2}$$

The quantity $\langle J\rangle$ consists of two contributions, the first of which corresponds to replacing the S_μ by their first-order values $S_\mu^{(2)}$. This part will be referred to as $\langle J^{(1)}\rangle$, the superscript on J referring to the order of the calculation rather than to the order of the multipole. The second part consists of terms one order higher than $\langle J^{(1)}\rangle$ and is obtained by collecting cross-product terms of the first-order terms and the correction terms to the $S_\mu^{(2)}$. The contribution to $\langle J\rangle$ due to these terms will be called $\langle J^{(2)}\rangle$. Thus

$$\langle J\rangle=\langle J^{(1)}\rangle+\langle J^{(2)}\rangle+\cdots. \tag{8.3}$$

One finds from Eq. (8.2)

$$\begin{aligned} \langle J^{(2)}\rangle=&-(5/14)(\sqrt{6})[S_0^{(2)}(S_2^i-S_{-2}^i) \\ &-S_0^i(S_2^{(2)}-S_{-2}^{(2)})](P_2-P_4)\sin2\delta \\ &-(5/2)(S_2^iS_{-2}^{(2)}-S_2^{(2)}S_{-2}^i) \\ &\times P_4\sin4\delta+\delta_r\langle J^{(1)}\rangle, \end{aligned} \tag{8.4}$$

where δ_r means the change caused by changing the $S_\mu^{(2)}$ in $J^{(1)}$ to $\mathrm{Re}\{S_\mu\}$, and where

$$S_\mu^i=\mathrm{Im}S_\mu, \tag{8.4'}$$

[7] See, for example, G. Breit, Revs. Modern Phys. 5, 91 (1933), Part VII.

while

$$\sin 2\delta = 2(\epsilon^2-1)^{\frac{1}{2}}/\epsilon^2, \quad \sin 4\delta = 4(2-\epsilon^2)(\epsilon^2-1)^{\frac{1}{2}}/\epsilon^4. \tag{8.5}$$

When one employs the last equation, the first-order contribution is

$$\langle J^{(1)}\rangle = (S_2^{(2)})^2 + (S_0^{(2)})^2 + (S_{-2}^{(2)})^2 + \frac{5}{14}\left\{(S_2^{(2)})^2 - (S_0^{(2)})^2 + (S_{-2}^{(2)})^2 + (\surd 6)\left(1-\frac{2}{\epsilon^2}\right)S_0^{(2)}(S_2^{(2)}+S_{-2}^{(2)})\right\}P_2 - \frac{3}{28}\left[(S_2^{(2)})^2 + 6(S_0^{(2)})^2 + (S_{-2}^{(2)})^2 + \frac{10}{3}6^{\frac{1}{2}}\left(1-\frac{2}{\epsilon^2}\right)S_0^{(2)}(S_2^{(2)}+S_{-2}^{(2)}) + \frac{70}{3}\left(1-\frac{8}{\epsilon^2}+\frac{8}{\epsilon^4}\right)S_2^{(2)}S_{-2}^{(2)}\right]P_4, \tag{9}$$

in agreement with Alder and Winther[6] provided the sign of the term in P_4 is changed as noted by Breit, Ebel, and Benedict.[8] In this comparison the explicit values of Alder and Winther's $B_2 = 5/14$, $B_4 = 8/7$ applying to the present $0\rightarrow 2$ case will be found helpful. In the evaluation of $c_\mu^{(2)}$, it is convenient to make use of the symmetry of various parts of the integrand in reducing the region $t>t'$ to the region $t>t'>0$. Introducing the abbreviations

$$P_0(w) = \frac{1}{(1+\epsilon\cosh w)^2}, \quad P_c(w) = P_0(w)\cos 2\varphi, \quad P_s(w) = P_0(w)\sin 2\varphi, \tag{9.1}$$

$$Q_0(w) = \int_0^w P_0(w)dw, \quad Q_c(w) = \int_0^w P_c(w)dw, \quad Q_s(w) = \int_0^w P_s(w)dw, \tag{9.2}$$

$$R_0 = Q_0(\infty), \quad R_c = Q_c(\infty), \quad R_s = Q_s(\infty), \tag{9.3}$$

$$T_0 = \int_0^\infty P_0(w)\cos\omega t dw, \quad T_c = \int_0^\infty P_c(w)\cos\omega t dw, \quad T_s = \int_0^\infty P_s(w)\sin\omega t dw, \tag{9.4}$$

and defining

$$A = \int_0^\infty (P_0Q_c + P_cQ_0)\sin\omega t dw, \quad B = \int_0^\infty (P_0Q_s + P_sQ_0)\cos\omega t dw, \quad C = \int_0^\infty (-P_0Q_0 + 3P_cQ_c + 3P_sQ_s)\sin\omega t dw, \tag{9.5}$$

one finds

$$S_0^{(2)} = 2T_0, \quad \tfrac{1}{2}(S_2^{(2)}+S_{-2}^{(2)}) = -(\surd 6)T_c, \quad \tfrac{1}{2}(S_{-2}^{(2)}-S_2^{(2)}) = (\surd 6)T_s, \tag{9.6}$$

while

$$S_0 = 2\mathcal{T}_0, \quad \tfrac{1}{2}(S_2+S_{-2}) = -(\surd 6)\mathcal{T}_c, \quad \tfrac{1}{2}(S_{-2}-S_2) = (\surd 6)\mathcal{T}_s, \tag{9.7}$$

$$S_2 = -(\surd 6)(\mathcal{T}_c+\mathcal{T}_s), \quad S_{-2} = (\surd 6)(\mathcal{T}_s-\mathcal{T}_c) \tag{9.7'}$$

with

$$\mathcal{T}_0 = T_0 + \lambda\{3R_sT_s + i(R_0T_0 - 3R_cT_c) - C\}, \quad \mathcal{T}_c = T_c + \lambda\{-i(R_cT_0 + R_0T_c) - A\}, \quad \mathcal{T}_s = T_s + \lambda\{-R_sT_0 - iR_0T_s + B\} \tag{9.8}$$

and

$$\lambda = Z_1e^2\langle r^2\rangle_{ff}/[7\hbar a'^2 v]. \tag{9.9}$$

In terms of these quantities, one finds on substitution into Eq. (8.2)

$$\langle J\rangle = 4\,\mathrm{Re}\{D+\delta D + (5/14)(N_2+\delta N_2)P_2 + (8/7)(N_4+\delta N_4)P_4\}, \tag{10}$$

with

$$D+\delta D = \mathrm{Re}\{3|\mathcal{T}_c|^2 + 3|\mathcal{T}_s|^2 + |\mathcal{T}_0|^2\}, \tag{10.1}$$

$$N_2+\delta N_2 = \mathrm{Re}\{3|\mathcal{T}_c|^2 + 3|\mathcal{T}_s|^2 - |\mathcal{T}_0|^2 + 6\mathcal{T}_0^*(\mathcal{T}_c\cos 2\delta - i\mathcal{T}_s\sin 2\delta)\}, \tag{10.2}$$

$$N_4+\delta N_4 = -\tfrac{3}{32}\,\mathrm{Re}\{3|\mathcal{T}_c|^2 + 3|\mathcal{T}_s|^2 + 6|\mathcal{T}_0|^2 + 20\mathcal{T}_0^*(\mathcal{T}_c\cos 2\delta - i\mathcal{T}_s\sin 2\delta) - 35[|\mathcal{T}_s|^2 - |\mathcal{T}_c|^2 + 2i\,\mathrm{Im}(\mathcal{T}_c\mathcal{T}_s^*)]e^{-4i\delta}\}. \tag{10.3}$$

Here D, N_2, N_4 correspond to $J^{(1)}$ and δD, δN_2, δN_4 to $J^{(2)}$ of Eq. (8.4). Explicitly

$$D = 3T_c^2 + 3T_s^2 + T_0^2, \tag{10.4}$$

$$N_2 = 3T_c^2 + 3T_s^2 - T_0^2 + 6T_0T_c\cos 2\delta, \quad \cos 2\delta = (2/\epsilon^2) - 1, \tag{10.5}$$

$$N_4 = -\tfrac{9}{32}(T_c^2+T_s^2) - \tfrac{9}{16}T_0^2 - (15/8)T_0T_c\cos 2\delta + (105/32)(T_s^2 - T_c^2)\cos 4\delta. \tag{10.6}$$

[8] Breit, Ebel, and Benedict, Phys. Rev. **100**, 429 (1955); see also Breit, Ebel, and Russell, Phys. Rev. **101**, 1504 (1956).

The parts corresponding to $J^{(2)}$ are

$$\delta D=\lambda[-6AT_c+6BT_s-2CT_0], \tag{10.7}$$

$$\begin{aligned}\delta N_2=\lambda\{&-6A[T_c+T_0\cos 2\delta]\\&-12R_sT_0T_s+18R_sT_cT_s\cos 2\delta\\&+[-12R_0T_0T_s+18R_cT_cT_s]\sin 2\delta\\&+6BT_s+C[2T_0-6T_c\cos 2\delta]\},\end{aligned} \tag{10.8}$$

$$\begin{aligned}\delta N_4=\lambda\Big\{&A[\tfrac{9}{16}T_c+(15/8)T_0\cos 2\delta+(105/16)T_c\cos 4\delta]\\&+B[-\tfrac{9}{16}T_s+(105/16)T_s\cos 4\delta]\\&+C[(9/8)T_0+(15/8)T_c\cos 2\delta]-(45/16)R_sT_0T_s\\&-(45/8)R_sT_cT_s\cos 2\delta+\frac{15}{4}\frac{(\epsilon^2-1)^{\frac{1}{2}}}{\epsilon^2}\\&\times[(-3R_cT_c+2R_0T_0)T_s-7R_cT_0T_s\cos 2\delta]\\&-(105/16)R_sT_0T_s\cos 4\delta\Big\}.\end{aligned} \tag{10.9}$$

For $\epsilon=1$, one has $\Phi=0$ and according to Eqs. (9.1)$\cdots$(9.5), $P_s=Q_s=R_s=T_s=0$, $P_c=P_0$, $Q_c=Q_0$, $R_c=R_0$, $T_c=T_0$, $B=0$, $A=C$. Consequently, in Eq. (9.8) one has

$$\mathcal{T}_0=T_0-\lambda(A+2iR_0T_0)=\mathcal{T}_c,\ \mathcal{T}_s=0,\ \delta=0,\ (\epsilon=1) \tag{11}$$

and hence

$$\langle J\rangle=4|\mathcal{T}_0|^2[4+(20/7)P_2-(48/7)P_4],\quad(\epsilon=1). \tag{11.1}$$

Since $|\mathcal{T}_0|^2$ enters as a common factor, there is no effect on the γ-angular distribution coefficients usually called $a_2(\xi)$ and $a_4(\xi)$. In this case $(\epsilon=1)$, the second-order effects are not detectable by an ordinary γ-angular distribution measurement although there is an effect on the absolute value entering through $|\mathcal{T}_0|^2$. A special case of this relationship has already been pointed out for $\epsilon=1$, $E_{fi}=0$ in connection with Eq. (6.1). On account of the vanishing of second-order effect in the γ-angular distribution coefficients $a_{2\epsilon}(\xi)$ and $a_{4\epsilon}(\xi)$ for $\epsilon=1$, the nearly head-on collisions are also especially unfavorable for the detection of the reorientation effects. This circumstance makes the reorientation effect smaller than might have been expected, because for head-on collisions the excitation effects are largest.

An interesting consequence of the linear variation of $\langle J^{(2)}\rangle$ with ξ is that the reorientation effect is, in first approximation, independent of bombarding energy. This can be seen from Eqs. (4.1′), (4.6), and (9.9) since the product of λ and ξ is

$$\lambda\xi\cong\frac{Z_1e^2}{\hbar va'^2}\frac{\langle r^2\rangle_{ff}}{7}\frac{\omega_{fi}a'}{v}=\frac{M\omega_{fi}\langle r^2\rangle_{ff}}{7Z_2\hbar}$$

and since in the range of ξ for which the linear approximation holds the reorientation effect depends on ξ only through $\lambda\xi$.

Since according to Eq. (10) the spherically symmetric part of $\langle J\rangle$ is represented by $D+\delta D$, the angular distribution coefficients are obtainable as

$$a_s(\xi)=(N_s+\delta N_s)/(D+\delta D)=N_s/D+\delta N_s/D-N_s(\delta D)/D^2,\quad(s=2,4). \tag{12}$$

According to Eq. (9.8), the number of excitations for $\epsilon=1$ is proportional to

$$|\mathcal{T}_0|^2=T_0^2\left[1-(4\lambda/T_0)\int_0^\infty P_0Q_0(\sin\omega t)dw\right],\quad(\epsilon=1). \tag{13}$$

The integral in this formula can be expressed as follows:

$$A/2=\int_0^\infty P_0Q_0(\sin\omega t)dw=\frac{4}{3}\int_1^\infty\frac{u}{(1+u)^4}\left(1-\frac{2(1+3u)}{(1+u)^3}\right)(\sin\omega t)du \tag{13.1}$$

with

$$\omega t=\xi\left[\ln u+\frac{1}{2}\left(u-\frac{1}{u}\right)\right]. \tag{13.2}$$

In terms of it one finds

$$\left(\frac{D+\delta D}{D}\right)_{\epsilon=1}=1-\frac{2A}{T_0}=1-4\lambda\frac{\int_0^\infty P_0Q_0\sin\omega t\,dw}{\int_0^\infty P_0\cos\omega t\,dw}. \tag{13.3}$$

By means of Eq. (13.1), this can be expressed as

$$\left(\frac{D+\delta D}{D}\right)_{\epsilon=1}=1-\frac{4\lambda}{3}\left\{\int_1^\infty\frac{u}{(1+u)^4}\left(1-\frac{2(1+3u)}{(1+u)^3}\right)\times(\sin\omega t)du\right\}\Big/\int_1^\infty\frac{u\cos\omega t}{(1+u)^4}du. \tag{13.4}$$

For 5-Mev protons on $_{78}$Pt, with an assumed $Q=7\times10^{-24}$ cm^2, $\Delta E=330$ kev, one has $\xi=0.1915$, $\lambda=-0.100$, and the effect on the collision for inelastic scattering may be estimated from (13.3) to be $\pm3.8\%$. Here the quadrupole moment Q is connected with the radial matrix element $\langle r^2\rangle_{ff}$ by the equation $|Q|=(4/7)\langle r^2\rangle_{ff}$.

By means of Eq. (12) combined with Eqs. (10.4), (10.5), (10.6), (10.7), (10.8), and (10.9), one obtains the change in the angular distribution of γ rays for a fixed value of ϵ but averaged over rotations of orbit planes around the direction of the incident beam. The change in the values of $a_2(\xi)$ and $a_4(\xi)$ is obtained by integration of the γ intensity in any direction over ϵ, taking the number of collisions as $2\pi v p dp = 2\pi v a'^2 \epsilon d\epsilon$. The formulas described so far suffice therefore for the calculation of γ-angular distributions in experiments on the angular distributions without coincidences and also with coincidences in which the inelastically scattered particles are collected in a cone with symmetry axis along the incident beam. Some information is lost in this type of experiment because of the averaging over orbit plane orientations.

This information is retained if the orbit plane is defined by the observation of the inelastically scattered particle and the direction of the γ ray is observed in coincidence. The angular distribution under these conditions is obtainable from Eq. (7.7) which gives J before this quantity is averaged over directions of orbit planes. Expressing the S_μ in terms of $\mathcal{T}_0$, $\mathcal{T}_c$, $\mathcal{T}_s$ by means of Eqs. (9.6) and (9.7), one obtains

$$\begin{aligned}(2/15)J = {} & 4\mathbf{c}^2\mathbf{s}^2|\mathcal{T}_0|^2 + 2\mathbf{s}^2(\mathbf{c}^2+1)[|\mathcal{T}_c|^2+|\mathcal{T}_s|^2] \\ & + 8\mathbf{c}^2\mathbf{s}^2 \operatorname{Re}\{\mathcal{T}_0^*[\mathcal{T}_c \cos 2\varphi + i\mathcal{T}_s \sin 2\varphi]\} \\ & + 2\mathbf{s}^4 \operatorname{Re}\{[|\mathcal{T}_s|^2 - |\mathcal{T}_c|^2 + 2i \operatorname{Im}(\mathcal{T}_s^*\mathcal{T}_c)]e^{4i\varphi}\}, \quad (14)\end{aligned}$$

and the division into first- and second-order parts is obtainable from Eq. (9.8). Averaging the first line over all directions and taking account of the factor 4 of Eq. (10) in the definition of D, one obtains $D+\delta D$ in agreement with Eq. (10.1).

The remainder of the first line contributes to $J/4$ the quantity

$$\begin{aligned}(5/7)[|\mathcal{T}_0|^2 - 3|\mathcal{T}_c|^2 - 3|\mathcal{T}_s|^2]P_2 \\ - (6/7)[|\mathcal{T}_c|^2 + |\mathcal{T}_s|^2 + 2|\mathcal{T}_0|^2]P_4. \quad (14.1)\end{aligned}$$

This quantity is related to but is not the same as the θ-dependent but δ-independent part of Eq. (10). The reason for the difference is that in the present consideration the directions of orbit planes have not been averaged over. *It should be observed that the argument of P_2 and P_4 in Eq. (14.1) is $\cos\theta$, with θ standing for the angle between the γ ray and the normal to the orbit plane while through all of the formulas for $\langle J \rangle$ the argument of P_2 and P_4 is the cosine of the angle with the incident wave.* The analysis of the angular distribution in the coincidence experiment is in principle capable of giving the coefficients of $\cos 2\varphi$, $\sin 2\varphi$, $\cos 4\varphi$, and $\sin 4\varphi$ in Eq. (14). It should be possible to verify the dependence of these coefficients on θ which appears in this equation. By doing so, the interpretation of the experiment in terms of the mechanism described in this paper would be more certain than in the axially symmetric type of coincidence experiment discussed in relation to $\langle J^{(2)} \rangle$. It is probably important to have such a verification in order to be sure that second-order effects caused by transitions to other levels have been sufficiently corrected for, and it may be helpful to have such a verification as a check on the experimental procedure. It would appear that for these reasons the coincidence experiment defining the orbit should be more informative than the axially symmetric type of experiment or the observation of the γ-intensity distribution without coincidences. A further reason for believing that the orbit-defining experiment is preferable is that some of the angle-independent quantities entering the coefficients of $\sin 2\varphi, \cdots \cos 4\varphi$ can be obtained in more than one way in this arrangement, providing an additional check as will be discussed presently. Before doing so, it should be pointed out, however, that the validity of the semiclassical theory has never been tested to the degree implied in the coincidence experiments. Since in the orbit-defining geometry lack of large diffuseness resulting from diffraction effects is presupposed both in the orbit plane and in a direction perpendicular to it, one may expect this circumstance to be more serious in the orbit-defining arrangement. The expressions derived in the present paper are applicable only approximately on account of the semiclassical nature of the treatment. The errors due to this cause are probably smaller, however, for bombardment by heavy ions such as that by N^{14} than for light particle projectiles such as protons.

In the coefficient of $\cos 2\varphi$, one has available

$$\begin{aligned}C_4 &\equiv \operatorname{Re}(\mathcal{T}_0^*\mathcal{T}_c) \\ &= T_0T_c + \lambda[-AT_0 + (3R_sT_s - C)T_c]; \quad (14.2)\end{aligned}$$

in that of $\sin 2\varphi$ there is present

$$C_5 \equiv \operatorname{Im}(\mathcal{T}_0^*\mathcal{T}_s) = \lambda(3R_cT_c - 2R_0T_0)T_s. \quad (14.3)$$

Similarly there is available in the coefficient of $\sin 4\varphi$

$$C_6 \equiv \operatorname{Im}(\mathcal{T}_s^*\mathcal{T}_c) = -\lambda R_cT_0T_s \quad (14.4)$$

and in the coefficient of $\cos 4\varphi$

$$\begin{aligned}C_7 &\equiv \operatorname{Re}[|\mathcal{T}_s|^2 - |\mathcal{T}_c|^2] \\ &= T_s^2 - T_c^2 + 2\lambda[AT_c + (B - R_sT_0)T_s]. \quad (14.5)\end{aligned}$$

Another way of stating the possibilities of the orbit-defining coincidence experiment is to observe that the coefficients of P_0, P_2, and P_4 of the φ independent parts of J give the combinations

$$\begin{aligned}C_1 &\equiv |\mathcal{T}_0|^2 + 3|\mathcal{T}_c|^2 + 3|\mathcal{T}_s|^2, \\ C_2 &\equiv |\mathcal{T}_0|^2 - 3|\mathcal{T}_c|^2 - 3|\mathcal{T}_s|^2, \quad (15) \\ C_3 &\equiv 2|\mathcal{T}_0|^2 + |\mathcal{T}_c|^2 + |\mathcal{T}_s|^2,\end{aligned}$$

and that through these quantities there is available $|\mathcal{T}_0|^2$, $|\mathcal{T}_c|^2 + |\mathcal{T}_s|^2$ with a check on experimental values through $7C_1 + 5C_2 = 6C_3$. The quantities

$$\Delta|\mathcal{T}_0|^2 = 2\lambda T_0(C - 3R_sT_s), \quad \Delta|\mathcal{T}_s|^2 + \Delta|\mathcal{T}_c|^2,$$

with

$$\Delta|\mathcal{T}_s|^2 = 2\lambda T_s(B - R_sT_0), \quad \Delta|\mathcal{T}_c|^2 = -2\lambda AT_c, \quad (15.1)$$

are thus available, and from the coefficient of $\cos 4\varphi$ there is available $\Delta|\mathcal{T}_s|^2 - \Delta|\mathcal{T}_c|^2$ so that $\Delta|\mathcal{T}_s|^2$ and $\Delta|\mathcal{T}_c|^2$ can both be obtained from experiment.

TABLE I. Values of ϵ and δ used for numerical computation.

ϵ	δ	ϵ	δ	ϵ	δ
1.000	0°	1.250	36.9°	2.000	60°
1.015	10°	1.414	45°	2.500	66.4°
1.064	20°	1.550	49.8°	3.000	70.5°
1.100	24.6°	1.700	54.0°	3.864	75°
1.155	30°				

The angles used in the description of the orbit-defining experiment are as in Fig. 1.

As has been mentioned in the discussion immediately preceding Eq. (14.2), the reorientation effect has to be separated from other second-order effects such as the change in the angular distribution caused by transitions to other levels. It may be pointed out, however, that in the usual 0, 2, 4 sequence of levels a quadrupole transition 0→4 is forbidden, and hence the populations of sublevels of $I=2$ are not affected by the 0→4→2 sequence of excitations if higher multipole effects are neglected. There may be effects caused by 0→2→2 sequences, but the intermediate level has to belong to another configuration of nucleons such as would be obtained by changing the vibrational quantum number. It may be expected to lie at a higher energy than the $I=4$ state and to be consequently less important in its effect on the γ-ray angular distribution; the transition quadrupole moments to states involving a change of vibrational quantum number are presumably also somewhat smaller than those in the normal 0→2→4 sequence.

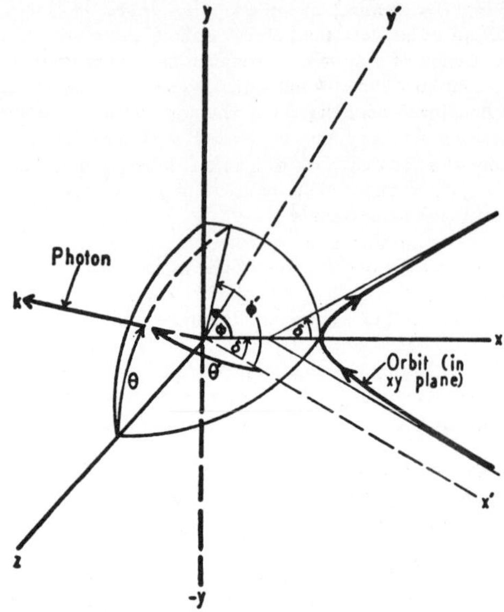

FIG. 1. Diagram of coordinate systems used, showing particle trajectory and direction of emitted photon.

It appears desirable to mention that even though the regular sequence of excitations 0→2→4 gives effects in populating the $I=4$ level, the associated depletion of the $I=2$ level does not produce an effect on the

TABLE II. Values of $\mathcal{T}_0$, $\mathcal{T}_c$, and $\mathcal{T}_s$ as a function of ξ and ϵ.

ξ	ϵ	$10\mathcal{T}_0$	$10\mathcal{T}_c$	$10\mathcal{T}_s$
0.0000	1.000	$3.33-(4.06\xi+2.22i)\lambda$	$3.33-(4.06\xi+2.22i)\lambda$	0
	1.015	$3.27-(3.65\xi+2.06i)\lambda$	$3.23-(3.86\xi+2.12i)\lambda$	$1.70\xi+(0-0.56i\xi)\lambda$
	1.064	$3.09-(2.56\xi+1.64i)\lambda$	$2.94-(3.30\xi+1.82i)\lambda$	$3.16\xi+(0-0.98i\xi)\lambda$
	1.155	$2.79-(1.23\xi+1.09i)\lambda$	$2.50-(2.51\xi+1.40i)\lambda$	$4.18\xi+(0-1.17i\xi)\lambda$
	1.414	$2.15+(0.27\xi-0.373i)\lambda$	$1.67-(1.25\xi+0.715i)\lambda$	$4.63\xi+(0-0.99i\xi)\lambda$
	2.000	$1.32+(0.549\xi-0.035i)\lambda$	$0.83-(0.372\xi+0.220i)\lambda$	$3.75\xi+(0-0.491i\xi)\lambda$
	3.864	$0.47+(0.158\xi+0.0068i)\lambda$	$0.22-(0.034\xi+0.0208i)\lambda$	$1.95\xi+(0-0.091i\xi)\lambda$
0.1915	1.000	$2.81-(0.534+1.88i)\lambda$	$2.81-(0.534+1.88i)\lambda$	0
	1.015	$2.76-(0.481+1.75i)\lambda$	$2.74-(0.508+1.79i)\lambda$	$0.21-(0.033+0.069i)\lambda$
	1.064	$2.59-(0.346+1.42i)\lambda$	$2.51-(0.435+1.54i)\lambda$	$0.39-(0.057+0.121i)\lambda$
	1.155	$2.32-(0.179+0.974i)\lambda$	$2.16-(0.333+1.18i)\lambda$	$0.52-(0.068+0.144i)\lambda$
	1.414	$1.74+(0.007-0.375i)\lambda$	$1.50-(0.171+0.610i)\lambda$	$0.56-(0.057+0.120i)\lambda$
	2.000	$1.00+(0.046-0.069i)\lambda$	$0.80-(0.054+0.189i)\lambda$	$0.43-(0.028+0.057i)\lambda$
	3.864	$0.28+(0.0101-0.0033i)\lambda$	$0.24-(0.0063+0.0175i)\lambda$	$0.19-(0.0050+0.0086i)\lambda$
0.4028	1.000	$1.97-(0.673+1.31i)\lambda$	$1.97-(0.673+1.31i)\lambda$	0
	1.015	$1.92-(0.610+1.23i)\lambda$	$1.92-(0.641+1.25i)\lambda$	$0.25-(0.064+0.083i)\lambda$
	1.064	$1.79-(0.445+1.02i)\lambda$	$1.78-(0.552+1.08i)\lambda$	$0.47-(0.111+0.145i)\lambda$
	1.155	$1.58-(0.243+0.732i)\lambda$	$1.56-(0.425+0.832i)\lambda$	$0.61-(0.131+0.171i)\lambda$
	1.414	$1.13-(0.016+0.320i)\lambda$	$1.12-(0.220+0.430i)\lambda$	$0.65-(0.107+0.139i)\lambda$
	2.000	$0.58+(0.037-0.078i)\lambda$	$0.62-(0.070+0.130i)\lambda$	$0.47-(0.048+0.061i)\lambda$
	3.864	$0.11+(0.0061-0.0056i)\lambda$	$0.16-(0.0069+0.0101i)\lambda$	$0.15-(0.0062+0.0069i)\lambda$
0.6882	1.000	$1.07-(0.546+0.716i)\lambda$	$1.07-(0.546+0.716i)\lambda$	0
	1.015	$1.05-(0.497+0.680i)\lambda$	$1.05-(0.521+0.684i)\lambda$	$0.19-(0.069+0.064i)\lambda$
	1.064	$0.97-(0.369+0.580i)\lambda$	$0.99-(0.450+0.592i)\lambda$	$0.36-(0.120+0.111i)\lambda$
	1.155	$0.84-(0.210+0.436i)\lambda$	$0.89-(0.348+0.458i)\lambda$	$0.46-(0.139+0.129i)\lambda$
	1.414	$0.56-(0.030+0.209i)\lambda$	$0.66-(0.181+0.235i)\lambda$	$0.47-(0.108+0.102i)\lambda$
	2.000	$0.25+(0.016-0.055i)\lambda$	$0.35-(0.055+0.067i)\lambda$	$0.30-(0.044+0.040i)\lambda$
	3.864	$0.03+(0.0019-0.0029i)\lambda$	$0.06-(0.0038+0.0037i)\lambda$	$0.06-(0.0035+0.0029i)\lambda$

TABLE III. Values of the coefficients, $C_1 \cdots C_7$, of the angular dependent terms available in a coincidence type experiment which defines the eccentricity and the plane of the orbit of the incident particle, in coincidence with the γ ray. The quantities C_i are defined in Eqs. (15), (14.2), (14.3), (14.4), (14.5).

ξ	ϵ	$100\,C_1$	$100\,C_2$	$100\,C_3$	$100\,C_4$	$100\,C_5$	$100\,C_6$	$100\,C_7$
0.0000	1.000	$44.4 - 108\xi\lambda$	$-22.2 + 54\xi\lambda$	$33.3 - 81\xi$	$11.11 - 27.1\xi\lambda$	0	0	$-11.11 + 27.1\xi\lambda$
	1.064	$35.5 - 74\xi\lambda$	$-16.4 + 42\xi\lambda$	$27.8 - 51\xi\lambda$	$9.10 - 17.7\xi\lambda$	$2.2\xi\lambda$	$-2.9\xi\lambda$	$-8.66 + 19.4\xi\lambda$
	1.414	$12.9 - 11.4\xi\lambda$	$-3.7 + 13.7\xi\lambda$	$12.0 - 1.88\xi\lambda$	$3.58 - 2.2\xi\lambda$	$-0.4\xi\lambda$	$-1.7\xi\lambda$	$-2.78 + 4.2\xi\lambda$
	2.000	$3.82 - 0.41\xi\lambda$	$-0.35 + 3.31\xi\lambda$	$4.17 + 2.28\xi\lambda$	$1.10 - 0.03\xi\lambda$	$-0.52\xi\lambda$	$-0.41\xi\lambda$	$-0.69 + 0.62\xi\lambda$
0.1915	1.000	$31.7 - 12.0\lambda$	$-15.8 + 6.0\lambda$	$23.8 - 9.0\lambda$	$7.92 - 3.00\lambda$	0	0	$-7.92 + 3.00\lambda$
	1.064	$26.1 - 8.5\lambda$	$-12.7 + 4.9\lambda$	$19.9 - 5.8\lambda$	$6.51 - 2.00\lambda$	0.24λ	-0.30λ	$-6.15 + 2.14\lambda$
	1.414	$10.7 - 1.70\lambda$	$-4.6 + 1.75\lambda$	$8.6 - 0.52\lambda$	$2.60 - 0.29\lambda$	0.002λ	-0.16λ	$-1.92 + 0.45\lambda$
	2.000	$3.47 - 0.24\lambda$	$-1.48 + 0.43\lambda$	$2.81 + 0.07\lambda$	$0.80 - 0.02\lambda$	-0.03λ	-0.04λ	$-0.45 + 0.06\lambda$
0.4028	1.000	$15.5 - 10.6\lambda$	$-7.7 + 5.3\lambda$	$11.6 - 7.9\lambda$	$3.86 - 2.65\lambda$	0	0	$-3.86 + 2.65\lambda$
	1.064	$13.4 - 7.8\lambda$	$-7.0 + 4.6\lambda$	$9.8 - 5.3\lambda$	$3.19 - 1.78\lambda$	0.22λ	-0.25λ	$-2.96 + 1.86\lambda$
	1.414	$6.3 - 1.9\lambda$	$-3.8 + 1.9\lambda$	$4.2 - 0.7\lambda$	$1.27 - 0.27\lambda$	0.05λ	-0.12λ	$-0.84 + 0.36\lambda$
	2.000	$2.14 - 0.35\lambda$	$-1.46 + 0.44\lambda$	$1.28 - 0.05\lambda$	$0.362 - 0.018\lambda$	0.000λ	-0.023λ	$-0.167 + 0.042\lambda$
0.6882	1.000	$4.62 - 4.70\lambda$	$-2.31 + 2.35\lambda$	$3.46 - 3.52\lambda$	$1.15 - 1.17\lambda$	0	0	$-1.15 + 1.17\lambda$
	1.064	$4.29 - 3.66\lambda$	$-2.42 + 2.23\lambda$	$2.98 - 2.41\lambda$	$0.96 - 0.80\lambda$	0.10λ	-0.10λ	$-0.86 + 0.81\lambda$
	1.414	$2.29 - 1.06\lambda$	$-1.66 + 0.99\lambda$	$1.29 - 0.41\lambda$	$0.37 - 0.12\lambda$	0.042λ	-0.044λ	$-0.21 + 0.136\lambda$
	2.000	$0.71 - 0.188\lambda$	$-0.59 + 0.204\lambda$	$0.34 - 0.049\lambda$	$0.088 - 0.008\lambda$	0.007λ	-0.006λ	$-0.032 + 0.012\lambda$

angular distribution of γ rays to within the order considered here. A similar effect has been noted in connection with the contribution of the term containing $c_0^{(1)}$ in Eq. (2.2) which turned out to be of a higher order.

IV. ESTIMATES AND VALUES OF INTEGRALS

Calculations for the three different types of observations discussed in Sec. III have been performed for the following values of $\xi = \eta_f - \eta_i$:

$\xi = 0$ (no excitation);
$\xi = 0.1915$ (5-Mev protons on Pt^{194}, 330-kev level);
$\xi = 0.4028$ (3.3-Mev protons on Cd^{114}, 555-kev level);
$\xi = 0.6882$ (2.4-Mev protons on Cd^{114}, 555-kev level).

The values of orbit eccentricity ϵ and the corresponding values of δ given by Eq. (7.8) used in the computations of the single and double integrals in Eqs. (9.1) to (9.5) are given in Table I. Computation of all integrals was performed in the form given in Eqs. (9.1) to (9.5), using Simpson's rule with w as the independent variable. The results for $\mathcal{T}_0$, $\mathcal{T}_c$, and $\mathcal{T}_s$, defined in Eqs. (9.6) to (9.8), are given in Table II for representative values of ϵ.

The quantities available from the coincidence type experiment discussed in the previous section are given in Eqs. (14.2)···(14.5), (15). In order to make their comparison with theory possible, the values were calculated making use of the numbers in Table II. The results are listed in Table III for a few of the values of ϵ in Table II.[9]

The limited-type coincidence experiment also discussed in the previous section, in which the eccentricity, but not the plane of the incident orbit is observed in coincidence with the γ rays, yields the quantities $D+\delta D$, $N_2+\delta N_2$, and $N_4+\delta N_4$ of Eqs. (10.1), ···(10.9) which are listed in Table IV for typical values of ϵ, so as to facilitate comparison with experiment.[9]

If only the angular correlation of the γ rays with respect to the incident beam direction is measured, one obtains only the quantities

$$\int_1^\infty \epsilon d\epsilon (D+\delta D,\, N_2+\delta N_2,\, N_4+\delta N_4).$$

TABLE IV. Values of the coefficients of the angular dependent terms available in a limited coincidence experiment which defines only the eccentricity of the orbit of the incident particle in coincidence with the γ ray.

ξ	ϵ	$100(D+\delta D)$	$100(N_2+\delta N_2)$	$100(N_4+\delta N_4)$
0.0000	1.000	$44.4 - 108\xi\lambda$	$88.9 - 217\xi\lambda$	$-66.7 + 162\xi\lambda$
	1.064	$35.5 - 74\xi\lambda$	$58.3 - 116\xi\lambda$	$-25.8 + 30\xi\lambda$
	1.414	$12.9 - 11.4\xi\lambda$	$3.73 - 16.1\xi\lambda$	$5.74 - 12.4\xi\lambda$
	2.000	$3.82 - 0.41\xi\lambda$	$-2.95 - 5.92\xi\lambda$	$1.00 + 1.50\xi\lambda$
0.1915	1.000	$31.7 - 12.0\lambda$	$63.4 - 24.0\lambda$	$-47.5 + 18.0\lambda$
	1.064	$26.1 - 8.5\lambda$	$42.6 - 13.2\lambda$	$-18.5 + 3.5\lambda$
	1.414	$10.7 - 1.70\lambda$	$4.6 - 1.74\lambda$	$3.9 - 1.32\lambda$
	2.000	$3.47 - 0.241\lambda$	$-0.91 - 0.515\lambda$	$0.70 + 0.108\lambda$
0.4028	1.000	$15.5 - 10.6\lambda$	$30.9 - 21.2\lambda$	$-23.2 + 15.9\lambda$
	1.064	$13.4 - 7.8\lambda$	$21.7 - 12.0\lambda$	$-9.0 + 3.2\lambda$
	1.414	$6.33 - 1.94\lambda$	$3.78 - 1.57\lambda$	$1.58 - 1.07\lambda$
	2.000	$2.14 - 0.352\lambda$	$0.37 - 0.383\lambda$	$0.25 + 0.055\lambda$
0.6882	1.000	$4.62 - 4.7\lambda$	$9.23 - 9.39\lambda$	$-6.93 + 7.04\lambda$
	1.064	$4.29 - 3.7\lambda$	$6.84 - 5.53\lambda$	$-2.71 + 1.51\lambda$
	1.414	$2.29 - 1.06\lambda$	$1.66 - 0.74\lambda$	$0.33 - 0.41\lambda$
	2.000	$0.714 - 0.188\lambda$	$0.325 - 0.145\lambda$	$0.038 + 0.0112\lambda$

TABLE V. Values of the coefficients of the angular dependent terms available from observation of the γ rays alone.

ξ	$100\int_1^\infty (D+\delta D)\epsilon d\epsilon$	$100\int_1^\infty (N_2+\delta N_2)\epsilon d\epsilon$	$100\int_1^\infty (N_4+\delta N_4)\epsilon d\epsilon$
0.0000	$28.4 - 21\xi\lambda$	$-1.54 - 50\xi\lambda$	$-0.022 - 0.24\xi\lambda$
0.1915	$23.5 - 3.4\lambda$	$8.1 - 5.0\lambda$	$0.50 - 0.078\lambda$
0.4028	$12.7 - 3.6\lambda$	$7.9 - 4.2\lambda$	$-0.12 - 0.105\lambda$
0.6882	$3.97 - 1.79\lambda$	$3.20 - 1.84\lambda$	$-0.20 - 0.027\lambda$

[9] The values listed in Tables III, IV, V for $\xi = 0$ are consistent with the vanishing of the reorientation effect for no excitation, as discussed in the previous section. The limit $\xi \to 0$ is implied in the terms listed as being proportional to ξ.

These values were obtained by numerical integration over ϵ and are listed in Table V.[9]

The computational error of the values listed in the tables is believed to be about 1 in the last figure listed. In Table V considerable cancellation has taken place in obtaining the values given in the last column. However, the quantity $N_4+\delta N_4$ is the most difficult of the three to obtain experimentally since it is most sensitive to the angular definition of the photon direction.

The values of $\int_1^\infty D\epsilon d\epsilon$, $\int_1^\infty N_2\epsilon d\epsilon$, and $\int_1^\infty N_4\epsilon d\epsilon$ listed for $\xi=0$ represent the exact values

$$\int_1^\infty D\epsilon d\epsilon = (\pi^2/16) - (1/3) \cong 0.284,$$

$$\int_1^\infty N_2\epsilon d\epsilon = (7\pi^2/16) - (13/3) \cong -0.0154,$$

$$\int_1^\infty N_4\epsilon d\epsilon = (17/9) - (49\pi^2/256) \cong -0.00022,$$

obtainable by direct analytic integration.

The quantity $\int_1^\infty (D+\delta D)\epsilon d\epsilon$ listed in Table V represents the angle-independent part of the correction due to the reorientation effect. As a result the quantity

$$\int_1^\infty (D+\delta D)\epsilon d\epsilon \Big/ \int_1^\infty D\epsilon d\epsilon,$$

represents the factor by which the presence of the reorientation effect increases the total cross section.

It is clear from examination of Tables III, IV, and V that the coincidence-type experiments offer opportunity for obtaining effects due to reorientation which are relatively larger than for the non-coincidence-type experiment. Specifically, one sees from Table III that a judicious choice of coincidence angles which selects the coefficients of $\sin 2\varphi$, $\sin 4\varphi$ ($\mathrm{Im}[\mathcal{T}_0{}^*\mathcal{T}_s]$, and $\mathrm{Im}[\mathcal{T}_s{}^*\mathcal{T}_c]$) may offer the possibility of measuring λ directly.

There seems to be an indication that the reorientation effect is relatively larger for higher excitation, but this will depend on the static quadrupole moment of the level in question. An additional inference which may be drawn from the tables is that the reorientation effect is larger for values of ϵ near 1. This is probably due to the fact that larger values of ϵ correspond to distant collisions for which the amplitude for excitation is small, and the relative importance of the reorientation effect can be expected to be reduced for large ϵ.

The authors would like to acknowledge the assistance of Mrs. Maureen Berry and Miss Jacqueline Gibson who performed much of the numerical work. They would also like to express their thanks to Dr. F. D. Benedict for the use of some of his numerical results on the SCT Coulomb excitation integrals.

The authors would like to express their thanks to the Computing Center of the IBM Research Laboratory at Poughkeepsie, New York, for the use of an IBM 650 machine and to Dr. Willard G. Bouricius of that laboratory for making the necessary arrangements.

At the time of submission for publication the numbers in the tables were less accurate than the final ones. The authors are indebted to the editorial offices of *The Physical Review* for making these changes possible.

EXPERIMENTAL OBSERVATION OF DOUBLE COULOMB EXCITATION*

J. O. Newton† and F. S. Stephens

Radiation Laboratory and Department of Chemistry, University of California, Berkeley, California
(Received June 11, 1958)

Reprinted with corrections by the authors.

In regions of the periodic table where nuclei have large spheroidal deformations, it is well known that there are, associated with the ground states of even-even nuclei, rotational bands in which the members have spins and parities of 0+ (ground), 2+, 4+, 6+, etc.[1] The 4+ state belonging to this rotational band can be Coulomb excited either by a direct $E4$ excitation or by a double $E2$ process via the 2+ member of the band.[2] (Other intermediate states can also give a contribution, but that from the 2+ state is likely to predominate.) When heavy ions are used as bombarding particles, the probability for the double $E2$ process is expected to become high, and to exceed considerably that for the $E4$ excitation.

In order to observe double $E2$ Coulomb excitation, we bombarded thick targets of natural tungsten with oxygen ions from the Berkeley heavy-ion linear accelerator. Natural tungsten consists almost entirely of three even-even isotopes W^{182}, W^{184}, and W^{186}, and of the odd-mass isotope W^{183} in 14% abundance. These nuclei are known to be highly deformed and to have rotational bands. The gamma rays arising from the decay of the excited states were observed with 1 in. × $1\frac{1}{2}$ in. diameter NaI (Tl) crystals together with a 50-channel and a single-channel pulse-height analyzer.

The pulse-height spectrum showed two broad peaks, having mean energies of 114 and 250 kev. The lower energy peak arises almost completely from the decay of the known first excited states of W^{182}, W^{184}, and W^{186}, which have energies of 100.07 kev, 111.13 kev, and 122.48 kev, respectively.[3] We believe that the second broad peak arises from the decay of the 4+ to the 2+ states of the even-even nuclei, and from the decay of the 292 kev level in[4] W^{183}. The energies of the 4+ to 2+ transitions in[5] W^{182} and[6] W^{184} are known to be 229.07 and 252.8 kev, and that in W^{186} is expected to be about 280 kev. The intensity arising from the excitation of the 292-kev state in W^{183} can be calculated from the meased $B(E2)$.[4] In order to verify that most of the 250-kev peak was in coincidence with a transition of about 114 kev, coincidence measurements were made with a resolving time $2\tau=10^{-7}$ sec. The results indicated that,at 66-Mev bombarding energy, essentially all the 250-kev peak was in coincidence with such a transition.

The excitation functions for the two peaks were also measured using bombarding energies between 30 and 80 Mev. The excitation function for the 114-kev peak is shown in Fig. 1, where it is compared with a theoretical thick-target excitation function for exciting a state at 114 kev.[2] The range-energy curves of Barrett[7] were used in calculating the theoretical curve. The normalization of the curve to the experimental points is arbitrary; nevertheless, the experimental thick-target yield of $(5.7 \pm 0.8) \times 10^6$ photons/μcoul. at 40 ± 2 Mev bombarding energy is consistent with that of 5.8×10^6 photons/μcoul. calculated from the previously measured $B(E2)$ values for the tungsten isotopes.[8,2] This corresponds to a cross section of 0.57 barn for the production of photons or, since the average conversion coefficient is about 2.3,[2] to an average cross section for exciting the 2+ states of about 1.9 barns. It can be seen that the points deviate from the theoretical curve by almost as much as a factor of two at the highest energy, which is close to that of the Coulomb barrier. The process of Coulomb de-excitation, estimated from a naive point of view, is too small to account for this deviation. Such a deviation from the theoretical curve at the highest energies is not particularly surprising since the probability for excitation is so high that the perturbation method used in the theoretical derivation is hardly a valid procedure.

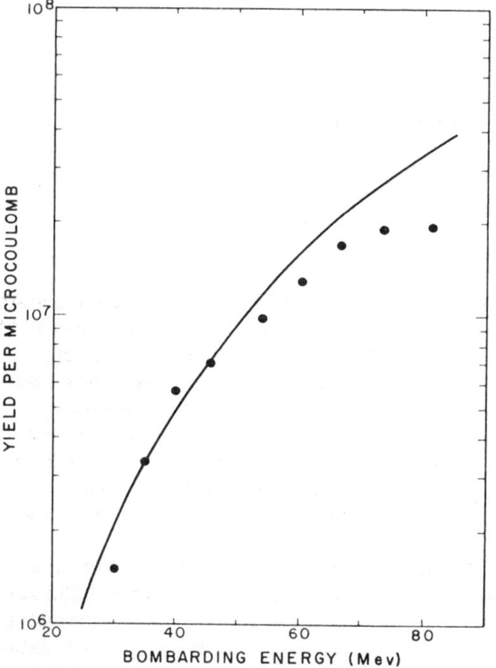

FIG. 1. Yield of 114-kev radiation from natural tungsten as a function of oxygen ion (charge +8) energy. The points are the experimental yields, and the line is the theoretical curve for $E2$ excitation of a 114-kev state in tungsten, arbitrarily normalized.

The excitation function for the 250-kev radiation is shown in Fig. 2. The calculated intensity of the 292-kev gamma ray of W^{183} is also shown, and this contribution has been subtracted from the experimental points before plotting. The resulting points are compared with an approximate theoretical excitation function for the double $E2$ process of the form[2]

$$\sigma_{E2\text{-}E2}(0 \to 4) = \alpha a^{-2}\, \sigma_{E2}(0 \to 2)\, \sigma_{E2}(2 \to 4), \quad (1)$$

where α is a constant and $2a$ is the distance of closest approach in a head-on collision. It can be seen that the agreement between this curve and the experimental points is good except at the highest energies where the points fall below the curve. This deviation is rather similar to that observed for the 114-kev peak, and it seems likely that the two are related.

If the theory of double $E2$ excitation were well established, we could in this case use it, together with the measured cross sections, to give a value for $B_{F2}(2 \to 4)$. We could then see whether that value was consistent with that expected from the theory of rotational states:

$$\frac{B_{E2}(2 \to 4)}{B_{E2}(0 \to 2)} = \left\{\frac{\langle 2200 \mid 40\rangle}{\langle 0200 \mid 20\rangle}\right\}^2 = \frac{18}{35} \quad (2)$$

Since the theory of double $E2$ excitation is not yet well established, we shall use the reverse procedure and, assuming that relation (2) is valid, compare our measured yield with that calculated from (1) using the theoretical value of

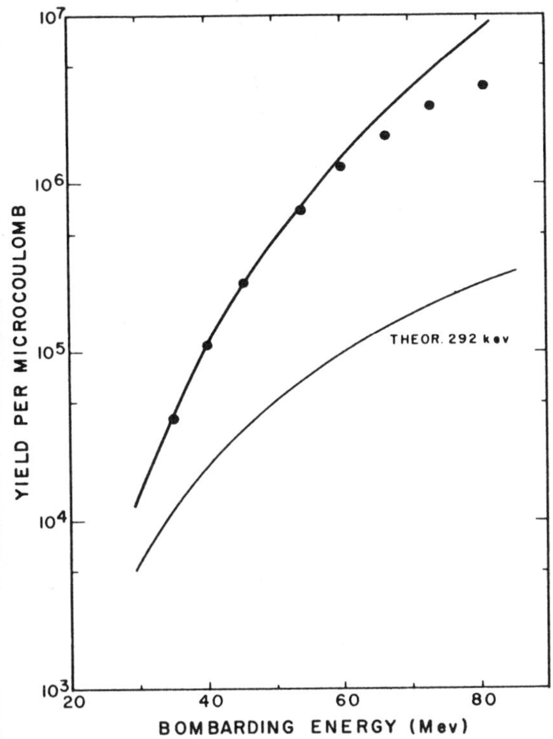

FIG. 2. Yield of 250-kev radiation from natural tungsten as a function of oxygen ion (charge +8) energy. The light line is the theoretical yield curve for excitation of the 292-kev state in W^{183} using the B_{E2} of reference 4. The heavy line is the theoretical yield curve for double $E2$ excitation in natural tungsten, arbitrarily normalized. The points are the experimental yields with 292-kev contribution from W^{183} subtracted.

0.0272 for α.[9] Our yield of $(1.10 \pm 0.17) \times 10^5$ photons/μcoul. at 40 ± 2 Mev bombarding energy compares well with the value of 1.19×10^5 photon/μcoul. obtained on the basis indicated above. This yield corresponds to an average cross section of 25 millibarns for the double excitation process. It should be noted here that in order to obtain this cross section from $E4$ excitation, the B_{E4} would have to be about 5000 times the single-particle value.[2] This does not seem to us very likely. It appears then that the present theory of double $E2$ excitation is not seriously in error.

We are grateful to Professor I. Perlman for his continued interest in this work. We are greatly indebted to the staff and operating crews of the heavy-ion linear accelerator for their help and cooperation.

*This work was performed under the auspices of the U. S. Atomic Energy Commission.

† At present on leave of absence from Atomic Energy Research Establishment, Harwell, England.

[1] A. Bohr and B. R. Mottelson, Kgl. Danske Videnskab. Selskab, Mat.-Fys. Medd. 27, No. 16 (1953).
[2] Alder, Bohr, Huus, Mottelson, and Winther, Revs. Modern Phys. 28, 432, (1956).
[3] Chupp, Clark, Du Mond, Gordon, and Mark, Phys Rev. 107, 745 (1957).
[4] P. H. Stelson and F. K. McGowan, Phys. Rev. 99 112 (1955).
[5] Murray, Boehm, Marmier, and Du Mond, Phys. Rev. 97, 1007 (1955).
[6] Gallagher, Strominger, and Unik, Phys. Rev. 110, 725 (1958).
[7] R. Barrett (private communication), recalculated from Aron, Hoffman, and Williams, Atomic Energy Commission Report AECU-663, 1949 (unpublished).
[8] McClelland, Mark, and Goodman, Phys. Rev. 97, 1191 (1955).
[9] Private communication from K. Alder to G. Breit of a correction to the formula in reference 2.

MULTIPLE COULOMB EXCITATION IN Th^{232} AND U^{238}

F. S. Stephens, Jr., R. M. Diamond, and I. Perlman
Lawrence Radiation Laboratory and Department of Chemistry, University of California, Berkeley, California
(Received October 12, 1959)

Reprinted with corrections by the authors.

Electromagnetic excitation or "Coulomb excitation" of a nucleus by the electric field of a passing ion occurs most readily in nuclei having low-lying states with the character of collective excitations of the ground-state configurations.[1] Electric quadrupole excitations are by far the most important. "Multiple Coulomb excitation" refers to excitation of higher states through successive quadrupole excitation steps. In an earlier paper,[2] the double Coulomb excitation of even-even tungsten isotopes by O^{16} ions was reported (excitation of the 4+ member of the ground-state rotational band through the 2+ state).

The probability of multiple Coulomb excitation should increase rather rapidly with increasing projectile charge, and Coulomb excitation in general is most favorable where the rotational level spacings are smallest. With these factors in mind, search was made for higher than second-order Coulomb excitation in Th^{232} and U^{238} by irradiating with Ne^{20}, S^{32}, and A^{40} ions produced by the Berkeley heavy-ion linear accelerator (Hilac). These particular target elements also have the virtue of being monoisotopic. (The uranium used was isotopically depleted in U^{235} and U^{234}.) In this paper evidence is presented for excitation up to sixth order in U^{238} and fifth order in Th^{232}.

Because of low beam intensities (1-5 micro-microamperes) it was necessary to employ thick targets of thorium and uranium metal (2-4 mils). The gamma rays resulting from Coulomb excitation were detected with a 3 in.$\times$3 in. NaI(Tl) crystal assembly mounted behind the target and recorded with a 100-channel pulse-height discriminator. In order to enhance the events from multiple Coulomb excitation and to reduce the background radiation, the gamma-ray detector was operated in fast coincidence (3×10^{-8} sec) with pulses produced by the back-scattered heavy ions in an argon scintillation chamber. These gating pulses were cut off below about 10 Mev.

The gamma-ray spectra obtained with A^{40} irradiation of U^{238} and Th^{232} are shown in Figs. 1(a) and 1(b). The gamma-ray energies observed are listed in columns 2 and 5 of Table I except that the energies for the 2+→0+ transitions were taken from other experiments. The 4+→2+ transition in U^{238} (103 kev) is somewhat uncertain both in energy and intensity because uranium K x-rays lie at 94 to 98 kev and accurate resolution was not possible. In Figs. 1(a) and 1(b) the small peaks which appear beyond energies of 311 kev and 273 kev can be reasonably well explained both in energy and intensity, as due to coincident arrival at the detector of lower energy photons. Such pile-up peaks, occasioned by the high acceptance angle of the detector, would make a negligible contribution to the gamma rays listed in Table I.

The interpretation of these data in the form of level schemes is shown in Fig. 2. The evidence is largely indirect since it would be experimentally difficult to determine all of the gamma-gamma coincidences in the de-excitation sequences. However, as a partial check, it was established that in U^{238} the 214-kev (8+→6+)

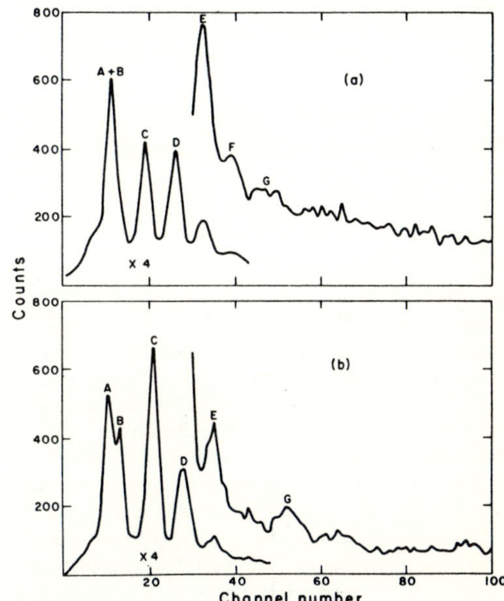

FIG. 1. De-excitation gamma-ray spectrum of (a) U^{238}, (b) Th^{232} irradiated with 190-Mev A^{40} ions. A, B, C, D, E, and F correspond to K x-rays, 4+→2+, 6+→4+, 8+→6+, 10+→8+, 12+→10+ transitions, respectively, and G is a pile-up peak (see text).

Table I. Transition energies between rotational states produced by multiple Coulomb excitation.

Transition	U^{238} transition energies (kev) Exper.	Calc. for $A = 7.45$	Calc. for $A = 7.46$, $B = -2.7 \times 10^{-3}$	Th^{232} transition energies (kev) Exper.	Calc. for $A = 8.33$	Calc. for $A = 8.38$, $B = -8.5 \times 10^{-3}$	Calc. for $A = 8.34$, $B = -1.2 \times 10^{-2}$, $C = 4 \times 10^{-5}$
2+→0+	(44.7 ± 0.2)[a]	44.7	44.7	(49.75 ± 0.25)[b]	50	50	50
4+→2+	103 ± 2	104	103	113 ± 2	117	114	113
6+→4+	161 ± 2	164	162	170 ± 3	183	173	170
8+→6+	214 ± 3	224	215	222 ± 4	250	222	221
10+→8+	264 ± 4	283	265	273 ± 5	317	259	272
12+→10+	311 ± 5	343	310				

[a]Taken from J. O. Newton, reference 5.
[b]This value was measured by the present authors from U^{236} α decay using a xenon-filled proportional counter.

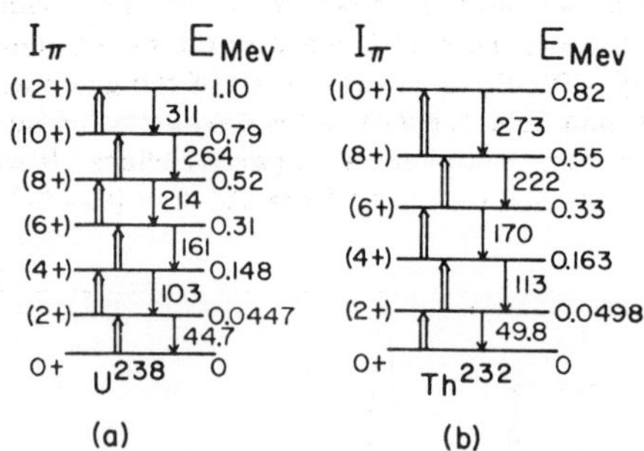

FIG. 2. Rotational energy level schemes proposed for (a) U^{238} and (b) Th^{232}.

gamma ray and part of the 103-kev (4+→2+) intensity are in coincidence with the 161-kev (6+→4+) transition. The principal pieces of evidence are the following: (1) The energies of the gamma rays are in agreement with expectations for excitation of rotational bands, (2) the yields of the photons for different bombarding ions and energies would be difficult to explain on any basis other than multiple Coulomb excitation of rotational bands, and (3) no other low-lying states are known in this region of the periodic system which would not at the same time give abundant crossover transitions of easily discernible energies.

It is now well known that levels of a rotational band of an even-even nucleus, such as those in Fig. 2, have energies given approximately by $E_I = AI(I+1)$, where E_I is the energy of the state with spin I and I is restricted to even integers.[3] The empirically determined constant A is related to the effective moment of inertia, $\mathcal{J}$, by $A = \hbar^2/2\mathcal{J}$. The first excited states (2+) have been established by other Coulomb-excitation work,[4,5] and the energies obtained have been used to determine A. With this value of A the energies of other states may be calculated. The corresponding energy spacings of the rotational bands are listed in columns 3 and 6 of Table I. It is not surprising that agreement with the experimental results is increasingly poor for the higher states since it is expected (and found in other cases) that higher order terms should be added to the expression for E_I to account for rotational-vibration interactions. The second term $BI^2(I+1)^2$ contains another constant, B, which can be related to vibrational energies. Columns 4 and 7 of Table I show that this second term is sufficient to bring into agreement the U^{238} data; but for Th^{232} a third term, $CI^3(I+1)^3$, is required (column 8).

Table II gives a comparison of these constants for several nuclei in the heavy-element region.

Table II. Rotational constants in the heavy elements.

Nuclide	A (kev)	$-B$ (kev)	C (kev)
Pu^{240}	7.17	3.9×10^{-3}	
Pu^{238}	7.37	3.6×10^{-3}	
U^{238}	7.46	2.7×10^{-3}	$< 2 \times 10^{-6}$
U^{234}	7.29	6.9×10^{-3}	$(\sim 3 \times 10^{-5})$
Th^{232}	8.34	1.2×10^{-2}	4×10^{-5}
Ra^{226}	11.74	8×10^{-2}	$(\sim 9 \times 10^{-4})$

As is well known, the value of A is rather constant for elements above thorium and increases rather sharply for lighter elements approaching the closed shells at Pb^{208}. Indeed, below radium the rotational picture is no longer a proper description of lowest-energy collective modes of motion. It is apparent that if accurate values for B and C are to be obtained and only a few members of a rotational band can be measured, great precision in the energy determination is required. However, if (as in the present study) a large number of states are discernible, high precision is not required. The values of C shown for Ra^{226} and U^{234} are only estimates which are entered to give some comparison with the more accurate values obtained in the present experiments.

Cross sections for Coulomb excitation have proved to be of great value because they are directly related to $E2$ transition probabilities and provide information for nuclear models. Such information on multiple Coulomb excitation would also be of interest as a guide in extending Coulomb excitation theory. Unfortunately, meaningful cross sections cannot be obtained from the present experiments because (1) it was necessary to use thick targets and the inherent difficulties in obtaining an excitation function are complicated by uncertainties in the dE/dx relation for argon ions; (2) only a portion of the excitations were observed, those for which the heavy ion was scattered through angles between 90° and 160° and emerged from the target with enough energy to enter the gas scintillation counter with greater than 5 to 10 Mev. Nevertheless some relative yields have been calculated for two different A^{40} energies (see Table III) and these will be explained presently.

There has been no published theoretical treatment applicable to high-order multiple excitation, but Alder and Winther[6] are now considering the problem. Their first results seem to be in excellent agreement with our observations in that qualitatively the predicted trends are being followed.

In columns 2 and 3 of Table III we have summarized the relative independent yields for excitation of the indicated levels in Th^{232} at two different A^{40} ion bombarding energies. These values were obtained from the integrated photopeaks corrected for absorption in the target and backing plate and for the counting efficiency of the NaI crystal. Correction was made for internal conversion (assuming $E2$ transitions), and allowance was made for the cascade from higher levels.

Table III. Relative yields from A^{40} on Th^{232}.

E_γ	Parent level	Direct yield of parent level $E(A^{40}) = 158$ Mev	$E(A^{40}) = 190$ Mev
113	4+	100	100
170	6+	46	88
222	8+	10	26
273	10+	0.7	5.9

The yield for the 4+ level has been normalized to 100 for each energy. It is seen that each successive level has an appreciably steeper excitation function so that at the higher bombarding energy the 10+ level is about eight times more intense relative to the 4+ than at the lower bombarding energy. This is explained in a natural way according to the scheme in Fig. 2, since here each higher state results from a higher order multiple excitation. Other explanations of this behavior are not so easy. If, for example, the gamma rays come from ordinary single $E2$ excitations from the ground state, then we would expect the 273-kev peak to increase only about 10% more than the 113-kev peak over this range of bombarding energy. To obtain the much greater variation in yield indicated in Table III, one would have to assume either a much higher (and different) energy of excitation for each transition, or excitations of higher multipole orders. Neither of these possibilities is very likely. The latter—the direct excitation of the rotational band members by $E4$, $E6$, etc., excitations instead of by the multiple process—was shown to be negligible in the case of double excitation in the tungsten isotopes using O^{16},[2] and should be even less important with a particle of higher charge. The yields seem entirely reasonable for the experimental conditions used (thick targets and back-scattered projectile ions). Had we used thin targets, the results of Winther and Alder would suggest that at the higher bombarding energy the primary population of the 6+ and perhaps the 8+ states would have been higher than for the 4+ state.[6] Such reversals in yield would be interesting to observe, but the difficulties associated with thin targets in our apparatus are rather severe. Data similar to those in Table III have been obtained for both U^{238} and Th^{232} with different bombarding ions. In all cases these data are consistent with the schemes in Fig. 2 insofar as it is possible to determine at the present time.

We are grateful to Dr. J. O. Newton who participated in many of the early phases of this work. We are also greatly indebted to Dr. E. Hubbard and the operating crews of the Hilac for the indispensable help they have given us in carrying out these experiments.

[1]For a comprehensive review of the subject, see: Alder, Bohr, Huus, Mottelson, and Winther, Revs. Modern Phys. 28, 432 (1956).

[2]J. O. Newton and F. S. Stephens, Phys. Rev. Letters 1, 63 (1958).

[3]A. Bohr and B. R. Mottelson, Kgl. Danske Videnskab. Selskab, Mat.-fys. Medd. 27, No. 16 (1953).

[4]G. M. Temmer and N. P. Heydenburg, Phys. Rev. 93, 351 (1954); Davis, Divatia, Lind, and Moffat, Phys. Rev. 103, 1801 (1956); P. H. Stelson and F. K. McGowan, Phys. Rev. 99, 112, 616A (1955).

[5]N. P. Heydenburg and G. M. Temmer, Phys. Rev. 93, 906 (1954); 94, 1252 (1954); J. O. Newton, Nuclear Phys. 3, 345 (1951).

[6]K. Alder and A. Winther, Federal Institute of Technology, Zurich, Switzerland (private communication, June, 1959).

ADDENDUM

Submitted by authors for this reprint edition.

New measurement with Ge detector gives the following improved values for the transition energies.

Transition	U^{238}	Th^{232}
$4 \rightarrow 2$	103.5 ± 1.0	112.7 ± 1.0
$6 \rightarrow 4$	159.4 ± 0.5	171.6 ± 0.5
$8 \rightarrow 6$	211.1 ± 0.5	223.8 ± 0.5
$10 \rightarrow 8$	258.3 ± 0.5	270.3 ± 0.5
$12 \rightarrow 10$	301.2 ± 0.5	310.7 ± 1.0
$14 \rightarrow 12$	339 ± 2	

Matematisk-fysiske Meddelelser
udgivet af
Det Kongelige Danske Videnskabernes Selskab
Bind **32**, nr. 8

Mat. Fys. Medd. Dan. Vid. Selsk. **32**, no. 8 (1960)

ON THE THEORY OF MULTIPLE COULOMB EXCITATION WITH HEAVY IONS

BY

KURT ALDER AND AAGE WINTHER

København 1960
i kommission hos Ejnar Munksgaard

CONTENTS

Synopsis

The present paper contains formulae and tables for the evaluation of multiple Coulomb excitation cross sections of rotational and vibrational states. For other cases, general calculational procedures have been developed and these are illustrated through examples. For the larger part of the work, the collision time is assumed to be short compared to the nuclear period. The investigation is furthermore simplified by an approximate treatment of the dependence of the cross section on the deflection angle of the projectile. The accuracy of the approximations is also discussed.

Printed in Denmark
Bianco Lunos Bogtrykkeri A/S

1. Introduction

In the last few years, the Coulomb excitation process has become a valuable tool for the investigation of low lying nuclear states. Several review articles on the experimental and theoretical aspects of Coulomb excitation have appeared, [(1), (2), (3), (4)] which contain bibliographies of the earlier work on this subject*.

The Coulomb excitation process has certain advantages over other nuclear reactions. The fact that the forces responsible for the process are well understood, and the theory is well developed, allows one from a careful analysis of the reaction to determine a number of quantities characteristic of the nuclear states. The main approximation in the existing calculations is the use of perturbation theory which is valid if the probability for nuclear excitation in a single encounter is small. If protons or α-particles are used as projectiles, and if the bombarding energy is kept so low that no nuclear reactions take place, this criterion will always be fulfilled. In these cases there is, however, a strong limitation on the number of states which can be investigated. The limitation lies, firstly, in the selection rules for the low multipole interactions which are important for the excitation process. Secondly, only low lying states are accessible, since the reaction for higher excitation energies soon becomes adiabatic. A way to overcome these difficulties is to use heavier ions as projectiles. The electric field exerted on the nucleus then becomes so large that higher order processes occur. While, e. g., a state with spin 4^+ in first-order perturbation treatment can only be reached from a ground state of spin 0^+ through an $E4$ interaction, it might already in second order be excited through a state of spin 2^+ by means of quadrupole interactions. In many cases, one might still use the perturbation expansion to calculate the excitation probabilities (see ref. 1, Chapt. II D, and ref. 2). If, however, the interaction becomes so strong that many levels are actively involved in the excitation process, one has to solve directly the set of coupled

* In the following, the notation of ref. 1 will always be used.

1*

equations which describes the population of the nuclear states during the collision.

The feasibility of such multiple excitations with heavy ions has recently been proved and, from these experiments as well as from the following calculations, it seems that a number of new possibilities are opened for the investigation of nuclear states[(5), (6)].

In the following we shall consider such multiple excitations. In Section 2, a discussion is given of the parameters which are important for the process. Section 3 contains the general formalism, while the following sections are concerned with special models and numerical tables.

2. Characteristic Parameters

The Coulomb excitation process is characterized by a number of parameters. These quantities describe the kind of approximations which are appropriate for the process in question.

A parameter which describes the motion of the projectile in the Coulomb field of the nucleus is η defined by

$$\eta = \frac{Z_1 Z_2 e^2}{\hbar v}, \tag{2.1}$$

where Z_1 and Z_2 are the charge numbers of the projectile and the target nucleus, respectively, while v is the relative velocity of the incident particle and the nucleus. While for protons this parameter may be as small as two, it is, for the heavy ions which are being considered in the following, always much larger than one. Since, furthermore, the projectile in a collision loses only a small part of its energy, one may to a very good approximation use a classical description for its path. The hyperbolic orbit of the particle will be described by the deflection angle ϑ (see Fig. 1).

The Coulomb interaction between the projectile and the nucleus is given by (see ref. 1, Eqs. (II A. 8) to (II A. 11))

$$\mathfrak{H}_E(t) = Z_1 e \int \varrho(\vec{r}) \frac{d\tau}{|\vec{r} - \vec{r}_p(t)|^2} - \frac{Z_1 Z_2 e^2}{r_p}, \tag{2.2}$$

where $\varrho(\vec{r})$ is the charge density operator at the position r of the nucleus and $\vec{r}_p(t)$ is the position vector of the projectile, which for a given hyperbolic orbit is a known function of time. The interaction can be expanded in multipole components

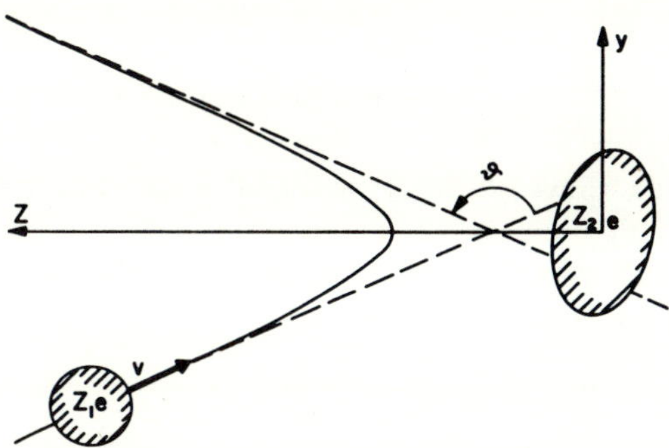

Fig. 1. Classical picture of the projectile orbit in the Coulomb field of the nucleus. The hyperbolic orbit of the projectile is shown in the frame of reference where the nucleus is at rest. The coordinate system which is employed in the present paper, with the z-axis along the axis of symmetry, is indicated. The charges of nucleus and projectile are denoted by $Z_2 e$ and $Z_1 e$, respectively, v is the initial relative velocity, and ϑ is the deflection angle.

$$\mathfrak{H}_E(t) = 4\pi Z_1 e \sum_{\lambda=1}^{\infty} \sum_{\mu=-\lambda}^{+\lambda} \frac{1}{2\lambda+1} r_p^{-\lambda-1} Y_{\lambda\mu}(\vartheta_p, \varphi_p) \mathfrak{M}^*(E\lambda, \mu), \tag{2.3}$$

where $\mathfrak{M}(E\lambda, \mu)$ is the electric multipole moment of order λ of the nucleus defined by

$$\mathfrak{M}(E\lambda, \mu) = \int r^\lambda Y_{\lambda\mu}(\vartheta, \varphi) \varrho(\vec{r})\, d\tau. \tag{2.4}$$

In first order perturbation treatment, one finds the following expression for the total probability P for the transition from the nuclear state 1 to the state 2 in a given collision with deflection angle ϑ (see ref. 1, Eqs. (II A. 4), (II A. 28), and (II A. 29)):

$$P_{1\to 2} = \frac{Z_1^2 \cdot e^2}{\hbar^2 v^2} \sum_{\lambda=1}^{\infty} \frac{16\pi^2}{(2\lambda+1)^3} \cdot \frac{1}{a^{2\lambda}} \cdot B(E\lambda) \sum_{\mu=-\lambda}^{+\lambda} \left| Y_{\lambda\mu}\left(\frac{\pi}{2}, 0\right)\right|^2 \cdot |I_{\lambda\mu}(\vartheta, \xi)|^2. \tag{2.5}$$

Here, a is half the distance of closest approach in a head-on collision

$$a = \frac{Z_1 Z_2 e^2}{mv^2}, \tag{2.6}$$

where m is the reduced mass of the projectile and the target nucleus. The reduced transition probability $B(E\lambda, I_1 \to I_2)$ is defined by

$$\left.\begin{aligned} B(E\lambda, I_1 \to I_2) &= \sum_{\mu M_2} |\langle I_1 M_1 | \mathfrak{M}(E\lambda, \mu) | I_2 M_2 \rangle|^2 \\ &= \frac{1}{2 I_1 + 1} |\langle I_1 \| \mathfrak{M}(E\lambda) \| I_2 \rangle|^2. \end{aligned}\right\} \tag{2.7}$$

For two states with spins I_1 and I_2, practically only one value of λ will give a contribution to the sum in (2.5).

The orbital integrals[7] $I_{\lambda\mu}(\vartheta, \xi)$ depend on the deflection angle and on the parameter ξ which is defined by

$$\xi_{1\to 2} = \eta_f - \eta_i \cong \frac{Z_1 Z_2 e^2}{\hbar v} \frac{E_2 - E_1}{2E}. \tag{2.8}$$

The quantities η_i and η_f are given by (2.1), substituting for the velocity the initial and final velocities, respectively. Similarly, E_1 and E_2 denote the energies of the nucleus in the states 1 and 2, while E is the energy of the projectile. The parameter ξ measures the suddenness of a head-on collission. In general, the suddenness is measured by the quantity

$$\xi(\vartheta) = \frac{\xi}{\sin\frac{\vartheta}{2}}. \tag{2.9}$$

If $\xi(\vartheta)$ is large, the process is essentially adiabatic and the excitation probability small. If $\xi(\vartheta)$ is small, the process has the character of a sudden impact and one may use a sudden approximation.

In the case of multiple Coulomb excitation, the parameter ξ is no more a characteristic of a nuclear state. A definite nuclear state can in this case be populated in different ways. The ξ which is important for the excitation of the state in question need not be the one corresponding to the excitation from the ground state, but is rather a set of ξ's corresponding to the transitions through which it is populated.

The validity of the perturbation treatment which leads to the result (2.5) is guaranteed if P is small compared to one. We may introduce the square root of the contribution to P from a definite multipole order as a measure of the strength with which the state 2 is coupled to the state 1 through the interaction with the projectile

$$\chi_{1\to 2}^{(\lambda)}(\vartheta, \xi) = \pm \sqrt{P_{1\to 2}^{(\lambda)}(\vartheta, \xi)}. \tag{2.10}$$

The sign is to be the same as the sign of the reduced matrix element $\langle I_1 \| \mathfrak{M}(E\lambda) \| I_2 \rangle$.

If all the parameters χ which connect the states of the nucleus are small compared to unity, one may use a first order perturbation treatment. This will practically always be the case when protons are used as projectiles. For α-particles and heavier ions, the χ's will also mostly be smaller than one if the matrix elements are of the order of the single-particle value (see ref. 1, Chapt. II A). One may, in such cases, still use the perturbation treatment, when necessary, to second or third order (see ref. 1, Chapt. II D). If, however, the nucleus possesses excited states of collective type[8] with strongly enhanced $B(E\,2)$ transition probabilities, $\chi^{(2)}$ might be as large as 5. Then, one has to use an approach which avoids the perturbation expansion. On the other hand, states with large quadrupole transition probabilities have usually small excitation energies, and one may use an expansion appropriate for small ξ.

The parameter $\chi\ (\vartheta, \xi)$ attains its largest value for $\xi = 0$ and $\vartheta = \pi$. It will be useful to introduce this value as the fundamental parameter, in the same way as ξ is used instead of $\xi\,(\vartheta)$. We thus define (see ref. 1, Chapt. II E. 4)

$$\chi_{1\to 2}^{(\lambda)} = \sqrt{16\,\pi}\,\frac{(\lambda-1)\,!}{(2\,\lambda+1)\,!!}\cdot\frac{Z_1 e}{\hbar v}\cdot\frac{\langle I_1\,||\,\mathfrak{M}(E\lambda)\,||\,I_2\rangle}{a^{\lambda}\sqrt{2\,I_1+1}}. \tag{2.11}$$

It will also be convenient sometimes to introducc the value of $\chi(\vartheta, \xi)$ for $\xi = 0$, but arbitrary ϑ as a parameter. We call this $\chi(\vartheta)$ and, according to (2.5), (2.10), and (2.11), it is defined by

TABLE 1

A survey of different limiting cases of the characteristic parameters η, ξ, and χ. In the table is indicated the kind of approximation which is appropriate for the different cases and the values of λ for which computations have been performed. The calculations mentioned under the heading "ξ arbitrary" are quoted in ref. 1. The computations for arbitrary η and $\xi \ll 1$ are given in ref. 11, while those mentioned in the last entry refer to the present work.

	η arbitrary	$\eta \gg 1$ semiclassical
ξ arbitrary	$\chi \ll 1$ 1. order perturbation $\lambda = 1,\ 2$	$\chi < 1$ 1. and 2. order perturbation $\lambda = 1,\ 2,\ 3,\ 4$
$\xi \ll 1$ sudden approximation	$\chi \ll 1$ 1. order perturbation $\lambda = 3,\ 4$	χ arbitrary multiple excitation $\lambda = 2$

$$\chi_{1\to 2}^{(\lambda)}(\vartheta) = \chi_{1\to 2}^{(\lambda)} \frac{(2\lambda-1)!!}{(\lambda-1)!}\left[\frac{\pi}{(2\lambda+1)}\sum_{\mu}' \left|Y_{\lambda\mu}\left(\frac{\pi}{2},0\right) I_{\lambda\mu}(\vartheta,0)\right|^2\right]^{\frac{1}{2}}. \tag{2.12}$$

In Table I, a survey is given of the different limiting cases for which computations on Coulomb excitation have until now been made, including the present work. In the following, we shall limit ourselves mainly to the case of quadrupole excitations ($\lambda = 2$).

3. General Theory

In this section, we investigate the equations which determine the multiple Coulomb excitation and discuss some general approximation methods. It will be shown that the special solution for $\xi = 0$ and $\vartheta = \pi$ is a convenient basic solution by means of which the excitation probability for small values of ξ and arbitrary angles may be expressed.

A. Expansion Methods

The Schrödinger equation for the nuclear state vector $|\psi\rangle$ is

$$i\hbar \frac{\partial}{\partial t}|\psi\rangle = [\mathfrak{H}_0 + \mathfrak{H}_E(t)]\,|\psi\rangle, \tag{3.1}$$

where $\mathfrak{H}_0$ is the Hamiltonian of the free nucleus and $\mathfrak{H}_E(t)$ the interaction energy given by (2.2), (2.3), and (2.4). It will be useful to introduce a new state vector $|\Phi\rangle$ defined by

$$|\psi\rangle = e^{-\frac{i}{\hbar}\mathfrak{H}_0 t}\,|\Phi\rangle. \tag{3.2}$$

Before and after the collision this state vector is time-independent and it satisfies the equation

$$i\hbar \frac{\partial}{\partial t}|\Phi\rangle = \tilde{\mathfrak{H}}(t)\,|\Phi\rangle, \tag{3.3}$$

where

$$\tilde{\mathfrak{H}}(t) = e^{\frac{i}{\hbar}\mathfrak{H}_0 t}\cdot \mathfrak{H}_E(t)\, e^{-\frac{i}{\hbar}\mathfrak{H}_0 t}. \tag{3.4}$$

The equation (3.3) may also be formulated as a set of coupled differential equations for the amplitudes on the nuclear eigenstates. If we thus define

$$a_n(t) = \langle n\,|\,\Phi\rangle, \tag{3.5}$$

where $|n\rangle$ is the time-independent eigenstate belonging to the eigenvalue E_n of $\mathfrak{H}_0$,

$$\mathfrak{H}_0 \,|\, n \rangle = E_n \,|\, n \rangle, \tag{3.6}$$

we obtain

$$i\hbar \dot{a}_n = \sum_m \langle n \,|\, \mathfrak{H}_E(t) \,|\, m \rangle \, e^{\frac{i}{\hbar}(E_n - E_m)t} \, a_m(t). \tag{3.7}$$

The solution of (3.3) and (3.7) can often conveniently be expressed as a series in powers of $\mathfrak{H}_E(t)$. This is the usual perturbation expansion which can be obtained by an iteration procedure. It can be written in a closed form due to Dyson

$$|\,\Phi(t)\rangle = T e^{-\frac{i}{\hbar}\int_{-\infty}^{t} \tilde{\mathfrak{H}}(t)\,dt} \,|\, \Phi(t = -\infty)\rangle, \tag{3.8}$$

where the symbol T stands for the time ordered product, i. e.,

$$\left.\begin{aligned} T e^{-\frac{i}{\hbar}\int_{-\infty}^{t} \tilde{\mathfrak{H}}(t')\,dt'} &= 1 - \frac{i}{\hbar}\int_{-\infty}^{t} \tilde{\mathfrak{H}}(t')\,dt' \\ +\left(\frac{-i}{\hbar}\right)^2 \int_{-\infty}^{t} \tilde{\mathfrak{H}}(t')\,dt' \int_{-\infty}^{t'} \tilde{\mathfrak{H}}(t'')\,dt'' &+ \left(\frac{-i}{\hbar}\right)^3 \int_{-\infty}^{t} \tilde{\mathfrak{H}}(t')\,dt' \int_{-\infty}^{t'} \tilde{\mathfrak{H}}(t'')\,dt'' \int_{-\infty}^{t''} \tilde{\mathfrak{H}}(t''')\,dt'''. \end{aligned}\right\} \tag{3.9}$$

If the nucleus before the collision is in the ground state $|\,0\rangle$, the solution (3.8) leads to the following expression for the amplitudes on the different excited states after the collision:

$$a_n(+\infty) = \langle n \,|\, T e^{-\frac{i}{\hbar}\int_{-\infty}^{+\infty} \tilde{\mathfrak{H}}(t)\,dt} \,|\, 0 \rangle. \tag{3.10}$$

When one inserts the series (3.9), one obtains exactly the usual perturbation expansion for the excitation amplitudes.

As has been mentioned above, the case where $\xi = 0$ for all states involved is of special importance for the problem of multiple Coulomb excitations. In this case, one has $E_n = E_m$ and $\tilde{\mathfrak{H}} = \mathfrak{H}$, and one can then leave out the time ordering in (3.8). The expression (3.10) now takes the simple form

$$a_n(+\infty) = \langle n \,|\, e^{-\frac{i}{\hbar}\int_{-\infty}^{\infty} \mathfrak{H}(t)\,dt} \,|\, 0 \rangle \tag{3.11}$$

characteristic of the sudden approximation. This formula is also applicable in cases where $\int_{-\infty}^{\infty} \mathfrak{H}(t)\,dt \gg \hbar$ since, for the evaluation, it is not necessary to perform a series expansion of the exponential function.

By means of formula (3.11) one may thus avoid the perturbation expansion. On the other hand, the effect of the motion of the nucleus during the collision has been neglected. The sudden approximation, however, forms a convenient starting point for a series expansion in powers of $\mathfrak{H}_0$. Such a series expansion is generated by the following substitution:

$$| \Phi \rangle = e^{-\frac{i}{\hbar}\int_{-\infty}^{t} \mathfrak{H}_E(t)\,dt} | \varphi \rangle . \tag{3.12}$$

The Schrödinger equation (3.3) then takes the form

$$i\hbar \frac{\partial | \varphi \rangle}{\partial t} = e^{\frac{i}{\hbar}\int_{-\infty}^{t} \mathfrak{H}_E(t)\,dt} [\mathfrak{H}(t) - \mathfrak{H}_E(t)]\, e^{-\frac{i}{\hbar}\int_{-\infty}^{t} \mathfrak{H}_E(t)\,dt} | \varphi \rangle . \tag{3.13}$$

In this expression, one can expand $\tilde{\mathfrak{H}}(t)$ in powers of $\mathfrak{H}_0$ in the following way:

$$\tilde{\mathfrak{H}}(t) = \mathfrak{H}_E(t) + \frac{i}{\hbar} t [\mathfrak{H}_0, \mathfrak{H}_E(t)] + \frac{1}{2}\left(\frac{i}{\hbar} t\right)^2 [\mathfrak{H}_0 [\mathfrak{H}_0, \mathfrak{H}_E(t)]] + \cdots \tag{3.14}$$

and (3.13) takes the form

$$i\hbar \frac{\partial | \varphi \rangle}{\partial t} = \left\{ \frac{i}{\hbar} t [\overline{\mathfrak{H}}_0, \mathfrak{H}_E(t)] + \frac{1}{2}\left(\frac{i}{\hbar} t\right)^2 [\overline{\mathfrak{H}}_0 [\overline{\mathfrak{H}}_0, \mathfrak{H}_E(t)]] + \cdots \right\} | \varphi \rangle , \tag{3.15}$$

where

$$\overline{\mathfrak{H}}_0 = e^{\frac{i}{\hbar}\int_{-\infty}^{t} \mathfrak{H}_E(t)\,dt}\, \mathfrak{H}_0\, e^{-\frac{i}{\hbar}\int_{-\infty}^{t} \mathfrak{H}_E(t)\,dt} . \tag{3.16}$$

The expression on the right-hand side of (3.15) is a series in powers of nuclear energy differences times the collision time, i. e., it is a series in powers of the ξ's involved. If we express the solution in a similar way,

$$| \varphi \rangle = | \varphi_0 \rangle + | \varphi_1 \rangle + | \varphi_2 \rangle + \cdots , \tag{3.17}$$

where $| \varphi_1 \rangle$ is of the order ξ times $| \varphi_0 \rangle$, $| \varphi_2 \rangle$ of the order ξ^2 times $| \varphi_0 \rangle$, etc., we obtain the following set of differential equations for these $| \varphi_n \rangle$'s:

$$\left.\begin{aligned}
& i\hbar \frac{\partial | \varphi_0 \rangle}{\partial t} = 0 \\
& i\hbar \frac{\partial | \varphi_1 \rangle}{\partial t} = \frac{i}{\hbar} t [\overline{\mathfrak{H}}_0, \mathfrak{H}_E(t)] | \varphi_0 \rangle \\
& i\hbar \frac{\partial | \varphi_2 \rangle}{\partial t} = \frac{i}{\hbar} t [\overline{\mathfrak{H}}_0, \mathfrak{H}_E(t)] | \varphi_1 \rangle + \frac{1}{2}\left(\frac{i}{\hbar} t\right)^2 [\overline{\mathfrak{H}}_0 [\overline{\mathfrak{H}}_0, \mathfrak{H}_E(t)]] | \varphi_0 \rangle \\
& \cdots\cdots\cdots\cdots\cdots\cdots\cdots\cdots
\end{aligned}\right\} \tag{3.18}$$

The initial condition determines $|\varphi_0\rangle$ to be

$$|\varphi_0\rangle = |0\rangle. \tag{3.19}$$

From $|\varphi_0\rangle$ one may determine $|\varphi_1\rangle, |\varphi_2\rangle$, etc. by means of quadratures

$$\left.\begin{aligned}
|\varphi_1\rangle &= \frac{1}{\hbar^2}\int_{-\infty}^{t} dt'\, t'\, [\overline{\mathfrak{H}}_0(t'), \mathfrak{H}_E(t')]\,|0\rangle \\
|\varphi_2\rangle &= \frac{1}{\hbar^4}\int_{-\infty}^{t} dt'\, t'\, [\overline{\mathfrak{H}}_0(t'), \mathfrak{H}_E(t')]\int_{-\infty}^{t'} dt''\, t''\, [\overline{\mathfrak{H}}_0(t''), \mathfrak{H}_E(t'')]\,|0\rangle \\
&+ \frac{1}{2}\frac{i}{\hbar^3}\int_{-\infty}^{t} dt'\, t'^2\, [\overline{\mathfrak{H}}_0(t'), [\overline{\mathfrak{H}}_0(t'), \mathfrak{H}_E(t')]]\,|0\rangle.
\end{aligned}\right\} \tag{3.20}$$

If the interaction energy $\mathfrak{H}_E(t)$ tends to zero sufficiently rapidly, all integrals converge, and (3.20) offers a systematic expansion in powers of the ξ's. In the case of quadrupole Coulomb excitation, however, $\mathfrak{H}_E(t)$ is of the order of $|t|^{-3}$ for large times, and already the second term in $|\varphi_2\rangle$ diverges. This difficulty is also encountered if one tries to expand the orbital integrals $I_{2,\mu}(\vartheta, \xi)$ in powers of ξ. The exact expression for these quantities in terms of confluent hypergeometric functions (see ref. 1, Eq. (II E. 50)) shows that the correct expansion is of the form

$$I_{2,\mu}(\vartheta, \xi) \approx a + b\xi + c\xi^2 + d\xi^2 \log \xi + \cdots. \tag{3.21}$$

In the following, we shall calculate only the first order terms of (3.20). For the evaluation of the higher terms a cut-off procedure might be used.

B. Choice of Coordinate System

In earlier calculations of Coulomb excitation, the orbital integrals were evaluated in the so-called focal system. In this coordinate system the z-axis is perpendicular to the plane of the orbit, and the x-axis is along the symmetry axis of the hyperbola. In this paper, another system will be used, where the z-axis is along the symmetry axis (see Fig. 1). This system is of special convenience for head-on collisions ($\vartheta = \pi$), where the invariance of the entire Hamiltonian for rotations around the z-axis ensures the conservation of the magnetic quantum number during the excitation process.

The time-dependence of the interaction energy (2.3) is, for $\lambda = 2$, given by the collision functions

$$\bar{S}_{2,\mu}(t) = r_p^{-3}(t)\, Y_{2,\mu}[\vartheta_p(t), \varphi_p(t)]. \tag{3.22}$$

In the new coordinate system these collision functions are explicitly given by

$$\left.\begin{aligned} \bar{S}_{2,2}(t) = \bar{S}_{2,-2}(t) &= -\sqrt{\frac{15}{32\pi}}\frac{1}{r_p(t)^3}\cdot\frac{y_p(t)^2}{r_p(t)^2} \\ \bar{S}_{2,1}(t) = \bar{S}_{2,-1}(t) &= -i\sqrt{\frac{15}{8\pi}}\frac{1}{r_p(t)^3}\cdot\frac{z_p(t)\cdot y_p(t)}{r_p(t)^2} \\ \bar{S}_{2,0}(t) &= \sqrt{\frac{5}{16\pi}}\frac{1}{r_p(t)^3}\cdot\frac{3z_p(t)^2 - r_p(t)^2}{r_p(t)^2} \end{aligned}\right\} \tag{3.23}$$

For the perturbation treatment, the important quantities are the orbital integrals defined by

$$S_{E2,\mu}(\vartheta, \xi) = \int_{-\infty}^{+\infty} \bar{S}_{2,\mu}(t)\, e^{i\xi\frac{v}{a}t}\, dt. \tag{3.24}$$

In the old focal system these orbital integrals were expressed by means of the tabulated functions $I_{2,\mu}$ (see ref. 7) in the following way:

$$S^{\text{old}}_{E2,\mu} = \frac{1}{va^2} Y_{2,\mu}\left(\frac{\pi}{2}, 0\right) I_{2,\mu}(\vartheta, \xi). \tag{3.25}$$

In the new coordinate system one can again express the orbital integrals in terms of the $I_{2,\mu}$.

$$\left.\begin{aligned} S_{E2,0}(\vartheta,\xi) &= \frac{1}{va^2}\sqrt{\frac{5}{16\pi}}\left\{\frac{1}{2}I_{2,0}(\vartheta,\xi) + \frac{3}{4}I_{2,2}(\vartheta,\xi) + \frac{3}{4}I_{2,-2}(\vartheta,\xi)\right\} \\ S_{E2,\pm1}(\vartheta,\xi) &= \frac{1}{va^2}\sqrt{\frac{15}{32\pi}}\left\{\frac{1}{2}I_{2,-2}(\vartheta,\xi) - \frac{1}{2}I_{2,2}(\vartheta,\xi)\right\} \\ S_{E2,\pm2}(\vartheta,\xi) &= \frac{-1}{va^2}\sqrt{\frac{15}{32\pi}}\left\{\frac{1}{2}I_{2,0}(\vartheta,\xi) - \frac{1}{4}I_{2,2}(\vartheta,\xi) - \frac{1}{4}I_{2,-2}(\vartheta,\xi)\right\} \end{aligned}\right\} \tag{3.26}$$

Since $\bar{S}_{2,\pm1}(t)$ is an odd function of time, one sees that $S_{2,\pm1}$ vanishes for $\xi = 0$. The two remaining orbital integrals can be expressed in a way similar to (3.25):

$$S_{E2,\mu}(\vartheta, \xi = 0) = -\frac{1}{va^2} Y_{2,\mu}\left(\frac{\pi}{2} 0\right) J_{2,\mu}(\vartheta), \tag{3.27}$$

where

$$\left.\begin{aligned} J_{2,\pm 2}(\vartheta) &= \frac{1}{2}\left[I_{2,0}(\vartheta,0) - I_{2,2}(\vartheta,0)\right] \\ J_{2,0}(\vartheta) &= \frac{3}{2} I_{2,2}(\vartheta,0) + \frac{1}{2} I_{2,0}(\vartheta,0). \end{aligned}\right\} \quad (3.28)$$

These quantities can be expressed by elementary functions, since (see ref. 1, Eq. (II E. 71))

$$\left.\begin{aligned} I_{2,\pm 2}(\vartheta,0) &= \frac{2}{3}\sin^2\frac{\vartheta}{2} \\ I_{2,0}\ \ (\vartheta,0) &= 2\tan^2\frac{\vartheta}{2}\left[1 - \frac{\pi-\vartheta}{2}\tan\frac{\vartheta}{2}\right] \end{aligned}\right\} \quad (3.29)$$

and they are tabulated in Table 2.

Table 2.

The classical orbital integrals for $\xi = 0$ in the coordinate system of Fig. 1. In the two first columns, the functions $J_{2,0}(\vartheta)$ and $J_{2,2}(\vartheta)$ (see Eq. (3.28)) are listed as functions of the deflection angle ϑ. The third column shows the ratio $J_{2,2}(\vartheta)/J_{20}(\vartheta)$ which is important for the $\chi(\vartheta)$ approximation, while the last two columns contain the quantities

$$\chi_{\text{eff}}(\vartheta)/\chi = J_{2,0}(\vartheta)\ /\ J_{2,0}(\pi) \text{ and } \chi(\vartheta)\ /\ \chi = \sqrt{(J_{20}(\vartheta))^2 + 3\,(J_{22}(\vartheta))^2}\ /\ J_{20}(\pi).$$

The entries are given in the form of a number followed (in paranthesis) by the power of ten by which it should be multiplied.

ϑ	$J_{2,0}(\vartheta)$	$J_{2,2}(\vartheta)$	$J_{2,2}/J_{2,0}$	χ_{eff}/χ	$\chi(\vartheta)/\chi$
0	0.0000	0.0000	3.333 (—1)	0.00000	0.00000
10	1.4257 (—2)	4.1288 (—3)	2.896 (—1)	0.01069	0.01196
20	5.3589 (—2)	1.3386 (—2)	2.498 (—1)	0.04019	0.04379
30	1.1360 (—1)	2.4285 (—2)	2.138 (—1)	0.08520	0.09085
40	1.9054 (—1)	3.4574 (—2)	1.814 (—1)	0.1429	0.1498
50	2.8102 (—1)	4.2878 (—2)	1.526 (—1)	0.2108	0.2180
60	3.8180 (—1)	4.8467 (—2)	1.269 (—1)	0.2864	0.2932
70	4.8973 (—1)	5.1078 (—2)	1.043 (—1)	0.3673	0.3732
80	6.0169 (—1)	5.0792 (—2)	8.442 (—2)	0.4513	0.4561
90	7.1460 (—1)	4.7935 (—2)	6.708 (—2)	0.5360	0.5396
100	8.2543 (—1)	4.2997 (—2)	5.209 (—2)	0.6191	0.6216
110	9.3125 (—1)	3.6570 (—2)	3.927 (—2)	0.6984	0.7000
120	1.0293	2.9301 (—2)	2.847 (—2)	0.7720	0.7729
130	1.1170	2.1831 (—2)	1.954 (—2)	0.8378	0.8382
140	1.1921	1.4772 (—2)	1.239 (—2)	0.8941	0.8943
150	1.2527	8.6659 (—3)	6.918 (—3)	0.9395	0.9396
160	1.2971	3.9600 (—3)	3.053 (—3)	0.9728	0.9728
170	1.3242	1.0393 (—3)	7.848 (—4)	0.9932	0.9932
180	1.3333	0.0000	0.000	1.0000	1.0000

For the special case $\vartheta = \pi$ one has $y_p = 0$, and the only non-vanishing collision function is $\bar{S}_{2,0}(t)$. In this case, $J_{2,\mu}(\pi) = 4/3\,\delta_{\mu,0}$. The special simplification for backward scattering is connected with the symmetry of the problem around the z-axis.

Also for $\vartheta \neq \pi$ one can obtain some general rules by symmetry considerations. The Hamiltonian is thus always invariant under a reflection in the plane of the orbit. This reflection brings a state vector $|I, M\rangle$ with spin I and magnetic quantum number M into a state $|I, -M\rangle$. One finds

$$|I, M\rangle \rightarrow (-1)^{p+M+I}\,|I, -M\rangle, \tag{3.30}$$

where p is the parity of the state. This rule implies that the excitation probabilities of states with magnetic quantum numbers M and $-M$ are equal, if the initial state is unoriented. The equality of $\bar{S}_{2,\mu}$ and $\bar{S}_{2,-\mu}$ follows also from this symmetry.

For $\xi = 0$, one has the additional symmetry that the Hamiltonian is invariant under a rotation of 180 degrees around the z-axis. This rotation gives rise to the following transformation:

$$|I, M\rangle \rightarrow e^{i\pi M}\,|I, M\rangle, \tag{3.31}$$

which implies

$$(-1)^{M_f - M_i} = 1, \tag{3.32}$$

where M_f and M_i are the magnetic quantum numbers in the final and initial states, respectively. The disappearance of $S_{1,\pm 1}$ for $\xi = 0$ is also a consequence of this symmetry.

C. Dependence on Deflection Angle

In the sudden approximation, the interaction energy $\mathfrak{H}_E(t)$ only enters through the expression (see Eqs. (2.3), (3.22), and (3.27))

$$\frac{1}{\hbar}\int_{-\infty}^{+\infty} \mathfrak{H}_E(t)\,dt = -\frac{4\pi Z_1 e}{5\,\hbar v a^2}\sum_{\mu} Y_{2,\mu}\left(\frac{\pi}{2}, 0\right) J_{2,\mu}(\vartheta)\,\mathfrak{M}^*(E2,\mu). \tag{3.33}$$

In this expression it will be convenient to collect the dependence on Z_1, v, and a in the parameter χ, which corresponds to the excitation from the ground state with spin I_0 to one of the excited states with spin I_1.

$$\chi_{0\rightarrow 1} = \frac{\sqrt{16\pi}}{15}\,\frac{Z_1 e}{\hbar v a^2}\cdot\frac{\langle I_0 \| \mathfrak{M}(E2) \| I_1\rangle}{\sqrt{2I_0+1}}. \tag{3.34}$$

The expression (3.33) then takes the form

$$\frac{1}{\hbar}\int_{-\infty}^{+\infty}\mathfrak{H}_E(t)\,dt = -\chi_{0\to 1}\sqrt{9\pi}\sum_{\mu} Y_{2,\mu}\left(\frac{\pi}{2},0\right)J_{2,\mu}(\vartheta)\frac{\mathfrak{M}^*(E2,\mu)\sqrt{2I_0+1}}{\langle I_0\|\mathfrak{M}(E2)\|I_1\rangle}. \quad (3.35)$$

The relative order of magnitude of the terms with $|\mu| = 2$ and $\mu = 0$ is given by the ratio $J_{2,2}(\vartheta)\,/\,J_{2,0}(\vartheta)$. This ratio, which vanishes for $\vartheta = \pi$, is given numerically in Table 2 and it is seen that it is very small for most angles.

This observation gives rise to the convenient approximation of neglecting the terms with $|\mu| = 2$. In this approximation, (3.35) has the same form for all angles and one may write it as follows:

$$\frac{1}{\hbar}\int_{-\infty}^{+\infty}\mathfrak{H}_E(t)\,dt = -\chi_{\rm eff}(\vartheta)\sqrt{9\pi}\,Y_{2,0}\left(\frac{\pi}{2},0\right)J_{2,0}(\pi)\frac{\mathfrak{M}^*(E2,0)\sqrt{2I_0+1}}{\langle I_0\|\mathfrak{M}(E2)\|I_1\rangle}, \quad (3.36)$$

where

$$\chi_{\rm eff}(\vartheta) = \chi_{0\to 1}\frac{J_{2,0}(\vartheta)}{J_{2,0}(\pi)} = \frac{3}{4}J_{2,0}(\vartheta)\,\chi_{0\to 1}. \quad (3.37)$$

If one uses the approximate interaction Hamiltonian (3.36), the final state vector for arbitrary deflection angles $|\Phi(\vartheta,\chi)\rangle$ is simply related to the state vector for backward scattering, i. e.,

$$|\Phi(\vartheta,\chi)\rangle \approx |\Phi(\pi,\chi_{\rm eff}(\vartheta))\rangle. \quad (3.38)$$

The accuracy of this approximation can easily be estimated by writing the state vector $|\Phi(\vartheta,\chi)\rangle$ in the form

$$|\Phi(\vartheta,\chi)\rangle = e^{i\chi_{\rm eff}(\vartheta)\sqrt{\frac{15}{2}}\frac{J_{22}(\vartheta)}{J_{20}(\vartheta)}\frac{\sqrt{2I_0+1}}{\langle I_0\|\mathfrak{M}(E2)\|I_1\rangle}\sum_{\mu=\pm 2}\mathfrak{M}^*(E2,\mu)}|\Phi(\pi,\chi_{\rm eff}(\vartheta))\rangle, \quad (3.39)$$

which follows from (3.11) and (3.35). In this expression, a series development of the exponential function may be performed, and one is thus led to the following expansion which contains (3.38) as the first term:

$$\left.\begin{aligned}
|\Phi(\vartheta,\chi)\rangle &= |\Phi(\pi,\chi_{\rm eff}(\vartheta))\rangle \\
&+ i\chi_{\rm eff}(\vartheta)\sqrt{\frac{15}{2}}\frac{J_{22}(\vartheta)}{J_{20}(\vartheta)}\frac{\sqrt{2I_0+1}}{\langle I_0\|\mathfrak{M}(E2)\|I_1\rangle}\sum_{\mu=\pm 2}\mathfrak{M}^*(E2,\mu)|\Phi(\pi,\chi_{\rm eff}(\vartheta))\rangle \\
&-(\chi_{\rm eff}(\vartheta))^2\frac{15}{4}\left(\frac{J_{22}(\vartheta)}{J_{20}(\vartheta)}\right)^2\frac{2I_0+1}{|\langle I_0\|\mathfrak{M}(E2)\|I_1\rangle|^2}\sum_{\mu\mu'}\mathfrak{M}^*(E2,\mu)\mathfrak{M}^*(E2,\mu')|\Phi(\pi,\chi_{\rm eff}(\vartheta))\rangle \\
&+\cdots
\end{aligned}\right\} \quad (3.40)$$

An indication of the accuracy of the approximation (3.36) can be obtained by applying it to the old perturbation calculation. For $\xi = 0$ one thus finds, by considering only the term with $\mu = 0$, a total cross section which only differs 5 per cent from the correct one, even though the forward angles, where the approximation is worst, here play a rather important role.

In the following, we shall apply the approximation (3.38) and in a number of cases also investigate the accuracy by calculating the correction terms in (3.40).

We have earlier, in Section 2, introduced a quantity $\chi(\vartheta)$ (see Eq. (2.12)) which is not very different from $\chi_{eff}(\vartheta)$; the connection between them is given by

$$\left.\begin{aligned} \chi(\vartheta) &= \chi \frac{3}{4} \sqrt{(J_{20}(\vartheta))^2 + 3(J_{22}(\vartheta))^2} \\ &\approx \chi_{eff}(\vartheta) \left\{ 1 + \frac{3}{2} \left(\frac{J_{22}(\vartheta)}{J_{20}(\vartheta)} \right)^2 + \cdots \right\}. \end{aligned}\right\} \quad (3.41)$$

As can be seen from Table 2, the two quantities $\chi(\vartheta)$ and $\chi_{eff}(\vartheta)$ differ at the most by 15 per cent, but for most angles the difference is much smaller. For foreward angles where the difference is largest, the excitation process can essentially be treated by the first order perturbation theory, where the excitation probability is $|\chi(\vartheta)|^2$. If we thus substitute $\chi(\vartheta)$ for $\chi_{eff}(\vartheta)$ in the approximation (3.38), we have made a change only of the order of $(J_{2,2}(\vartheta)/J_{2,0}(\vartheta))^2$, but on the other hand obtained an expression which leads to the correct result for the excitation probability for foreward angles.

In the more general case where the sudden approximation is not applicable, the interaction energy $\mathfrak{H}_E(t)$ enters in a more complicated way into the problem. For $\vartheta = \pi$, again only the term with $\mu = 0$ will appear. For other angles, however, the order of magnitude of the terms with $\mu \neq 0$ depends directly on the collision functions (see Eq. (3.23)). These, especially $\bar{S}_{2,1}(t)$, are in general not very small compared to $\bar{S}_{2,0}(t)$, and the approximation is thus only valid in the neighbourhood of $\vartheta = \pi$. One may here investigate the angular dependence by considering the terms with $\mu \neq 0$ in $\mathfrak{H}_E(t)$ as a perturbation in the Hamiltonian. In the sudden approximation, this method would just lead to the result (3.40).

For the sake of completeness it should be mentioned that, once the final amplitudes $a_{I,M}(\infty)$ on the states with spin I and magnetic quantum number M are known, one may easily obtain all the quantities which are important for the experiments. Thus, the total excitation probability of a level of spin I_f is given by

$$P_{I_f I_i} = \frac{1}{2 I_i + 1} \sum_{M_i, M_f} | a_{I_f M_f} |^2. \tag{3.42}$$

The differential cross section $d\sigma$ is obtained by multiplying P with the Rutherford cross section, i. e.,

$$\left. \begin{aligned} d\sigma &= P_{I_f I_i} d\sigma_R \\ &= \frac{a^2}{4} P_{I_f I_i} \sin^{-4}\left(\frac{\vartheta}{2}\right) d\Omega. \end{aligned} \right\} \tag{3.43}$$

The angular distribution of γ-quanta emitted after the excitation is also calculable from the amplitudes. One must here take into account that a level which emits the γ-quantum under consideration may be populated not only through an excitation, but also through the deexcitation by cascade γ's from higher excited states.

4. Diagonalization Method

In this section, we shall discuss a method of evaluating the multiple Coulomb excitation which does not use any specific nuclear model. We shall thus consider the properties of the nuclear states, i. e., energies and transition matrix elements as empirically determined quantities. Since, in this case, we have a very large number of parameters in the problem, it is not practically possible to give a systematic numerical tabulation of cross sections, etc., and we shall therefore confine ourselves to a few numerical examples which illustrate some important aspects of the problem.

A. Sudden Approximation

In the sudden approximation (3.11), we have the following expression for the final amplitude a_n on the state $| n \rangle$:

$$a_n = \langle n | e^{-\frac{i}{\hbar}\int_{-\infty}^{\infty} \mathfrak{H}_E(t)\, dt} | 0 \rangle, \tag{4.1}$$

where the exponent is given by (3.33) or (3.35). If the wave functions of the nuclear states are known, the problem is reduced to calculate matrix elements of a known operator. Usually one will be interested, however, in calculating the cross sections from a knowledge of the matrix elements of the multipole operators themselves. These matrix elements enter in other processes also and are often determined from nuclear spectroscopy.

In order to perform this calculation, we introduce a unitary transformation U which diagonalizes the hermitian operator (3.33) and which is thus defined by the equations

$$U^{\dagger} U = UU^{\dagger} = 1 \tag{4.2}$$

and

$$\left.\begin{aligned} &\langle n \mid U^{\dagger} \frac{1}{\hbar} \int_{-\infty}^{+\infty} \mathfrak{H}_E(t)\, dt\, U \mid q \rangle = \delta_{nq} \cdot \lambda_q \\ &= \sum_{m,p} \langle n \mid U^{\dagger} \mid m \rangle \langle m \mid \frac{1}{\hbar} \int_{-\infty}^{+\infty} \mathfrak{H}_E(t)\, dt \mid p \rangle \langle p \mid U \mid q \rangle. \end{aligned}\right\} \tag{4.3}$$

The result (4.1) can then be expressed in terms of U and the eigenvalues λ_q in the following way:

$$\left.\begin{aligned} a_n &= \langle n \mid UU^{\dagger} e^{-\frac{i}{\hbar}\int_{-\infty}^{\infty} \mathfrak{H}_E(t)\, dt} UU^{\dagger} \mid 0 \rangle \\ &= \sum_m \langle n \mid U \mid m \rangle e^{-i\lambda_m} \langle m \mid U^{\dagger} \mid 0 \rangle \\ &= \sum_m \langle n \mid U \mid m \rangle \langle 0 \mid U \mid m \rangle^* e^{-i\lambda_m}. \end{aligned}\right\} \tag{4.4}$$

The determination of U requires the knowledge of the matrix elements of the operator (3.33). If we specify the nuclear states by means of the spin I_n and magnetic quantum number M_n, these matrix elements are expressible by the reduced multipole matrix elements* (9) defined by

$$\langle I_m \| \mathfrak{M}(E2) \| I_n \rangle = (-1)^{I_m - M_m} \begin{pmatrix} I_m & 2 & I_n \\ -M_m & \mu & M_n \end{pmatrix}^{-1} \langle I_m M_m \mid \mathfrak{M}(E2, \mu) \mid I_n M_n \rangle. \tag{4.5}$$

For the diagonalization it will be convenient to apply the $\chi(\vartheta)$ approximation. In this approximation, the operator $\int_{-\infty}^{\infty} \mathfrak{H}_E(t)\, dt$ (see Eq. (3.36)) is diagonal in M, and one may write the matrix elements in the form

$$\langle I_m M \mid \frac{1}{\hbar} \int_{-\infty}^{+\infty} \mathfrak{H}_E(t)\, dt \mid I_n M' \rangle \approx \chi_{\text{eff}}(\vartheta)\, \varrho_{mn}^{M} \delta_{MM'}, \tag{4.6}$$

where the (symmetric) matrix ϱ_{mn}^{M} is defined by

* For the angular momentum algebra we use throughout this paper the notation of ref. 9.

$$\varrho_{mn}^{M} = (-1)^{I_m - M} \sqrt{5\,(2\,I_0+1)} \begin{pmatrix} I_m & 2 & I_n \\ -M & 0 & M \end{pmatrix} \frac{\langle I_m \,||\, \mathfrak{M}(E\,2) \,||\, I_n \rangle}{\langle I_0 \,||\, \mathfrak{M}(E\,2) \,||\, I_1 \rangle}. \tag{4.7}$$

The number of states which have to be included in the matrix (4.6) by the diagonalization depends of course on the χ's. Only those states which are mutually connected with large (collective) matrix elements must be taken into account. One may furthermore classify these in different groups where states within a group are strongly coupled, while states from two different groups are weakly coupled. A group consists, e. g., of the states in a rotational band, and the different groups are the bands belonging to different single-particle states. For each group one must perform the diagonalization and must here take into account a number of those states which are most directly coupled to the ground state. This number will depend on the χ's and must be determined so that the inclusion of still more states would not change the result. The weak interplay between the groups can be treated by a perturbation calculation.

Since the energy of the projectile and the deflection angle enter only through the common factor $\chi_{\text{eff}}(\vartheta)$, the diagonalization can be used for all energies and all ϑ's.

The deviation from the $\chi_{\text{eff}}(\vartheta)$ approximation is given by the expression (3.40) which we may write explicitly in the form

$$\left.\begin{aligned}
a_{I_f M_f} &= a^{(0)}_{I_f M_f}(\chi_{\text{eff}}(\vartheta))\,\delta_{M_i M_f} \\
&+ i\chi_{\text{eff}}(\vartheta)\frac{J_{22}(\vartheta)}{J_{20}(\vartheta)} \sum_{z\,\mu=\pm 2} \sqrt{\frac{3}{2}} \begin{pmatrix} I_z & 2 & I_f \\ -M_i & \mu & M_f \end{pmatrix} \begin{pmatrix} I_z & 2 & I_f \\ -M_i & 0 & M_i \end{pmatrix}^{-1} \varrho_{zf}^{M_i}\, a^{(0)}_{I_z M_i}(\chi_{\text{eff}}(\vartheta)) \\
&- \left(\chi_{\text{eff}}(\vartheta)\frac{J_{22}(\vartheta)}{J_{20}(\vartheta)}\right)^2 \sum_{zz'\mu\mu'=\pm 2} \frac{3}{4} \begin{pmatrix} I_z & 2 & I_z' \\ -M_i & \mu & M_z' \end{pmatrix} \begin{pmatrix} I_z & 2 & I_z' \\ -M_i & 0 & M_i \end{pmatrix}^{-1} \varrho_{zz'}^{M_i} \\
&\quad \times \begin{pmatrix} I_z' & 2 & I_f \\ -M_z' & \mu & M_f \end{pmatrix} \begin{pmatrix} I_z' & 2 & I_f \\ -M_i & 0 & M_i \end{pmatrix}^{-1} \varrho_{z'f}^{M_i}\, a^{(0)}_{I_z M_i}(\chi_{\text{eff}}(\vartheta)) \\
&+ \cdots,
\end{aligned}\right\} \tag{4.8}$$

where $a^{(0)}$ indicates the amplitude in the $\chi_{\text{eff}}(\vartheta)$ approximation. Since, in this approximation, $M_i = M_f$, it is seen that, while the first term only contributes to this substate, the second term proportional to $J_{22}(\vartheta) \,/\, J_{20}(\vartheta)$ only contributes to the states with $M_f = M_i \pm 2$. The third term proportional to $(J_{22}(\vartheta) \,/\, J_{20}(\vartheta))^2$ contributes to both $M_f = M_i$ and $M_f = M_i \pm 4$. The excitation probability will thus contain no terms linear in $J_{22}(\vartheta) \,/\, J_{20}(\vartheta)$. The

2*

terms quadratic in this quantity will arise partly from the square of the second term, and partly from an interference between the first and third term.

One may also avoid the $\chi(\vartheta)$ approximation and the expansion (4.8) by directly diagonalizing the matrix of the complete Hamiltonian (3.35). This matrix is no more diagonal in the magnetic quantum number M and is essentially different for different angles so that the diagonalization will have to be performed for all angles.

For not too large values of $\chi(\vartheta)$ it may be advantageous to use the perturbation expansion to higher order instead of the diagonalization method. The power series expansion of (3.11) leads to the following expression for the excitation amplitude:

$$\left.\begin{aligned} a_n &= \langle I_n M \,|\, e^{-\frac{i}{\hbar}\int_{-\infty}^{\infty} \mathfrak{H}_E(t)\,dt} \,|\, I_0 M\rangle \\ &= \delta_{no} - i\chi(\vartheta)\,\varrho_{no}^M - \frac{1}{2!}\chi(\vartheta)^2 \sum_r \varrho_{nr}^M \varrho_{ro}^M \\ &\quad + \frac{i}{3!}\chi(\vartheta)^3 \sum_{rs} \varrho_{nr}^M \varrho_{rs}^M \varrho_{so}^M + \cdots. \end{aligned}\right\} \quad (4.9)$$

This expansion can also be useful for the discussion of small changes in the matrix elements, e. g., from a rotational model.

B. Examples

In this section, we shall consider some examples of the methods discussed above. They will mainly be given in order to illustrate how many levels one has to take into account in the diagonalization method, and secondly to illustrate the accuracy of the perturbation expansion and the $\chi(\vartheta)$ approximation. For the sake of comparison with the exact treatment (see Section 5), we shall use the matrix elements characteristic of a rotational band.

For a pure rotational band, one may express the reduced matrix elements entering in (4.7) by means of the constant intrinsic quadrupole moment Q_0. One finds

$$\left.\begin{aligned} \langle I_m \| \mathfrak{M}(E2) \| I_n\rangle &= \sqrt{\frac{5}{16\pi}}(-1)^{I_m-K}(2I_m+1)^{1/2}(2I_n+1)^{1/2} \\ &\quad \begin{pmatrix} I_m & 2 & I_n \\ -K & 0 & K \end{pmatrix} e\,Q_0, \end{aligned}\right\} \quad (4.10)$$

where K is the (constant) projection of the total angular momentum on the nuclear symmetry axis. We shall consider only the case of an even-even nucleus with ground state spin $I_0 = K = 0$. In this case, the matrix ϱ_{mn} (see Eq. (4.7)) takes the form

$$\varrho_{mn} = \sqrt{5}\,(2\,I_m + 1)^{1/2}\,(2\,I_n + 1)^{1/2} \begin{pmatrix} I_m & 2 & I_n \\ 0 & 0 & 0 \end{pmatrix}^2 . \tag{4.11}$$

We shall now successively take more and more states into account. If we include only the ground state and the first excited state $I_1 = 2$, we have to diagonalize the matrix

$$\varrho_{mn} = \begin{Bmatrix} 0 & 1 \\ 1 & \frac{2\sqrt{5}}{7} \end{Bmatrix} . \tag{4.12}$$

The eigenvalues of this matrix are

$$\lambda_0 = \frac{\sqrt{5} - 3\sqrt{6}}{7} \quad \text{and} \quad \lambda_1 = \frac{\sqrt{5} + 3\sqrt{6}}{7} . \tag{4.13}$$

The unitary matrix which diagonalizes (4.12) is then found to be

$$\langle m \,|\, U \,|\, n \rangle = \begin{Bmatrix} \frac{\sqrt{18 + \sqrt{30}}}{6} & \frac{\sqrt{18 - \sqrt{30}}}{6} \\ -\frac{\sqrt{18 - \sqrt{30}}}{6} & \frac{\sqrt{18 + \sqrt{30}}}{6} \end{Bmatrix} . \tag{4.14}$$

According to (4.4) we thus obtain the result

$$\left.\begin{aligned} a_0 &= \frac{18 + \sqrt{30}}{36}\, e^{-i\chi(\vartheta)\frac{\sqrt{5} - 3\sqrt{6}}{7}} + \frac{18 - \sqrt{30}}{36}\, e^{-i\chi(\vartheta)\frac{\sqrt{5} + 3\sqrt{6}}{7}} \\ &= e^{-i\frac{\sqrt{5}}{7}\chi(\vartheta)} \left[\cos\frac{3\sqrt{6}}{7}\chi(\vartheta) + i\sqrt{\frac{5}{54}}\sin\frac{3\sqrt{6}}{7}\chi(\vartheta) \right] \\ a_1 &= -\frac{7\sqrt{6}}{36}\, e^{-i\chi(\vartheta)\frac{\sqrt{5} - 3\sqrt{6}}{7}} + \frac{7\sqrt{6}}{36}\, e^{-i\chi(\vartheta)\frac{\sqrt{5} + 3\sqrt{6}}{7}} \\ &= -i\frac{7\sqrt{6}}{18}\, e^{-i\frac{\sqrt{5}}{7}\chi(\vartheta)} \sin\frac{3\sqrt{6}}{7}\chi(\vartheta) . \end{aligned}\right\} \tag{4.15}$$

The excitation probability $P_2 = |a_1|^2$ is then

$$P_2 = \frac{49}{54}\sin^2\frac{3\sqrt{6}}{7}\chi(\vartheta). \tag{4.16}$$

This quantity and the probability that the nucleus is left in the ground state $P_0 = |a_0|^2 = 1 - P_2$ are illustrated in Fig. 2 as a function of $\chi(\vartheta)$.

In Fig. 3 and Fig. 4 are shown the extensions of the above calculation to include two and four excited states in the rotational band. The matrix ϱ_{mn}, in the latter case, is explicitly given by

$$\varrho_{mn} = \begin{Bmatrix} 0 & 1.0000 & 0 & 0 & 0 \\ 1.0000 & 0.6389 & 0.8571 & 0 & 0 \\ 0 & 0.8571 & 0.5808 & 0.8457 & 0 \\ 0 & 0 & 0.8457 & 0.5692 & 0.8423 \\ 0 & 0 & 0 & 0.8423 & 0.5649 \end{Bmatrix} \tag{4.17}$$

In the case that one includes only two of the excited states, one finds the eigenvalues

$$\lambda_0 = -0.9270, \quad \lambda_1 = 0.3484 \quad \text{and} \quad \lambda = 1.7984 \tag{4.18}$$

and the matrix U is then

$$\langle m|U|n\rangle = \begin{Bmatrix} 0.6840 & -0.6006 & 0.4139 \\ -0.6341 & -0.2093 & 0.7444 \\ 0.3605 & 0.7717 & 0.5240 \end{Bmatrix} \tag{4.19}$$

The final amplitudes on the three states are thus, according to (4.4),

$$\left.\begin{aligned} a_0 &= 0.4679\, e^{i0.9270\chi} + 0.3608\, e^{-i0.3484\chi} + 0.1713\, e^{-i1.7984\chi} \\ a_1 &= -0.4338\, e^{i0.9270\chi} + 0.1257\, e^{-i0.3484\chi} + 0.3081\, e^{-i1.7984\chi} \\ a_2 &= 0.2466\, e^{i0.9270\chi} - 0.4635\, e^{-i0.3484\chi} + 0.2169\, e^{-i1.7984\chi} \end{aligned}\right\} \tag{4.20}$$

Similarly, one finds for the complete matrix (4.17) the eigenvalues

$$\left.\begin{aligned} \lambda_0 &= -1.0437 \\ \lambda_1 &= -0.4880 \\ \lambda_2 &= 0.4302 \\ \lambda_3 &= 1.3920 \\ \lambda_4 &= 2.0633 \end{aligned}\right\} \tag{4.21}$$

and the matrix U

$$\langle m \mid U \mid n \rangle = \left\{ \begin{array}{rrrrr} 0.5436 & -0.5189 & 0.4681 & -0.3866 & 0.2582 \\ -0.5674 & 0.2532 & 0.2014 & -0.5381 & 0.5328 \\ 0.4795 & 0.2725 & -0.5951 & -0.0218 & 0.5841 \\ -0.3461 & -0.6010 & -0.0981 & 0.5245 & 0.4840 \\ 0.1812 & 0.4808 & 0.6137 & 0.5342 & 0.2721 \end{array} \right\} \quad (4.22)$$

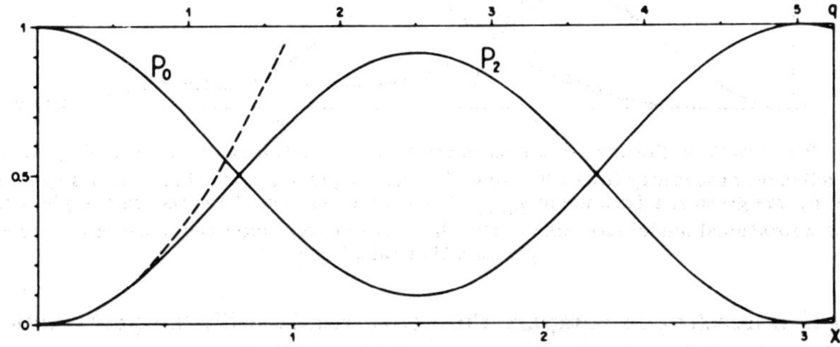

Fig. 2. The result of the two-state calculation for a rotational band on a 0^+ ground state. The probability for the excitation in the 2^+ state, P_2, and the probability for no excitation, P_0, are given as functions of $\chi_{0\to 2}(\vartheta)$ and as a function of the parameter $q(\vartheta)$ characteristic of the rotational model (see Eq. (5.11)). The broken curve shows the result of the first order perturbation calculation.

Figs. 2, 3, and 4 show a very general feature of the multiple excitation process. The excitation probability for a definite state has a maximum as a function of χ. Where this maximum is reached depends on how directly the state is connected with the ground state. The more intermediate states that have to be passed, the higher is the value of χ for which the maximum is attained. For the rotational band on the 0^+ ground state, the 2^+ state is maximally excited for $\chi \simeq 1$, the 4^+ for $\chi \simeq 2$, the 6^+ for $\chi \simeq 3$, etc. The heights of the maxima decrease as one passes to higher excited states, partly because a small tail is left in the excitation probability of the lower states. If the band is broken off as in the above calculation, the maximum in the excitation of the last state is much higher than that of any of the others.

A comparison of the curves shows that the deletion of higher states practically does not change the excitation probability of the lower states. This is true at least as long as the last state included is not strongly excited. In Fig. 2, the curve for P_0 is thus essentially correct until $\chi = 0.8$. In Fig. 3, the curves for P_0 and P_2 are similarly correct until $\chi = 1.5$ and, in Fig. 4, one expects P_0, P_2, P_4, and P_6 to be correct until $\chi = 3$.

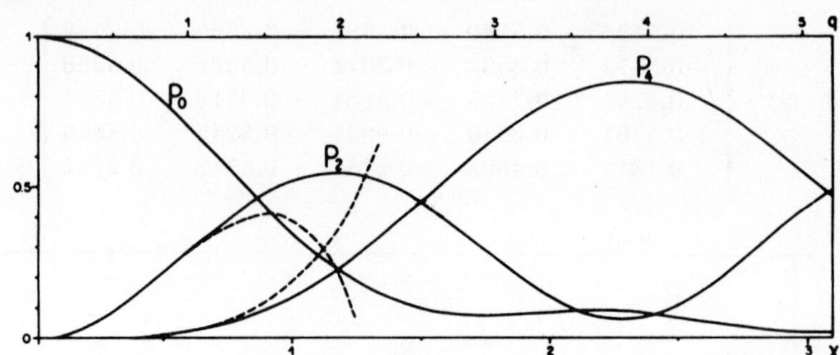

Fig. 3. The result of the three-state calculation for a rotational band on a 0^+ ground state. The excitation probability in the 2^+ and 4^+ states, P_2 and P_4, and the probability for no excitation, P_0, are given as a function of $\chi_{0\to 2}(\vartheta)$ and as a function of the parameter $q(\vartheta)$ characteristic of a rotational model (see Eq. (5.11)). The broken curves show the result of the second order perturbation calculation.

It is interesting to compare the above results with the perturbation calculation. According to the equation (4.9) one finds

$$\left.\begin{aligned} a_0 &= 1 \quad -0.5000\,\chi(\vartheta)^2 + i\,0.1065\,\chi(\vartheta)^3 + 0.0893\,\chi(\vartheta)^4 - i\,0.0242\,\chi(\vartheta)^5 \\ &\qquad -0.0092\,\chi(\vartheta)^6 \\ a_1 &= -i\,\chi(\vartheta) - 0.3195\,\chi(\vartheta)^2 + i\,0.3571\,\chi(\vartheta)^3 + 0.1210\,\chi(\vartheta)^4 - i\,0.0551\,\chi(\vartheta)^5 \\ a_2 &= \qquad -0.4286\,\chi(\vartheta)^2 + i\,0.1742\,\chi(\vartheta)^3 + 0.1274\,\chi(\vartheta)^4 \\ a_3 &= \qquad + i\,0.1208\,\chi(\vartheta)^3 \end{aligned}\right\} \quad (4.23)$$

The power expansion in χ of the excitation probabilities contains (since $\xi = 0$) only even powers of χ. It is noted that, e. g., a third order perturbation calculation leads to the correct answer for P_6 to terms of the order of χ^6, while P_2 and P_4 are correct only to terms of the order of χ^4, and P_0 only to terms of the order of χ^2.

In the comparison of the perturbation expansion with the more exact treatment given above, we have calculated the excitation probabilities to second, fourth, and sixth order. In Fig. 2 is shown the calculation to second order (first order perturbation). This gives a good approximation only up to $\chi \cong 0.4$. In Fig. 3 is shown the calculation to fourth order in χ. This is good up to $\chi \cong 0.7$. Similarly, the calculation to sixth order in χ shown on Fig. 4 is seen to be correct only up to $\chi \cong 1.0$. It thus seems that the perturbation expansion only offers a poor approximation for large values of χ.

The accuracy of the $\chi(\vartheta)$ approximation which we have used can be

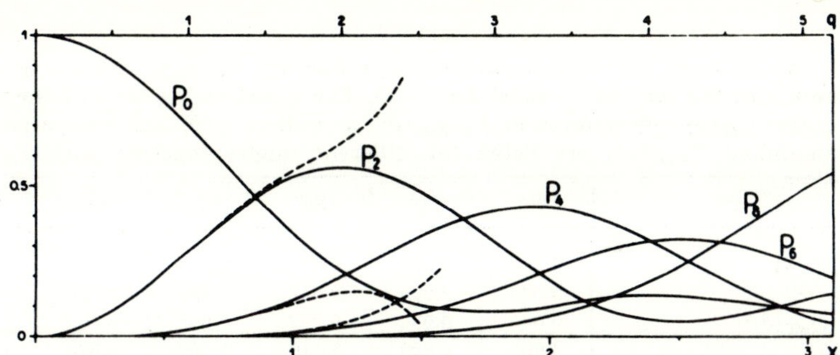

Fig. 4. The result of the five-state calculation for a rotational band on a 0^+ ground state. The excitation probabilities for the 2^+, 4^+, 6^+ and 8^+ states P_2, P_4, P_6 and P_8, and the probability for no excitation, P_0, are given as a function of $\chi_{0\to 2}(\vartheta)$ and as a function of the parameter $q(\vartheta)$ characteristic of a rotational model (see Eq. (5.11)). The broken curves show the result of the third order perturbation calculation.

evaluated by means of Eq. (4.8). We shall only do the explicit calculation in the case of the two-state model (4.12) to (4.16). One finds directly from (4.8) the following expressions to second order in $J_{22}(\vartheta)/J_{20}(\vartheta)$:

$$\left.\begin{aligned}
a_{22} &= i\chi_{eff}(\vartheta)\frac{J_{22}(\vartheta)}{J_{20}(\vartheta)}e^{-i\frac{\sqrt{5}}{7}\chi_{eff}(\vartheta)}\left[\sqrt{\frac{3}{2}}\cos\frac{3\sqrt{6}}{7}\chi_{eff}(\vartheta)+i\frac{\sqrt{5}}{2}\sin\frac{3\sqrt{6}}{7}\chi_{eff}(\vartheta)\right]\\
a_{20} &= a_1^{(0)}-\left(\chi_{eff}(\vartheta)\frac{J_{22}(\vartheta)}{J_{20}(\vartheta)}\right)^2 e^{-i\frac{\sqrt{5}}{7}\chi_{eff}(\vartheta)}\left[-\frac{3\sqrt{5}}{7}\cos\frac{3\sqrt{6}}{7}\chi_{eff}(\vartheta)-i\frac{5\sqrt{6}}{14}\sin\frac{3\sqrt{6}}{7}\chi_{eff}(\vartheta)\right]\\
a_{00} &= a_0^{(0)}-\left(\chi_{eff}(\vartheta)\frac{J_{22}(\vartheta)}{J_{20}(\vartheta)}\right)^2 e^{-i\frac{\sqrt{5}}{7}\chi_{eff}(\vartheta)}\left[\frac{3}{2}\cos\frac{3\sqrt{6}}{7}\chi_{eff}(\vartheta)+i\frac{\sqrt{30}}{4}\sin\frac{3\sqrt{6}}{7}\chi_{eff}(\vartheta)\right].
\end{aligned}\right\}\quad (4.24)$$

In these expressions, $a^{(0)}$ are the amplitudes in the $\chi_{eff}(\vartheta)$ approximation (4.15).

From (4.24) one obtains the excitation probability

$$P_2(\vartheta,\chi)=\frac{49}{54}\sin^2\frac{3\sqrt{6}}{7}\chi_{eff}(\vartheta)+\left(\chi_{eff}(\vartheta)\frac{J_{22}(\vartheta)}{J_{20}(\vartheta)}\right)^2\left[\frac{5}{6}+\frac{13}{6}\cos^2\frac{3\sqrt{6}}{7}\chi_{eff}(\vartheta)\right]. \quad (4.25)$$

One observes that the correction term is an oscillating function of $\chi_{eff}(\vartheta)$ which has its maxima where $P_2^{(0)}$ shows its minima. The tendency of the correction is thus to fill out the minima of the excitation probability.

To illustrate the magnitude of the correction we have evaluated (4.25)

TABLE 3.

Comparison between the correct excitation probability $P_2(\vartheta, \chi)$ and the $\chi(\vartheta)$ approximation in the two-state model for $\chi = 3$. The quantity $\chi_{eff}(\vartheta)$ and the probability in the $\chi_{eff}(\vartheta)$ approximation $P_2(\chi_{eff}(\vartheta))$ as well as $\chi(\vartheta)$ and the corresponding probability $P_2(\chi(\vartheta))$ are listed for different angles together with $P_2(\vartheta, \chi)$.

	180°	150°	120°	90°	60°	30°
$P_2(\vartheta, \chi)$	0.000	0.031	0.396	0.880	0.578	0.072
$\chi_{eff}(\vartheta)$	3.000	2.820	2.315	1.698	0.859	0.256
$P_2(\chi_{eff}(\vartheta))$	0.000	0.030	0.387	0.868	0.558	0.064
$\chi(\vartheta)$	3.000	2.820	2.318	1.709	0.880	0.272
$P_2(\chi(\vartheta))$	0.000	0.030	0.384	0.863	0.578	0.072

numerically in the case of $\chi = 3$, and the result is given in Table 3. One observes here that the maximum correction (~ 0.020) appears for angles between 60 and 90 degrees. This is connected with the fact that $J_{22}(\vartheta)$ is maximal in this range. In the two last rows we have made a comparison with the approximation where $\chi(\vartheta)$ is used instead of $\chi_{eff}(\vartheta)$. It is seen that this approximation reproduces the correct excitation probability for small angles until the angle which gives the maximum probability. On the other side of the maximum, the $\chi(\vartheta)$ approximation is no improvement over the $\chi_{eff}(\vartheta)$ approximation.

C. First Order Expansion in ξ

In this paragraph, we shall consider the first order corrections in ξ to the results which were derived earlier in this section.

In Section 3A, we obtained the following expression for the amplitude to first order in ξ:

$$\left.\begin{aligned} a_n &= a_n(\xi = 0) \\ &+ \frac{1}{\hbar^2}\langle n \,|\, e^{-\frac{i}{\hbar}\int_{-\infty}^{\infty}\mathfrak{H}_E(t)\,dt}\int_{-\infty}^{\infty} dt\, t\, e^{\frac{i}{\hbar}\int_{-\infty}^{t}\mathfrak{H}_E(t')\,dt'}[\mathfrak{H}_0, \mathfrak{H}_E(t)]\, e^{-\frac{i}{\hbar}\int_{-\infty}^{t}\mathfrak{H}_E(t')\,dt'}\,|\,0\rangle. \end{aligned}\right\} \quad (4.26)$$

We shall here make the simplifying assumption that $\vartheta = \pi$. One may use the result which we shall obtain, for other angles also, by the usual substitution $\chi \to \chi_{eff}(\vartheta)$. As was discussed earlier, the approximation is here less accurate than it was in the case of the sudden approximation.

The simplification by considering only terms with $\mu = 0$ in $\mathfrak{H}_E(t)$ is that

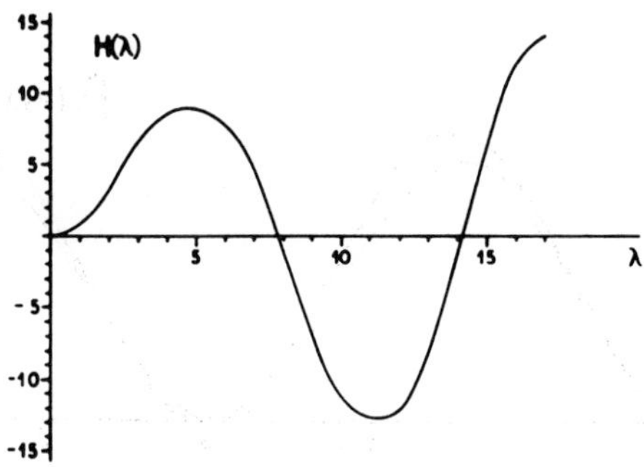

Fig. 5. The function $H(\lambda)$. This function is of importance for the evaluation of the deviation from the sudden approximation in the diagonalization method (see Eq. (4.31)).

the unitary matrix U (see Eqs. (4.2) and (4.3)) diagonalizes not only $\int_{-\infty}^{\infty}\mathfrak{H}_E(t)\,dt$, but also $\int_{-\infty}^{t}\mathfrak{H}_E(t)\,dt$ and $\mathfrak{H}_E(t)$. We thus have

$$\langle n\,|\,U^{\dagger}\frac{1}{\hbar}\int_{-\infty}^{t}\mathfrak{H}_E(t)\,dt\,U\,|\,q\rangle = \lambda_q\,\delta_{qn}\left(\frac{1}{2}+h(t)\right) \tag{4.27}$$

and

$$\langle n\,|\,U^{\dagger}\frac{1}{\hbar}\mathfrak{H}_E(t)\,U\,|\,q\rangle = \lambda_q\,\delta_{qn}\,g(t), \tag{4.28}$$

where

$$h(t) = (S_{E\,2,0}(\vartheta,0))^{-1}\int_{0}^{t}\bar{S}_{20}(t)\,dt \tag{4.29}$$

and

$$g(t) = (S_{E\,2,0}(\vartheta,0))^{-1}\,\bar{S}_{20}(t). \tag{4.30}$$

The functions $S_{E\,2,0}$ and $\bar{S}_{2,0}$ are defined in Eqs. (3.23) and (3.24), and one sees that $h(t)$ is an odd function while $g(t)$ is an even function of t.

By introducing an appropriate number of factors $UU^{\dagger}$ in (4.26) one may write it in the form

$$\left.\begin{aligned} a_n &= a_n(\xi=0)\\ &-i\sum_{lm}\langle n\,|\,U\,|\,l\rangle\, e^{-i\frac{1}{2}(\lambda_l+\lambda_m)}\,H(\lambda_l-\lambda_m)\,\Xi_{lm}\langle m\,|\,U^{\dagger}\,|\,0\rangle,\end{aligned}\right\} \tag{4.31}$$

where the Ξ matrix is the transformed energy matrix

$$\left.\begin{aligned}\Xi_{lm} &= \frac{a}{\hbar v}(U^{\dagger}\,\mathfrak{H}_0\,U)_{lm}\\ &= \sum_{p}\langle l\,|\,U^{\dagger}\,|\,p\rangle\frac{aE_p}{\hbar v}\langle p\,|\,U\,|\,m\rangle.\end{aligned}\right\} \tag{4.32}$$

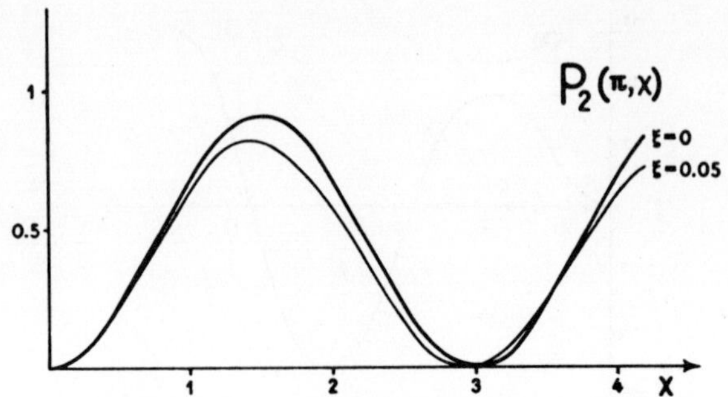

Fig. 6. The excitation probability $P_2(\pi, \lambda)$ for backward scattering in the two-state model is shown as a function of χ for $\xi = 0$ and $\xi = 0.05$.

Since a constant energy will give no contribution, one may replace the energies in (4.32) by the ξ's corresponding to the excitation of the p s state from the ground state, i. e.,

$$\xi_p = \frac{a(E_p - E_0)}{\hbar v}. \tag{4.33}$$

One thus finds

$$\Xi_{lm} = \sum_p \xi_p \langle p \mid U \mid l \rangle^* \langle p \mid U \mid m \rangle. \tag{4.34}$$

The function $H(\lambda)$ which appears in (4.32) is defined by

$$H(\lambda) = \frac{v}{a} \lambda \int_{-\infty}^{\infty} dt\, t\, g(t) \sin(\lambda h(t)). \tag{4.35}$$

One observes that $H(\lambda)$ is a symmetric function of λ, i.e.,

$$H(-\lambda) = H(\lambda) \tag{4.36}$$

and that the Ξ matrix is symmetric also in the indices l and m.

The function $H(\lambda)$ has been evaluated numerically and is given in Fig. 5. For small values of λ it is quadratic in λ, as may be seen from (4.35), and one finds

$$H(\lambda) \approx 0.9172\, \lambda^2 \quad (\lambda \ll 1). \tag{4.37}$$

For larger values of λ, $H(\lambda)$ is an oscillating function whose amplitude increases slowly. From formula (4.31) one may thus draw the general conclusion that the first order correction in ξ is only a slowly increasing (and oscillating) function of χ.

As an illustration we shall apply the result (4.31) to the two-state model.

Using the Eqs. (4.12) to (4.16) one finds the following expression for the excitation probability to first order in ξ:

$$P_2 = \frac{49}{54}\sin^2\frac{3\sqrt{6}}{7}\chi - \xi\frac{49\sqrt{30}}{972}\sin\left(\frac{3\sqrt{6}}{7}\chi\right)H\left(\frac{6\sqrt{6}}{7}\chi\right). \qquad (4.38)$$

This result is illustrated in Fig. 6, where the excitation probability for $\xi = 0.05$ is compared to the earlier calculated excitation probability for $\xi = 0$.

5. Excitation of Rotational States

In this section we shall treat the excitation of a rotational band. It will be shown that, in the sudden approximation, one can obtain a closed expression for the cross section including all (infinitely many) states in the band. The problem is analogous to the classical problem of a charged ellipsoid which is set in motion by a fast projectile. At the end of the section we shall make some comments on this classical treatment.

A. Sudden Approximation

We shall assume that we have a pure rotational band and that only this band is involved in the excitation process. The Schrödinger equation (3.1) for the rotational motion may then be written in the form

$$i\hbar\frac{\partial\overline{\psi}}{\partial t} = H_0\overline{\psi} + H'(t)\overline{\psi}, \qquad (5.1)$$

where $\overline{\psi}$ only depends on the Eulerian angles α and β describing the orientation of the nuclear symmetry axis. The complete wave function ψ is connected with $\overline{\psi}$ through the equation

$$\psi = e^{-\frac{i}{\hbar}E_{\mathrm{intr}}t}\,\overline{\psi}(\alpha, \beta, t)\,\chi(x'), \qquad (5.2)$$

where $\chi(x')$ is the intrinsic wave function and E_{intr} is the intrinsic energy. The free Hamiltonian for the rotation H_0 is given by

$$H_0 = -\frac{\hbar^2}{2\mathfrak{J}}\left\{\frac{\partial^2}{\partial\beta^2} + \cot\beta\frac{\partial}{\partial\beta} + \frac{1}{\sin^2\beta}\frac{\partial^2}{\partial\alpha^2}\right\}, \qquad (5.3)$$

where $\mathfrak{J}$ is the moment of inertia. Since the quadrupole operator can be expressed in terms of the intrinsic quadrupole moment Q_0 in the following way:

$$\mathfrak{M}(E\,2,\mu) = \frac{1}{2} Q_0 Y_{2\mu}(\beta,\alpha), \tag{5.4}$$

the interaction Hamiltonian $H'(t)$ is given by

$$H'(t) = \frac{2\pi Z_1 e}{5} Q_0 \sum_{\mu} \bar{S}_{2\mu}(t) Y^*_{2\mu}(\beta,\alpha). \tag{5.5}$$

The evaluation of the excitation amplitudes in the sudden approximation has now been reduced to the calculation of matrix elements of a known operator. We shall specify the eigenstates of H_0 by means of the spin I, the magnetic quantum number M, and the (constant) projection K of the total angular momentum on the nuclear symmetry axis. The wave function may then be written

$$\bar{\psi}_{IMK} = \sqrt{\frac{2I+1}{4\pi}} D^I_{MK}(\alpha,\beta,0), \tag{5.6}$$

where D^I_{MK} is the rotation matrix. The excitation amplitude on the state specified by I_f, M_f, and K is then, according to (3.11), (3.24), and (3.27),

$$\left.\begin{aligned} a_{I_f M_f} &= \sqrt{\frac{(2I_i+1)(2I_f+1)}{(4\pi)^2}} \int_0^{2\pi} d\alpha \int_0^{\pi} d\beta \sin\beta \,(D^{I_f}_{M_f K}(\alpha,\beta,0))^* \\ &\times D^{I_i}_{M_i K}(\alpha,\beta,0) \exp\left\{ i \frac{2\pi Z_1 e Q_0}{5\hbar v a^2} \sum_{\mu} Y_{2\mu}\left(\frac{\pi}{2},0\right) J_{2\mu}(\vartheta) Y^*_{2\mu}(\beta,\alpha)\right\}. \end{aligned}\right\} \tag{5.7}$$

We shall now first show that the excitation of any rotational band with ground state spin I_i and final state spin I_f can be expressed by means of the amplitudes for the excitation of a band with ground state spin 0. This follows from (5.7) by expanding the product of the two D-functions on D-functions. The amplitude (5.7) may then be expressed in the following way:

$$\left.\begin{aligned} a_{I_f M_f} &= \sum_I (2I_i+1)^{1/2} (2I_f+1)^{1/2} (2I+1)(-1)^{M_f-K} \\ &\times \begin{pmatrix} I_f & I_i & I \\ -M_f & M_i & M_f-M_i \end{pmatrix} \begin{pmatrix} I_f & I_i & I \\ -K & K & 0 \end{pmatrix} A_{I,\,M_f-M_i}(\vartheta,q), \end{aligned}\right\} \tag{5.8}$$

where we have introduced the functions

$$\left.\begin{aligned}A_{IM}(\vartheta, q) &= [4\pi(2I+1)]^{-1/2}\\ &\times\int_0^{2\pi} d\alpha \int_0^{\pi} d\beta \sin\beta\, Y_{IM}(\beta,\alpha)\, e^{i\frac{8\pi}{5} q \sum_{\mu} Y_{2\mu}\left(\frac{\pi}{2},0\right) J_{2\mu}(\vartheta)\, Y^*_{2\mu}(\beta,\alpha)}\end{aligned}\right\} \quad (5.9)$$

One observes that these functions are proportional to the amplitudes on the state I, M in a rotational band with ground state spin 0, i. e.,

$$A_{IM}(\vartheta, q) = \frac{1}{\sqrt{2I+1}} a_{IM}(I_i = M_i = 0). \quad (5.10)$$

The quantity q is defined by

$$q = \frac{Z_1 e Q_0}{4\hbar v a^2}. \quad (5.11)$$

This quantity is independent of the spins in the rotational band and plays the role of a common χ. It is connected with the χ corresponding to the first excitation in an even-even nucleus with the same intrinsic quadrupole moment by the relation

$$q = \sqrt{\frac{45}{16}}\, \chi_{0\to 2}. \quad (5.12)$$

The calculation of the excitation cross sections of any rotational band is then reduced to the determination of the functions $A_{I,M}(\vartheta, q)$. From Eq. (5.8) one obtains, e. g., according to (3.42), the following formula for the excitation probability of the state of spin I_f:

$$P_{I_f I_i} = (2I_f+1)\sum_{IM}(2I+1)\begin{pmatrix} I_f & I_i & I \\ -K & K & 0\end{pmatrix}^2 |A_{IM}(\vartheta, q)|^2. \quad (5.13)$$

The functions $A_{I,M}$ can most easily be evaluated in the $\chi(\vartheta)$ approximation where the terms with $|\mu| = 2$ in the exponential function are neglected. The integration over α in (5.9) shows then that $A_{I,M}$ vanishes except for $M = 0$, where one finds

$$A_{I0}(\vartheta, q) \approx A_{I0}(\pi, q_{\text{eff}}(\vartheta)) = e^{i\frac{2}{3}q_{\text{eff}}(\vartheta)} \int_0^1 dx\, P_I(x)\, e^{-2iq_{\text{eff}}(\vartheta)x^2}. \quad (5.14)$$

We have here introduced a quantity

$$q_{\text{eff}}(\vartheta) = \frac{3}{4} J_{20}(\vartheta)\, q \quad (5.15)$$

which corresponds to the $\chi_{\text{eff}}(\vartheta)$ introduced in paragraph 3C. The function

$P_I(x)$ is the Legendre polynomial of order I. The integral (5.14) can be expressed in terms of a confluent hypergeometric function ${}_1F_1$ with the following result (see ref. 10, Vol. I, p. 171):

$$A_{I0}(\pi, q) = \frac{\Gamma\left(\frac{I+1}{2}\right)}{2\,\Gamma\left(\frac{2I+3}{2}\right)} e^{-i\frac{4}{3}q} (-2iq)^{\frac{I}{2}} {}_1F_1\left(\frac{I+2}{2}, \frac{2I+3}{2}, 2iq\right). \tag{5.16}$$

The confluent hypergeometric function which appears here can always be expressed by means of Fresnel integrals. For $I = 0$,the expression (5.16) thus takes the simple form (see ref. 10, Vol. I, p. 266)

$$A_{00}(\pi, q) = \sqrt{\frac{\pi}{4q}}\, e^{i\frac{2}{3}q} \left[C(2q) - iS(2q)\right], \tag{5.17}$$

where $C(x)$ and $S(x)$ are the Fresnel integrals which are tabulated in refs. 12 and 13.

The functions A_{I0} for higher values of I are most easily obtained by means of recursion formulae. The existence of such relations is guaranteed by the theorem that three confluent hypergeometric functions with parameters differing only by integer numbers are linear dependent. Accordingly, one finds the following recursion formula for the functions A_{I0}:

$$\left.\begin{aligned}(I+2)(2I-1)A_{I+2,0}(\pi, q) = &\left[\frac{(2I-1)(2I+1)(2I+3)}{4iq} + 2I + 1\right] A_{I0}(\pi, q) \\ &+ (I-1)(2I+3)A_{I-2,0}(\pi, q).\end{aligned}\right\} \tag{5.18}$$

For the application of this formula one needs two consecutive A's. Instead of $A_{2,0}$ it is practical to use the non-physical function $A_{-2,0}$ which, according to (5.16), is a simple exponential function

$$A_{-2,0}(\pi, q) = -\frac{1}{4iq} e^{-i\frac{4}{3}q}. \tag{5.19}$$

The functions $A_{I,0}(\pi, q)$ have been computed numerically in this way. The result is given in Table 4.

The excitation probabilities in a rotational band with ground state spin 0 are easily found from these numbers. They are tabulated in Table 5 and the result is shown in Fig. 7.

TABLE 4.

The functions $A_{I,0}(\pi, q)$ for backward scattering. The real part, Re $A_{I,0}$, and the imaginary part, Im $A_{I,0}$, are tabulated as functions of the parameter q for spin values up to 22.

q	Re $A_{0,0}$	Im $A_{0,0}$	Re $A_{2,0}$	Im $A_{2,0}$	Re $A_{4,0}$	Im $A_{4,0}$
0.0	1.00000	0.00000	0.00000	0.00000	0.00000	0.00000
0.5	0.95625	0.00276	−0.01228	−0.12916	−0.01236	0.00150
1.0	0.83311	0.02081	−0.04427	−0.23461	−0.04567	0.01123
1.5	0.65268	0.06336	−0.08317	−0.29888	−0.08980	0.03372
2.0	0.44537	0.12960	−0.11265	−0.31506	−0.13173	0.06754
2.5	0.24234	0.20825	−0.11852	−0.28773	−0.15978	0.10530
3.0	0.06838	0.28073	−0.09370	−0.23062	−0.16731	0.13599
3.5	−0.06252	0.32687	−0.04075	−0.16163	−0.15447	0.14879
4.0	−0.14819	0.33131	0.02890	−0.09722	−0.12756	0.13695
4.5	−0.19556	0.28834	0.09852	−0.04786	−0.09629	0.10036
5.0	−0.21596	0.20372	0.15132	−0.01611	−0.06998	0.04592
5.5	−0.22000	0.09300	0.17549	−0.00242	−0.05425	−0.01454
6.0	−0.21388	−0.02302	0.16734	−0.01578	−0.04929	−0.06786
6.5	−0.19840	−0.12396	0.13171	0.03154	−0.05024	−0.10350
7.0	−0.17046	−0.19497	0.07968	0.05327	−0.04959	−0.11644
7.5	−0.12642	−0.22953	0.02453	0.07913	−0.04044	−0.10807
8.0	−0.06545	−0.22954	−0.02240	0.10270	−0.01943	−0.08498
8.5	0.00816	−0.20297	−0.05466	0.11574	0.01177	−0.05610
9.0	0.08504	−0.16012	−0.07155	0.11168	0.04694	−0.02948
9.5	0.15274	−0.11018	−0.07677	0.08834	0.07745	−0.00979
10.0	0.19925	−0.05912	−0.07582	0.04909	0.09538	0.00255

(to be continued)

It is interesting to compare these curves with the excitation probabilities which were obtained for the same situation by means of the diagonalization method. It is seen that the excitation curves for the five-state model (see Fig. 4) are in good agreement with the exact calculation for χ values up to 3. It is interesting that also the secondary maxima of the excitation curves are present in the calculation with infinitely many levels. These secondary maxima which for P_2 appear for $q = 5.5$ and $q = 9$ must be understood as rudiments of the secondary maxima in the calculation with a finite number of states. In the two-state calculation the secondary maxima of P_2 appear at $q = 7.5$, 12.5, etc. When more states are introduced, these maxima are decreased (and shifted) due to the possibility of exciting the higher states which are introduced. One must expect that the secondary maxima are rather characteristic of the multiple Coulomb excitation of a pure rotational

TABLE 4 (continued).

q	Re $A_{6,0}$	Im $A_{6,0}$	Re $A_{8,0}$	Im $A_{8,0}$	Re $A_{10,0}$	Im $A_{10,0}$
0.0	0.00000	0.00000	0.00000	0.00000	0.00000	0.00000
0.5	0.00012	0.00087	0.00006	−0.00003		
1.0	0.00176	0.00647	0.00074	−0.00020		
1.5	0.00813	0.01934	0.00324	−0.00143	−0.00014	−0.00045
2.0	0.02255	0.03862	0.00872	−0.00546	−0.00104	−0.00160
2.5	0.04638	0.06013	0.01724	−0.01443	−0.00352	−0.00400
3.0	0.07751	0.07799	0.02728	−0.03002	−0.00899	−0.00767
3.5	0.11025	0.08694	0.03601	−0.05229	−0.01881	−0.01192
4.0	0.13660	0.08451	0.04021	−0.07890	−0.03361	−0.01518
4.5	0.14878	0.07217	0.03764	−0.10527	−0.05272	−0.01549
5.0	0.14179	0.05491	0.02823	−0.12555	−0.07385	−0.01113
5.5	0.11535	0.03947	0.01454	−0.13437	−0.09335	−0.00157
6.0	0.07426	0.03174	0.00120	−0.12857	−0.10705	0.01195
6.5	0.02704	0.03448	−0.00650	−0.10847	−0.11151	0.02635
7.0	−0.01664	0.04613	−0.00457	−0.07794	−0.10518	0.03733
7.5	−0.04883	0.06126	0.00809	−0.04338	−0.08909	0.04069
8.0	−0.06558	0.07250	0.02898	−0.01177	−0.06674	0.03370
8.5	−0.06765	0.07320	0.05252	0.01150	−0.04311	−0.01632
9.0	−0.05967	0.05992	0.07176	0.02413	−0.02313	−0.00852
9.5	−0.04797	0.03368	0.08053	0.02740	−0.01011	−0.03548
10.0	−0.03808	−0.00022	0.07552	0.02535	−0.00474	−0.05831

(to be continued)

band, and sensitive to any deviation. The maxima are also, as we shall see, less pronounced for finite ξ, and the deviation from the $q(\vartheta)$ approximation will also tend to wash out the oscillations.

The deviation from the $q(\vartheta)$ approximation can be treated by means of the expansion discussed in paragraph 3C. From (3.40), (3.34), (5.4), and (5.12) one finds the amplitude a_{IM} to second order in $J_{22}(\vartheta)/J_{20}(\vartheta)$

$$\left.\begin{aligned} a_{IM} &= a_{IM}^{(0)} \\ &+ i\sqrt{\frac{32\pi}{15}}\, q_{\text{eff}}(\vartheta)\frac{J_{22}(\vartheta)}{J_{20}(\vartheta)} \sum_{I_z M_z \mu=\pm 2}{}' \langle IM \mid Y_{2\mu}^*(\beta,\alpha) \mid I_z M_z \rangle a_{I_z M_z}^{(0)} \\ &- \frac{16\pi}{15}\left(q_{\text{eff}}(\vartheta)\frac{J_{22}(\vartheta)}{J_{20}(\vartheta)}\right)^2 \sum_{I_z M_z \mu, \mu'=\pm 2}{}' \langle IM \mid Y_{2\mu}^*(\beta,\alpha)\, Y_{2\mu'}^*(\beta,\alpha) \mid I_z M_z \rangle a_{I_z M_z}^{(0)} \end{aligned}\right\} \quad (5.20)$$

where $a^{(0)}$ are the amplitudes in the $q_{\text{eff}}(\vartheta)$ approximation.

The formula (5.20) for the special case of $I_i = M_i = 0$ may, according

TABLE 4 (continued).

q	Re $A_{12,0}$	Im $A_{12,0}$	Re $A_{14,0}$	Im $A_{14,0}$	Re $A_{16,0}$	Im $A_{16,0}$
0.0						
0.5						
1.0						
1.5						
2.0	—0.00025	0.00013				
2.5	—0.00074	0.00072				
3.0	—0.00180	0.00220	0.00040	0.00032		
3.5	—0.00330	0.00551	0.00137	0.00079	0.00018	0.00031
4.0	—0.00478	0.01151	0.00331	0.00130	0.00033	—0.00080
4.5	—0.00531	0.02089	0.00691	0.00157	0.00041	—0.00197
5.0	—0.00358	0.03361	0.01263	0.00096	0.00022	—0.00406
5.5	0.00172	0.04865	0.02065	—0.00145	—0.00071	—0.00746
6.0	0.01128	0.06390	0.03052	—0.00667	—0.00305	—0.01228
6.5	0.02462	0.07658	0.04106	—0.01541	—0.00758	—0.01833
7.0	0.03983	0.08387	0.05050	—0.02764	—0.01492	—0.02496
7.5	0.05373	0.08389	0.05682	—0.04231	—0.02530	—0.03100
8.0	0.06255	0.07637	0.05836	—0.05723	—0.03822	—0.03508
8.5	0.06303	0.06298	0.05444	—0.06944	—0.05230	—0.03587
9.0	0.05349	0.04698	0.04574	—0.07585	—0.06541	—0.03266
9.5	0.03460	0.03232	0.03437	—0.07411	—0.07500	—0.02564
10.0	0.00945	0.02241	0.02338	—0.06344	—0.07872	—0.01619

(to be continued)

to (5.10), be interpreted as an expansion of the function $A_{I,M}$. The specialization $I_i = M_i = 0$ may thus be done without any loss of generality, since the amplitudes for other ground state spins can be computed by means of (5.8). Introducing this simplification we obtain

$$\left.\begin{aligned} A_{I0}(\vartheta, q) = A_{I0}(\pi, q_{\text{eff}}(\vartheta)) - \frac{8}{3}\left(q_{\text{eff}}(\vartheta)\frac{J_{22}(\vartheta)}{J_{20}(\vartheta)}\right)^2 \sum_{I_z I_z'} (2I_z+1)(2I_z'+1) \\ \times \begin{pmatrix} I & 2 & I_z' \\ 0 & 0 & 0 \end{pmatrix}\begin{pmatrix} I & 2 & I_z' \\ 0 & -2 & 2 \end{pmatrix}\begin{pmatrix} I_z' & 2 & I_z \\ 0 & 0 & 0 \end{pmatrix}\begin{pmatrix} I_z' & 2 & I_z \\ -2 & 2 & 0 \end{pmatrix} A_{I_z 0}(\pi, q_{\text{eff}}(\vartheta)) \end{aligned}\right\} \quad (5.21)$$

$$\left.\begin{aligned} A_{I,\pm 2}(\vartheta, q) = i\sqrt{\frac{8}{3}}\, q_{\text{eff}}(\vartheta)\frac{J_{22}(\vartheta)}{J_{20}(\vartheta)} \sum_{I_z} (2I_z+1) \\ \begin{pmatrix} I_z & 2 & I \\ 0 & 0 & 0 \end{pmatrix}\begin{pmatrix} I_z & 2 & I \\ 0 & -2 & 2 \end{pmatrix} A_{I_z 0}(\pi, q_{\text{eff}}(\vartheta)) \end{aligned}\right\} \quad (5.22)$$

3*

TABLE 4 (continued).

q	Re $A_{18,0}$	Im $A_{18,0}$	Re $A_{20,0}$	Im $A_{20,0}$	Re $A_{22,0}$	Im $A_{22,0}$
0.0						
0.5						
1.0						
1.5						
2.0						
2.5						
3.0						
3.5						
4.0						
4.5	–0.00051	–0.00009				
5.0	–0.00114	–0.00004	0.00000	0.00027		
5.5	–0.00236	0.00027	0.00009	0.00069		
6.0	–0.00429	0.00116	0.00037	0.00133		
6.5	–0.00706	0.00312	0.00111	0.00240	0.00071	–0.00034
7.0	–0.01056	0.00672	0.00261	0.00392	0.00130	–0.00090
7.5	–0.01435	0.01248	0.00530	0.00580	0.00209	–0.00199
8.0	–0.01768	0.02072	0.00958	0.00774	0.00301	–0.00389
8.5	–0.01958	0.03125	0.01575	0.00922	0.00384	–0.00692
9.0	–0.01904	0.04330	0.02382	0.00952	0.00420	–0.01133
9.5	–0.01540	0.05540	0.03334	0.00788	0.00354	–0.01718
10.0	–0.00863	0.06564	0.04338	0.00371	0.00128	–0.02423

and

$$\left.\begin{aligned} A_{I,\pm 4}(\vartheta, q) &= -\frac{4}{3}\left(q_{\mathrm{eff}}(\vartheta)\frac{J_{22}(\vartheta)}{J_{20}(\vartheta)}\right)^2 \sum_{I_z I_z'} (2\,I_z+1)(2\,I_z'+1) \\ &\times \begin{pmatrix} I & 2 & I_z' \\ 0 & 0 & 0 \end{pmatrix}\begin{pmatrix} I & 2 & I_z' \\ 4 & -2 & -2 \end{pmatrix}\begin{pmatrix} I_z' & 2 & I_z \\ 0 & 0 & 0 \end{pmatrix}\begin{pmatrix} I_z' & 2 & I_z \\ 2 & -2 & 0 \end{pmatrix} A_{I_z 0}(\pi, q_{\mathrm{eff}}(\vartheta)). \end{aligned}\right\} \quad (5.23)$$

In the excitation probabilities (see Eq. (5.13)) only the squares of the $A_{I,M}$ appear and to second order in $J_{22}(\vartheta)/J_{20}(\vartheta)$ only $A_{I,0}$ and $A_{I,2}$ contribute. The seamplitudes have been calculated numerically for $I = 0,2$, and 4 by means of the known functions $A_{I,0}\,(\pi, q_{\mathrm{eff}}(\vartheta))$. The probability for the excitation of the states in an even-even nucleus can be written

$$P_I = P_I^{(0)}(q_{\mathrm{eff}}(\vartheta)) + \left(\frac{J_{22}(\vartheta)}{J_{20}(\vartheta)}\right)^2 \Delta_I(q_{\mathrm{eff}}(\vartheta)), \quad (5.24)$$

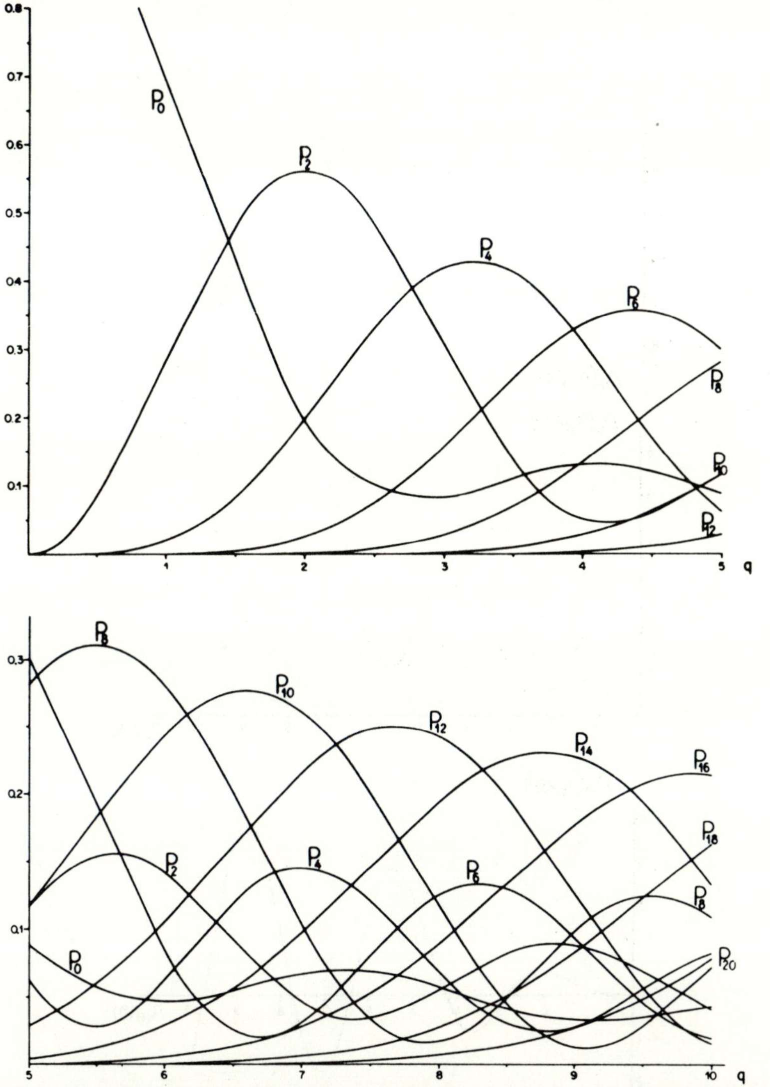

Fig. 7. The multiple Coulomb excitation of a pure rotational band in an even-even nucleus. The excitation probability P_I of the state with spin I is given as a function of the parameter q for backward scattering. The excitation probabilities for other deflection angles and other ground state spins can also be inferred from these curves (see Sect. 5).

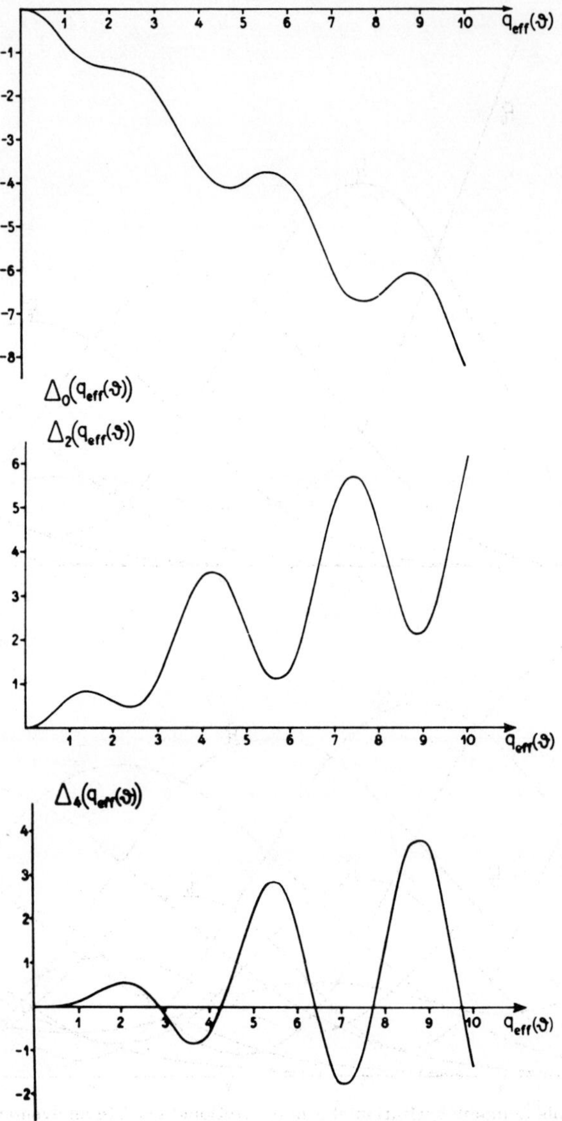

Fig. 8. The coefficients Δ_I for the correction of the excitation probabilities in the $q_{eff}(\vartheta)$ approximation. The coefficients Δ_I (which are defined in Eq. (5.24)) are plotted as functions of $q_{eff}(\vartheta)$ for the three lowest states $I = 0$, 2, and 4 in a rotational band in an even-even nucleus.

where $P_I^{(0)}$ are the probabilities given in Fig. 7. The coefficients Δ_I receive a contribution, partly from $A_{I,0}$ partly from $A_{I,2}$. The numerical result for Δ_I in the cases of $I = 0,2$, and 4 are given in Table 6 and are illustrated in Fig. 8.

It is seen that Δ_I is an oscillating function of $q_{\text{eff}}(\vartheta)$, and a rough estimate shows that the correction may amount to about 0.1, but usually it will be much smaller. If one compares the curves for Δ_I and P_I, one sees that the tendency of the correction is to fill out the minima of P_I. The relative error at the minima of P_I might be rather considerable, as is shown on Fig. 9. The relative error in P_2 is here plotted as a function of $q_{\text{eff}}(\vartheta)$ for different angles. The curves end at the value of $q_{\text{eff}}(\vartheta)$ where q reaches the value 10. The maxima on the curves appear as expected at the points where $P_2^{(0)}$ is minimal.

One sees also that for small angles the relative error is rather considerable in the whole range of q. This discrepancy can, as was discussed in paragraph 3C, be removed by applying $q(\vartheta)$ defined by

$$q(\vartheta) = q\frac{3}{4}\left[(J_{20}(\vartheta))^2 + 3(J_{22}(\vartheta))^2\right]^{1/2} \tag{5.25}$$

instead of $q_{\text{eff}}(\vartheta)$. This approximation will lead to the correct result for angles where the perturbation calculation is applicable. In Fig. 10 we have plotted the relative error of this latter approximation. It is seen that for $q(\vartheta)$ less than 2 one obtains a considerable improvement over the $q_{\text{eff}}(\vartheta)$ approximation (compare Fig. 9). For $q(\vartheta)$ larger than 2, the error is mostly larger than the error of the $q_{\text{eff}}(\vartheta)$ approximation, but it is here not very different from this. As a net result the $q(\vartheta)$ approximation is preferable.

In the case of a pure rotational band, one may calculate the excitation amplitude for arbitrary angles directly from (5.8). This can be done in the following way. We write the exponent in (5.9) explicitly in the form

$$\left.\begin{aligned} &\frac{8\pi}{5} q \sum_{\mu} Y_{2\mu}\left(\frac{\pi}{2}, 0\right) J_{2\mu}(\vartheta)\, Y_{2\mu}^{*}(\beta,\alpha) \\ &= q\left[\frac{1}{2} J_{20}(\vartheta)\,(1 - 3\cos^2\beta) + \frac{3}{2} J_{22}(\vartheta)\sin^2\beta\cos 2\alpha\right]. \end{aligned}\right\} \tag{5.26}$$

For the spherical harmonics we use the definition

$$Y_{IM}(\beta,\alpha) = (-1)^M\left[\frac{(2I+1)(I-M)!}{4\pi(I+M)!}\right]^{\frac{1}{2}} P_I^M(\cos\beta)\, e^{iM\alpha}. \tag{5.27}$$

The integration over α may then be expressed by means of a Bessel function $J_{M/2}$ of order $M/2$ with the following result[10]:

$$\left.\begin{aligned} A_{IM}(\vartheta, q) &= i^{\frac{M}{2}}\left[\frac{(I-M)!}{(I+M)!}\right]^{\frac{1}{2}} e^{i\frac{1}{2}qJ_{20}(\vartheta)} \\ &\times \int_0^1 dx\, P_I^M(x)\, J_{\frac{M}{2}}\left(\frac{3}{2}qJ_{22}(\vartheta)(1-x^2)\right) e^{-i\frac{3}{2}qJ_{20}(\vartheta)x^2}. \end{aligned}\right\} \quad (5.28)$$

This formula applies to I and M even. For I or M odd, $A_{I,M}$ vanishes. One observes furthermore the symmetry relation

$$A_{I,-M}(\vartheta, q) = A_{I,M}(\vartheta, q). \quad (5.29)$$

One may then proceed by using the following integral representation of the Bessel function (see ref. 10, Vol. II, p. 81)

$$J_n(y) = \frac{(y/2)^n}{\Gamma\left(n+\frac{1}{2}\right)\sqrt{\pi}} \int_{-1}^{1} dt\, e^{iyt}(1-t^2)^{n-\frac{1}{2}}. \quad (5.30)$$

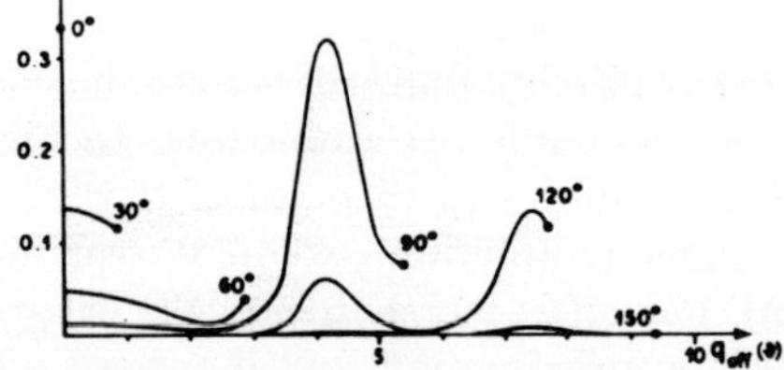

Fig. 9. The relative error of the $q_{eff}(\vartheta)$ approximation for the excitation of the 2^+ rotational state in an even-even nucleus. The error $[P_2(\vartheta, q) - P_2(\pi, q_{eff}(\vartheta))]/P_2(\pi, q_{eff}(\vartheta))$ is plotted as a function of $q_{eff}(\vartheta)$ for different angles. The curves end at a value of $q_{eff}(\vartheta)$ where q reaches the value 10.

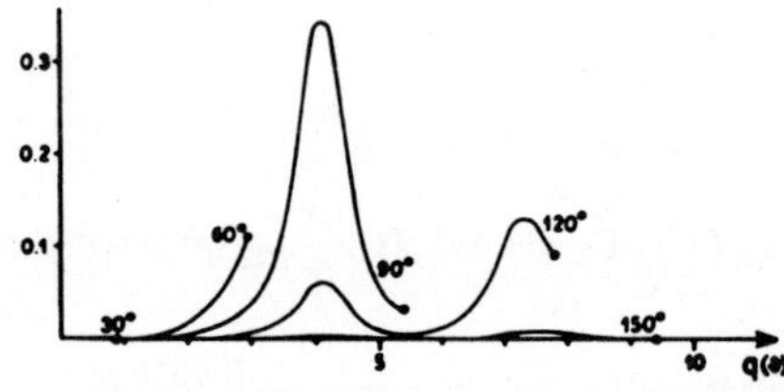

Fig. 10. The relative error of the $q(\vartheta)$ approximation for the excitation of the 2^+ rotational state in an even-even nucleus. The error $[P_2(\vartheta, q) - P_2(\pi, q(\vartheta))]/P_2(\pi, q(\vartheta))$ is plotted as a function of $q(\vartheta)$ for different angles. The curves end at a value of $q(\vartheta)$ where q reaches the value 10.

The integration over x can be done by expanding the exponential function in power series in $q\,[J_{2,0}(\vartheta)+tJ_{2,2}(\vartheta)]$ (see ref. 10, Vol. I, p. 172). The integration over t can finally be done when the powers of this quantity are expanded according to the binomial formula. The result is a double series

$$\left.\begin{aligned} A_{IM}(\vartheta, q) &= \left[\frac{(I+M)!}{(I-M)!}\right]^{\frac{1}{2}} \frac{\Gamma\left(\frac{I-M+1}{2}\right)(-1)^{\frac{I-M}{2}}}{2^{M+1}\,\Gamma\left(\frac{2I+3}{2}\right)\Gamma\left(\frac{M+2}{2}\right)} e^{-i\frac{4}{3}q_{\mathrm{eff}}(\vartheta)} \\ &\times \sum_{m,n}{}' \frac{\left(\frac{I+M+2}{2}\right)_m}{\left(\frac{2I+3}{2}\right)_m} \frac{\left(-\frac{I-M}{2}-m\right)_{2n}}{\left(\frac{M+2}{2}\right)_n} \frac{(2\,iq_{\mathrm{eff}}(\vartheta))^{m+\frac{I}{2}}}{m!} \frac{b^{2n+\frac{M}{2}}}{n!}, \end{aligned}\right\} \quad (5.31)$$

where

$$b = \frac{J_{22}(\vartheta)}{2\,J_{20}(\vartheta)}. \quad (5.32)$$

We have furthermore used the notation

$$\alpha_m = \alpha(\alpha+1)\ldots.(\alpha+m-1). \quad (5.33)$$

The formula (5.31) holds for $M > 0$. The functions $A_{I,M}$ for negative M are determined by means of (5.29).

It is useful to perform the summation over M whereby (5.31) may be written in the form

$$\left.\begin{aligned} A_{IM}(\vartheta, q) &= \left[\frac{(I+M)!}{(I-M)!}\right]^{\frac{1}{2}} \frac{\Gamma\left(\frac{I-M+1}{2}\right)}{2^{M+1}\,\Gamma\left(\frac{2I+3}{2}\right)\Gamma\left(\frac{M+2}{2}\right)} e^{-i\frac{4}{3}q_{\mathrm{eff}}(\vartheta)} \\ &\times (-2\,iq_{\mathrm{eff}}(\vartheta))^{\frac{I}{2}}(-b)^{\frac{M}{2}} \sum_{n}{}' \frac{b^{2n}}{\left(\frac{M+2}{2}\right)^n n!} (q_{\mathrm{eff}}(\vartheta))^{2n-\frac{I-M}{2}} \\ &\times \frac{d^{2n}}{(dq_{\mathrm{eff}}(\vartheta))^{2n}}\left[(q_{\mathrm{eff}}(\vartheta))^{\frac{I-M}{2}}\,{}_1F_1\left(\frac{I+M+2}{2}, \frac{2I+3}{2}, 2\,iq_{\mathrm{eff}}(\vartheta)\right)\right]. \end{aligned}\right\} \quad (5.34)$$

If one sets $b = 0$ the expressions (5.31) and (5.34) reduce to the simple result (5.16). The expression (5.34) is, similar to (5.20), a systematic expansion in powers of $J_{22}(\vartheta)/J_{20}(\vartheta)$.

B. First Order Correction in ξ

In paragraph 3 A it was outlined how one may calculate the deviation of the excitation amplitudes from the sudden approximation. The result was expressed in a power series in the ξ's which enter into the excitation process. For rotational bands, one may define a common ξ in terms of the moment of inertia in a similar way as, for such spectra, we defined a common χ in terms of the intrinsic quadrupole moment. We shall use the notation

$$\xi = \frac{3\,\hbar a}{v\mathfrak{J}}, \tag{5.35}$$

where $\mathfrak{J}$ is the moment of inertia entering in (5.3). The quantity (5.35) is identical to the ξ corresponding to the excitation of the lowest rotational state in an even-even nucleus.

The excitation amplitudes which were evaluated in the previous paragraph are essentially complex numbers. The first order corrections must also be expected to be complex, and it follows therefore that the excitation probabilities have linear terms in ξ. This is in contrast to the first order perturbation theory which is independent of ξ to first order in this quantity.

To first order, the excitation amplitude a_n may be written in the form

$$a_n = a_n^{(0)} + a_n^{(1)}, \tag{5.36}$$

where $a_n^{(0)}$ is the amplitude (5.7) in the sudden approximation. The first order correction $a_n^{(1)}$ is, according to (3.20), given by

$$\left.\begin{aligned} a_n^{(1)} &= \langle n\,|\,e^{-\frac{i}{\hbar}\int_{-\infty}^{\infty} H'(t)\,dt}\,|\,\varphi_1\rangle \\ &= \frac{\xi v}{6\,\hbar a}\langle n\,|\,e^{-\frac{i}{\hbar}\int_{-\infty}^{\infty} H'(t)\,dt}\int_{-\infty}^{\infty} e^{\frac{i}{\hbar}\int_{-\infty}^{t} H'(t')\,dt'}\,t\,[L^2, H'(t)]\,e^{-\frac{i}{\hbar}\int_{-\infty}^{t} H'(t')\,dt'}\,dt\,|\,0\rangle. \end{aligned}\right\} \tag{5.37}$$

In this equation, $H'(t)$ is given by (5.5) and (5.11)

$$H'(t) = \frac{8\,\pi\,\hbar v a^2}{5}\,q \sum_{\mu} \bar{S}_{2\mu}(t)\,Y_{2\mu}^{*}(\beta,\alpha), \tag{5.38}$$

while H_0 (see Eq. (5.3)) has been expressed by means of ξ and the angular momentum operator L through

$$H_0 = \frac{\hbar v}{6\,a}\,\xi\,L^2. \tag{5.39}$$

For the evaluation of (5.37) we shall proceed in the following way: Firstly, the differentiations of the operator L^2 are performed. Hereby the two exponential functions in $|\varphi_1\rangle$ will cancel. The expression for $|\varphi_1\rangle$ will then be suitable for an expansion in terms of the eigenfunctions $|m\rangle$ of H_0, and the problem is reduced to the already performed calculation of matrix elements in the sudden approximation.

The result of the first step in this program can be written in the form

$$\left.\begin{aligned} |\varphi_1\rangle = & -i\frac{8\pi}{5}q\xi\sum_\mu f_\mu Y^*_{2\mu}(\beta,\alpha)|0\rangle \\ & + i\frac{8\pi}{15}q\xi\sum_\mu f_\mu\left(\frac{\partial Y^*_{2\mu}}{\partial\beta}\cdot\frac{\partial}{\partial\beta}+\frac{1}{\sin^2\beta}\frac{\partial Y^*_{2\mu}}{\partial\alpha}\frac{\partial}{\partial\alpha}\right)|0\rangle \\ & + i\frac{64\pi^2}{75}q^2\xi\sum_{\mu\mu'} f_{\mu\mu'}\left(\frac{\partial Y^*_{2\mu}}{\partial\beta}\frac{\partial Y^*_{2\mu'}}{\partial\beta}+\frac{1}{\sin^2\beta}\frac{\partial Y^*_{2\mu}}{\partial\alpha}\frac{\partial Y^*_{2\mu'}}{\partial\alpha}\right)|0\rangle, \end{aligned}\right\} \quad (5.40)$$

where the coefficients f_μ and $f_{\mu\mu'}$ are defined by

$$f_\mu = iv^2 a\int_{-\infty}^{\infty} t\bar{S}_{2\mu}(t)\,dt \quad (5.41)$$

and

$$f_{\mu\mu'} = v^3 a^3\int_{-\infty}^{\infty}\bar{S}_{2\mu}(t)\,t\int_{-\infty}^{t}\bar{S}_{2\mu'}(t')\,dt'\,dt. \quad (5.42)$$

From the symmetries of the $\bar{S}_{2,\mu}(t)$ (see Eq. (3.23)) one sees immediately that $f_1 = f_{-1}$ is the only non-vanishing f_μ. The first two terms of (5.40) also appear in the first order perturbation treatment while the third, which is proportional to q^2, is characteristic of higher order excitations. The second term arises from the initial motion of the rotator, and it disappears for the ground-state spin $I = 0$.

For the evaluation of (5.40) one has to use the properties of the eigenfunctions $|0\rangle$ which are given in terms of the D-function in (5.6). We note the following formula

$$\left.\begin{aligned} & 2\left[\frac{\partial D^I}{\partial\beta}\frac{\partial D^{I'}}{\partial\beta}+\frac{1}{\sin^2\beta}\frac{\partial D^I}{\partial\alpha}\frac{\partial D^{I'}}{\partial\alpha}\right] \\ & = D_I L^2(D^{I'}) + D^{I'}L^2(D^I) - L^2(D^I D^{I'}) \\ & = (I(I+1) + I'(I'+1) - L^2)\,D^I D^{I'}, \end{aligned}\right\} \quad (5.43)$$

where we have suppressed the lower indices M and K on the D functions. By means of this formula the problem of expanding $|\varphi_1\rangle$ in terms of $|m\rangle$ is reduced to the problem of expanding a product of D functions in terms of D functions.

The result can be expressed by means of the functions A_{IM} defined in (5.9) in the following way:

$$\left.\begin{aligned}
a^{(1)}_{I_f M_f} = -\sqrt{\frac{16\pi}{5}}\, q\xi \sum_{II'M\mu=\pm 1} f_\mu(\vartheta)\left[1+\frac{1}{6}I_i(I_i+1)-\frac{1}{6}I'(I'+1)+i\right] \\
\times(2I_i+1)^{\frac{1}{2}}(2I_f+1)^{\frac{1}{2}}(2I+1)(2I'+1) \\
\times\begin{pmatrix} 2 & I_i & I' \\ -\mu & M_i & \mu-M_i\end{pmatrix}\begin{pmatrix} 2 & I_i & I' \\ 0 & K & -K\end{pmatrix}\begin{pmatrix} I_f & I' & I \\ -M_f & \mu-M_i & M\end{pmatrix}\begin{pmatrix} I_f & I' & I \\ -K & K & 0\end{pmatrix} A_{IM}(\vartheta,q) \\
+i\frac{8\pi}{15}q^2\xi\sum_{II'Mlm}(-1)^m f^l_m(\vartheta)(2I_i+1)^{\frac{1}{2}}(2I_f+1)^{\frac{1}{2}}(2I+1)(2I'+1) \\
\times\begin{pmatrix} l & I_i & I' \\ -m & M_i & m-M_i\end{pmatrix}\begin{pmatrix} l & I_i & I' \\ 0 & K & -K\end{pmatrix}\begin{pmatrix} I_f & I' & I \\ -M_f & m-M_i & M\end{pmatrix}\begin{pmatrix} I_f & I' & I \\ -K & K & 0\end{pmatrix} A_{IM}(\vartheta,q)
\end{aligned}\right\} \tag{5.44}$$

We have here introduced the notation

$$f^l_m(\vartheta) = (2l+1)\begin{pmatrix} 2 & 2 & l \\ 0 & 0 & 0\end{pmatrix}[12-l(l+1)]\sum_{\mu\mu'}\begin{pmatrix} 2 & 2 & l \\ \mu & \mu' & m\end{pmatrix} f_{\mu\mu'} \tag{5.45}$$

for the tensors of rank 0,2, and 4 which can be built up of the $f_{\mu\mu'}$, given by (5.42).

The coefficients $f_\mu(\vartheta)$ and $f^l_m(\vartheta)$ which are necessary for the evaluation of $a^{(1)}$ have been computed for a few angles and the result is given in Table 7. We note the following property of the f^l_m functions:

$$f^l_{-m}(\vartheta) = f^l_m(\vartheta). \tag{5.46}$$

The f^l_m functions for odd values of m may be expressed by means of f_1 and the functions $J_{2,\mu}(\vartheta)$ defined by (3.27):

$$f^2_1(\vartheta) = i\frac{3\sqrt{5}}{28\sqrt{\pi}} f_1[3J_{22}(\vartheta)+J_{20}(\vartheta)] \tag{5.47}$$

$$f^4_1(\vartheta) = -i\frac{3\sqrt{3}}{7\sqrt{2\pi}} f_1[2J_{20}(\vartheta)-J_{22}(\vartheta)] \tag{5.48}$$

$$f^4_3(\vartheta) = i\frac{3\sqrt{3}}{\sqrt{14\pi}} f_1 J_{22}(\vartheta). \tag{5.49}$$

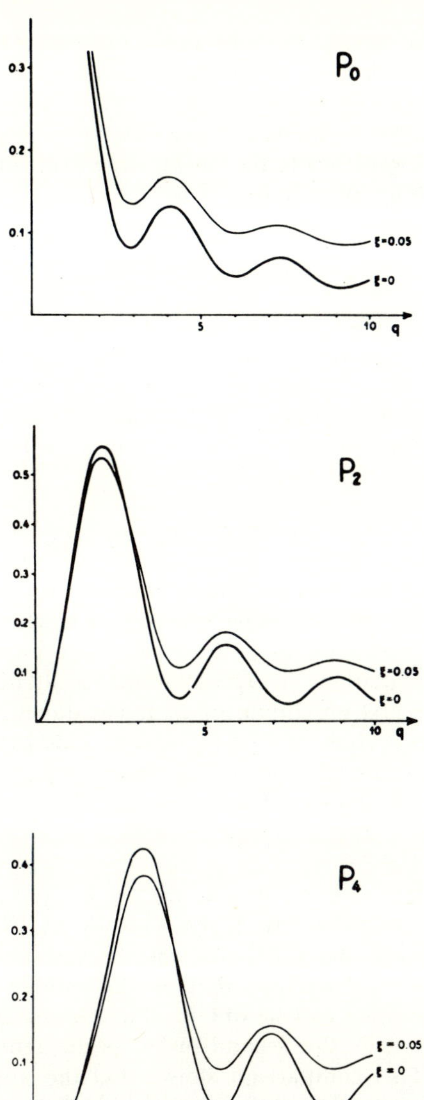

Fig. 11. The first order correction in ξ for the excitation of a rotational band in an even-even nucleus. The curves show the excitation probabilities of the states of spin 2 (P_2) and of spin 4 (P_4) and the probability that the nucleus stays in its ground state (P_0). The probabilities are given as functions of q for backward scattering and for the cases $\xi = 0$ and $\xi = 0.05$.

For backward scattering the only non-vanishing f-functions are f_0^0, f_0^2, and f_0^4. This is a consequence of the fact that all f_μ and $f_{\mu\mu'}$ of Eqs. (5.41) and (5.42) vanish, except f_{00}.

We shall here illustrate the first order correction in ξ by considering the special case of backward scattering on an even-even nucleus. In this case, the only non-vanishing amplitude $a^{(1)}$ is

$$\left.\begin{aligned} a_{I_f 0}^{(1)} &= i\frac{8\pi}{15} q^2 \xi (2I_f+1)^{\frac{1}{2}} f_0^0(\pi) \sum_I (2I+1) \\ &\times \left\{ \begin{pmatrix} 0 & I_f & I \\ 0 & 0 & 0 \end{pmatrix}^2 + \frac{5}{7} \begin{pmatrix} 2 & I_f & I \\ 0 & 0 & 0 \end{pmatrix}^2 - \frac{12}{7} \begin{pmatrix} 4 & I_f & I \\ 0 & 0 & 0 \end{pmatrix}^2 \right\} A_{I0}(\pi, q). \end{aligned}\right\} \quad (5.50)$$

We have here used that $f_0^2(\pi) = 5/7\, f_0^0(\pi)$ and $f_0^4(\pi) = -12/7\, f_0^0(\pi)$ which follows from the definition (5.45). The excitation probabilities to first order in ξ may be written in the form

$$P_I(\pi, q) = P_I(\xi = 0) + \Lambda_I(q)\,\xi. \quad (5.51)$$

The coefficient $\Lambda_I(q)$ has been evaluated numerically and is given in Table 8. It is seen that Λ_I is an oscillating function of q which is of the order of magnitude 1. The corrections for $\xi \neq 0$ are thus not dominated by the factor q^2 in (5.50). The oscillations in Λ_I follow the oscillations of $P_I(\xi = 0)$ in such a way that the first maximum of P_I is cut down, while the excitation probability for larger values of q is increased. This increment is largest at the minima of P_I and the effect of $\xi \neq 0$ is thus essentially to smooth out the whole excitation curve. This is clearly seen on Fig. 11 where the excitation probabilities for $I = 0$, 2, and 4 and $\xi = 0.05$ are compared with the excitation probabilities for $\xi = 0$.

For other deflection angles one may use the $q_{\mathrm{eff}}(\vartheta)$ approximation. One must here substitute only the q in $A_{I0}(\pi, q)$ with $q_{\mathrm{eff}}(\vartheta)$. Furthermore, the $f_0^0(\pi)$ should be replaced by $f_0^0(\vartheta)$. As was discussed earlier, this approximation is much less accurate here than in the sudden approximation. An indication of the accuracy can be obtained by comparing the limiting case of (5.50) for $q \ll 1$ with the second order perturbation calculation performed in ref. 14. This comparison shows that the approximation should not be applied for angles less than 90 degrees.

The failure of the $q_{\mathrm{eff}}(\vartheta)$ approximation for $\xi \neq 0$ is due to the fact that the relative importance of the different coefficients $f_m^l(\vartheta)$ in (5.44) for angles smaller than 90 degrees is completely different from the relative importance in the neighbourhood of 180 degrees where only f_{00} is different from zero. This means that one has to take into account also the amplitudes on the states with magnetic quantum number $M \neq 0$.

For $\xi \neq 0$, one observes from (5.44) that also the states with magnetic quantum numbers which differ by an odd integer from the magnetic quantum number of the ground state are populated. The amplitude on these states will be proportional to $\xi f_1(\vartheta)$ and the excitation probability will thus only receive a contribution of the order ξ^2 from such terms.

C. Numerical Results

In this paragraph we shall collect the numerical results which have been obtained for the excitation of rotational states together with some formulae which facilitate the application of these results to the experiments.

It is thus convenient to write the important parameters directly as functions of the energy of the incident projectile in the laboratory system (see ref. 1, Chapter II C). We shall here quote the expression for half the distance of closest approach in a head-on collision

$$a = 0.07199\left(1 + \frac{A_1}{A_2}\right)\frac{Z_1 Z_2}{E_{\mathrm{MeV}}} \times 10^{-12}\,\mathrm{cm}. \tag{5.52}$$

Here, A_1, Z_1 and A_2, Z_2 are the mass numbers and charges of projectile and target nucleus, respectively. The quantity E_{MeV} is the bombarding energy expressed in MeV.

The parameter ξ is similarly given by

$$\xi_{1\to 2} = \frac{Z_1 Z_2 A_1^{\frac{1}{2}}\,\Delta E'_{\mathrm{MeV}}}{12.65\left(E_{\mathrm{MeV}} - \frac{1}{2}\Delta E'_{\mathrm{MeV}}\right)^{3/2}}, \tag{5.53}$$

where $\Delta E'$ is connected with the energy difference $E_2 - E_1$ by the relation

$$\Delta E' = \left(1 + \frac{A_1}{A_2}\right)(E_2 - E_1). \tag{5.54}$$

An expression for the parameter χ (in the case $\lambda = 2$) is found by inserting (5.52) in (2.11), i. e.,

$$\chi_{1\to 2} = 14.36 \frac{A_1^{\frac{1}{2}} [B(E\,2, I_1 \to I_2)]^{1/2}}{(1 + A_1/A_2)^2 Z_1 Z_2^2} E_{\text{MeV}}^{3/2}. \tag{5.55}$$

The reduced transition probability $B(E\,2)$ is here measured in units of $e^2 \cdot 10^{-48}\,\text{cm}^4$.

For the excitation of rotational states we have introduced two parameters which are related to $\chi_{1\to 2}$ and $\xi_{1\to 2}$ and are defined in terms of the nuclear moments, so that they are independent of the spin sequence in the rotational bands. We have thus (see Eq. (5.35)) defined a common ξ in terms of the moment of inertia $\mathfrak{J}$ by means of (5.53) where

$$\Delta E' = \left(1 + \frac{A_1}{A_2}\right)\left(\frac{3\,\hbar}{\mathfrak{J}}\right)_{\text{MeV}}. \tag{5.56}$$

For an even-even nucleus this ξ is identical with the $\xi_{0\to 2}$ for the excitation of the lowest rotational state.

We have furthermore defined a quantity q by means of the intrinsic quadrupole moment Q_0 in the following way (see Eq. (5.11)):

$$q = 7.6241 \frac{A_1^{\frac{1}{2}} Q_0}{(1 + A_1/A_2)^2 Z_1 Z_2^2} E_{\text{MeV}}^{3/2}, \tag{5.57}$$

where Q_0 is measured in units of $e \cdot 10^{-24}\,\text{cm}^2$. The quantity q is related to the χ for the excitation of the lowest state in an even-even nucleus by Eq. (5.12).

The differential Coulomb excitation cross section is given by (3.43) through the excitation probability $P_{I_f,\,I_i}(\vartheta, q, \xi)$ which is the probability that the nucleus is excited from the ground state with spin I_i into the state with spin I_f when the projectile moves in an orbit with deflection angle ϑ in the center of mass system.

The probabilities P as well as other quantities interesting for the experiments can be obtained from the excitation amplitudes (see Eq. (3.42)). For $\xi \ll 1$ the amplitudes $a_{I_f,\,M_f}(\vartheta, q, \xi)$ are easily obtainable from the functions $A_{IM}(\vartheta, q)$ (see Eq. (5.8)). These functions can in turn be expressed by the functions $A_{I0}(\pi, q)$ which are related to the amplitudes for the excitation of rotational states in an even-even nucleus for $\vartheta = \pi$ and $\xi = 0$ by means of (5.10). These fundamental quantities have been calculated according to the formulae given in paragraph 5 A, and the result is given in Table 4.

For other deflection angles, the functions $A_{IM}(\vartheta, q)$ can be obtained to a good approximation from those tabulated by means of Eqs. (5.21) to (5.23) which we shall quote here for the cases $M = 0$ and $M = \pm 2$:

$$\left.\begin{aligned} A_{I0} = A_{I0}(\pi, q_{\text{eff}}) - \left(q_{\text{eff}}\frac{J_{22}(\vartheta)}{J_{20}(\vartheta)}\right)^2 & \left\{\frac{(I-3)(I-2)(I-1)I}{(2I-5)(2I-3)(2I-1)(2I+1)} A_{I-4,0}(\pi, q_{\text{eff}})\right. \\ & - \frac{4(I-1)I[(I-1)I-4]}{(2I-5)(2I-1)(2I+1)(2I+3)} A_{I-2,0}(\pi, q_{\text{eff}}) \\ & + \frac{2[3I^2(I+1)^2 - 14I(I+1) + 12]}{(2I-3)(2I-1)(2I+3)(2I+5)} A_{I,0}(\pi, q_{\text{eff}}) \\ & - \frac{4(I+1)(I+2)[(I+1)(I+2)-4]}{(2I-1)(2I+1)(2I+3)(2I+7)} A_{I+2,0}(\pi, q_{\text{eff}}) \\ & \left. + \frac{(I+1)(I+2)(I+3)(I+4)}{(2I+1)(2I+3)(2I+5)(2I+7)} A_{I+4,0}(\pi, q_{\text{eff}})\right\} \end{aligned}\right\} \quad (5.58)$$

and

$$\left.\begin{aligned} A_{I,\pm 2} = iq_{\text{eff}}\frac{J_{22}(\vartheta)}{J_{20}(\vartheta)}[(I-1)I(I+1)(I+2)]^{\frac{1}{2}} & \left\{\frac{1}{(2I-1)(2I+1)} A_{I-2,0}(\pi, q_{\text{eff}})\right. \\ - \frac{2}{(2I-1)(2I+3)} A_{I,0}(\pi, q_{\text{eff}}) + & \left.\frac{1}{(2I+1)(2I+3)} A_{I+2,0}(\pi, q_{\text{eff}})\right\}. \end{aligned}\right\} \quad (5.59)$$

In these equations, $q_{\text{eff}}(\vartheta)$ is given by (5.15). The ratio $q_{\text{eff}}(\vartheta)/q$ is shown in Table 2 where also the ratio $J_{22}(\vartheta)/J_{20}(\vartheta)$ has been tabulated.

The excitation amplitudes for arbitrary spin sequence in the rotational band is given by Eq. (5.8). The first order correction in the amplitude for $\xi \neq 0$ is expressed by means of the $A_{I,M}(\vartheta, q)$ in Eq. (5.44). We shall in this paragraph only consider the application of $A_{I,M}(\vartheta, q)$ for the evaluation of the excitation probability $P_{I_f, I_i}(\vartheta, q, \xi)$.

In the simplest case of the excitation of a rotational band in an even-even nucleus for $\xi = 0$ and $\vartheta = \pi$, the excitation probabilities $P_I(q) = P_{I,0}(\pi, q, 0)$ are given by

$$P_I(q) = (2I+1)\,|A_{I,0}(\pi, q)|^2. \quad (5.60)$$

These probabilities have been evaluated in Table 5, and they are plotted in Fig. 7.

For other deflection angles the probability can be obtained from (5.58) and (5.59). To second order in $J_{22}(\vartheta)/J_{20}(\vartheta)$ it may be written

$$P_{I,0}(\vartheta, q, 0) = P_I(q_{\text{eff}}(\vartheta)) + \left(\frac{J_{22}(\vartheta)}{J_{20}(\vartheta)}\right)^2 \Delta_I(q_{\text{eff}}(\vartheta)). \quad (5.61)$$

TABLE 5.

The probabilities for excitation of the rotational states in an even-even nucleus. The result which is given for backward scattering and $\xi = 0$ is tabulated as a function of q and of the spin of the excited state. The excitation probability for other deflection angles and other spins can easily be inferred from these numbers.

q	P_0	P_2	P_4	P_6	P_8	P_{10}	P_{12}
0.0	1.0000	0.0000	0.0000	0.0000	0.0000	0.0000	0.0000
0.5	0.9144	0.0842	0.0014				
1.0	0.6945	0.2850	0.0199	0.0006			
1.5	0.4300	0.4812	0.0828	0.0057	0.0002		
2.0	0.2152	0.5597	0.1972	0.0260	0.0018	0.0001	
2.5	0.1021	0.4842	0.3296	0.0750	0.0086	0.0006	
3.0	0.0835	0.3098	0.4184	0.1572	0.0280	0.0029	0.0002
3.5	0.1108	0.1389	0.4140	0.2563	0.0685	0.0104	0.0010
4.0	0.1317	0.0514	0.3152	0.3354	0.1333	0.0286	0.0039
4.5	0.1214	0.0600	0.1741	0.3555	0.2125	0.0634	0.0116
5.0	0.0881	0.1158	0.0630	0.3006	0.2815	0.1171	0.0285
5.5	0.0571	0.1540	0.0284	0.1932	0.3105	0.1831	0.0593
6.0	0.0463	0.1412	0.0633	0.0848	0.2810	0.2436	0.1053
6.5	0.0547	0.0917	0.1191	0.0250	0.2007	0.2757	0.1618
7.0	0.0671	0.0459	0.1442	0.0312	0.1036	0.2616	0.2155
7.5	0.0687	0.0343	0.1198	0.0798	0.0331	0.2014	0.2481
8.0	0.0570	0.0552	0.0684	0.1242	0.0166	0.1174	0.2436
8.5	0.0413	0.0819	0.0296	0.1292	0.0491	0.0446	0.1985
9.0	0.0329	0.0879	0.0276	0.0930	0.0974	0.0128	0.1267
9.5	0.0355	0.0685	0.0549	0.0447	0.1230	0.0286	0.0560
10.0	0.0432	0.0408	0.0819	0.0189	0.1079	0.0719	0.0148

(to be continued)

The coefficient $\Delta_I(\vartheta)$ has been evaluated numerically for $I = 0,2$, and 4 and the result is given in Table 6 and Fig. 8.

In many cases a simpler approximation for $P_{I,0}(\vartheta, q, 0)$ will be quite adequate, namely the $q(\vartheta)$ approximation. In this approximation the excitation probability is given by

$$P_{I,0}(\vartheta, q, 0) \approx P_I(q(\vartheta)), \qquad (5.62)$$

where the quantity $q(\vartheta)$ is defined by (5.25). The ratio $q(\vartheta) / q$ is given in Table 2. The accuracy of the approximation (5.62) is illustrated in Fig. 10.

As an illustration of the application of (5.62) the differential and total cross sections have been evaluated for the case $q = 3$, and the result is given in Fig. 12. While the cross sections for all higher states tend towards zero for small deflection angles, the excitation of the $I = 2$ state reaches a finite

TABLE 5 (continued).

q	P_{14}	P_{16}	P_{18}	P_{20}	P_{22}	P_{24}	P_{26}
0.0	0.0000	0.0000	0.0000	0.0000	0.0000	0.0000	0.0000
0.5	0.0000						
1.0	0.0000						
1.5	0.0000						
2.0	0.0000						
2.5	0.0000						
3.0	0.0000						
3.5	0.0001						
4.0	0.0004						
4.5	0.0014	0.0001					
5.0	0.0046	0.0007					
5.5	0.0124	0.0018	0.0002				
6.0	0.0283	0.0053	0.0007	0.0001			
6.5	0.0558	0.0130	0.0022	0.0003			
7.0	0.0961	0.0279	0.0058	0.0009	0.0001		
7.5	0.1455	0.0529	0.0134	0.0025	0.0004		
8.0	0.1938	0.0888	0.0274	0.0062	0.0011	0.0002	
8.5	0.2258	0.1327	0.0503	0.0136	0.0028	0.0004	0.0001
9.0	0.2275	0.1764	0.0828	0.0270	0.0066	0.0013	0.0002
9.5	0.1935	0.2073	0.1223	0.0481	0.0138	0.0031	0.0005
10.0	0.1326	0.2132	0.1622	0.0777	0.0265	0.0069	0.0014

TABLE 6.

The coefficient $\Delta_I(q)$ for the correction of the $q_{\text{eff}}(\vartheta)$ approximation (see Eq. (5.61)), in the case of a rotational band in an even-even nucleus for $\xi = 0$. The result is given for the states of spin $I = 0$, 2, and 4 as a function of q.

q	Δ_0	Δ_2	Δ_4
0.0	0.000	0.000	0.000
0.5	−0.247	0.239	0.008
1.0	−0.787	0.678	0.103
1.5	−1.225	0.821	0.347
2.0	−1.381	0.585	0.561
2.5	−1.477	0.490	0.404
3.0	−1.887	1.120	−0.206
3.5	−2.712	2.390	−0.796
4.0	−3.608	3.441	−0.610
4.5	−4.084	3.787	0.630
5.0	−4.000	2.263	2.214
5.5	−3.756	1.178	2.844
6.0	−3.960	1.387	1.766
6.5	−4.841	3.078	−0.349
7.0	−5.993	5.048	−1.746
7.5	−6.712	5.700	−1.046
8.0	−6.654	4.538	1.409
8.5	−6.195	2.733	3.639
9.0	−6.120	2.156	3.664
9.5	−6.898	3.655	1.329
10.0	−8.213	6.188	−1.379

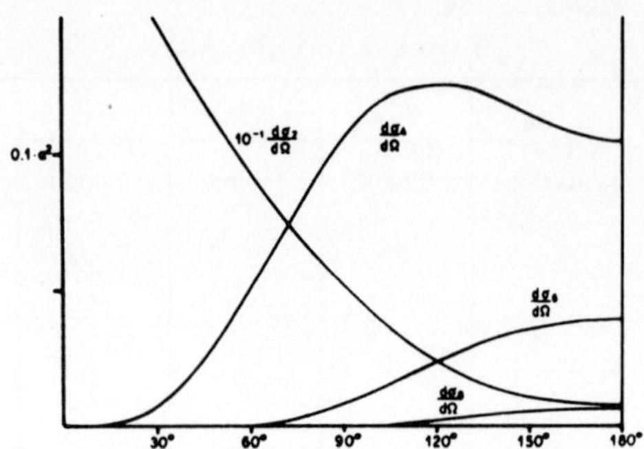

Fig. 12. The differential cross sections for multiple Coulomb excitation of a rotational band in an even-even nucleus for $q = 3$. The curves show the cross sections $d\sigma_I/d\Omega$ for the excitation of the state with spin I in the sudden approximation in units of a^2. The cúrve for the first excited state has been scaled down by a factor 10.

value for $\vartheta = 0°$. This is seen from the perturbation expression which is valid in this region. One thus finds

$$\left(\frac{d\sigma_2}{d\Omega}\right)_{\vartheta=0} = \frac{4}{15} q^2 a^2. \tag{5.63}$$

From the differential cross sections the following values for the total cross sections have been obtained

$$\left.\begin{aligned} \sigma_{I=2} &= 7.93\ a^2 \\ \sigma_{I=4} &= 1.06\ a^2 \\ \sigma_{I=6} &= 0.160\, a^2 \\ \sigma_{I=8} &= 0.016\, a^2. \end{aligned}\right\} \tag{5.64}$$

For other ground state spins the excitation probabilities of the rotational band can be obtained by means of (5.13). Since $K = I_i$ this equation may be written

$$\left.\begin{aligned} P_{I_f I_i}(\vartheta, q, 0) &= (2I_f+1)\sum_I \begin{pmatrix} I_f & I_i & I \\ -I_i & I_i & 0 \end{pmatrix}^2 P_{I,0}(\vartheta, q, 0) \\ &= \frac{(2I_f+1)\ (2I_i)!\,(I_i+I_f)!}{(I_f-I_i)!} \sum_I \frac{(I_f-I_i+I)!\,P_{I,0}(\vartheta, q, 0)}{(I_i+I_f+I+1)!\,(I_i-I_f+I)!\,(I_i+I_f-I)!} \end{aligned}\right\} \tag{5.65}$$

As an illustration, the case of $I_i = 5/2$ is shown in Fig. 13 for $\vartheta = 180°$.

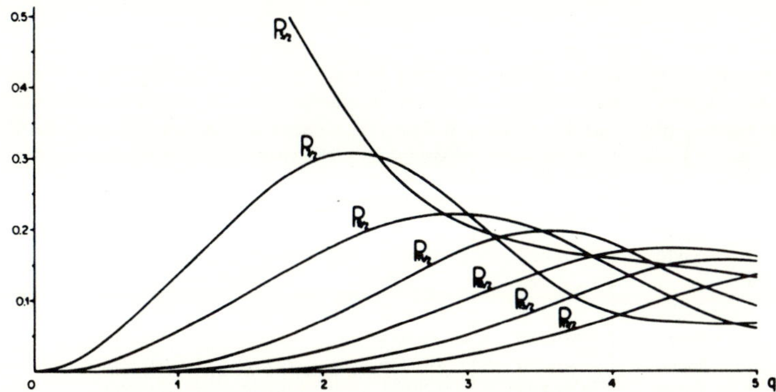

Fig. 13. The multiple Coulomb excitation of a pure rotational band in an odd A nucleus with ground state spin 5/2. The excitation probability P_I of the state with spin I is given as a function of the parameter q for backward scattering.

TABLE 7.

The coefficients $f_1(\vartheta)$ and $f_m^l(\vartheta)$ for the first order corrections in ξ to the excitation of rotational states (see Eq. (5.44)). The coefficients are given as functions of the deflection angle ϑ (in degrees) for even values of m. For odd values of m the $f_m^l(\vartheta)$ are easily obtained from $f_1(\vartheta)$ and the functions $J_{2,\mu}(\vartheta)$ given in Table 2 by means of the Eqs. (5.47) to (5.49). The entries are given in the form of a number followed by the power of ten by which it should be multiplied.

θ	f_1	f_0^0	f_0^2	f_2^2
180	0.000	3.893 (—1)	2.781 (—1)	0.000
150	1.917 (—1)	3.423 (—1)	2.423 (—1)	9.405 (—3)
120	3.242 (—1)	2.188 (—1)	1.495 (—1)	2.672 (—2)
90	3.586 (—1)	9.615 (—2)	5.915 (—2)	3.181 (—2)
60	2.904 (—1)	2.370 (—2)	9.662 (—3)	1.920 (—2)
30	1.513 (—1)	1.442 (—3)	—1.047 (—3)	4.078 (—3)

θ	f_0^4	f_2^4	f_4^4
180	—6.673 (—1)	0.000	0.000
150	—5.935 (—1)	1.015 (—3)	—4.062 (—5)
120	—3.942 (—1)	4.773 (—3)	—4.713 (—4)
90	—1.880 (—1)	8.893 (—3)	—1.373 (—3)
60	—5.511 (—2)	8.255 (—3)	—1.654 (—3)
30	—5.787 (—3)	2.628 (—3)	—6.296 (—4)

TABLE 8.

The coefficients $\Lambda_l(q)$ for the first order correction in ξ to the excitation probability of even-even nuclei. The coefficient which is defined in Eq. (5.51) is given for backward scattering on even-even nuclei. An approximate expression for other deflection angles can be obtained from the table by means of Eq. (5.67).

q	Λ_0	Λ_2	Λ_4	Λ_6	Λ_8	Λ_{10}
0.0	0.000	0.000	0.000	0.000	0.000	0.000
0.5	0.030	−0.029	−0.001			
1.0	0.204	−0.186	−0.017			
1.5	0.537	−0.422	−0.106	−0.008		
2.0	0.890	−0.512	−0.325	−0.050	−0.004	
2.5	1.084	−0.250	−0.636	−0.175	−0.022	−0.002
3.0	1.044	0.337	−0.861	−0.427	−0.083	−0.009
3.5	0.861	0.963	−0.773	−0.775	−0.234	−0.038
4.0	0.714	1.282	−0.278	−1.073	−0.509	−0.117
4.5	0.728	1.159	0.451	−1.107	−0.880	−0.288
5.0	0.880	0.783	1.053	−0.735	−1.224	−0.578
5.5	1.033	0.510	1.216	−0.025	−1.344	−0.966
6.0	1.061	0.580	0.919	0.721	−1.080	−1.341
6.5	0.955	0.930	0.465	1.136	−0.433	−1.526
7.0	0.821	1.266	0.249	1.039	0.370	−1.351
7.5	0.780	1.321	0.456	0.579	0.969	−0.779
8.0	0.865	1.082	0.922	0.148	1.087	0.031
8.5	0.995	0.785	1.278	0.099	0.728	0.751
9.0	1.055	0.701	1.259	0.480	0.199	1.063
9.5	0.996	0.901	0.920	1.006	0.089	0.855
10.0	0.880	1.204	0.580	1.282	−0.094	0.322

(to be continued)

The first order corrections for $\xi \neq 0$ must be calculated by means of Eq. (5.44) from the quantities $A_{IM}(\vartheta, q)$ and from the functions $f_m^l(\vartheta)$ and $f_1(\vartheta)$. The latter have been evaluated numerically for some angles and the result is given in Table 7.

The excitation probability may be written in the form (5.51)

$$P_{I_f I_i}(\vartheta, q, \xi) \approx P_{I_f I_i}(\vartheta, q, 0) + \Lambda_{I_f I_i}(\vartheta, q)\,\xi. \qquad (5.66)$$

The functions Λ have been calculated for the special case of $\vartheta = \pi$ and $I_i = 0$, and the result is given in Table 8. The effect of the correction in the excitation probability is illustrated in Fig. 11. The result (5.66) may be applied for angles in the neighbourhood of 180 degrees by the following substitution:

TABLE 8 (continued).

q	A_{12}	A_{14}	A_{16}	A_{18}	A_{20}
0.0	0.000	0.000	0.000	0.000	0.000
0.5					
1.0					
1.5					
2.0					
2.5					
3.0	—0.001				
3.5	—0.004				
4.0	—0.017	—0.001			
4.5	—0.055	—0.007			
5.0	—0.149	—0.025	—0.003		
5.5	—0.336	—0.073	—0.016		
6.0	—0.640	—0.180	—0.034	—0.005	
6.5	—1.038	—0.380	—0.092	—0.016	—0.001
7.0	—1.436	—0.694	—0.210	—0.044	—0.007
7.5	—1.669	—1.101	—0.421	—0.110	—0.021
8.0	—1.571	—1.516	—0.744	—0.238	—0.055
8.5	—1.072	—1.788	—1.163	—0.459	—0.128
9.0	—0.284	—1.753	—1.586	—0.789	—0.266
9.5	0.511	—1.324	—1.888	—1.210	—0.495
10.0	0.981	—0.571	—1.907	—1.648	—0.832

$$A_{I_f0}(\vartheta, q) \approx \frac{q^2 f_0^0(\vartheta)}{(q_{\rm eff}(\vartheta))^2 f_0^0(\pi)} A_{I_f0}(\pi, q_{\rm eff}(\vartheta)). \tag{5.67}$$

This equation only holds as long as f_0^0, f_0^2, and f_0^4 dominate over the coefficients f_2^2, f_2^4, and f_4^4.

The collision between the target nucleus and the projectile may also lead to an excitation of the projectile. The results which we have obtained for target excitation can also be used for projectile excitations, since we have worked in a relative coordinate system. The parameter ξ (see Eq. (2.8)) is thus given by the Eqs. (5.53) and (5.54) where one must insert for $E_2 - E_1$ the excitation energy of the projectile. Similarly the expression for χ (see Eq. (2.11)) is given by

$$\chi_{1\to 2}^{\rm proj} = 14.36 \frac{A_1^{1/2}[B(E2, I_1 \to I_2)]^{1/2}}{(1 + A_1/A_2)^2 Z_2 Z_1^2} E_{\rm MeV}^{3/2}, \tag{5.68}$$

where $B(E2)$ now refers to the projectile. The formula for $q^{\rm proj}$ has the same relation to q in (5.57) as $\chi^{\rm proj}$ has to χ in (5.55).

D. Classical Treatment

We shall make a few comments about the classical limit of the excitation of rotational states, which can be used for large angular momenta and large q. The classical problem of a collision between a charged particle and a charged symmetric top leads to a non-linear equation of motion which, like in the quantum mechanical problem, can only be solved in closed form in the limit where the collision time is short compared to the time of rotation of the top.

The classical Hamiltonian can be written in the form

$$H = \frac{v}{6\,a\hbar}\xi\left[p_\beta^2 + \frac{1}{\sin^2\beta}p_\alpha^2\right] + \frac{8\,\pi}{5}\hbar v a^2 q \sum_\mu \bar{S}_{2\mu}(t)\, Y_{2\mu}^*(\beta, \alpha), \qquad (5.69)$$

where p_β and p_α are the momenta which are conjugate to the Eulerian angles β and α, describing the orientation of the axis of the top.

We shall here consider only the case where one may neglect the terms with $\mu \neq 0$, i.e., we limit ourselves to the case of backward scattering or the $q(\vartheta)$ approximation. In this case the angle α is a cyclic variable. For the angle β one obtains from (5.69) the following equation of motion:

$$\ddot{\beta} = \sqrt{\frac{4\,\pi}{5}}\, q\,\xi\, a v^2 \bar{S}_{20}(t) \sin 2\,\beta. \qquad (5.70)$$

In the sudden approximation one assumes β on the right-hand side to be unchanged (equal to β_0) during the collision, and the final angular velocity $\dot{\beta}_f$ is thus given by

$$\dot{\beta}_f = \dot{\beta}_i + \frac{2\,v}{3\,a} q_{\mathrm{eff}}(\vartheta)\, \xi \sin 2\,\beta_0. \qquad (5.71)$$

We have here used Eqs. (5.15), (3.24), and (3.27), and have denoted the angular velocity $\dot{\beta}$ before the collision by $\dot{\beta}_i$.

From (5.71) we obtain the following simple expression for the transfer of angular momentum $\Delta L_\perp$ perpendicular to the symmetry axis of the orbit.

$$\Delta L_\perp = 2\, q_{\mathrm{eff}}(\vartheta)\, \hbar \sin 2\,\beta_0, \qquad (5.72)$$

while the component of L parallel to the axis is unchanged.

In the classical treatment one thus finds that the angular momentum transfer depends on the initial orientation of the top and one sees that the projectile can transfer at most (for $\beta_i = \pi/4$) an angular momentum of magnitude

$$\Delta L_{\max} = 2\, q_{\mathrm{eff}}(\vartheta)\, \hbar. \qquad (5.73)$$

If one considers all initial orientations of the top to be equal probable, one may evaluate the classical energy distribution of the top after the collision. In the simplest case where the nucleus is at rest before the collision, one finds corresponding to (5.73) also a maximum energy transfer $E_{\max} = 2(q_{\rm eff}(\vartheta))^2 \hbar^2/\mathfrak{J}$. In this case, the energy distribution can be written in the form

$$P(E)\,dE = \frac{d\varepsilon}{4\sqrt{\varepsilon}\sqrt{1-\sqrt{\varepsilon}}}, \tag{5.74}$$

where

$$\varepsilon = \frac{E}{E_{\max}} = \frac{\mathfrak{J}}{2(q_{\rm eff}(\vartheta))^2 \hbar^2} E. \tag{5.75}$$

This energy distribution (5.74) is illustrated in Fig. 14.

The classical treatment gives a qualitative understanding of the result of the quantum mechanical calculations of Fig. 7. In the classical limit, the excitation probability of a state of spin I is zero until q reaches the value $I/2$. As a function of q the excitation probability thereafter goes through a maximum and finally decreases slowly. The quantum mechanical energy distribution is a function of both q and the discrete excitation energies. For a fixed value of q the points corresponding to the different energies oscillate around the classical curve. On Fig. 14 we have illustrated the case of q = 10, and we have here, for illustrative purposes, connected the points (indicated by circles) by a smooth curve. It is seen that the result is still far from the classical limit.

Like in the quantum mechanical treatment, the case of $\xi \neq 0$ can be solved for small values of ξ. One must then take into account that the nucleus is moving during the collision time. In first order one may consider the change in the right-hand side of (5.70) to be linear in β. One is thereby led to a hypergeometric differential equation which can be solved explicitly.

The result for the angular momentum transfer perpendicular to the z axis can be written in the form

$$\Delta L_{\perp} = 2\, q_{\rm eff}(\vartheta)\, \hbar \sin 2\beta_0\, F(q\xi \cos 2\beta_0), \tag{5.76}$$

where the correction factor F to the result for $\xi = 0$ is given by

$$F(x) = \frac{3\cos\left(\pi\sqrt{\frac{9}{4} - 4x}\right)}{4\pi x(2x-1)} \approx 1 + \frac{22}{9}x \quad \text{for} \quad x \ll 1. \tag{5.77}$$

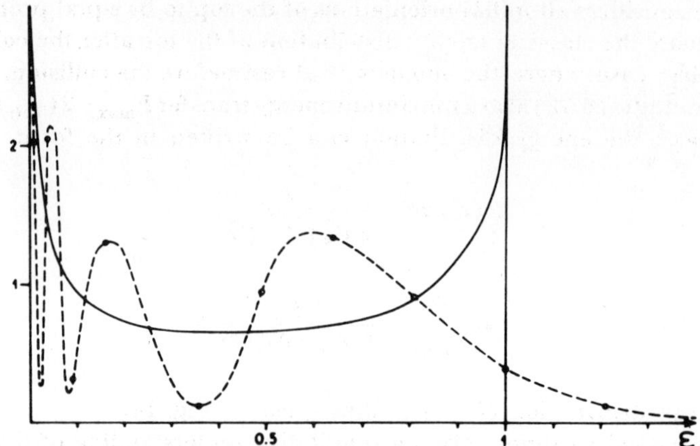

Fig. 14. The classical energy distribution of a charged symmetric top after a head-on impact of a charged particle (full drawn curve). The top is assumed to be at rest before the collision and the impact is assumed to be sudden. The scale of the abscissa is the ratio between the energy of the top E and the maximum energy E_{max} which can be transferred. The circles, which are connected by a broken curve, show the corresponding quantum mechanical result for $q = 10$.

It is illustrative to evaluate the average excitation energy of the nucleus after the collision. If we assume an isotropic distribution for the orientation of the nucleus before the collision, we find from (5.76) to first order in ξ (and backward scattering)

$$\langle E \rangle = \frac{16}{15} \frac{q^2 \hbar^2}{\Im} \left(1 - \frac{44}{63} q \xi\right). \tag{5.78}$$

This result must be correct also in the quantum mechanical treatment. In the limit of $q \ll 1$ where only the lowest state in a rotational band in even-even nuclei is excited, one may use it to calculate the excitation probability. Since the energy of the state of spin 2 is $E_2 = 3\,\hbar^2/\Im$, one finds

$$P_2 = \frac{16}{45} q^2 \left(1 - \frac{44}{63} q \xi\right). \tag{5.79}$$

If one compares this result with the result (5.51) in the limit of $q \ll 1$, one finds

$$f_0^0(\pi) = \frac{11}{9\pi} = \frac{7}{5} f_0^2(\pi) = -\frac{7}{12} f_0^4(\pi) \tag{5.80}$$

in agreement with Table 7.

6. Excitation of Vibrational States

Another important kind of collective excitations in nuclei is that connected with the vibrational degree of freedom. In even-even nuclei a number of low-lying states have been identified as vibrational levels but, in general, the spectra are not as well understood as the corresponding rotational states in deformed nuclei. A survey of the experimental and theoretical status is given in ref. 1, Chapt. V C.

The excitation of pure vibrational states can be solved exactly not only in the sudden approximation, but also for arbitrary ϑ, ξ, and χ. The problem is analogous to the classical problem of a forced vibration which can also be solved in an explicit form.

For a pure quadrupole vibration, the Hamiltonian of the free nucleus is given by

$$H_0 = \frac{1}{2} B \sum_{\mu} |\dot{\alpha}_{2\mu}|^2 + \frac{1}{2} C \sum_{\mu} |\alpha_{2\mu}|^2, \tag{6.1}$$

where B is the inertial parameter and C the restoring force. The parameters $\alpha_{2,\mu}$, where $\mu = -2, -1, 0, 1, 2$ describe the shape of the nuclear surface. In the idealized case where the surface is sharply defined and where the nuclear density is constant, the nuclear shape is given by

$$R(\vartheta, \varphi) = R_0 \left[1 + \sum_{\mu} \alpha_{2\mu} Y_{2\mu}(\vartheta, \varphi)\right]. \tag{6.2}$$

The eigenstates of the Hamiltonian (6.1) can be classified according to the five vibrational quantum numbers n_μ, where $n_\mu = 0, 1, 2 \ldots$. The energy of a state $|n_\mu\rangle$ can thus be written in the form

$$E = \hbar\omega \sum_{\mu} \left(n_\mu + \frac{1}{2}\right) = \hbar\omega \left(N + \frac{5}{2}\right). \tag{6.3}$$

Here, the frequency ω is given by

$$\omega = \sqrt{\frac{C}{B}}, \tag{6.4}$$

while the principal quantum number N is defined by

$$N = \sum_{\mu} n_\mu . \tag{6.5}$$

The degenerate nuclear states can also be labelled by this principal quantum number together with the total angular momentum I and the mag-

5*

netic quantum number M. For $N < 3$ these numbers are sufficient to specify the state completely, while for $N > 3$ one needs additional quantum numbers[17]. The connection between the two labellings n_μ and N, I, M is given in refs. 15 and 16 for a number of cases.

In the following, it will be convenient to introduce a dimensionless coordinate x_μ defined by the equation

$$\alpha_{2\mu} = \sqrt{\frac{\hbar}{\sqrt{BC}}}\, x_\mu \tag{6.6}$$

instead of $\alpha_{2,\mu}$. If we introduce furthermore ξ by means of the equation

$$\xi = \omega \frac{a}{v}, \tag{6.7}$$

we may write H_0 in the form

$$H_0 = \frac{\hbar v \xi}{2\,a} \sum_\mu [\,|\,p_\mu/\hbar\,|^2 + |\,x_\mu\,|^2\,], \tag{6.8}$$

where the momenta p_μ are defined by

$$p_\mu = -B\,\dot{\alpha}^*_{2\mu} = -\sqrt{\hbar\sqrt{\frac{B}{C}}}\;\dot{x}^*_\mu. \tag{6.9}$$

The nuclear multipole moments $\mathfrak{M}(E\,2, \mu)$ are related to the deformation parameters $\alpha_{2,\mu}$ by the following expression (see ref. 1, Eq. (V. 24)),

$$\mathfrak{M}(E2, \mu) = \frac{3}{4\pi} Z_1 e R_0^2 \alpha^*_{2\mu}. \tag{6.10}$$

By evaluating the reduced matrix element of (6.10) between the ground state and the first excited state one finds the following expression for the parameter χ (see Eq. (2.12)):

$$\chi = \frac{Z_1 Z_2 e^2 R_0^2}{v a^2 \sqrt{10\,\pi\hbar\sqrt{BC}}} \tag{6.11}$$

and one may therefore write the interaction Hamiltonian (2.3) in the following form:

$$H'(t) = \sqrt{\frac{18\pi}{5}}\,\hbar v a^2 \chi \sum_\mu \bar{S}_{2\mu}(t)\, x_\mu, \tag{6.12}$$

where $\bar{S}_{2\mu}(t)$ is given by (3.23).

The eigenstates of the free Hamiltonian H_0 are given in terms of the Hermite polynomials $H_n(x)$, i.e.,

$$\left.\begin{aligned}\psi_{n_{-2}\,n_{-1}\,n_0\,n_1\,n_2}(x) &= \prod_{\mu} \psi_{n_\mu}(x_\mu)\\ &= \prod_{\mu} \frac{1}{\sqrt{\sqrt{\pi}\,2^{n_\mu}\,n_\mu!}}\, e^{-\frac{1}{2}x_\mu^2} H_{n_\mu}(x_\mu).\end{aligned}\right\} \quad (6.13)$$

The excitation amplitude in the sudden approximation (3.11) is now very easily found.

The result for the distribution on the different energy levels (N) turns out to be a Poisson distribution, where the mean excitation energy is the same as that one would find in the perturbation calculation (see Eq. (6.27) below). This result can be understood by noting that the excitation of an harmonic oscillator can always be interpreted as the collective motion of a large number of mutually uncoupled harmonic oscillators which are, each of them, only weakly excited. The weak excitation of these oscillators can be treated by a perturbation calculation and, since they are mutually uncoupled, the resulting total energy distribution must be a Poisson distribution.

Since the above argument is independent of the sudden approximation, we shall in the following give the details of the calculation in the more general case of $\xi \neq 0$.

The first step in the program will be to introduce a number of auxiliary variables $x^{(i)}$, where $i = 1, 2, \ldots\ldots \mathfrak{N}$ and where $x_\mu^{(1)} = x_\mu$. The 5 $(\mathfrak{N}-1)$ new degrees of freedom are supposed to be coordinates for free vibrations which have the same frequency ω as the x_μ oscillators. Furthermore, we take them to be coupled, neither to each other nor to the old x_μ. Under these circumstances they will be left undisturbed in the Coulomb excitation process and will only change the problem in a trivial way. The total Hamiltonian will thus be

$$H = \frac{\hbar v \xi}{2a} \sum_{\mu, i} \left[\left| \frac{p_\mu^{(i)}}{\hbar} \right|^2 + |x_\mu^{(i)}|^2 \right] + \sqrt{\frac{18\pi}{5}}\, \hbar v a^2 \chi \sum_\mu \bar{S}_{2\mu}(t)\, x_\mu^{(1)}, \quad (6.14)$$

while the eigenstates which are of physical interest will be

$$\psi(x) = \psi_{n_{-2},n_{-1},n_0,n_1,n_2}(x^{(1)}) \prod_{\substack{\mu \\ i \neq 1}} \psi_0(x_\mu^{(i)}) \quad (6.15)$$

where ψ_0 is the ground state wave function (6.13) with $n_\mu^{(i)} = 0$.

We perform now a linear transformation on the coordinates $x_\mu^{(i)}$ and introduce hereby new coordinates $Q_\mu^{(i)}$ and new momenta $P_\mu^{(i)}$. The transformation matrix U is supposed to be unitary, i. e.

$$UU^\dagger = 1 \tag{6.16}$$

and is assumed to be diagonal in μ, i. e.,

$$Q_\mu^{(i)} = \sum_j U_{ij}^{(\mu)} x_\mu^{(j)}. \tag{6.17}$$

Furthermore, we prescribe the first row of $U^{(\mu)}$ for all values of μ to be given by

$$U_{1,j}^{(\mu)} = \frac{1}{\sqrt{\mathfrak{N}}}. \tag{6.18}$$

The new Hamiltonian in the variables $Q_\mu^{(i)}$ and $P_\mu^{(i)}$ is found from (6.14) (6.16), and (6.18) to be

$$H = \frac{\hbar v \xi}{2a} \sum_{\mu i} [\,|\,P_\mu^{(i)}/\hbar\,|^2 + |\,Q_\mu^{(i)}\,|^2] + \sqrt{\frac{18\pi}{5}}\,\hbar v a^2 \frac{\chi}{\sqrt{\mathfrak{N}}} \sum_\mu \bar{S}_{2\mu}(t) \sum_i Q_\mu^{(i)}. \tag{6.19}$$

The Schrödinger equation for the variables $Q_\mu^{(i)}$ is again separable and the eigenstates of the free Hamiltonian are

$$\varphi_n(Q) = \prod_{i\mu} \psi_{n_\mu^{(i)}}(Q_\mu^{(i)}), \tag{6.20}$$

where all $n_\mu^{(i)}$ can take the values $n_\mu^{(i)} = 0, 1, 2 \ldots\ldots$

In the new Hamiltonian, however, the interaction term can be made very small for all values of μ and i by choosing $\mathfrak{N}$ to be a large number. In the new variables, the excitation process can therefore be treated by a perturbation calculation, and one obtains for each of the oscillators (μ, j) only a very small probability that the oscillator is excited. In the perturbation treatment one finds, for each oscillator the following excitation amplitude on the first excited state:

$$a_\mu^{(j)} = \frac{1}{i\hbar} \langle 1 \,|\, \int_{-\infty}^{\infty} \sqrt{\frac{18\pi}{5}}\,\hbar v a^2 \frac{\chi}{\sqrt{\mathfrak{N}}} \bar{S}_{2\mu}(t)\, Q_\mu^{(j)} e^{i\omega t} dt \,|\, 0 \rangle. \tag{6.21}$$

Since the matrix element of $Q_\mu^{(j)}$ between the states of one phonon $|\,1\,\rangle$ and the ground state $|\,0\,\rangle$ is given by

$$\langle 1 \mid Q_\mu^{(j)} \mid 0 \rangle = \frac{(-1)^\mu}{\sqrt{2}}, \tag{6.22}$$

one may write (6.21) in the form

$$a_\mu^{(j)} = -\frac{i}{\sqrt{\mathfrak{N}}} \chi_\mu (\vartheta, \xi), \tag{6.23}$$

where (see also Eq. (3.24))

$$\chi_\mu (\vartheta, \xi) = \chi \sqrt{\frac{9\pi}{5}} v a^2 (-1)^\mu S_{E2,\mu} (\vartheta, \xi). \tag{6.24}$$

We can now easily determine the total excitation probability. We ask first for the probability that any one of the five oscillators belonging to a definite value of j is excited. This probability is

$$P_{(j)} = \sum_\mu \mid a_\mu^{(j)} \mid^2 = \frac{1}{\mathfrak{N}} \mid \chi (\vartheta, \xi) \mid^2, \tag{6.25}$$

where $\chi (\vartheta, \xi)$ is given by Eq. (2.12), i. e.,

$$\chi (\vartheta, \xi) = \chi^2 \frac{9\pi}{5} \sum_\mu \mid Y_{2\mu} \left(\frac{\pi}{2}, 0 \right) I_{2\mu} (\vartheta, \xi) \mid^2. \tag{6.26}$$

We can then calculate the probability that all the $\mathfrak{N}$ groups of five oscillators have together the total excitation energy $N \hbar \omega$. Since all groups have the same probability (6.25) of having the energy $\hbar \omega$, we obtain in the limit of $\mathfrak{N} \to \infty$ a Poisson distribution for this probability P_N

$$P_N = \frac{1}{N!} e^{-[\chi (\vartheta, \xi)]^2} [\chi (\vartheta, \xi)]^{2N}. \tag{6.27}$$

In the old variables $x_\mu^{(i)}$, this result must be interpreted as the total excitation probability of the vibrational state with principal quantum number N.

We shall be interested also in the amplitudes on the eigenstates (6.13). We shall evaluate these by calculating the amplitudes $\langle \varphi_n (Q) \mid \Phi (Q) \rangle$ of the final wave function, Φ, on the eigenstates (6.20) as well as the amplitudes $\langle \psi (x (Q)) \mid \varphi_n (Q) \rangle$ of (6.13) on these eigenstates.

From (6.23) one finds directly the amplitude on the states (6.20)

$$\langle \varphi_n(Q) \mid \Phi (Q) \rangle = \prod_{\mu, i} [a_\mu^{(i)}]^{n_\mu^{(i)}} [1 - \mid a_\mu^{(i)} \mid^2]^{(1 - n_\mu^{(i)})/2}, \tag{6.28}$$

where the quantum numbers $n_\mu^{(i)}$ are all 0 or 1.

The amplitude of (6.13) on (6.20) follows from the expansion of a Hermite polynomial of a linear function of $Q_\mu^{(i)}$ in terms of a product of Hermite polynomials of $Q_\mu^{(i)}$. One finds from ref. 10 (Vol. 2, p. 196)

$$\langle \psi(x(Q)) \mid \varphi_n(Q) \rangle = \prod_\mu \sqrt{\frac{n_\mu!}{\mathfrak{N}^{n_\mu}(n_\mu^{(1)})!(n_\mu^{(2)})!\cdots(n_\mu^{(\mathfrak{N})})!}}. \tag{6.29}$$

According to (6.28) and (6.29), the expression for the excitation amplitude is

$$\left.\begin{aligned} a_{n_{-2}n_{-1}n_0n_1n_2} &= \sum_n \langle \psi(x(Q)) \mid \varphi_n(Q) \rangle \langle \varphi_n(Q) \mid \Phi(Q) \rangle \\ &= \prod_\mu \sqrt{\frac{(n_\mu)!}{\mathfrak{N}^{n_\mu}}} \left(\frac{-i\chi_\mu(\vartheta,\xi)}{\sqrt{\mathfrak{N}}}\right)^{n_\mu} \left(1 - \frac{(\chi_\mu(\vartheta,\xi))^2}{\mathfrak{N}}\right)^{\frac{\mathfrak{N}}{2}-\frac{n_\mu}{2}} \binom{\mathfrak{N}}{n_\mu}. \end{aligned}\right\} \tag{6.30}$$

We have here utilized that $n_\mu^{(i)} = 0$ or 1 and have performed the summation over $n_\mu^{(i)}$ with the restriction $\Sigma\, n_\mu^{(i)} = n_\mu$ by multiplying with the number of ways in which n_μ objects may be chosen among $\mathfrak{N}$ objects. When we let $\mathfrak{N} \to \infty$ the expression (6.30) takes the form

$$a_{n_{-2}n_{-1}n_0n_1n_2} = \prod_\mu \frac{(-i)^{n_\mu}}{\sqrt{(n_\mu)!}} (\chi_\mu(\vartheta,\xi))^{n_\mu} e^{-\frac{1}{2}(\chi_\mu(\vartheta,\xi))^2}. \tag{6.31}$$

This equation offers the complete solution of the excitation of pure vibrational states.

It is interesting to observe that the total excitation probability (6.27) depends on ϑ and ξ only through the quantity $\chi(\vartheta, \xi)$. This means that the excitation probability for arbitrary ϑ and ξ can be obtained from the probabilities for $\vartheta = \pi$ and $\xi = 0$ by substituting $\chi(\vartheta, \xi)$ for χ, i. e.,

$$P_N(\vartheta, \xi, \chi) = P_N(\pi, 0, \chi(\vartheta, \xi)). \tag{6.32}$$

In the special case of $\xi = 0$ this equation shows that the $\chi(\vartheta)$ approximation (see Eq. (5.62)) in the case of vibrational states is exactly fulfilled.

The function $P_N(\pi, 0, \chi)$ is illustrated on Fig. 15, as a function of χ. It is interesting to compare this result with the corresponding result for the excitation of a rotational band which is illustrated on Fig. 7. The maximum excitation probabilities are larger for rotational states than for vibrational states. However, in the latter case, higher lying states are reached for a definite value of χ.

The ground state and the first excited state have the definite angular momenta 0 and 2. The second excited state, however, is a triplet with spins 0, 2 and 4. Since the vibrational states in nuclei are not pure, the degeneracy

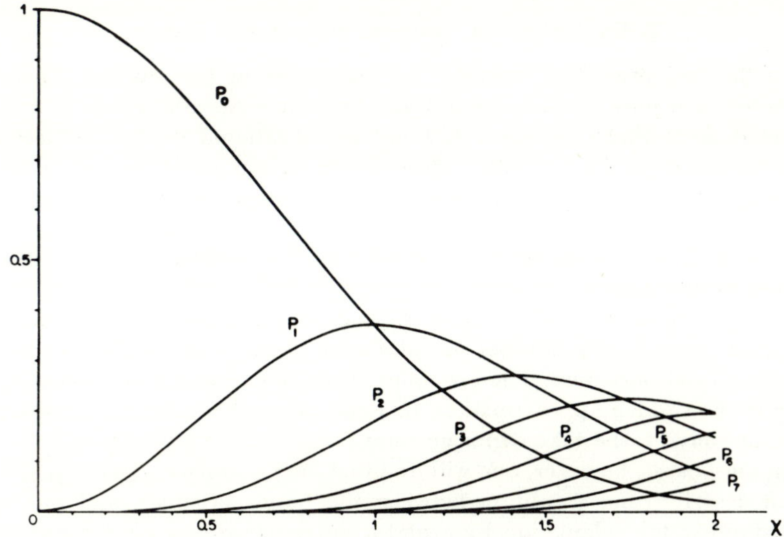

Fig. 15. The multiple Coulomb excitation of a pure vibrational band in an even-even nucleus. The excitation probability P_N of the state with principal quantum number N is given as a function of the parameter $\chi(\vartheta, \xi)$.

is in actual cases removed, and it is thus interesting to find how the total excitation probability (6.27) is distributed on each of the substates.

To perform this calculation one needs the expression (6.31) for the amplitudes on the states specified by $n_{-2}, n_{-1}, \ldots\ldots n_2$. Furthermore, one needs the coefficients for the transformation between the n_μ and the N, I, M labelling (see refs. 15 and 16).

One finds in the case of $N = 2$ the following result

$$\left.\begin{aligned} P_{2,\,I=0} &= 1/5\ P_2 \\ P_{2,\,I=2} &= 2/7\ P_2 \\ P_{2,\,I=4} &= 18/35\,P_2 \end{aligned}\right\} \quad (6.33)$$

only for $\xi = 0$ or $\xi \neq 0$ and $\vartheta = \pi$

where P_2 is given by (6.27) for $N = 2$. In this case the rule (6.32) thus also holds for the excitation of the substates with $I = 0$, 2 and 4. This is, however, not true any more for the excitation of the substates of the state with principal quantum number $N = 3$.

7. Excitation of Coupled Rotational Bands

In the two preceding sections, we have treated the multiple Coulomb excitation of a pure rotational band and of a pure vibrational band, and we assumed there that only the rotational or vibrational degrees of freedom were involved in the excitation process. In actual cases several different degrees of freedom can be excited. One might have cases, such as in most deformed nuclei, where both rotational and vibrational degrees of freedom are involved, or one might have to consider the excitation of the intrinsic degrees of freedom.

In this section, we shall consider a situation where the low energy nuclear spectrum consists of a number of rotational bands which differ in internal (or vibrational) structure. The excitation of these bands can be treated rigorously in the sudden approximation, in some cases when only a finite number of bands have to be taken into account.

In most cases, however, one will find that the parameter χ (see Eq. (2.12)) which describes the transition between the bands is small, and one may then simplify the calculation by a perturbation expansion for the transition from one band to the others. The transitions within any one of the bands must in any case be treated rigorously.

We shall assume that the nuclear states are described by state vectors of the form

$$|\psi\rangle = |n, K\rangle \, |I, K, M\rangle, \tag{7.1}$$

where $|I, K, M\rangle$ stands for a rotational wave function of the form (5.6) which only depends on the Eulerian angles describing the orientation of the nuclear axis, while $|n, K\rangle$ describes a state of the intrinsic and vibrational degrees of freedom, which has a component of angular momentum K along the nuclear symmetry axis. The state vector $|n, K\rangle$ depends only on relative coordinates measured with respect to a coordinate system which has its z-axis along the nuclear axis.

In actual cases, the nuclear state vector will be a linear combination of state vectors of the form (7.1). Firstly, it will always contain a term identical with (7.1), except for a change of sign on K. Secondly, it may often contain admixtures from other bands with different values of K and n. The actual excitation probabilities can, however, easily be evaluated once the excitation probabilities for states of the simple type (7.1) are known.

It is convenient to transform the interaction energy to the rotating coordinate system which has its z-axis along the nuclear axis.

For the multipole operator (2.4) one finds

$$\mathfrak{M}(E\,2,\mu) = \sum_{\nu} D^2_{\mu,\nu}(\alpha,\beta,0)\,\mathfrak{M}_{\text{int}}(E\,2,\nu), \tag{7.2}$$

where the intrinsic multipole operator $\mathfrak{M}_{\text{int}}(E\,2,\nu)$ is independent of the Eulerian angles α and β.

If we adopt the $\chi(\vartheta)$ approximation, the excitation amplitude in the sudden approximation (3.11) may be written

$$a^{(f)}_{I_f M_f K_f} = \langle I_f K_f M_f \,|\, \langle f K_f \,|\, e^{-i\sqrt{\frac{64\pi}{45}}\sum_{\nu} \mathfrak{q}_{\nu}(\vartheta)\, Y^*_{2\nu}(\beta,0)} \,|\, i K_i \rangle \,|\, I_i K_i M_i \rangle \tag{7.3}$$

where the operator $\mathfrak{q}_{\nu}(\vartheta)$ is defined by

$$\mathfrak{q}_{\nu}(\vartheta) = \sqrt{\frac{\pi}{5}}\,\frac{Z_1 e^2 \mathfrak{M}^*_{\text{int}}(E2,\nu)}{\hbar v a^2}\,\frac{3}{4} J_{2,0}(\vartheta). \tag{7.4}$$

The expectation value of $\mathfrak{q}_{\nu}$ for states within a band is exactly the earlier defined quantity of $q_{\text{eff}}(\vartheta)$ for this band (see Eq. (5.15)),viz.,

$$\langle nK \,|\, \mathfrak{q}_{\nu}(\vartheta) \,|\, nK \rangle = q^{(n)}_{\text{eff}}(\vartheta)\,\delta_{\nu 0} \approx q^{(n)}(\vartheta)\,\delta_{\nu 0}. \tag{7.5}$$

If the matrix elements of $\mathfrak{q}_{\nu}$ between the bands are small, and if the expectation value of $\mathfrak{q}_{\nu}$ or the intrinsic quadrupole moment is not very different in the initial and final band, one may use a perturbation expansion to evaluate (7.3). We may write the amplitude (7.3) in the form

$$\left.\begin{aligned} a^{(f)}_{I_f K_f M_f} &= \langle I_f K_f M_f \,|\, e^{-i\sqrt{\frac{64\pi}{45}}\, q(\vartheta)\, Y_{2,0}(\beta,0)} \\ &\times \langle f K_f \,|\, e^{-i\sqrt{\frac{64\pi}{45}}\sum_{\nu}[\mathfrak{q}_{\nu}(\vartheta) - \delta_{\nu 0}\, q(\vartheta)]\, Y^*_{2\nu}(\beta,0)} \,|\, i K_i \rangle \,|\, I_i K_i M_i \rangle, \end{aligned}\right\} \tag{7.6}$$

where $q(\vartheta)$ is the q for the initial band. We then perform a series expansion of the second exponential whereby we obtain the following expression for the excitation amplitude for a band different from the ground state band:

$$\left.\begin{aligned} a^{(f)}_{I_f K_f M_i} &\approx -i\sqrt{\frac{64\pi}{45}}\,\langle f K_f \,|\, \mathfrak{q}_{K_i - K_f}(\vartheta) \,|\, i K_i \rangle \\ &\times \langle I_f K_f M_f \,|\, Y^*_{2,\,K_i - K_f}(\beta,0)\, e^{-i\sqrt{\frac{64\pi}{45}}\, q(\vartheta)\, Y_{2,0}(\beta,0)} \,|\, I_i K_i M_i \rangle. \end{aligned}\right\} \tag{7.7}$$

The excitation probability in the perturbation treatment can thus be written in the form

$$P^{(f)}_{I_f} = r^2_{if}\, N^{(f)}_{I_f K_f}(q(\vartheta)) \tag{7.8}$$

where we have introduced the notation

$$r_{if} = \frac{\langle fK_f \,|\, q_{K_f - K_i}(\vartheta) \,|\, iK_i \rangle}{q(\vartheta)} = \frac{\langle fK_f \,|\, \mathfrak{M}(E\,2,\nu) \,|\, iK_i \rangle}{\langle iK_i \,|\, \mathfrak{M}(E\,2,0) \,|\, iK_i \rangle}. \tag{7.9}$$

From (7.7), one observes that the total probability of exciting any member of the final band is given by the simple expression

$$P^{(f)} = \sum_I P_{I_f}^{(f)} = \frac{16}{45} r_{if}^2 \,[q(\vartheta)]^2. \tag{7.10}$$

This result is identical with the expression which would be obtained in the ordinary perturbation theory for $q \ll 1$, and one sees that the strong coupling within the bands give rise only to a redistribution of the single probabilities $P_{I_f}^{(f)}$.

In the perturbation treatment which we have used here one may also approximately take into account the effect of finite ξ. If the energy difference between the ground states of the two bands is larger than the energy of the lowest states in the bands, the ξ corresponding to the possible transitions between the two bands will be approximately constant. One may then take ξ into account by calculating the total transition probability (7.10) for the finite ξ in the ordinary perturbation treatment and apply the same distribution $P_{I_i}^{(f)}/P^{(f)}$ as for $\xi = 0$.

The matrix element (7.7) can be expressed in terms of the functions $A_{I,M}(\pi, q)$ (see Eq. (5.9)) by expanding the product of D-functions in terms of D-functions (compare also Eq. (5.20)). The result has been evaluated in the special case of $I_i = 0$. For a band with $K_f = 2$, one finds

$$\left.\begin{aligned} a_{I,2,0}^{(f)} &= -i\frac{2}{3} r_{if}\, q(\vartheta) \sqrt{6(I-1)I(I+1)(I+2)(2I+1)} \\ &\times \left\{ \frac{1}{2(2I+1)(2I+3)} A_{I+2,0}(q(\vartheta)) - \frac{1}{(2I-1)(2I+3)} A_{I,0}(q(\vartheta)) \right. \\ &\left. + \frac{1}{2(2I-1)(2I+1)} A_{I-2,0}(q(\vartheta)) \right\}. \end{aligned}\right\} \tag{7.11}$$

For a band with $K_f = 0$ one finds similarly

$$\left.\begin{aligned} a_{I,0,0}^{(f)} &= -i\frac{2}{3} r_{if} \cdot q(\vartheta) \sqrt{2I+1} \\ &\times \left\{ \frac{3(I+1)(I+2)}{(2I+1)(2I+3)} A_{I+2,0}(q(\vartheta)) + \frac{2I(I+1)}{(2I-1)(2I+3)} A_{I,0}(q(\vartheta)) \right. \\ &\left. + \frac{3(I-1)I}{(2I-1)(2I+1)} A_{I-2,0}(q(\vartheta)) \right\}. \end{aligned}\right\} \tag{7.12}$$

The resulting excitation probabilities are given in Figs. 16 and 17 in terms of the coefficients $N^f_{I_f,K_f}$. This coefficient is plotted for $K_f = 0$, $I = 0, 2$ and 4 and for $K_f = 2$, $I = 2$ and 4. The states with odd spins in the $K = 2$ band are not excited in the $q(\vartheta)$ approximation.

The perturbation treatment (7.7) is only correct if the quantity $r_{if}\, q(\vartheta)$ is smaller than one (compare Fig. 2).

If the bands are strongly coupled through the interaction with the projectile one may evaluate the matrix element (7.3) by a diagonalization method. We thus introduce a unitary transformation U which diagonalizes the matrix elements of the exponent in (7.3), i. e.,

$$\langle aK_a \mid U^\dagger \sum_\nu q_\nu(\vartheta)\, Y^*_{2,\nu}(\beta, 0)\, U \mid bK_b \rangle = \delta_{ab} \cdot \lambda_a. \tag{7.13}$$

The result (7.3) can then be written in the form

$$a_{I_f K_f M_f} = \sum_z \langle I_f K_f M_f \mid \langle f \mid U \mid z \rangle e^{-i\sqrt{\frac{64\pi}{45}}\lambda_z} \langle z \mid U^\dagger \mid i \rangle \mid I_i K_i M_i \rangle. \tag{7.14}$$

Since λ and the unitary matrix U in the general case depend on the Eulerian angle β in a rather complex way, this result is of practical interest only in some special cases. We shall consider the case where only two bands are involved in the excitation process. We assume that they have the same intrinsic quadrupole moment and that $K_i = K_f$. The matrix diagonalization is then easily performed and one finds the result

$$\left.\begin{aligned} a^{(i)}_{IKM} &= \frac{1}{2}\{a_{IM}[q(1+r_{if})] + a_{IM}[q(1-r_{if})]\} \\ a^{(f)}_{IKM} &= \frac{1}{2}\{a_{IM}[q(1+r_{if})] - a_{IM}[q(1-r_{if})]\}. \end{aligned}\right\} \tag{7.15}$$

It is interesting to compare this result with the result (7.7) from the perturbation treatment. In Fig. 18 we have plotted the excitation probability for the state $\mid 2,0,0 \rangle$ as a function of q for different values of r. It is seen that the perturbation treatment is correct if $rq \lesssim 1$, as was to be expected. This condition will in actual cases usually be fulfilled.

If $\mid K_f - K_i \mid$ is larger than 2 the transition between the bands is K forbidden. A possible transition between the bands can then occur only if the wave function (7.1) contains admixtures from other bands. Such admixtures can also play an important role in K allowed transitions[18].

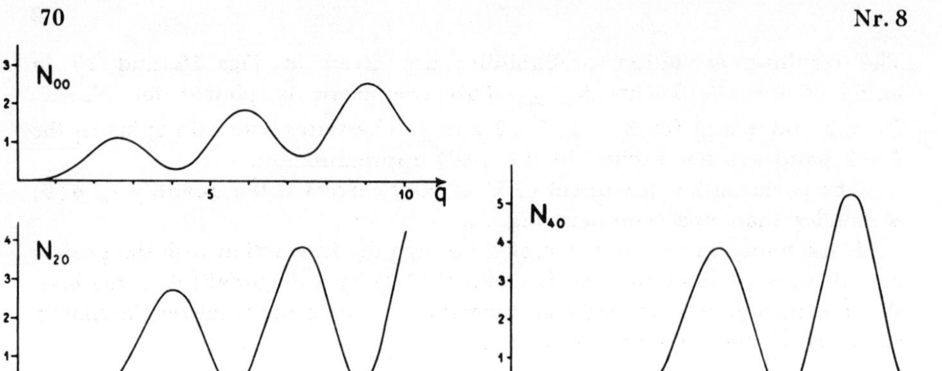

Fig. 16. The excitation probability of a weakly coupled pure rotational band with $K = 0$ in an even-even nucleus. The figure shows the coefficient $N^{(f)}_{I,0}$ in the perturbation treatment (see Eq. (7.8)) for $I = 0,2$ and 4 as a function of $q(\vartheta)$, which is assumed to be the same for the two bands.

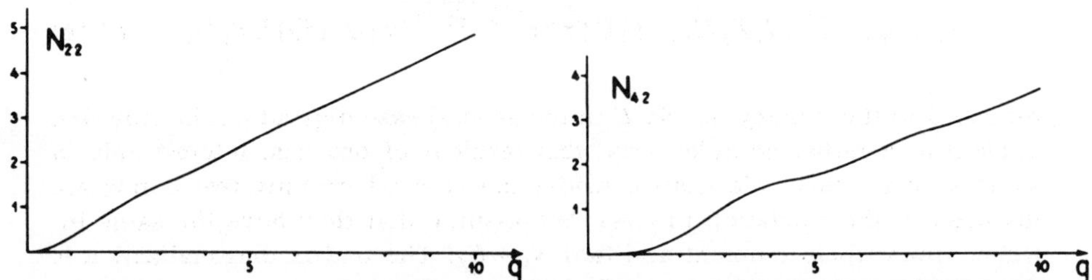

Fig. 17. The excitation probability of a weakly coupled, pure rotational band with $K = 2$ in an even-even nucleus. The figure shows the coefficient $N^{(f)}_{I,2}$ in the perturbation treatment (see Eq. (7.8)) for $I = 2$ and 4 as a function of $q(\vartheta)$ which is assumed to be the same for the two bands. In the $q(\vartheta)$ approximation, which has been used for the evaluation of N, the states with odd I are not excited.

8. Conclusion

The multiple Coulomb excitation has until now been observed only in a few cases, but, from these observations (see refs. 5 and 6) as well as from the survey given in the present paper, it seems that a large number of experimental possibilities are offered. Especially it seems promising to investigate the excitation of the vibrational degrees of freedom of the nucleus, since our present knowledge on such states is rather limited. It is known that the vibrational states are mostly rather impure and for a quantitative comparison one may need some modification of the present theory. Simi-

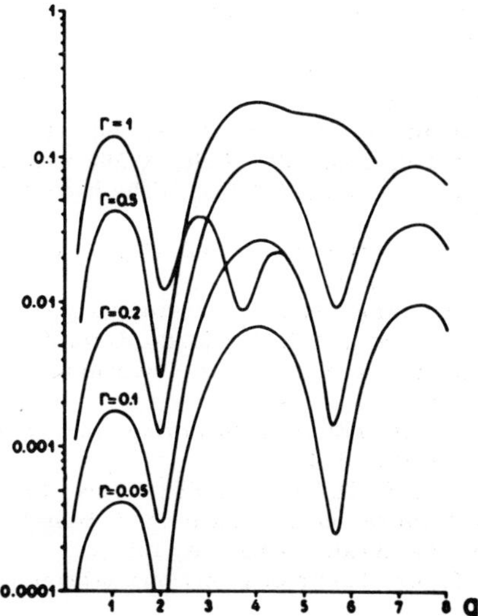

Fig. 18. The excitation probability of a strongly coupled pure rotational band with $K = 0$. The figure shows the excitation probability of the $I = 2$ state for different values of the coupling r (see Eq. (7.9)). The probabilities are given as functions of $q\,(\vartheta)$ which is assumed to be equal in the two bands.

larly deviations from the pure rotational model have been observed. These deviations introduce a number of new parameters in the theory, and these can, in turn, be determined by a comparison of the experimental cross sections with the theory. On the other hand the increasing number of parameters make a systematic tabulation of cross sections increasingly difficult.

The larger part of the present paper was written during a stay of one of the authors (A.W.) at the Federal Institute of Technology in Zürich, Switzerland. We wish to thank Professor Paul Scherrer for making this stay possible. Sincere thanks are due Professor Aage Bohr for his continual interest in our work.

Federal Institute of Technology, Zürich, Switzerland
and
Institute for Theoretical Physics, University of Copenhagen, Denmark.

References

1) K. Alder, A. Bohr, T. Huus, B. Mottelson and A. Winther, Revs. Mod. Phys. **28**, 432 (1956).

2) G. Breit and R. L. Gluckstern, Encyclopedia of Physics (Springer, Berlin, 1959) Vol. XLI/1 p. 496.

3) N. P. Heydenburg and G. M. Temmer, Annual Review of Nuclear Science (Annual Reviews Inc., Palo Alto, Calif. 1956) Vol. 6, p. 77.

4) R. Huby, Reports on Progress in Physics, (The Physical Society, London 1958) Vol. XXI p. 59.

5) J. O. Newton and F. S. Stephens, Phys. Rev. Let. **1**, 63 (1958).

6) F. S. Stephens and R. Diamont, Private communieation.

7) K. Alder and A. Winther, Mat. Fys. Medd. Dan. Vid. Selsk. **31**, no. 1 (1956).

8) A. Bohr and B. Mottelson, Mat. Fys. Medd. Dan. Vid. Selsk. **27**, no. 16 (1953).

9) A. R. Edmonds, Angular Momentum in Quantum Mechanics (Princeton University Press, Princeton, 1957).

10) A. Erdélyi et al., Higher Transcendental Functions, (Mc Graw-Hill Book Company, Inc., New York, 1953).

11) L. C. Biedenharn and R. M. Thaler, Phys. Rev. **104**, 1643 (1956).

12) G. N. Watson, A Treatise on the Theory of Bessel Functions, (Cambridge University Press, Cambridge, England, 1952).

13) E. Jahnke and F. Emde, Tables of Functions, (Dover Publications, New York, 1945).

14) G. Breit, R. L. Gluckstern and J. E. Russel, Phys. Rev. **103**, 727 (1956).

15) D. C. Choudhury, Mat. Fys. Medd. Dan. Vid. Selsk. **28**, no. 4 (1954).

16) K. W. Ford and C. Levinson, Phys. Rev. **100**, 1 (1955).

17) G. Racavy, Nucl. Phys. **4**, 289 (1957).

18) P. Gregers Hansen, O. B. Nielsen and R. K. Sheline, Nucl. Phys. **12**, 389 (1959).

Indleveret til Selskabet den 5. december 1959
Færdig fra trykkeriet den 25. juli 1960.

2.H Nuclear Physics 36 (1962) 452—456; © North-Holland Publishing Co., Amsterdam

A NOTE ON THE ANGULAR DISTRIBUTION OF INELASTICALLY SCATTERED PARTICLES IN COULOMB EXCITATION

T. A. GRIFFY and L. C. BIEDENHARN
Duke University, Durham, North Carolina †

Received 13 February 1962

Abstract: The angular distribution of the inelastically scattered particles following Coulomb excitation is calculated for electric quadrupole transitions. The quantum mechanical calculations are performed for the case of finite energy loss and a comparison is made between the semi-classical and quantum mechanical angular distributions.

1. Introduction

The angular distribution of the inelastically scattered particles following Coulomb excitation has recently been measured by several authors [1-3] †† and has proven to be a useful tool in the identification of the multipolarity of the nuclear transition induced in the excitation process. Measurements of the cross section at backward angles, combined with theoretical predictions of the angular distribution, can be used to determine the total cross section and hence the value of the reduced nuclear matrix element $B(E\lambda)$. This method of obtaining the total cross section is more advantageous than the more usual method – measuring the yield of internal conversion electrons following the excitation – in that one does not require a knowledge of the internal conversion coefficients in order to obtain the cross section.

The angular distributions of the inelastically scattered particles may be characterized by two parameters [4]),

$$\eta_i = \frac{Z_1 Z_2 e^2}{\hbar v_i}, \qquad \text{and} \qquad \xi = \eta_f - \eta_i,$$

where Z_1 is the charge of the projectile, Z_2 is the charge of the target nucleus and v_i, v_f are the asymptotic initial and final velocities of the projectile. Previous calculations of the differential cross section for Coulomb excitation have been performed in the semi-classical approximation [4]) ($\eta_i = \infty$) and in the quantum mechanical case where the energy loss of the projectile is assumed to be zero [5]) (η_i = finite, $\xi = 0$). In the present calculation we have evaluated the differential cross section for electric quadrupole transitions for both η_i and ξ finite. A comparison of these calculations with the

† Work supported in part by the Army Research Office (Durham) and the National Science Foundation.

†† Ref. [1]) contains references to earlier work of the same laboratory.

previous semi-classical calculations indicates that the *shapes* of the angular distributions differ by at most about 10% for any reasonable values of the parameters η_i and ξ. The *magnitude* of the cross section may differ by as much as 40 % however.

2. Method of Calculation

The differential cross section for Coulomb excitation of a given electric multipolarity is given by [4])

$$\frac{d\sigma_{E\lambda}}{d\Omega} = \left(\frac{Z_1 e}{\hbar v_i}\right)^2 a^{-2\lambda+2} B(E\lambda) \frac{df_{E\lambda}}{d\Omega}(\theta, \eta_i, \xi), \tag{1}$$

where Z_1 is the charge number of the projectile, a is given by $a = Z_1 Z_2 e^2/M_0 v_i v_f$, and $B(E\lambda)$ is the reduced nuclear transition probability. As is seen from eq. (1), the angular distribution of the inelastically scattered particles is determined by the dimensionless function $df_{E\lambda}/d\Omega$ which is defined by

$$\frac{df_{E\lambda}}{d\Omega} = \frac{4k_i k_f a^{2\lambda-2}}{(2\lambda+1)^3} \sum_\mu |\langle \boldsymbol{k}_f | r_p^{-\lambda-1} Y_\lambda^\mu(\theta_p, \phi_p) | \boldsymbol{k}_i \rangle|^2, \tag{2}$$

where

$$\langle \boldsymbol{k}_f | r_p^{-\lambda-1} Y_\lambda^\mu(\theta_p, \phi_p) | \boldsymbol{k}_i \rangle = (4\pi)^{\frac{1}{2}} \sum_{l_i l_f m_i m_f} i^{l_i - l_f} (-)^\mu$$
$$\times e^{i[\sigma_{li}(\eta_i)+\sigma_{lf}(\eta_f)]} [(2l_i+1)(2l_f+1)(2\lambda+1)]^{\frac{1}{2}} \begin{pmatrix} l_i & l_f & \lambda \\ 0 & 0 & 0 \end{pmatrix} \tag{3}$$
$$\begin{pmatrix} l_i & l_f & \lambda \\ m_i & -m_f & \mu \end{pmatrix} Y_{l_i}^{-m_i}(\hat{\boldsymbol{k}}_i) Y_{l_f}^{m_f}(\hat{\boldsymbol{k}}_f) M_{l_i l_f}^{-\lambda-1}.$$

Here, λ is the multipolarity of the transition, the quantities

$$\begin{pmatrix} a & b & c \\ \alpha & \beta & \gamma \end{pmatrix}$$

are Clebsch-Gordan (Wigner) coefficients, $\sigma_l(\eta)$ is the Coulomb phase shift and is given by $\sigma_l(\eta) = \arg \Gamma(l+1+i\eta)$. The quantities $M_{l_i l_f}^{-\lambda-1}$ are the radial matrix elements:

$$M_{l_i l_f}^{-\lambda-1} = \frac{1}{k_i k_f} \int_0^\infty F_{l_f}(k_f r) r^{-\lambda-1} F_{l_i}(k_i r) dr. \tag{4}$$

If the beam axis is taken as the axis of quantization, the differential cross section may be written in the form

$$\frac{d\sigma}{d\Omega} \propto \sum_\mu |A_\lambda^\mu(\theta)|^2, \tag{5}$$

where

$$A^{\mu}_{\lambda}(\theta) = \sum_{l_i l_f} i^{l_i - l_f} e^{i[\sigma_{l_i}(\eta_i)+\sigma_{l_f}(\eta_f)]}[(2l_i+1)(2l_f+1)]^{\frac{1}{2}}$$
$$\times \begin{pmatrix} l_i & l_f & \lambda \\ 0 & 0 & 0 \end{pmatrix} \begin{pmatrix} l_i & l_f & \lambda \\ 0 & \mu & \mu \end{pmatrix} Y^{\mu}_{l_f}(\theta\phi) M^{-\lambda-1}_{l_i l_f}. \tag{6}$$

In eqs. (5) and (6) we have suppressed all factors independent of the scattering angle.

Eqs. (5) and (6) have been calculated using the radial matrix elements tabulated by Goldstein, Thaler and Biedenharn [6]). The sum on l_i was performed from $0 \leqq l_i \leqq 15$. The Clebsch-Gordan coefficients were taken from the algebraic tables of Condon and Shortley [7]). The Coulomb phase shifts were calculated by means of the relation [8])

$$\sigma_0 = 0.57721\eta + \sum_{s=1} \left\{ \frac{\eta}{s} - \text{arctg} \frac{\eta}{s} \right\} \tag{7}$$

and the recurrence relation

$$\sigma_{l+1} = \sigma_l + \text{arctg} \frac{\eta}{l+1}.$$

The results of the calculation are shown in figs. 1, 2 and 3. It might be useful to note once again that the result given in eq. (2) is completely specified by ξ, η_i and θ for a fixed multipolarity – there are, for example, no arbitrary normalizations used in the figures.

3. Discussion of Results

Figs. 1 and 2 show the comparison between the semi-classical angular distributions and the results of the present calculation for $\eta_i = 1.0$ and various values of ξ. As these figures indicate, the shapes of the various angular distributions are quite similar. The magnitudes of the quantum mechcnical calculations are considerably smaller, reflecting the decrease in the total cross section with decreasing η_i. Fig. 3 shows a comparison of the semi-classical and quantum mechanical calculation for $\xi = 0.8$ and $\eta_i = 1$, 2 and 3. This comparison indicates even for this rather large value of ξ that the cross sections differ by less than 10% at the peak for $\eta_i > 3$ and are essentially the same for backward angles ($\theta > 120°$). *These comparisons indicate that for most cases of experimental interest* ($\eta_i \geqq 3$, $\xi \leqq 0.5$) *the semi-classical treatment gives an adequate description of the differential cross section.*

The accuracy of the present calculations were checked by comparison with previous calculations – the $\xi = 0$ results of Bang [5]) and the semi-classical calculations of Alder and Winther [4]). The Coulomb phase shifts and the Legendre polynomials which enter in the calculation were also compared with existing tables. The differential cross sections were integrated numerically and compared with the previous calculation of the total cross section [9]) (see also ref. [4])). These comparisons indicate that the calculations are accurate to about 3%, the major sources of error being small inaccuracies

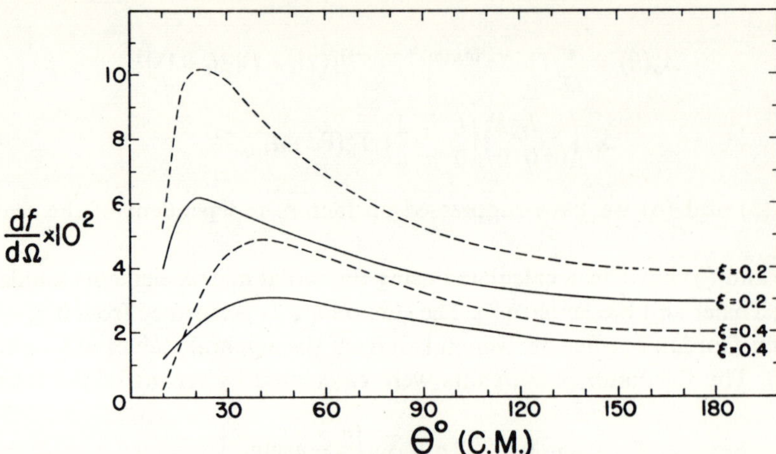

Fig. 1. The differential cross section $df_{E2}/d\Omega$ (defined in equation 2) is plotted versus the centre of mass angle θ for the values $\eta_1 = 1$ and ξ as indicated. The corresponding semi-classical results are shown by the dashed curves.

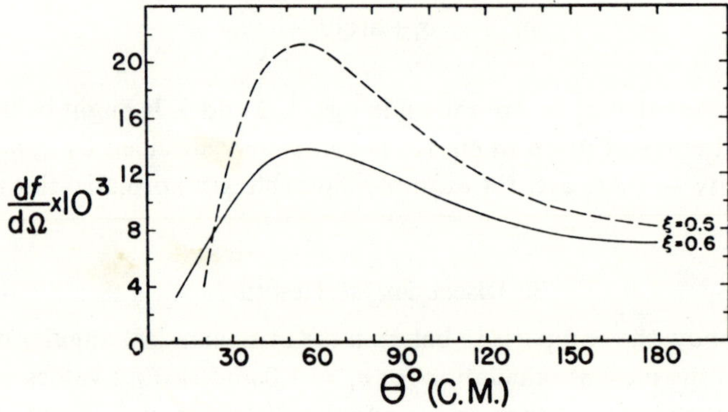

Fig. 2. The differential cross section $df_{E2}/d\Omega$ (defined in equation 2) is plotted versus the centre of mass angle θ for the values $\eta_1 = 1$ and $\xi = 0.6$. The corresponding semi-classical result is shown by the dashed curve.

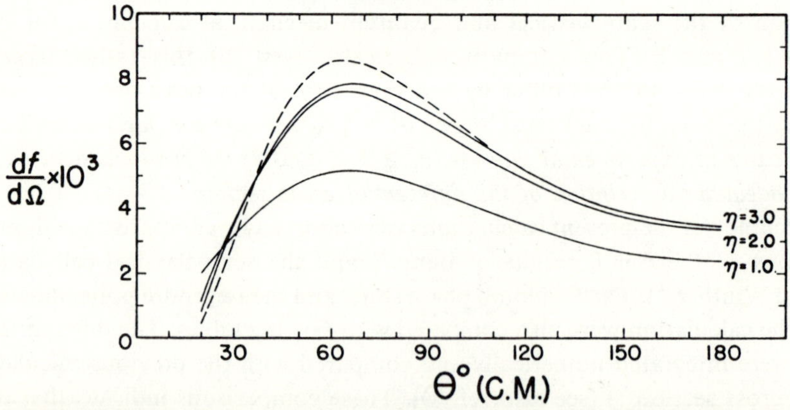

Fig. 3. The differential cross section is shown as a function of the centre of mass angle θ for $\xi = 0.8$ and the indicated values of η_1. The semi-classical result is shown as a dashed curve.

in the tabulated radial integrals and the fact that the sum over l_i was terminated at $l_i = 15$.

The present calculations were performed as the initial step in a more general investigation of the angular distribution of relativistic projectiles used in Coulomb excitation [10]). The paper of Bernstein and Skurnik indicated, however, the desirability of publishing the present results separately.

Note added in proof: The semi-classical results for electric quadrupole excitation can be improved very considerably by employing a new symmetrization [11]) procedure based upon Wentzel's [12]) use of the correspondence principle, rather than the customary "ad hoc" symmetryzation used in the present paper. With this procedure the semi-classical and the quantum-mechanical cross-sections agree to within 10 %, rather than the ca. 40 % deviation shown here in figs. 1, 2 and 3.

The computations reported here were carried out at the Duke University Computing Center on the IBM-7070 digital computer. We are indebted to this staff for their cooperation. Helpful discussions with Drs. P. J. Brussaard and D. P. Saylor are acknowledged with thanks.

References

1) B. Elbek, M. C. Olesen and O. Skilbreid, Nuclear Physics **19** (1960) 523
2) W. R. Wisseman and R. M. Williamson, Nuclear Physics **21** (1960) 688
3) E. M. Bernstein and E. Z. Skurnik, Phys. Rev. **121** (1961) 841
4) Alder, Bohr, Huus, Mottelson and Winther, Rev. Mod. Phys. **28** (1956) 432
5) J. Bang, Mat. Fys. Medd. Dan. Vid. Selsk, **32**, No. 5 (1960)
6) Goldstein, Thaler and Biedenharn, Los Alamos Scientific Laboratory Report (LA-2106) 1957, unpublished
7) E. U. Condon and G. H. Shortley, Theory of atomic spectra (Cambridge University Press, New York, 1935)
8) G. Breit and M. Abramowitz, Tables of Coulomb wave functions, National Bureau of Standards Applied Mathematics Series, Vol. **17** (1952)
9) Biedenharn, Goldstein, McHale and Thaler, Phys. Rev. **101** (1956) 662
10) T. A. Griffy, D. S. Onley and J. T. Reynolds, Bull. Am. Phys. Soc., Sec. Series, **7** (1962) 41
11) P. J. Brussaard, private communication (to appear in Biedenharn and Brussaard, Coulomb excitation, Oxford, 1963)
12) G. Wentzel, Z. Physik **27** (1924) 257

Multiple Coulomb Excitation of Rotational Levels in Even-Even Nuclei*

J. DE BOER, G. GOLDRING,† AND H. WINKLER
California Institute of Technology, Pasadena, California

(Received 20 January 1964)

An experimental study of excitation probabilities of rotational states with spins 2, 4, 6, and 8 in 22 even-even nuclei in the region $150 \leq A \leq 192$ has been carried out by means of multiple Coulomb excitation with oxygen ions of energies up to 44 MeV. In the region of strong nuclear deformation, the results are in good agreement with excitation probabilities calculated on the basis of the rotational model. It has, however, been observed, that the intrinsic quadrupole moment Q_0, derived from the observed excitation probabilities, increases for higher rotational states, especially in the transition region. For Sm^{152}, the dependence of the probabilities for excitation of the 2^+ and 4^+ states on the bombarding energy are compared with calculations based on higher order perturbation theory and on multiple Coulomb-excitation theory. In the nuclei Sm^{152}, Gd^{160}, W^{186}, Os^{188}, and Os^{192}, nonrotational transitions were observed. The excitation probabilities of the second excited 2^+ states are compared with the predictions of the Davydov model.

I. INTRODUCTION

THE electric quadrupole transition probabilities between the ground state and the lowest excited states of most nuclei have been determined by Coulomb excitation. If light particles (protons, deuterons, alpha particles) are used as projectiles, the excitation probability is usually so small that a first-order perturbation calculation describes the process quite accurately.[1] For heavy ions of sufficiently high bombarding energies, however, the excitation probability can become comparable to unity and the perturbation treatment breaks down. In an intermediate situation, higher than first-order perturbation expansions may be used which involve an increasingly larger number of parameters, and the calculation of the process becomes more and more complex. The sudden approximation[2-4] avoids the perturbation expansion but has instead to assume a model for the nucleus which gives all the nuclear quantities used in the calculations in terms of a small number of parameters, which are characteristic for the particular nuclear model.

The Coulomb excitation of a symmetric rotator is described by two parameters, namely the moment of inertia $\mathfrak{J}$ which gives the energy spectrum as

$$E_I = E_0 + (\hbar^2/2\mathfrak{J}) I(I+1) \tag{1}$$

and the intrinsic quadrupole moment Q_0 which gives the $B(E2)$ values and the static quadrupole moments as

$$B(E2, I \to I+2) = \frac{15}{32\pi} e^2 Q_0^2 \frac{(I+1)(I+2)}{(2I+1)(2I+3)}, \tag{2}$$

$$Q_I = -Q_0 I/(2I+3); \quad I = 0, 2, 4, \cdots. \tag{3}$$

For nuclei whose proton and neutron numbers are both far from closed-shell numbers, the energies and spin sequence of the lowest states are fairly accurately given by Eq. (1). Deviation from Eq. (1) can approximately be expressed in terms of a parameter B, defined by

$$E_I - E_0 = (\hbar^2/2\mathfrak{J}) I(I+1) - B[I(I+1)]^2. \tag{4}$$

The value of B, determined from the energies of the 4^+ levels is found to be of the order of $\hbar^2/2\mathfrak{J} \times 10^{-3}$ for even-even nuclei with $154 \leq A \leq 180$. (See also Fig. 11.)

Much less information is available on the static and dynamic electric quadrupole moments of these nuclei. Double Coulomb-excitation experiments[5] and measurements of the lifetimes of 4^+ rotational states[6] have shown that the $B(E2)$ ratios for the transitions $4^+ \to 2^+$ and $2^+ \to 0^+$ are correctly given by Eq. (2).

The present work was performed in order to obtain systematic information on the higher rotational states of strongly deformed nuclei and on the lowest 2^+ and 4^+ states of nuclei in the transition region between strong deformations and closed-shell configurations. Some of the results presented in this work have previously been communicated.[7,8] Small differences between the results given in Refs. 7 and 8 and the present ones are mainly due to an improved determination of the efficiencies of the gamma counters. Twenty-two nuclei with $150 \leq A \leq 192$ have been Coulomb-excited with oxygen ions of 18–44 MeV. The results are analyzed in terms of calculations by Alder and Winther, in which the sudden approximation is applied to the rotational model.

* Supported in part by the U. S. Atomic Energy Commission and in part by the Office of Naval Research.

† Present address: The Weizmann Institute of Science, Rehovoth, Israel.

[1] K. Alder, A. Bohr, T. Huus, B. Mottelson, and A. Winther, Rev. Mod. Phys. **28**, 432 (1956).

[2] K. Alder and A. Winther, Kgl. Danske Videnskab. Selskab, Mat. Fys. Medd. **32**, No. 8 (1960).

[3] K. Alder, in *Proceedings of the Third Conference on the Reactions between Complex Nuclei, Asilomar, California, 1963* (University of California Press, Berkeley and Los Angeles, 1963), p. 253; and private communication.

[4] K. Alder, W. Bierter, M. Simonius, and A. Zwicky (private communication).

[5] R. Graetzer and E. M. Bernstein, Phys. Rev. **129**, 1772 (1963).

[6] A. C. Li and A. Schwarzschild, Phys. Rev. **129**, 2664 (1963).

[7] G. Goldring, J. de Boer, and H. Winkler, in *Proceedings of the Third Conference on the Reactions between Complex Nuclei, Asilomar, California, 1963* (University of California Press, Berkeley and Los Angeles, 1963), p. 278.

[8] J. de Boer, G. Goldring and H. Winkler, in *Proceedings of the Third Conference on the Reactions between Complex Nuclei, Asilomar, California* (University of California Press, Berkeley and Los Angeles, 1963), p. 317.

Several transitions which do not belong to the ground-state rotational band were observed in Sm^{152}, Gd^{160}, W^{186}, Os^{188}, and Os^{192}. The excitation probabilities for these nonrotational transitions are analyzed in terms of second-order perturbation calculations for the Coulomb-excitation process. For Gd^{160}, W^{186}, Os^{188}, and Os^{192}, the $B(E2)$ values, extracted from this analysis, are compared to the theory of Davydov-Filippov[9] for a nonaxially symmetric rotator.

The Coulomb-excitation process was studied by observing the gamma radiation depopulating the rotational levels. The gamma spectra were recorded in coincidence with projectiles backscattered from the target.

II. EXPERIMENTAL PROCEDURE

A. Apparatus

The experiments described here were performed with the ONR-CIT tandem accelerator. For the beam production, the same procedure as the one described by Graetzer and Bernstein[5] was used and similar performances were achieved.

The energy of the incident beam is known to 0.2%. The bombarding energy is obtained by subtracting the mean energy loss of the oxygen ions in the thin targets, assuming a value of 3 MeV mg^{-1} cm^2.[10] This correction was always smaller than 300 keV.

A schematic view of the setup is shown in Fig. 1. The beam, collimated to 1 mm diam, passes through the hole in an annular shaped solid-state counter of 8-mm-outside and 4-mm-inside diameter. The target is placed at about 7 mm from the counter surface, so that the counter accepts oxygen ions, backscattered through $152°<\theta<165°$. The spectrum seen by the ring counter is shown in Fig. 2 for 31-MeV oxygen ions impinging on a 70-μg/cm² Sm^{152} target evaporated onto a 1000-Å nickel foil. The width of the peaks of about 1 MeV arises from the combined effects of the energy loss in the target and the kinematic spread. The peak comprises elastic scattering and all significant inelastic

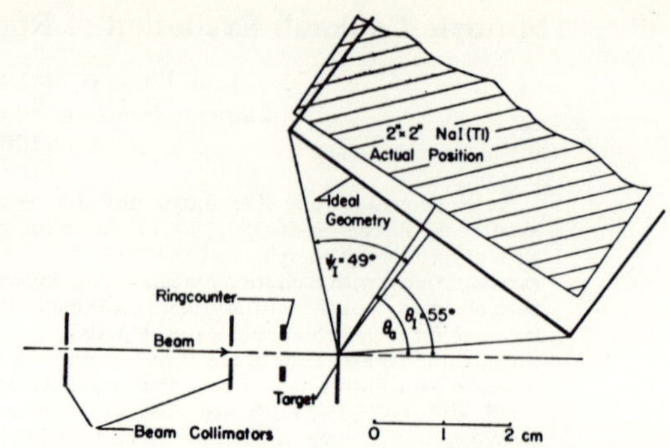

FIG. 1. Schematic view of the apparatus. The beam passes through the hole in an annular shaped solid-state detector which accepts particles backscattered through $152°<\theta<165°$. The NaI counter is placed at such an angle with respect to the beam and at such a distance from the target as to average out the angular distributions of the de-excitation gamma rays (see text).

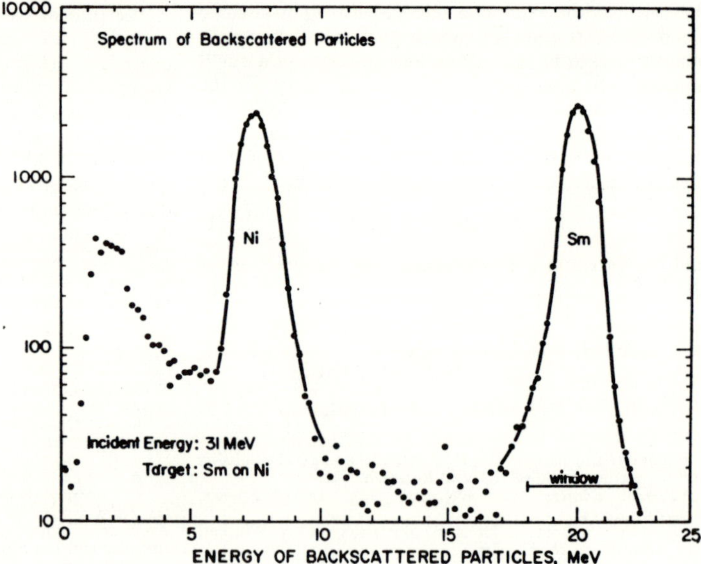

FIG. 2. Energy spectrum of the backward scattered oxygen ions. The window is set so as to accept only ions backscattered from the target material.

[9] A. S. Davydov and G. F. Filippov, Nucl. Phys. 8, 237 (1958).
[10] L. C. Northcliffe (unpublished).

scattering processes. A positive potential of about 300 V was applied to the target. Without this precaution, the performance of the solid-state detector was found to deteriorate (high noise, slow rise time, smaller pulses) within a counting period of about one hour.

The gamma-ray detector consists of a cylindrically symmetric 2-in.×2-in. NaI crystal placed so that its symmetry axis goes through the beam spot on the target and forms an angle φ of about 58° with respect to the beam direction. The distance is such, that the opening angle 2ψ for gamma rays hitting the front face of the crystal is about 86°. With this geometry, the fraction of gamma rays seen by the crystal is almost independent of the (strongly anisotropic) angular distributions of the gamma rays. The solid angle subtended by the counter is about 10% of the whole sphere. For the runs on osmium and on Sm^{150}, a 3-in. ×3-in. sodium-iodide crystal was used in a similar geometry.

B. Targets

The targets were prepared by vacuum-evaporation of 5–20 mg of the isotopically enriched material onto nickel backings of about 1000-Å thickness. The high temperatures required for the evaporation of rare-earth oxides were obtained by electron bombardment of small quantities of the samples contained in carbon crucibles of 1.5–3 mm diameter.[11] Most targets had a thickness of about 100 μg/cm^2. Osmium targets turned out to be very difficult to produce in adequate thicknesses, the thickest being less than 1 μg/cm^2.

The use of thin targets avoids overloading of the junction detector which is sensitive to high counting rates[12] and also gives an energy definition of the bombarding particles close to the energy definition of the accelerator.

C. Gamma-Ray Angular Distributions

Because of the $E2$ nature of all transitions involved, and the symmetry of the particle counter around the beam axes, the gamma angular distribution can be written in the form

$$W(\Omega)=1+A_2P_2(\cos\Omega)+A_4P_4(\cos\Omega)\,, \qquad (5)$$

where Ω is the angle between the direction of emission of the gamma ray and the incident beam. The counting rate, measured by a counter of finite angular aperture 2ψ and position φ can be written in the form

$$W(\varphi,\psi,E_\gamma;A_2,A_4)=1+\epsilon_2(\psi E_\gamma)\times A_2\times P_2(\cos\varphi) + \epsilon_4(\psi E_\gamma)\times A_4\times P_4(\cos\varphi)\,, \quad (6)$$

where ϵ_2 and ϵ_4 depend only on the angular opening 2ψ and on the gamma energy E_γ. If $\varphi=\varphi_I=55°$ then $P_2(\cos\varphi_I)=0$ and W is independent of A_2 for all ψ and E_γ. The opening ψ can now be chosen in such a way that $\epsilon_4=0$ for one particular E_γ. Gamma rays of sufficiently low energy are all counted in the front face of the crystal. In this case, $\epsilon_4=0$ for $\psi=\psi_I=49°$ and W is independent of A_4 for all φ.

This "ideal geometry" could not exactly be realized in our apparatus, because it would have brought the edge of the scintillator housing too close to the beam (see Fig. 1). In the "actual position," a compromise was made by taking $\varphi=\varphi_a=58.5°$ and $\psi=\psi_a=43°$. In this geometry, P_2 has a value of -0.09. The magnitude of $P_4(\varphi_a)\times\epsilon_4(\psi=43°,\,E_\gamma)$ for the 2-in.×2-in. crystal is estimated to be -0.03 for $E_\gamma<100$ keV, -0.08 for $E_\gamma=200$ keV, -0.14 for $E_\gamma=300$ keV, and -0.20 for $E_\gamma=600$ keV. The signs and magnitudes of A_2 and A_4 in formula (5) are such that the contributions due to the P_2 and P_4 terms almost cancel each other. The size of the term $\epsilon_2\times A_2\times P_2+\epsilon_4\times A_4\times P_4$ in Eq. (6) is estimated to be less than 0.05 for the deexcitation of a 2^+ state and less than 0.02 for all other transitions.

D. Evaluation of Coincidence Spectra

The pulses from the photomultiplier, gated by the output of a fast-slow coincidence system, were analyzed in a multichannel kicksorter. The occurrence of a gate signal required a coincidence between any multiplier pulse and a pulse from the particle counter corresponding to an oxygen ion backscattered from the target nucleus. The resolving time 2τ of the fast coincidence system was set to 48 nsec to ensure 100% coincidence efficiency for all gamma-ray energies between 50 and 500 keV. Examples of coincidence spectra are given in Figs. 3, 4, and 8–11.

In order to obtain the number of gamma rays emitted per backscattered oxygen ion from the areas under the photopeaks in the coincidence spectra, the efficiencies of the counters were determined by placing calibrated sources at the location of the beam spot on the target. This calibration is believed to be accurate to about 8%. In the runs with high bombarding energies, a Cu+Sn absorber was placed between target and gamma counter. This absorber served two purposes. First it reduced the counting rate in the K x-ray peak and in the $2^+\rightarrow 0^+$ transition by a factor of 10–20 so that the pileup of pulses was appreciably reduced. Second, the absorber almost eliminates the sum peak, which occurs when both members of a gamma cascade depopulating the 4^+ state are counted in the NaI crystal. According to the $I(I+1)$ rule, the energy of this sum peak is 10/11 of the energy of the $6^+\rightarrow 4^+$ transition. (Compare Figs. 4 and 11.) The uncertainty in the attenuation factor made it impossible to obtain accurate values of the $2^+\rightarrow 0^+$ intensities in the runs in which absorbers were used.

The number of gamma rays per backscattered oxygen ion was finally obtained by applying corrections for the fractional dead time of the multichannel analyzer, for random coincidences, for counting losses

[11] Y. Dar, H. M. Loebenstein, and J. de Boer, Nucl. Instr. Methods (to be published).

[12] D. Eccleshall, B. M. Hinds, and M. J. L. Yates, Nucl. Phys. 32, 190 (1962).

in the scaler monitoring the number of backscattered particles, for Compton-tails of higher lines and for the loss of counts due to the addition of simultaneous members of a cascade into the sum peak. From this number, the number of de-excitations was derived by taking into account internal conversion. The K and L conversion coefficients were taken from the tables of Sliv and Band[13] and the contributions of the higher shells were accounted for by adding $\frac{1}{3}$ of the L coefficient.

We define the de-excitation probability R_I as

$$R_I = \frac{\text{number of de-excitations of a level with spin } I}{\text{number of backscattered particles}}. \tag{7}$$

For rotational levels in even-even nuclei, which decay by $E2$ cascades only, the transition $I \rightarrow I-2$ will occur whenever a level with spin $J \geq I$ has been excited. We therefore have

$$R_I = \sum_{J \geq I} P_J, \tag{8}$$

where P_J denotes the probability for Coulomb excitation of a level with spin J. The excitation probabilities are normalized so that

$$\sum_{J=0}^{\infty} P_J = 1. \tag{9}$$

The probabilities R_I, calculated from the theoretical values for P_J, are the quantities that are compared with the experiment in the present analysis.

III. EXCITATION FUNCTIONS FOR Sm^{152}

The first group of experiments consists of a series of precision measurements on Sm^{152} for a wide range of bombarding energies, in order to study the dependence of the Coulomb-excitation probabilities on the excitation parameters.

A. Coincidence Spectra

Examples of coincidence gamma spectra from the bombardment of Sm^{152} with 18- and 42-MeV oxygen ions are shown in Figs. 3 and 4. The prominent peaks

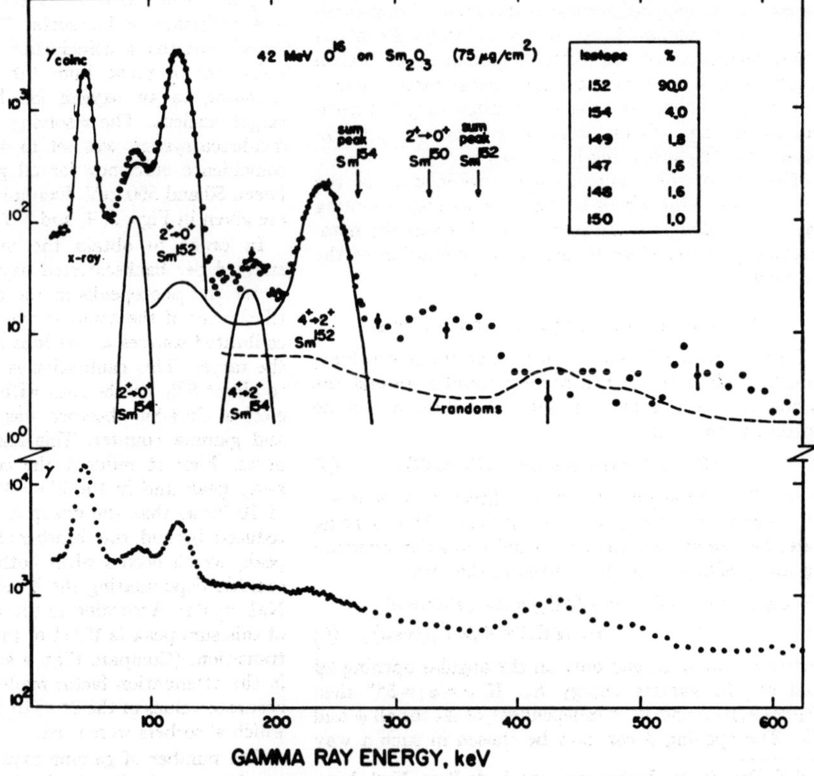

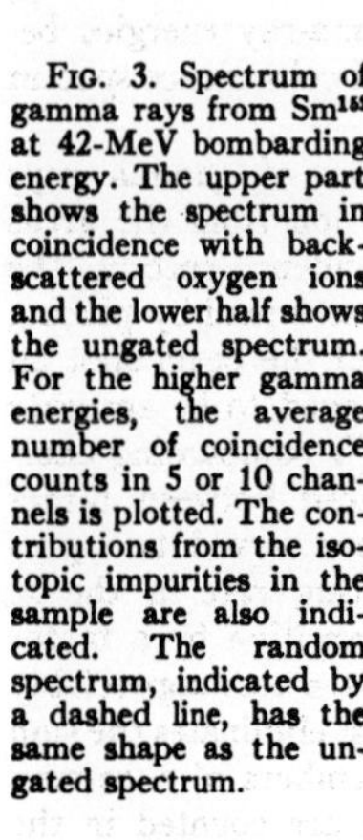
FIG. 3. Spectrum of gamma rays from Sm^{152} at 42-MeV bombarding energy. The upper part shows the spectrum in coincidence with backscattered oxygen ions and the lower half shows the ungated spectrum. For the higher gamma energies, the average number of coincidence counts in 5 or 10 channels is plotted. The contributions from the isotopic impurities in the sample are also indicated. The random spectrum, indicated by a dashed line, has the same shape as the ungated spectrum.

[13] L. A. Sliv and M. I. Band, *Coefficients for Internal Conversion of Gamma Radiation* (Academy of Sciences of the USSR, Moscow, 1956, 1958), Parts I and II.

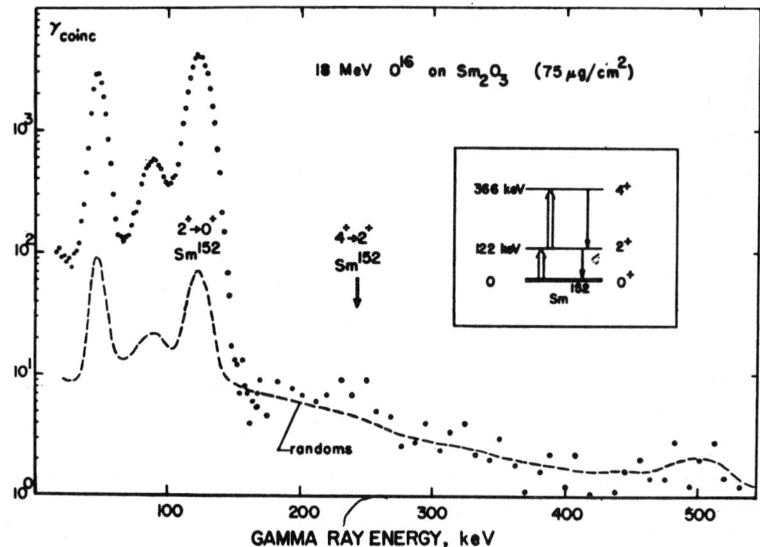

FIG. 4. Coincidence spectrum of gamma rays from Sm^{152} at 18-MeV bombarding energy. The two most prominent peaks are the K x-ray and the $2^+\to0^+$ transition in Sm^{152}. A peak corresponding to the $2^+\to0^+$ transition in Sm^{154} is visible. At 244 keV, an indication of the $4^+\to2^+$ transition can be seen. The dashed line shows the random coincidences.

are due to K x rays and gamma rays of 122 and 245 keV from the $2^+\to0^+$ and $4^+\to2^+$ transitions in Sm^{152}. In the lower part of Fig. 3, the singles gamma spectrum is shown for the same bombarding conditions. The spectrum of the random coincidences, indicated by a dashed line in the coincidence spectrum, was found to have the same shape as the singles spectrum. Almost all counts in the coincidence spectra can be assigned to Coulomb excitation of Sm^{152} and its isotopic impurities. The random coincidences which range to high gamma-ray energies, are probably due to nuclear reactions of the projectiles with the nickel backing and light elements in the target.

B. Excitation Parameters and Experimental Results

The experimental values of the excitation probabilities for Sm^{152} are listed in Table I. The first column gives the bombarding energy, reduced by the mean energy loss of the oxygens in the target (about 80 keV for the Sm^{152} target). The next two columns list the Coulomb excitation parameters ξ and χ, defined by

$$\xi_{if}=\frac{Z_1Z_2A_1^{1/2}(1+A_1/A_2)\Delta E_{if}}{12.65E^{3/2}}, \qquad (10)$$

and

$$\chi_{i\to f}=14.36\frac{\pm[B(E2;i\to f)]^{1/2}\times A_1^{1/2}}{(1+A_1/A_2)^2Z_1Z_2^2}E^{3/2}. \qquad (11)$$

The indices 1 and 2 refer to projectiles and target nucleus, respectively. The charge numbers are denoted by Z. The laboratory energy E and the transition energy ΔE_{if} are both measured in MeV, the reduced transition probability $B(E2)$ in $e^2\times10^{-48}$ cm^4 and the masses A in amu.[14] The sign of $[B(E2,i\to f)]^{1/2}$ is to be the same as the sign of the reduced matrix element $\langle i||E2||f\rangle$. For the calculation of the adiabaticity parameter ξ, ΔE is taken as the energy of the $2^+\to0^+$ transition. The parameter $\chi_{0\to2}$ is the transition amplitude for a $0^+\to2^+$ transition in first-order perturbation theory for $\xi=0$ and $\theta=180°$. The χ values, given in column 2 of Table I, are calculated for $B(E2)=3.43$, a weighted average[15] of previous determinations of this quantity. The last two columns list the experimentally determined values for R_2 and R_4. The uncertainties in the absolute efficiencies of the gamma counter, which

TABLE I. Results of the precision measurements of the probabilities for Coulomb excitation of the 2^+ and 4^+ rotational states in Sm^{152}. ξ and χ are excitation parameters defined by Eqs. (10) and (11). The value of the bombarding energy E is known to $\pm0.2\%$. The values R_2 and R_4 are the experimental probabilities for observing the decay of the 2^+ and 4^+ rotational states, respectively, in a collision in which the oxygen ion was backscattered. The uncertainty in the efficiency of the gamma counter is absorbed in the constants K and F. $K=1.00\pm0.08$ and $F=1.00\pm0.08$.

E MeV	$\chi_{0\to2}$	$\xi_{2\to0}$	$R_{2\,exp}$	$R_{4\,exp}$
17.94	0.216	0.277	$K\times0.0236\pm0.0004$	$F\times0.000048\pm0.000015$
23.06	0.315	0.190	$K\times0.0645\pm0.001$	$F\times0.00048\pm0.00005$
28.07	0.423	0.141	$K\times0.1305\pm0.001$	$F\times0.00247\pm0.00015$
31.15	0.494	0.121	$K\times0.182\pm0.002$	$F\times0.00525\pm0.00015$
32.95	0.537	0.111	$K\times0.216\pm0.002$	$F\times0.0084\pm0.0002$
36.08	0.616	0.097	$K\times0.289\pm0.004$	$F\times0.0152\pm0.0003$
40.46	0.731	0.082	$K\times0.395\pm0.005$	$F\times0.0292\pm0.0006$
42.10	0.776	0.077	$K\times0.426\pm0.005$	$F\times0.0373\pm0.0007$

[14] The numerical factor in relation (11) differs from the one given in Ref. 2 due to a different choice of the mass unit.

[15] E. M. Bernstein and E. Z. Skurnik, Phys. Rev. 121, 841 (1961).

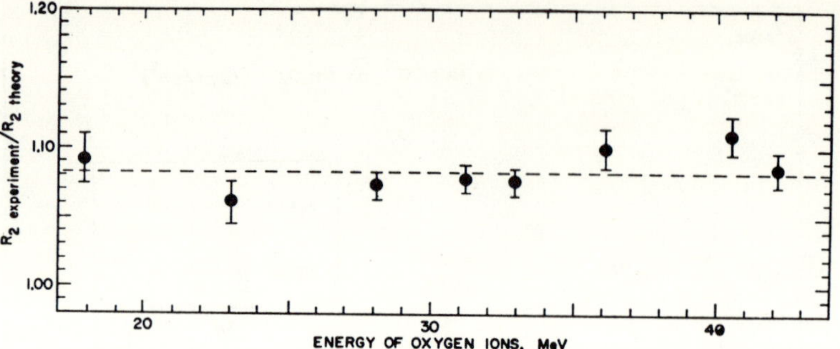

FIG. 5. Comparison of the experimental values from Table I for the de-excitation probability of the 2^+ rotational state in Sm^{152} with the calculations of Ref. 3. The points are consistent with a constant value for $R_{2\,\mathrm{exp}}/R_{2\,\mathrm{theor}}$.

are independent of the bombarding conditions, are absorbed in the constants K and F. The error, quoted for each experimental value, is due to the errors in the corrections, the statistical error and the uncertainty in the position of the beam spot. (In Ref. 7, this last source of error was not taken into account.)

In Figs. 5 and 6, the experimental values are compared with theoretical calculations of the multiple Coulomb excitation of rotational bands. In Refs. 2 and 16, P_I is given for $\xi=0$ and $\theta=180°$ as a function of q which is related to $\chi_{0\to2}$ by the expression

$$q=(45/16)^{1/2}\times\chi_{0\to2}. \tag{12}$$

More recently, Alder[3] has calculated $P_I(q,\xi\neq0,\theta=180°)$. The theoretical values, which were used in the comparison of Figs. 5 and 6, are based on these calculations. The dependence on the deflection angle $\theta\neq180°$ was taken into account by using an "effective" value for χ or q [see Ref. 2, formula (5.15)]. For the geometry used in this experiment

$$\chi_{\mathrm{eff}}(\theta)=0.965\chi(180°). \tag{13}$$

For backward scattered particles ($\theta\approx160°$) this approximation is expected to yield quite accurate theoretical values (see Ref. 2, pp. 14, 15). The accuracy may be estimated by comparing the values, obtained by the χ_{eff} approximation with calculations by Alder *et al.*,[4] in which the exact dependence of P_I on θ was taken into account for $\xi=0$. The deviations between the two calculations are smaller than 0.5% for all χ values encountered in the present experiments. For $\xi\neq0$, the deviations are somewhat larger.

From the comparison in Fig. 5 it can be seen that the experimental dependence of $R_2=1-P_0$ on the bombarding energy is very well reproduced by the theory. In Fig. 6, the de-excitation probability of the 4^+ state in Sm^{152} is compared with the calculation of Ref. 3. The theoretical values for the solid points in

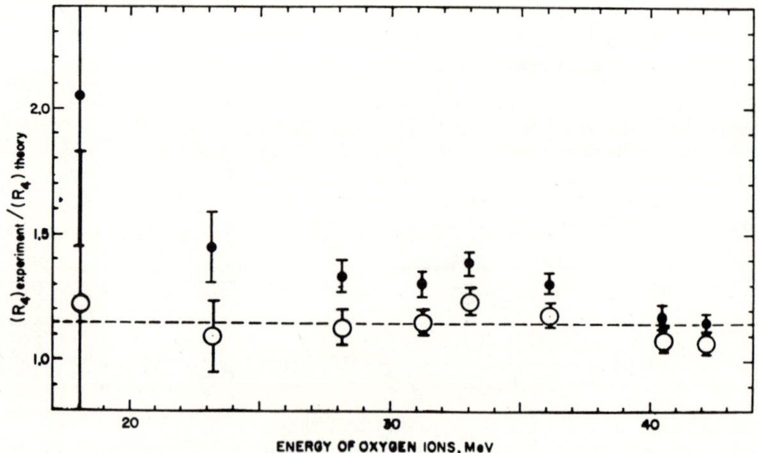

FIG. 6. Comparison of the experimental values from Table I for the de-excitation probability of the 4^+ state in Sm^{152} with the calculations of Ref. 3. For the solid points, ξ values corresponding to the moment of inertia representing the $2\to0$ transition energy were employed in the calculations of the theoretical values. The open circles were calculated for a moment of inertia that represents the $4^+\to2^+$ transition energies. The open circles are consistent with a constant value for $R_{4\,\mathrm{exp}}/R_{4\,\mathrm{theor}}$.

[16] R. Graetzer, R. Hooverman and E. M. Bernstein, Nucl. Phys. 39, 124 (1962).

Fig. 6 were obtained by taking $\xi = \xi_{2\to 0}$. If, however, a ξ value is used that corresponds to a moment of inertia which reproduces the experimentally determined transition energy $\Delta E_{4\to 2}$ rather than $\Delta E_{2\to 0}$ (see discussion below) the points marked by open circles are obtained. These points, which have the same experimental error as the solid ones, are consistent with a horizontal line in Fig. 6.

C. Higher Order Perturbation

In Ref. 7 it was attempted to relate the dependence of R_2 on E to the magnitude of the static quadrupole moment of the 2^+ state. Theoretically, the dependence of the Coulomb-excitation probability on Q_{2^+} can be calculated, if higher than first-order perturbation expansions are used. This "reorientation effect" has been considered by various authors.[17–19] Lin and Masso[19] have pointed out that in the case of Sm^{152}, the bombarding energies must be very low in order to prevent the higher than second-order terms in the perturbation expansion from being comparable in size to the reorientation effect,[20] so that the present experiments are not conclusive in this respect. The situation is

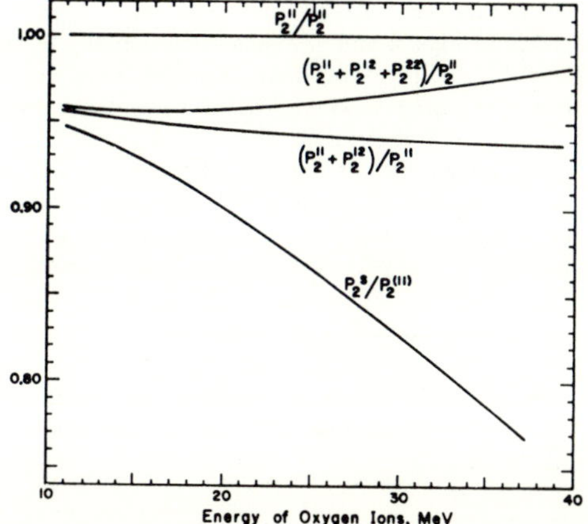

FIG. 7. Perturbation calculations and diagonalization calculations of the excitation probability P_2 of the 2^+ state in Sm^{152}. All probabilities are given relative to the first order perturbation treatment. P^{11} and P^{22} denote first- and second-order perturbation, respectively, and P^{12} denotes the interference term between first and second order. P^S is the excitation probability calculated by the diagonalization method of Ref. 3.

illustrated in Fig. 7. The contributions of the second-order terms in the perturbation expansion were calculated using the tables of Douglas.[21] The ratios $P_2^{(12)}/P_2^{(11)}$, $(P_2^{(12)}+P_2^{(22)})/P_2^{(11)}$ and $P_2^{(S)}/P_2^{(11)}$ are plotted as a function of the bombarding energy E for Sm^{152} and for $\theta = 160°$. The symbols $P^{(11)}$ and $P^{(22)}$ denote the excitation probabilities in first and second order, $P^{(12)}$ is the interference between first and second order and $P^{(S)}$ is the excitation probability in the calculation of Ref. 3. The numerical formulas for these probabilities are

$$P_2^{(11)} = 17.90\chi_{0\to 2}{}^2 \times \sin^4(\theta/2) \times df(\xi,\theta)/d\Omega, \tag{14}$$

$$P_2^{(12)} = 9.49\chi_{0\to 2}{}^3 \times \frac{\langle 2||E2||2\rangle}{[B(E2;0\to 2)]^{1/2}} \times\{(\tfrac{2}{7})^{1/2}[I_{22}(\xi,\theta)\beta_{2-2}(2,2,\xi,0,\theta) + I_{2-2}(\xi,\theta)\beta_{22}(2,2,\xi,0,\theta)] - I_{20}(\xi,\theta)\beta_{20}(2,2,\xi,0,\theta)\}, \tag{15}$$

$$P_2^{(22)} = 16.03\chi_{0\to 24}\times\sin^4(\theta/2) \times\frac{|\langle 2||E2||2\rangle|^2}{B(E2,0\to 2)}\times dF(\xi,0,2,\theta)/d\Omega. \tag{16}$$

In these expressions, $\chi_{0\to 2}$ is defined by (11) and $\langle 2||E2||2\rangle$ is related to the quadrupole moment of the 2^+ state by

$$eQ_{2^+} = \tfrac{4}{5}(2\pi/7)^{1/2}\langle 2||E2||2\rangle. \tag{17}$$

In the rotational model, $\langle 2||E2||2\rangle/[B(E2,0\to 2)]^{1/2} = -(10/7)^{1/2}$. This value was used for the calculations in Fig. 7. The functions $df/d\Omega$ and I are tabulated in Ref. 1 and β and $dF/d\Omega$ are given in Ref. 21.

From Fig. 7 it can be seen that for energies higher than 22 MeV, $P_2^{(12)}$ accounts for less than half of the difference between $P_2^{(S)}$ and $P_2^{(11)}$. The measurement of P_2 would have to be extended to very low bombarding energies in order to obtain accurate values for $P_2^{(12)}$, which is proportional to the static quadrupole moment of the 2^+ state. This shows that other terms of order χ^4 (interference between first- and second-order perturbation) are, in this case, more important than $P_2^{(22)}$.

On the other hand the multiple Coulomb-excitation theory of Alder[3] accounts for these higher order effects, provided they all conform to the rotational patterns. The remarkable degree of accuracy with which the theory[3] fits the strong energy dependence of the excitation probabilities (Figs. 5 and 6, Table I) shows that the set of nuclear quantities, given by the rotational model, describes the multiple process well. In order to assess the accuracy with which a specific nuclear quantity may be determined in this way, we have to

[17] G. Breit, R. L. Gluckstern and J. E. Russell, Phys. Rev. **103**, 727 (1956); **105**, 1121 (1957).

[18] D. Beder, Phys. Letters **3**, 206 (1963); Can. J. Phys. **41**, 547 (1963).

[19] D. L. Lin and J. F. Masso, in *Proceedings of the Third Conference on the Reactions between Complex Nuclei, Asilomar, California, 1963* (University of California Press, Berkeley and Los Angeles, 1963), p. 267.

[20] From the work by J. Eichler, Phys. Rev. **133**, B1162 (1964), it follows that the effects due to virtual $E1$ transitions are negligible in this case.

[21] A. C. Douglas, Nucl. Phys. **42**, 428 (1963) and Atomic Weapons Research Establishment Report NR/P-2/62, Aldermaston, England (unpublished).

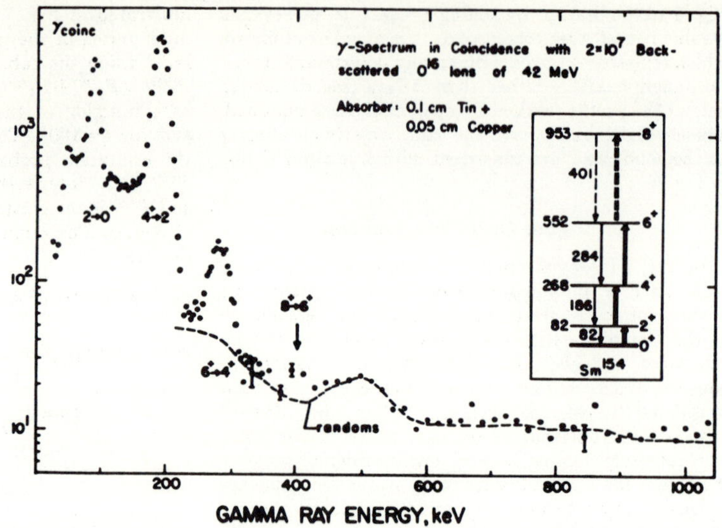

FIG. 8. Coincidence gamma spectrum of Sm^{154}. The peaks corresponding to the $2^+ \rightarrow 0^+$ and $4^+ \rightarrow 2^+$ transitions are attenuated with an absorber by factors of 12 and 1.5, respectively. The assignment of the weak line at ≈ 401 keV as the $8^+ \rightarrow 6^+$ transition is based on the energy and intensity of the gamma ray and on the fact that it disappears at lower bombarding energies.

know how sensitively the calculations depend on each quantity. Such calculations are, however, not yet available.

IV. DEFORMED EVEN-EVEN NUCLEI

A. Coincidence Spectra

This group of experiments deals with a large number of nuclei in the region of stable equilibrium deformation. Examples of coincidence gamma-ray spectra are given in Figs. 8–11. The nuclei in the region of strong deformation ($154 \leq A \leq 180$) all have similar rotational parameters and therefore exhibit similar spectra. The strong dependence of the Coulomb-excitation probabilities on Z_2, the charge number of the target, is illustrated in Figs. 8 and 9. Peaks corresponding to the first, second, and third rotational transitions can clearly be seen. Compared to ${}_{62}Sm_{92}{}^{154}$, the coincidence spectrum for ${}_{72}Hf_{106}{}^{178}$ shows a much weaker $6^+ \rightarrow 4^+$ transition. The assignment of the weak line at 401 keV in the Sm^{154} spectrum as the $8^+ \rightarrow 6^+$ transition is based on the energy and intensity of the gamma ray and on the fact that it disappears at lower bombarding energies.

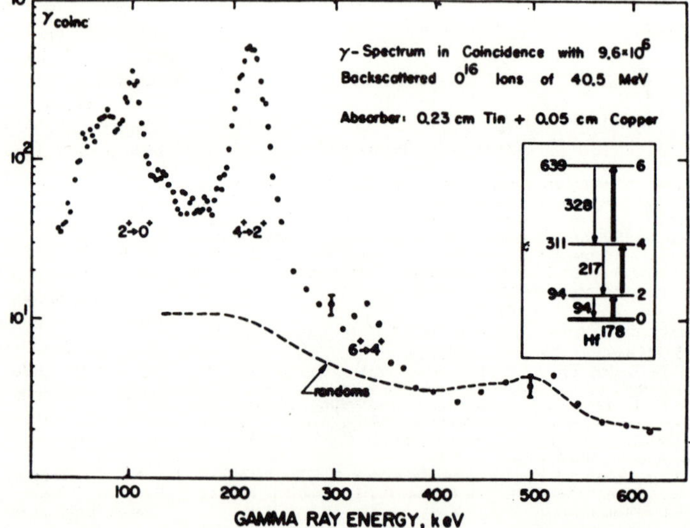

FIG. 9. Coincidence gamma spectrum of Hf^{178}. The peak corresponding to the $2^+ \rightarrow 0^+$ transition is attenuated by a factor of 24.

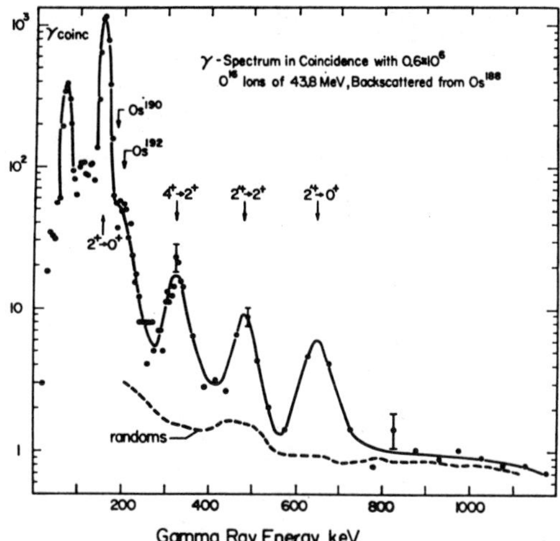

FIG. 10. Coincidence gamma spectrum of Os^{188}. In addition to the rotational transitions, the $2'^{+}\rightarrow 0^{+}$ and the $2'^{+}\rightarrow 2^{+}$ peaks can be seen.

The spectrum of Os^{188} is shown in Fig. 10 as an example of a nucleus in the transition region between spheroidal and spherical equilibrium shape. The transitions corresponding to the cascade and crossover decay of the second 2^{+} state have intensities which are comparable to the intensity of the $4^{+}\rightarrow 2^{+}$ transition.

In Fig. 11, the coincidence spectrum from Sm^{152} bombarded with 43.1-MeV oxygen is shown. The effect of the gamma absorber in reducing the intensity of the sum peak can be seen by comparison with Fig. 3. The prominent peak at 550 keV is assigned to the $0'^{+}\rightarrow 2^{+}$ transition. The isotopic impurities that show up in Fig. 3, are not visible in Fig. 11 because a 99% enriched sample was used for this run.

Except for osmium, where the thinness of the targets required extremely long exposure times, each target has been bombarded with oxygen ions of at least two different energies in order to confirm the multiplicity of the $E2$ excitation process assumed by the rotational description. For the lowest bombarding energies, the spectra were usually taken without a gamma absorber so that the intensities of the $2^{+}\rightarrow 0^{+}$ and $4^{+}\rightarrow 2^{+}$ transitions could be determined. At the higher bombarding energies, the use of absorbers was necessary in order to determine the energies and intensities of the $6^{+}\rightarrow 4^{+}$ transitions without contributions from the sum peaks.

B. Energies of Rotational Transitions

The transition energies, determined in our experiments, are listed in Table II. They are in good agreement with energy determinations by other investigators. From a fit of the weighted averages of the level energies with an expression of the form (4) the values of $B\times 2\mathfrak{J}/\hbar^{2}$ have been determined and are plotted in Fig. 12 as a function of the mass number. We can see from Fig. 12 that deviations from the $I(I+1)$ rule [Eq. (1)] are encountered. They are particularly large in the transition region ($A\approx 150$ and $A\lesssim 188$).

Alternatively, the experimental energy values can be fitted with a formula of type (1) if an effective moment of inertia $\mathfrak{J}_{\text{eff}}(I-2\rightarrow I)$ is introduced for each transition. Because B in (4) is positive, $\mathfrak{J}_{\text{eff}}$ increases with increasing spin. On the other hand, the theoretical calculations for the multiple Coulomb excitation of a rotational band $P_I(\chi,\xi,\theta=\pi)$ have been carried out[8] for an energy spectrum given by Eq. (1), i.e., a constant value for $\mathfrak{J}$ has been assumed. Thus, all $\xi_{J-2\rightarrow J}$-values for the higher transitions which enter implicitly in Alder's calculation of P_I, are expressed in terms of $\xi_{0\rightarrow 2}$ with the help of Eqs. (1) and (10). Since P_I

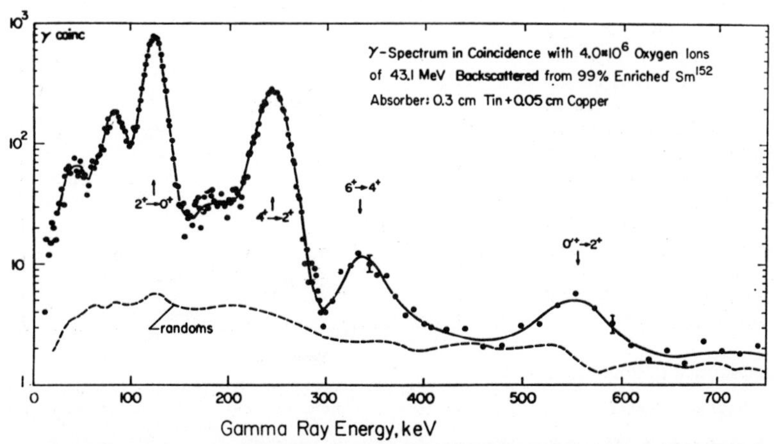

FIG. 11. Coincidence gamma spectrum of Sm^{152}. The use of a gamma-ray absorber eliminates the sum peak, visible in Fig. 4. The peak at 550 keV is assigned to the $0'^{+}\rightarrow 2^{+}$ transition in Sm^{152}.

depends strongest on the largest transition energy, i.e., the energy $\Delta E_{I-2\rightarrow I}$, a value for ξ corresponding to $\mathfrak{J}_{eff}(I-2\rightarrow I)$ has been used in the comparison of theory and experiment. (See also discussion of Fig. 6.) For rotational bands whose $\mathfrak{J}_{eff}$ increases strongly with I, this consideration is of importance if ξ is large.

C. Results on Rotational Transitions

The measured values of R_I are listed in Table III. The ξ values given in the third column of this table are determined by Eq. (10) from the energy of the $2^+\rightarrow 0^+$ transitions. Column 4 lists the value of $\chi_{0\rightarrow 2}$ defined by Eq. (11). The $B(E2)$ values used for the calculation of $\chi_{0\rightarrow 2}$ were determined by inelastic scattering (Refs. 22 and 23). For the osmium isotopes, where no inelastic scattering data were available, the $B(E2)$ values were determined from the present experiments in such a way as to reproduce the theoretical values[3] for the de-excitation probabilities of the 2^+ states. Columns 6–9 list the de-excitation probabilities R_I for $I=2$, 6, and 8. The number in parenthesis following each experimental value of R_I gives the error in percent of the measured value. In columns 10–13, the ratios of experimental to theoretical values of R_I are given. The theoretical de-excitation probabilities were obtained by interpolation of the tables by Alder[3] which give P_I for $I=2$–8 as a function of χ and ξ and for $\theta=180°$. An effective χ_{eff} was used to account for the dependence on θ [see Eq. (13)]. For the theoretical excitation probability of a state with spin I, a ξ value corresponding to a moment of inertia that represents the energy of the $I\rightarrow I-2$ transition was employed.

We notice that the average value of the ratio $R_{2\,\mathrm{exp}}/R_{2\,\mathrm{theor}}$ is higher than unity. This can have various reasons: (a) The assumed values for $B(E2, 0^+\rightarrow 2^+)$ from Refs. 22 and 23 may be too small, (b) the assumed value for the total conversion coefficient $\alpha(2^+\rightarrow 0^+)$ may be too high, and (c) the assumed efficiency of the gamma counter may be too low. The discrepancies are, however, too small to allow a definite decision on the basis of the present experiments. The conversion coefficient, required to make $R_{2\,\mathrm{exp}}/R_{2\,\mathrm{theor}}=1$ would be smaller than the theoretical value of Sliv and Band.[11] Some recent experiments, in particular the work of Fossan and Herskind[24] indicates, however, an experimental value of $\alpha(2^+\rightarrow 0^+)$ about 12% larger than the theoretical one. As can be seen from columns 10–13, $R_{I\,\mathrm{exp}}/R_{I\,\mathrm{theor}}$ is independent of the bombarding energy within the limits of error. This is consistent with the multiple excitation of a rotational state of spin I.

The results are graphically represented in Fig. 13. The solid curves show the calculations of Ref. 3 for $R_I(\chi,\ \xi,\ \theta=180°)$; and the broken curve gives $R_I(\chi, \xi=0, \theta=160°)$ of Ref. 4. The experimental values for R_I are plotted as a function of χ, and for four ranges of ξ values, as indicated by different points. The points clearly group into clusters corresponding to the respective rotational states.

TABLE II. Energies of the rotational transitions, determined in the present experiments. Italic values denote transitions that were not previously reported.

Isotope	Energies of rotational transitions, keV $2^+\rightarrow 0^+$	$4^+\rightarrow 2^+$	$6^+\rightarrow 4^+$	$8^+\rightarrow 6^+$
Sm^{150}	335 ±3	450±5		
Sm^{152}	122 ±2	243±2	*337±5*	
Sm^{154}	82 ±1	186±2	*284±2*	*401±6*
Gd^{156}	88.5±1	200±2	300±3	*380±10*
Gd^{158}	79.5±1	183±2	282±3	
Gd^{160}	75 ±1	174±2	*265±2*	354±10
Dy^{162}	80 ±1	185±2	282±6	370±6
Dy^{164}	74 ±1	*169±2*	*258±2*	*332±8*
Er^{166}	81 ±1	183±3	278±4	
Er^{168}	80.5±1	185±2	282±4	
Er^{170}	79 ±1	180±2	*275±4*	
Yb^{172}	80 ±1	181±3	272±4	
Yb^{174}	76 ±1	*175±2*	*271±4*	*345±8*
Yb^{176}	82 ±1	187±2	*294±3*	*383±6*
Hf^{178}	94 ±1	217±3	328±9	
Hf^{180}	93 ±1	215±2	338±4	
W^{182}	101 ±1	229±3	350±4	
W^{184}	112 ±1	254±2		
W^{186}	122.5±1	276±3	*419±6*	
Os^{188}	155 ±3	325±6		
Os^{190}	188 ±3	360±15		
Os^{192}	205 ±2	380±10		

D. Results on Nonrotational Transitions

The excitations of the second 2^+ states in the nuclei Gd^{160}, W^{186}, Os^{188}, and Os^{192}, and of an excited $0'^+$ state in Sm^{152} (see Fig. 14) are analyzed on the basis of first- and second-order perturbation calculations. For the bombarding energies used in the present work,

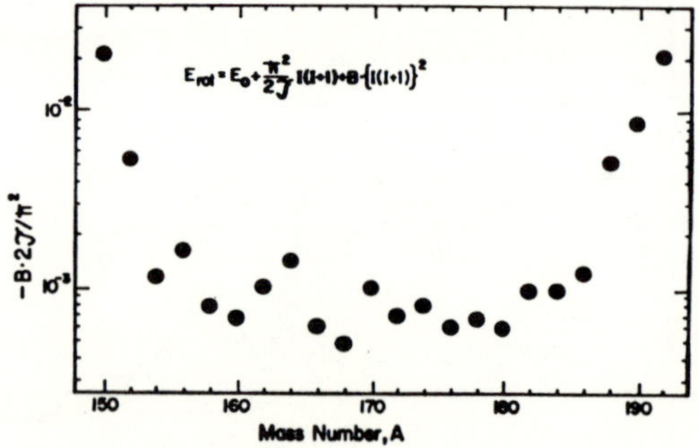

FIG. 12. The energy parameter $B\times 2\mathfrak{J}/\hbar^2$ as a function of the mass number A. The parameter was not determined exclusively from the values of Table III but from an average of all data available.

[22] O. Hansen, M. C. Olesen, O. Skilbreid, and B. Elbek, Nucl. Phys. **25**, 634 (1961).
[23] B. Elbek, M. C. Olesen, and O. Skilbreid, Nucl. Phys. **19**, 523 (1960).
[24] D. B. Fossan and B. Herskind, Phys. Letters **2**, 155 (1962).

the $2'^{+}$ states are populated with about equal strength by the direct $E2$ excitation from the ground state and by the double $E2$ transition via the 2^{+} rotational state. In second-order perturbation, the excitation probability contains three terms, corresponding to the two mentioned modes of excitation and the interference between the two.

The sign of the interference term depends on the relative sign of the reduced $E2$ matrix elements connecting the 0^{+} and 2^{+} states with the $2'^{+}$ state. It can in principle be determined by Coulomb-excitation experiments, if the excitation probability of the $2'^{+}$ state is measured for several bombarding energies. The perturbation analysis of the experiments then yields two sets of $B(E2)$ values for each choice of the sign of the interference term, and the decision between the two can be made by the requirement that the $B(E2)$ values of one set be equal for all bombarding energies.

The numerical formula, used in the analysis of the $2'^{+}$ excitation probabilities, is given by

$$\begin{aligned}P_{2'} = \chi_{0\to2}{}^{2}\times\frac{B(E2,\,0\to2')}{B(E2,\,0\to2)} &\times[17.90\sin^{4}(\theta/2)df(\xi_{0\to2'},\theta)/d\Omega\\ &\pm9.49\chi_{0\to2}\times(\lambda(\Delta E_{2'\to0}/\Delta E_{2'\to2})^{5})^{1/2}\\ &\times\{\xi_{0\to2'},\xi_{0\to2},\xi_{2\to2'},\theta\}\\ &+16.03\chi_{0\to2}{}^{2}\times\lambda(\Delta E_{2'\to0}/\Delta E_{2'\to2})^{5}\\ &\times\sin^{4}(\theta/2)dF(\xi_{0\to2'},\xi_{2\to2'},2,\theta)/d\Omega].\end{aligned}\tag{18}$$

The quantities $df/d\Omega$ and $dF/d\Omega$ are tabulated in Refs. 1 and 21. The branching ratio λ is defined by

$$\lambda=\Gamma(E2,\,2'\to2)/\Gamma(E2,\,2'\to0)\,,\tag{19}$$

where $\Gamma(E2)$ denotes the electric quadrupole gamma-ray transition probability. The sign of the second term

TABLE III. Summary of experimental results and comparison with theory. The number in parenthesis, following each experimental value, gives the relative error in percent.

1	2	3	4	5	6	7	8	9	10	11	12	13
				$B(E2)^{d}$	Experimental de-excitation probabilities R_I				$R_{I\,\mathrm{exp}}/R_{I\,\mathrm{theory}}{}^{e}$			
Isotope	$E_{0x}{}^{a}$ MeV	ξ^{b} $0\to2$	χ^{c} $0\to2$	$0\to2$ $e^{2}\times10^{-48}$ cm^{4}	$R_2\times10$	$R_4\times10^{2}$	$R_6\times10^{3}$	$R_8\times10^{4}$	$2\to0$	$4\to2$	$6\to4$	$8\to6$
Sm^{150}	34.8	0.284	0.357	1.30	0.50(15)	···	···	···	0.89			
	43.6	0.202	0.500		1.71(10)	0.89 (20)	···	···	1.27	2.0	···	···
Sm^{152}	29.9	0.130	0.470	3.43	1.45(12)	0.36 (22)	···	···	0.98	1.00	···	···
	39.0	0.087	0.700		···	2.38 (17)	0.59(40)	···	···	1.05	1.03	···
	43.1	0.075	0.820		···	4.27 (9)	2.48(15)	···	···	0.99	1.57	···
Sm^{154}	31.6	0.080	0.635	4.54	···	1.19 (10)	0.26(22)	···	···	0.77	0.97	···
	37.7	0.062	0.824		···	3.80 (9)	1.28(15)	0.36(50)	···	0.84	0.82	1.4
	41.5	0.053	0.952		···	5.79 (9)	3.06(15)	1.1 (40)	···	0.75	0.75	1.1
	41.5	0.053	0.952		5.09(10)	5.60 (10)	···	···	0.91	0.72	···	···
Gd^{156}	34.8	0.077	0.629	4.50	3.48(10)	1.59 (11)	···	···	1.22	1.04	···	···
	40.3	0.062	0.784		···	4.39 (9)	1.75(25)	···	···	1.15	1.42	···
	43.8	0.055	0.890		···	6.31 (8)	3.57(15)	1.3 (35)	···	1.02	1.30	2.0
Gd^{158}	34.8	0.069	0.687	5.36	3.70(10)	2.04 (11)	···	···	1.11	0.92	···	···
	36.9	0.064	0.752		···	2.92 (9)	0.94(15)	···	···	0.91	1.03	···
	40.3	0.056	0.858		···	5.18 (9)	2.13(15)	···	···	0.88	1.00	···
Gd^{160}	33.9	0.068	0.718	5.71	···	1.99 (10)	0.50(22)	···	···	0.76	0.76	···
	39.1	0.054	0.892		5.07(9)	4.09 (10)	···	···	1.00	0.66	···	···
	39.1	0.054	0.892		···	4.67 (10)	1.92(18)	···	···	0.75	0.71	···
	42.0	0.049	0.993		···	7.10 (10)	3.87(17)	···	···	0.77	0.74	···
	43.9	0.046	1.060		···	8.83 (8)	5.20(15)	3.4 (50)	···	0.76	0.65	1.3
Dy^{162}	34.6	0.072	0.630	5.03	2.92(11)	1.77 (9)	0.17(50)	···	1.02	1.14	0.60	···
	41.5	0.055	0.819		···	5.05 (10)	2.20(30)	0.4 (60)	···	1.10	1.34	1.3
Dy^{164}	34.9	0.066	0.667	5.55	3.68(11)	2.15 (11)	···	···	1.15	1.04	···	···
	41.6	0.051	0.864		···	6.52 (10)	2.35(26)	1.5 (50)	···	1.14	1.01	2.8
Er^{166}	34.9	0.074	0.634	5.58	3.23(11)	1.76 (11)	···	···	1.04	0.98	···	···
	41.5	0.055	0.815		···	4.98 (10)	1.63(22)	···	···	1.10	1.04	···
Er^{168}	34.8	0.074	0.634	5.64	3.21(11)	1.90 (11)	···	···	1.11	1.19	···	···
	41.5	0.055	0.822		···	4.50 (11)	1.49(32)	···	···	0.97	0.87	···
Er^{170}	34.8	0.072	0.617	5.35	3.36(12)	1.82 (12)	···	···	1.22	1.13	···	···
	34.8	0.072	0.617		···	1.98 (11)	0.37(40)	···	···	1.23	1.40	···
	41.7	0.054	0.809		···	5.19 (10)	1.96(22)	···	···	1.18	1.25	···

TABLE III. (*Continued*).

1	2	3	4	5	6	7	8	9	10	11	12	13
Isotope	E_{0x}[a] MeV	ξ[b] $0\rightarrow2$	χ[c] $0\rightarrow2$	$B(E2)$[d] $0\rightarrow2$ $e^2\times10^{-48}$ cm^4	Experimental de-excitation probabilities R_I $R_2\times10$	$R_4\times10^2$	$R_6\times10^3$	$R_8\times10^4$	$R_{I\,\text{exp}}/R_{I\,\text{theory}}$[e] $2\rightarrow0$	$4\rightarrow2$	$6\rightarrow4$	$8\rightarrow6$
Yb^{172}	34.8	0.077	0.608	5.80	3.02(11)	1.60 (11)	···	···	1.13	1.15	···	···
	41.6	0.058	0.792		···	4.21 (9)	1.51(22)	···	···	1.04	1.12	···
Yb^{174}	34.7	0.072	0.607	5.80	3.15(11)	1.51 (11)	···	···	1.18	1.10	···	···
	38.8	0.061	0.717		···	3.06 (10)	0.81(22)	···	···	1.10	1.14	···
	41.5	0.055	0.791		···	4.19 (9)	1.58(17)	0.5 (60)	···	1.03	1.15	2.0
Yb^{176}	34.7	0.077	0.602	5.69	2.94(10)	1.37 (9)	···	···	1.13	1.06	···	···
	34.7	0.077	0.602		···	1.44 (9)	0.17(31)	···	···	1.11	0.85	···
	41.6	0.059	0.790		···	3.88 (9)	1.29(22)	0.6 (50)	···	0.98	1.03	2.8
Hf^{178}	34.8	0.091	0.521	4.66	2.19(11)	0.78 (12)	···	···	1.14	1.20	···	···
	40.4	0.073	0.625		···	2.07 (9)	0.68(32)	···	···	1.36	2.56	···
Hf^{180}	34.9	0.090	0.502	4.30	2.25(11)	0.71 (12)	···	···	1.24	1.25	···	···
	38.1	0.079	0.572		···	1.26 (9)	0.30(32)	···	···	1.21	2.07	···
	41.2	0.070	0.644		···	2.26 (10)	0.60(50)	···	···	1.28	1.85	···
W^{182}	34.2	0.103	0.451	4.10	1.72(11)	0.48 (11)	···	···	1.18	1.42	···	···
	41.5	0.077	0.600		···	1.23 (17)	0.32(35)	···	···	0.96	1.56	···
	43.0	0.073	0.633		···	1.66 (12)	0.19(50)	···	···	1.04	0.64	···
W^{184}	34.7	0.112	0.441	3.74	1.64(10)	0.35 (12)	···	···	1.20	1.10	···	···
	37.9	0.099	0.503		···	0.73 (11)	···	···	···	1.34	···	···
W^{186}	28.0	0.167	0.304	3.40	0.73(13)	0.025(35)	···	···	1.19	0.62	···	···
	43.8	0.086	0.594		···	1.40 (11)	0.41(60)	···	···	1.23	2.3	···
Os^{188}	43.3	0.113	0.463[f]	3.1[f]	1.49(17)	0.44 (36)	···	···	1.00[f]	1.20	···	···
	43.8	0.112	0.455[f]		1.44(12)	0.51 (30)	···	···	1.00[f]	1.48	···	···
Os^{190}	43.8	0.135	0.414[f]	2.5[f]	1.15(22)	0.48 (50)	···	···	1.00[f]	2.18	···	···
Os^{192}	43.8	0.148	0.414[f]	2.5[f]	1.12(17)	0.20 (40)	···	···	1.00[f]	0.98	···	···

[a] Corrected for energy loss in the target.
[b] Adiabaticity parameter, defined in Eq. (10).
[c] Transition amplitude for the $0^+\rightarrow2^+$ excitation in first order perturbation ($\xi=0$ and $\theta=180°$).
[d] Values taken from Refs. 22 and 23.
[e] $R_{I\,\text{theory}}$ was determined for a moment of inertia which reproduces the energy of the transition $I\rightarrow I-2$.
[f] For the osmium isotopes $B(E2,\,0\rightarrow2)$ values determined by inelastic scattering experiments, are not available. The quoted values are adjusted in such a way as to make $R_{2\,\text{exp}}/R_{2\,\text{theor}}=1.00$.

in the square brackets of Eq. (18) is the same as the relative sign of the two reduced matrix elements $\langle2'||E2||0\rangle$ and $\langle2'||E2||2\rangle$. The curly bracket in the second term in defined as

$$\{\xi_{0\rightarrow2'},\xi_{0\rightarrow2},\xi_{2\rightarrow2'},\theta\}$$
$$=(\tfrac{3}{2})^{1/2}[I_{22}(\cdots)\times\beta_{2-2}(\cdots)+I_{2-2}(\cdots)\times\beta_{22}(\cdots)]$$
$$-I_{20}(\cdots)\beta_{20}(\cdots). \quad (20)$$

The arguments of $I_{2\mu}(\cdots)$ in Eq. (20) are $(\xi_{0\rightarrow2'},\theta)$, and of $\beta_{2\kappa}(\cdots)$ they are $(2,2,\xi_{0\rightarrow2},\xi_{2\rightarrow2'},\theta)$. The functions $I_{2\mu}$ and $\beta_{2\kappa}$ are tabulated in Refs. 1 and 21, respectively.

In Sm^{152}, a $0'^+$ level at 685 keV was excited by double Coulomb excitation. The numerical formula for the excitation probability is for this case

$$P_{0'}{}^{(22)}=4.48\chi_{0\rightarrow2}{}^4\times\frac{B(E2,\,2\rightarrow0')}{B(E2,\,0\rightarrow2)}$$
$$\times\sin^4(\theta/2)dF(\xi_{0\rightarrow2},\xi_{2\rightarrow0'},0,\theta)/d\Omega. \quad (21)$$

The results on the excitation of nonrotational levels and the $B(E2)$ values extracted from the experimental excitation probabilities are summarized in Table IV. The number in parenthesis following each experimental quantity gives its relative error in percent. The relative errors in the $B(E2)$ values are the same as in the excitation probabilities. The multipolarity of the $2'^+\rightarrow2^+$ transition has been assumed to be pure $E2$, the $M1$ contributions being in most cases smaller than 2%.[25] For the de-excitation of the $2'^+$ states, two sets of $B(E2)$ values are given, corresponding to the two signs of the interference term in Eq. (18). The relative sign of the reduced matrix elements, which was used in the determination of the $B(E2)$ values, is indicated in parenthesis following the experimental number for $B(E2)$. In the case of Gd^{160}, where the excitation of the 1010-keV state was observed at two bombarding energies, the experimental accuracy is not sufficient to allow a decision between the two signs. In the nuclei W^{186}, Os^{188}, and Os^{112}, the interference term is so small that the two sets of $B(E2)$'s are almost equal. The

[25] F. K. McGowan and P. H. Stelson, Phys. Rev. **122**, 1274 (1961).

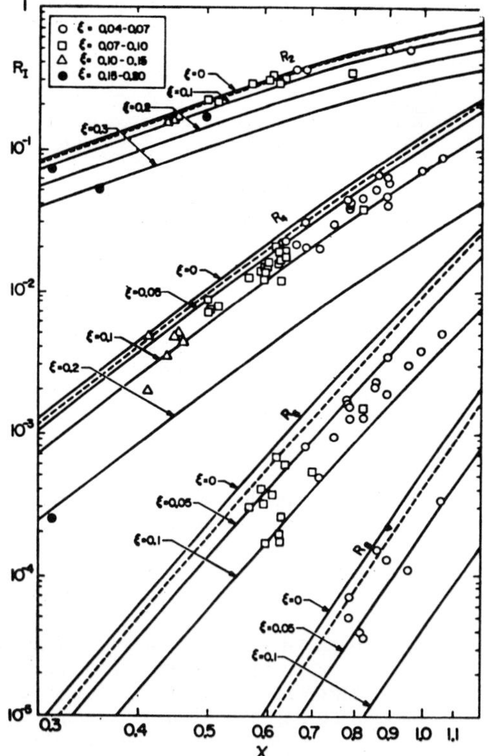

FIG. 13. De-excitation probabilities R_I as a function of χ and ξ. The solid curves represent $R_I(\chi, \xi, \theta=180°)$, calculated in Ref. 3 and the dashed lines show $R_I(\chi, \xi=0, \theta=160°)$ from Ref. 4. The experimental points are plotted for the $\chi_{0\to2}$ and $\xi_{0\to2}$ values quoted in Table IV.

$B(E2)$ values, reported in the present paper, are in fair agreement with the ones quoted in Ref. 25. In all cases, our values are somewhat smaller.

V. DISCUSSION

The aim of the present experiments was to study in detail the multiple Coulomb excitation of the ground-state rotational bands in deformed even-even nuclei. In the experiments some transitions to other states were also observed.

A. Rotational Transitions

As was shown in Sec. IIIC, the perturbation treatment fails to describe the excitation of the ground-state rotational bands under the bombarding conditions of the present experiments. A realistic description of the process must take into account a large number of nuclear quantities which can not be determined independently from the measured excitation probabilities. Instead we have compared our results with a calculation for a pure rotational band[3] and a good agreement has been found (cf. Figs. 5, 6, and 13). We believe, however, that the accuracy of the experiments allows one to interpret the small deviations from the calculations in terms of small deviations from the pure rotational description. The effects arising from the deviations of the energies from the $I(I+1)$ rule are discussed in Sec. IVB. In the following it is attempted to interpret the remaining discrepancies in terms of a variation of the intrinsic quadrupole moment Q_0.

We can best illustrate this approximative procedure by an example: In lowest order perturbation (l.o.p.), R_2 depends only on $B(E2, 0\to2)$ and R_4 depends only on the product $B(E2, 0\to2)\times B(E2, 2\to4)$, so that $(R_4/R_2)_{\text{l.o.p.}} \propto B(E2, 2\to4)$. We now calculate $R_4^{(\text{l.o.p.})}/$

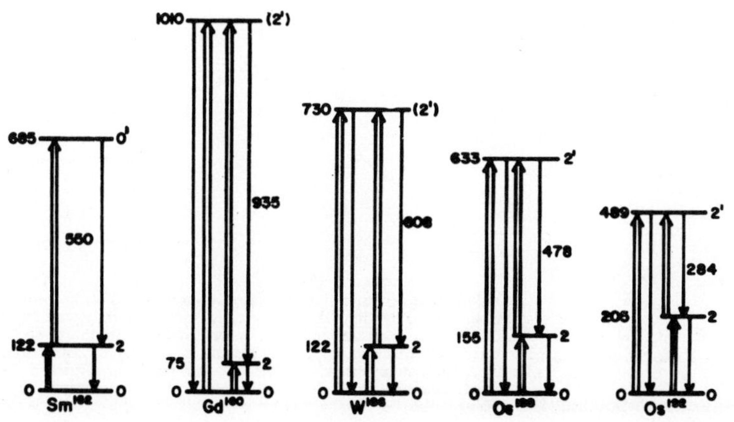

FIG. 14. Level diagrams, depicting the nonrotational transitions observed in the present experiments.

TABLE IV. Results on the excitation of nonrotational levels. The numbers in parenthesis, following each experimental figure, are the relative errors in percent. The plus (or minus) sign following the $B(E2)$ values denotes that the two reduced $E2$-matrix elements connecting the $2'^+$ state with the 2^+ and 0^+ rotational states have the same (or the opposite) sign.

Isotope	E_{0z} MeV	E_γ keV	Level energy keV	Assignment of decay	Branching ratio $(E2)$ $\frac{2'\to 2}{2'\to 0}$	Excitation probability 10^{-3}	$B(E2)$ for de-excitation $e^2\times 10^{-48}$ cm^4
Sm^{152}	39.0	550±15	685	$0'\to 2^a$	...	0.6(30)	1.9
	43.1	550±15	685	$0'\to 2^a$	...	1.7(25)	2.5
Gd^{160}	39.1	1010±15	1010	$2'\to 0$	1.6(30)	1.3(40)	0.026(−); 0.015(+)
		930±15	1010	$2'\to 2$			0.060(−); 0.035(+)
	43.9	1010±15	1010	$2'\to 0$	1.4(30)	2.9(35)	0.020(−); 0.013(+)
		930±15	1010	$2'\to 2$			0.046(−); 0.030(+)
W^{186}	43.8	732±10	730	$2'\to 0^b$	0.73(25)	3.1(30)	0.019(−); 0.017(+)
		605±10	730	$2'\to 2^b$			0.035(−); 0.032(+)
Os^{188}	43.8	640±10	633	$2'\to 0^b$	0.68(25)	6.4(30)	0.030(−); 0.028(+)
		475±10	633	$2'\to 2^b$			0.083(−); 0.077(+)
Os^{192}	43.8	495±10	489	$2'\to 0^b$	0.57(30)	6.9(40)	0.013(−); 0.012(+)
		285±10	489	$2'\to 2^b$			0.11(−); 0.10(+)

[a] Assignment taken from Ref. 27 and 28.
[b] Assignments taken from Ref. 25.

$R_2^{(\text{l.o.p.})}$ with $B(E2,\,2\to 4)_{\text{theor}}$ given by Eq. (2), whereby Q_0 is taken from Refs. 22 and 23. In l.o.p. we have

$$(R_4/R_2)_{\text{exp}}/(R_4/R_2)_{\text{l.o.p.}} = B(E2,\,2\to 4)_{\text{exp}}/B(E2,\,2\to 4)_{\text{theor}}. \quad (22)$$

The proportionality between (R_4/R_2) and $B(E2,\,2\to 4)$ does no longer hold exactly, if higher than lowest order terms are included in the calculations. We can, however, estimate the effect of the inclusion of higher order terms by comparing R_4/R_2 for the lowest order perturbation calculations with the value of the calculation of Ref. 3. This comparison is illustrated in Fig. 15. The de-excitation probabilities R_2 and R_4 are calculated for a symmetric rotator by the diagonalization method[3] (solid lines) and in l.o.p. (broken lines). In the doubly logarithmic plot, the first- and second-order perturbation calculations give straight lines with slopes of χ^2 and χ^4, respectively. It can be seen from the calculated curves that, in the range of χ and ξ encountered in the present experiments, the ratios R_4/R_2 are indeed very closely the same for the perturbation calculations and for the diagonalization method. If we now assume that corresponding relations also hold for higher excitations, we can write for the double ratios

$$D_I \equiv \frac{(R_I/R_{I-2})_{\text{exp}}}{(R_I/R_{I-2})_{\text{diag. method}}} \approx \frac{(R_I/R_{I-2})_{\text{exp}}}{(R_I/R_{I-2})_{\text{l.o.p.}}} \approx \frac{B(E2,\,I-2\to I)_{\text{exp}}}{B(E2,\,I-2\to I)_{\text{theor}}}. \quad (23)$$

The double ratios D_I, calculated for Q_0-values of Refs. 22 and 23, are plotted in Fig. 16 for $I=4$, 6 and 8, as a function of the mass number A. The experimental points lie fairly close to a line corresponding to $D_I=1$ which indicates that the intrinsic quadrupole moment Q_0 is to a good approximation a constant for all transitions within the ground state rotational band. Li and Schwarzschild[6] came to the same conclusion from their measurements of the lifetimes of 2^+ and 4^+ states of strongly deformed nuclei. The present measurements, however, show a slight increase of D_I for nuclei in the transition region, where Q_0 is smaller than in the region of strong deformation. Also, the deviation of D_I to values greater than 1 is more pronounced for the higher rotational states. These observations are in

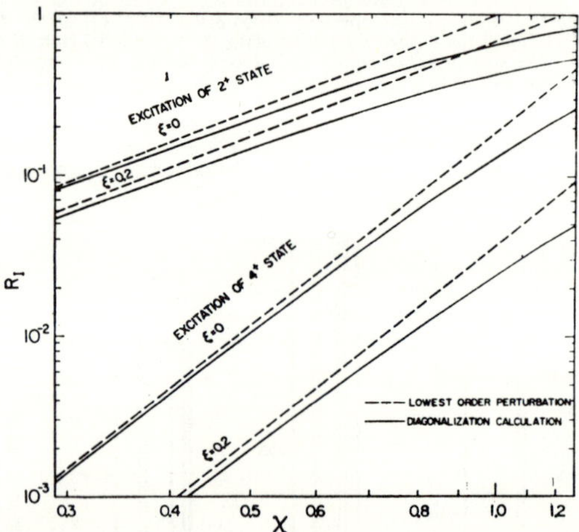

FIG. 15. Comparison of the calculations in Ref. 3 of the multiple Coulomb excitation of a symmetric rotator with lowest order perturbation calculations (first order for the excitation of the 2^+ state, second order for the 4^+ state). In the doubly logarithmic plot, the perturbation calculations are straight lines with slopes corresponding to χ^2 for first-order and χ^4 for second-order perturbation.

TABLE V. Comparison of the $B(E2)$ ratios with the predictions of the asymmetric-rotator model by Davydov *et al.* (Ref. 9).

Isotope	$B(E2, 2'\to 2)/B(E2, 2\to 0)$ Experiment	Theory	$B(E2, 2'\to 0)/B(E2, 2\to 0)$ Experiment	Theory	$B(E2, 2'\to 2)/B(E2, 2'\to 0)$ Experiment	Theory
Gd^{160}	0.053±0.016	0.064	0.023±0.007	0.031	2.3±0.5	1.78
W^{186}	0.051±0.015	0.18	0.028±0.009	0.061	1.9±0.4	3.05
Os^{188}	0.13 ±0.04	0.32	0.048±0.014	0.071	2.2±0.6	4.60
Os^{192}	0.22 ±0.09	0.94	0.026±0.008	0.042	8.6±1.7	23.5

qualitative agreement with the fact that the moment of inertia becomes larger with higher spin[26] and that its relative change increases at the border of the region of strong deformations (see Fig. 11). A quantitative treatment would require the inclusion of nuclear degrees of freedom in addition to those of a symmetric top.

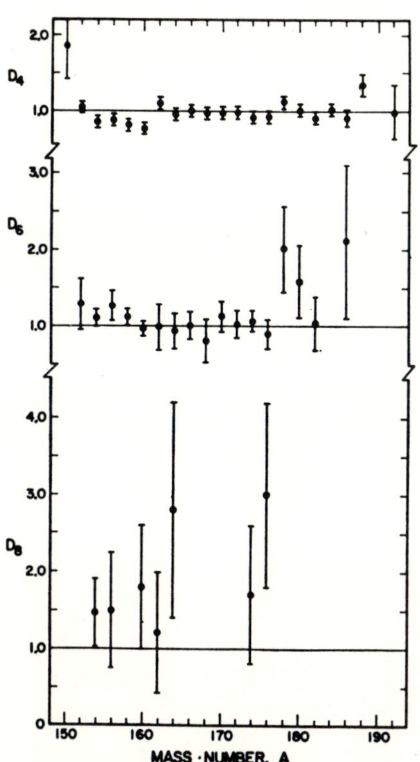

FIG. 16. The quantities D_I, defined as $(R_I/R_{I-2})_{exp}$: $(R_I/R_{I-2})_{diag.\ method}$ are plotted for $I=4$, 6, and 8 as a function of the mass number A. In an approximative way (see text), the deviation of D_I from 1 is proportional to the deviation of $B(E2, I\to I-2)/B(E2, 2\to 0)$ from the value predicted by the symmetric-rotator model [Eq. (2)].

[26] H. Morinaga and P. C. Gugelot, Nucl. Phys. **46**, 210 (1963).

B. Nonrotational Transitions

An excited $0'^+$ state in Sm^{152} has been observed in the decay of Eu^{152} by Marklund *et al.*[27] and in Coulomb-excitation experiments by Greenberg *et al.*[28] The assignment of the 550±15-keV gamma ray as a $0'^+\to 2^+$ transition is based on the work of these authors. Our experiments, however, do not show a gamma ray of about 690 keV, which is reported in Ref. 28. The intensity of this line is, according to Ref. 28, about equal to the intensity of the $0'^+\to 2^+$ transition, and and the line should therefore have been observed in the present investigations.[29]

The nature of the $2'^+$ states has been extensively discussed by McGowan and Stelson.[25] In analogy with these authors, we present in Table V a comparison of our results for the ratios of $B(E2)$ values with the asymmetric rotator model of Davydov and Filippov.[9] For this comparison, the sign of the interference term has been chosen in such a way as to give the best agreement with Ref. 9. In each case, this leads to the assumption that the two matrix elements connecting the $2'^+$ state with the 2^+ and 0^+ rotational states have opposite sign. The agreement of our values with the model of Davydov and Filippov is in most cases not as good as the agreement obtained by McGowan and Stelson.[25,30]

ACKNOWLEDGMENTS

The authors want to thank Dr. Zeev Vager and Yacov Dar at the Weizmann Institute of Science for the preparation of some of the targets used in these experiments. We are most grateful to Clyde Zaidins for his assistance during the measurements. We are indebted to Dr. K. Alder, D. Beder, Dr. F. Boehm, Dr. J. Eichler, Dr. T. Lauritsen, and Dr. A. Winther for many valuable discussions. It was of great help to us that Dr. K. Alder made the results of his calculations available prior to their publication.

[27] I. Marklund, O. Nathan, and O. B. Nielsen, Nucl. Phys. **15**, 199 (1960).

[28] J. S. Greenberg, G. G. Seaman, E. V. Bishop, and D. A. Bromley, Phys. Rev. Letters **11**, 211 (1963).

[29] *Note added in proof*. The relative intensity of the 690 keV line was found to be much higher in the *direct* gamma spectrum than in coincidence with *backscattered* particles [J. S. Greenberg (private communication)].

[30] *Note added in proof*. The work of Lutken and Winther shows that the perturbation treatment might not be adequate in this case (to be published in Kgl. Danske Videnskab. Selskab, Mat. Fys. Medd.).

California Institute of Technology, Technical Report, November 18, 1965

*A Computer Program for Multiple Coulomb Excitation**

AAGE WINTHER† AND JORRIT DE BOER

California Institute of Technology, Pasadena, California
and Rutgers, The State University, New Brunswick, New Jersey

(Reprinted with corrections by the authors.)

A computer program is presented by which one may calculate the multiple quadrupole Coulomb excitation with heavy ions. The excitation amplitudes are evaluated by direct numerical integration of the coupled differential equations for the time-dependent amplitudes. The program applies to an arbitrary nucleus, specified by the spins and energies of the levels and by all E2 matrix elements. For given bombarding conditions, the differential cross sections and the gamma-ray angular distributions are computed.

I. Introduction

The use of heavy ions as projectiles in Coulomb excitation experiments makes it possible to populate a large number of nuclear levels. The present report describes a computer program for the calculation of multiple excitation among a finite number of nuclear states. The system of coupled differential equations for the time-dependent amplitudes of the eigenstates of the free nucleus is solved numerically for electric quadrupole interactions. All nuclear quantities, either known from experiments or calculated from a model, as well as the conditions realized in the experiment, are explicitly specified as input parameters. The program then computes the Coulomb-excitation probabilities and differential cross sections as well as the angular distribution of the de-excitation γ quanta. The number of nuclear levels which can be incorporated in the calculation and the accuracy with which the differential equations are integrated depend on the memory capacity of the computer and on the time allowed for the calculation.

II. Solution of the Time-Dependent Schrödinger Equation

In pure Coulomb excitation where the projectile does not enter into the range of nuclear forces, the parameter $\eta = Z_1Z_2e^2/\hbar v$ is larger than one. For heavy ions

* Supported in part by the Office of Naval Research [Nonr-220(47)], the U.S. Atomic Energy Commission [AT(04-3)-63], and the National Science Foundation.

† On leave from Niels Bohr Institute, Copenhagen, Denmark.

one finds $\eta \gg 1$ and since the energy loss of the projectile is small compared to the bombarding energy, one is justified in using a semiclassical description. While the first-order corrections to the semiclassical approximation are discussed in Sec. III, we here adhere to the pure semiclassical picture where the Coulomb excitation is caused by a well-defined time-dependent electromagnetic field acting on the nucleus as the projectile moves along the classical hyperbola.

The nuclear wave function ψ satisfies the Schrödinger equation

$$i\hbar\dot{\psi} = [H_0 + H_E(t)]\psi, \tag{1}$$

where H_0 is the Hamiltonian of the free nucleus and $H_E(t)$ is the time-dependent electromagnetic interaction. Expanding the wave function on the eigenstates ψ_s of the free nucleus, one obtains the following set of coupled differential equations for the expansion amplitudes $a_r(t)$ (see Ref. 1):

$$i\hbar\dot{a}_r(t) = \sum_s \langle r|H_E(t)|s\rangle \exp[i(E_r - E_s)t/\hbar]a_s(t), \tag{2}$$

where E_i denotes the energy of the ith state. Equation (2) must be solved with initial conditions $a_s(t = -\infty)$ describing the target nucleus in its ground state. The final amplitudes $a_s(t = +\infty)$, describing the target nucleus after the collision, can be found by integrating (2) numerically.

In the following we neglect the interaction via the transverse electromagnetic field. The interaction energy is then the Coulomb interaction, which may be written in the form[2]

$$H_E(t) = \sum_{\lambda,\mu} \frac{4\pi Z_1 e}{2\lambda + 1} r_p^{-\lambda-1}(t)\, Y_{\lambda\mu}[\theta_p(t), \varphi_p(t)]\, \mathfrak{M}^*(E\lambda,\mu), \tag{3}$$

where r_p, θ_p, φ_p denote the time-dependent spherical coordinates of the projectile and $\mathfrak{M}(E\lambda,\mu)$ is the nuclear electric multipole moment of order $\lambda\mu$,

$$\mathfrak{M}(E\lambda,\mu) = \int r^\lambda Y_{\lambda\mu}(\theta,\varphi)\rho(\mathbf{r})\, d^3r, \tag{4}$$

ρ being the nuclear charge density.

For the solution of the differential equations (2) it is convenient to use, instead of the time t, a parameter w defined by[2]

$$t = (a/v)(\epsilon \sinh w + w) \tag{5}$$

$$r_p(t) = a(\epsilon \cosh w + 1), \tag{6}$$

where ϵ, given by

$$\epsilon = 1/\sin \vartheta/2 \tag{7}$$

[1] K. Alder and A. Winther, *Kgl. Danske Videnskab. Selskab, Mat. Fys. Medd.* **32,** No. 8 (1960).
[2] K. Alder, A. Bohr, T. Huus, B. Mottelson, and A. Winther, *Rev. Mod. Phys.* **28,** 432 (1956).

is the eccentricity of the hyperbolic orbit and ϑ is the scattering angle in the center of mass system. Half the distance of closest approach in a head-on collision a is given by

$$a = Z_1 Z_2 e^2 / M_0 v^2, \tag{8}$$

where v is the relative velocity of the projectile and the target at large distances and M_0 is the reduced mass. Inserting this into Eq. (2) one obtains

$$\frac{da_r(w)}{dw} = \frac{4\pi Z_1 e}{i\hbar v} \sum_{\lambda\mu s} \frac{Y_{\lambda\mu}(\theta_p,\varphi_p)\exp[i\xi_{rs}(\epsilon \sinh w + w)]}{a^\lambda(\epsilon \cosh w + 1)^\lambda (2\lambda + 1)} \langle s|\mathfrak{M}(E\lambda,\mu)|r\rangle\, a_s(w) \tag{9}$$

where ξ_{rs} is defined by

$$\xi_{rs} = (a/\hbar v)(E_r - E_s). \tag{10}$$

To simplify expression (9) we introduce the dimensionless parameter $\psi_{rs}^{(\lambda)}$ by the relation

$$\psi_{rs}^{(\lambda)} = \frac{(16\pi)^{1/2}(\lambda - 1)!}{(2\lambda + 1)!!} \frac{Z_1 e}{\hbar v} \frac{\langle I_s || i^\lambda M(E\lambda) || I_r \rangle}{a^\lambda}, \tag{11}$$

where the reduced multipole matrix element is defined by

$$\langle I_s M_s | i^\lambda \mathfrak{M}(E\lambda,\mu) | I_r M_r \rangle = (-1)^{I_s - M_s} \begin{pmatrix} I_s & \lambda & I_r \\ -M_s & \mu & M_r \end{pmatrix} \langle I_s || i^\lambda \mathfrak{M}(E\lambda) || I_r \rangle. \tag{12}$$

The nuclear states are specified by the spin quantum numbers I and M. If the phases of the nuclear wave functions are chosen in such a way that the reduced matrix elements are real, they satisfy the following symmetry relation:

$$M_{rs}^{(\lambda)} = \langle I_s || i^\lambda \mathfrak{M}(E\lambda) || I_r \rangle = (-1)^{\lambda + I_r - I_s} \langle I_r || i^\lambda \mathfrak{M}(E\lambda) || I_s \rangle. \tag{13}$$

The parameters $\psi_{rs}^{(\lambda)}$ are, except for a different phase convention, related to the parameters $\chi_{s\to r}^{(\lambda)}$ defined in Ref. 1 by the expression

$$\psi_{rs}^{(\lambda)} = (2I_s + 1)^{1/2} \chi_{s\to r}^{(\lambda)}. \tag{14}$$

Introducing, finally, the quantities $Q_{\lambda\mu}(\xi,w)$ defined by

$$Q_{\lambda\mu}(\xi,w) = \frac{(2\lambda - 1)!!}{(\lambda - 1)!} \left(\frac{\pi}{2\lambda + 1}\right)^{1/2} \frac{Y_{\lambda\mu}[\theta_p(w),\varphi_p(w)]\exp[i\xi(\epsilon \sinh w + w)]}{(\epsilon \cosh w + 1)^\lambda}, \tag{15}$$

we may write Eq. (9) in the form

$$\frac{da_r(w)}{dw} = -i \sum_{\lambda\mu s} i^{-\lambda} Q_{\lambda\mu}(\xi_{rs},w) \times \zeta_{sr}^{(\lambda)} \times a_s(w), \tag{16}$$

where

$$\zeta_{rs}^{(\lambda)} = (2\lambda + 1)^{1/2}(-1)^{I_s - M_s}\begin{pmatrix} I_s & \lambda & I_r \\ -M_s & \mu & M_r \end{pmatrix} \times \psi_{rs}^{(\lambda)}. \tag{17}$$

For the solution of the differential equations (16) we chose a coordinate system in which the z axis bisects the angle between the vectors $\mathbf{v}_f$ and $-\mathbf{v}_i$ where $\mathbf{v}_i$ and

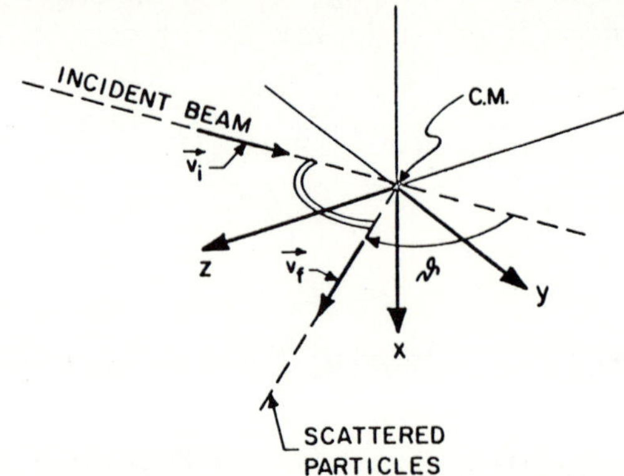

FIG. 1. The coordinate system, used in the solution of the differential equations. The vectors $\mathbf{v}_i$ and $\mathbf{v}_f$ denote the initial and final velocities of the projectile in the center-of-mass system. The z axis bisects the angle between $-\mathbf{v}_i$ and $\mathbf{v}_f$. The x axis is perpendicular to the plane of the orbit and the y axis points to the side to which the particle is deflected.

$\mathbf{v}_f$ are the initial and final relative velocities. The x axis is perpendicular to the plane of the orbit such that it points in the direction of $\mathbf{v}_i \times \mathbf{v}_f$ (see Fig. 1). In this system we have

$$\begin{aligned} x &= 0 \\ y &= a(\epsilon^2 - 1)^{1/2} \sinh w \\ z &= a(\cosh w + \epsilon) \end{aligned} \tag{18}$$

Inserting (18) into Eq. (15) one finds the following explicit expressions for $Q_{\lambda\mu}(\xi,w)$ for $\lambda = 1$, 2, and 3:

$$\left.\begin{aligned} Q_{10} &= \frac{1}{2}\frac{\cosh w + \epsilon}{(\epsilon \cosh w + 1)^2} \exp[i\xi(\epsilon \sinh w + w)] \\ Q_{1\pm1} &= -i\frac{1}{2}\left(\frac{1}{2}\right)^{1/2} \frac{(\epsilon^2 - 1)^{1/2} \sinh w}{(\epsilon \cosh w + 1)^2} \exp[i\xi(\epsilon \sinh w + w)] \end{aligned}\right\} \tag{19}$$

$$\left.\begin{aligned} Q_{20} &= \frac{3}{4}\frac{2(\cosh w + \epsilon)^2 - (\epsilon^2 - 1)\sinh^2 w}{(\epsilon \cosh w + 1)^4} \exp[i\xi(\epsilon \sinh w + w)] \\ Q_{2\pm1} &= -i\frac{3}{2}\left(\frac{3}{2}\right)^{1/2} \frac{(\epsilon^2 - 1)^{1/2} \sinh w(\cosh w + \epsilon)}{(\epsilon \cosh w + 1)^4} \exp[i\xi(\epsilon \sinh w + w)] \\ Q_{2\pm2} &= -\frac{3}{4}\left(\frac{3}{4}\right)^{1/2} \frac{(\epsilon^2 - 1)\sinh^2 w}{(\epsilon \cosh w + 1)^4} \exp[i\xi(\epsilon \sinh w + w)] \end{aligned}\right\} \tag{20}$$

$$
\left.
\begin{aligned}
Q_{30} &= \frac{15}{8}\,\frac{[2(\cosh w + \epsilon)^2 - 3(\epsilon^2 - 1)\sinh^2 w](\cosh w + \epsilon)}{(\epsilon\cosh w + 1)^6} \\
&\qquad\qquad \times \exp[i\xi(\epsilon\sinh w + w)] \\
Q_{3\pm1} &= -i\,\frac{15}{16}\,(3)^{1/2}\,\frac{[4(\cosh w + \epsilon)^2 - (\epsilon^2 - 1)\sinh^2 w](\epsilon^2 - 1)^{1/2}\sinh w}{(\epsilon\cosh w + 1)^6} \\
&\qquad\qquad \times \exp[i\xi(\epsilon\sinh w + w)] \\
Q_{3\pm2} &= -\,\frac{15}{8}\,(15)^{1/2}\,\frac{(\cosh w + \epsilon)(\epsilon^2 - 1)\sinh^2 w}{(\epsilon\cosh w + 1)^6}\,\exp[i\xi(\epsilon\sinh w + w)] \\
Q_{3\pm3} &= i\,\frac{15}{16}\,(5)^{1/2}\,\frac{(\epsilon^2 - 1)^{3/2}\sinh^3 w}{(\epsilon\cosh w + 1)^6}\,\exp[i\xi(\epsilon\sinh w + w)].
\end{aligned}
\right\}
\tag{21}
$$

It is noted that in the above coordinate system

$$Q_{\lambda\mu}(\xi,w) = Q_{\lambda-\mu}(\xi,w) \tag{22}$$

and that the integral of $Q_{\lambda\mu}$ over w is equal to the integrals $R_{\lambda\mu}(\vartheta,\xi)$ defined in Ref. 3.

$$R_{\lambda\mu}(\vartheta,\xi) = \int_{-\infty}^{\infty} Q_{\lambda\mu}(\xi,w)\,dw. \tag{23}$$

The reaction matrix $\langle r|T|l\rangle$ for multiple Coulomb excitation corresponding to an initial state l and a final state r is proportional to the excitation amplitude, which is obtained from the differential equations (16) as the solution $a_r(+\infty)$ for the initial condition $a_s(-\infty) = \delta_{sl}$, where l is one of the magnetic substates of the nuclear ground state. We denote this excitation amplitude by $a_{r,l}$ or $a_{I_rM_r}(M_l)$, where in the latter notation the dependence on the magnetic substates has been explicitly specified. From the symmetry of the process with respect to reflections in the plane of the orbit,[4] one finds in the above coordinate system the following relation between the elements of the reaction matrix:

$$\langle I_f,M_f|T|I_i,M_i\rangle = (-1)^{\Delta\pi+I_i-I_f}\langle I_f,-M_f|T|I_i,-M_i\rangle \tag{24}$$

or

$$a_{I_fM_f}(M_i) = (-1)^{\Delta\pi+I_i-I_f}a_{I_f-M_f}(-M_i) \tag{25}$$

where the parity change $(-1)^{\Delta\pi}$ is equal to $(-1)^{\lambda}$.

From the time reversal invariance of the Hamiltonian in (1) one finds in the coordinate system used here

$$\langle I_i,M_i|T|I_f,M_f\rangle = \langle I_f,M_f|T|I_i,M_i\rangle. \tag{26}$$

or

$$a_{I_iM_i}(M_f) = a_{I_fM_f}(M_i). \tag{27}$$

which may also be shown directly by solving the diffential equations (2) and (16) with the nucleus initially in the excited state I_f,M_f.

[3] H. Lütken and A. Winther, *Kgl. Danske Videnskab. Selskab, Mat. Fys. Skrifter* **2**, No. 6 (1964).
[4] K. Alder and A. Winther, *Nucl. Phys.* **37**, 194 (1962).

The probability for Coulomb excitation of the Nth state (spin I_N) from an unpolarized ground state can be determined from the amplitudes $a_{I_rM_r}(M_l)$ by the expression

$$P(N) = \frac{1}{2I_1 + 1} \sum_{M_1, M_N} |a_{I_NM_N}(M_1)|^2. \tag{28}$$

The differential cross sections in the center-of-mass system are obtained by multiplying $P(N)$ with the Rutherford cross section,

$$d\sigma(N)/d\Omega = P(N) \times \tfrac{1}{4}a^2 \sin^{-4} \vartheta/2. \tag{29}$$

III. Symmetrization

In the semiclassical description of Coulomb excitation used above, the energy loss of the projectile in the collision is neglected, and only one asymptotic velocity v is used in the expressions (5), (6), (8), (10), and (11). The change in results which would be obtained by inserting for v instead of the bombarding velocity the relative velocity after the excitation, gives rise to an uncertainty in the results which in principle lies outside the scope of the semiclassical approximation. The uncertainty may be especially important for high excited states since the cross sections then depend rather sensitively on ξ.

In first-order perturbation theory a systematic improvement of the semiclassical theory can be obtained by a WKB treatment. One finds in this treatment that the parameter ξ arises from the difference in wave number, between initial and final state, and that the parameter in the semiclassical limit approaches

$$\xi_{MN} = (Z_1Z_2e^2/\hbar)(1/v_M - 1/v_N) \tag{30}$$

instead of (10), where v_N and v_M are the asymptotic relative velocities in initial and final state, respectively. One finds, furthermore, that one should use an orbit with the following symmetrized expression

$$a_{MN} = Z_1Z_2e^2/M_0v_Nv_M \tag{31}$$

for half the distance of closest approach instead of (8). From the excitation probabilities P computed with the help of these parameters, one obtains the differential cross section in the center-of-mass system by the expression

$$d\sigma(N \to M) = (v_M/v_N) \cdot P(N \to M, \xi, \vartheta) \cdot \tfrac{1}{4}a_{MN}^2 \sin^{-4} (\vartheta/2)\, d\Omega. \tag{32}$$

It was found[1] that this procedure leads to results which are in very close agreement with the results of a quantum mechanical description of the first-order Coulomb excitation.

In multiple Coulomb excitation, where several nuclear states are excited, a substitution analogous to Eq. (31) would introduce a different value of a_{MN} for each final velocity v_M. For each level, the excitation would then be calculated for a different orbit of the projectile, violating hereby the condition of unitarity,

$$\sum_M P(N \to M) = 1. \tag{33}$$

The fundamental rules, which have to be observed in the construction of a symmetrization are, besides (33), the relation

$$(2I_N + 1)v_N^2\, d\sigma(N \to M) = (2I_M + 1)v_M^2\, d\sigma(M \to N) \tag{34}$$

between the cross sections for the two processes $N \to M$ and $M \to N$. The rule (34) must hold for any inelastic scattering according to time reversal invariance. Equation (34) is fulfilled with expression (32) since in perturbation theory

$$P(N \to M)(2I_N + 1) = P(M \to N)(2I_M + 1). \tag{35}$$

A symmetrization can be realized in multiple Coulomb excitation without violating the two rules (33) and (34) if, instead of modifying the projectile orbit, we symmetrize all parameters ξ_{rs} in (10) according to the prescription

$$\xi_{rs}^{(\mathrm{symm})} = (Z_1 Z_2 e^2/\hbar) = (1/v_r - 1/v_s) \tag{36}$$

and all parameters $\psi_{rs}^{(\lambda)}$ in (11) according to the prescription

$$\psi_{rs}^{(\mathrm{symm},\,\lambda)} = \frac{(16\pi)^{1/2}(\lambda - 1)!}{(2\lambda + 1)!!}\,\frac{Z_1 e}{\hbar(v_r \cdot v_s)^{1/2}}\,\frac{M_{rs}^{(\lambda)}}{a_{rs}^{\lambda}}. \tag{37}$$

The velocities v_r, expressed in terms of energies and masses, are given by

$$v_r = [2(\mathcal{E}_p - E'_r)/M_p]^{1/2}, \tag{38}$$

where

$$E'_r = (1 + A_1/A_2)(E_r - E_i) \tag{39}$$

and $\mathcal{E}_p$ and M_p are the laboratory energy and the mass of the projectile, respectively, and E_i is the energy of the initial state.

Since these symmetrizations may be thought of as (slight) renormalizations of the nuclear energies and matrix elements, the symmetrized equations (16) will lead to solutions fulfilling unitarity.

Furthermore, if we consider the problem of reversed motion, where the nucleus in the state f is bombarded with projectiles of energy $\mathcal{E}_p - E'_f$, we find according to (38) and (39):

$$v_r^{(\mathrm{reversed})} = \left\{\frac{2[(\mathcal{E}_p - E'_f) - (E'_r - E'_f)]}{M_p}\right\}^{1/2} = v_r \tag{40}$$

The differential equations which have to be solved are therefore the same for the two problems except for the initial conditions. Under these circumstances the amplitudes for the transitions from the state $I_i M_i$ to the state $I_f M_f$ are equal to the amplitudes for the opposite transition according to (27).

It thus follows that the quantities (28) for the two problems satisfy (35) and that the following expression for the differential cross sections

$$d\sigma(N \rightarrow M,\lambda) = \frac{v_M}{v_N} \cdot \frac{a_{MN}^2}{4} \cdot P[N \rightarrow M, \psi_{sr}^{(\text{symm},\lambda)}, \xi_{sr}^{(\text{symm})}] \sin^{-4}\left(\frac{\vartheta}{2}\right) d\Omega \qquad (41)$$

satisfies condition (34).

IV. Gamma-Ray Angular Distributions

In the calculation of gamma-ray angular distributions complications arise from the fact that the nuclear levels are not only populated by Coulomb excitation but also by conversion and gamma transitions cascading down from higher states. This situation is illustrated in Fig. 2.

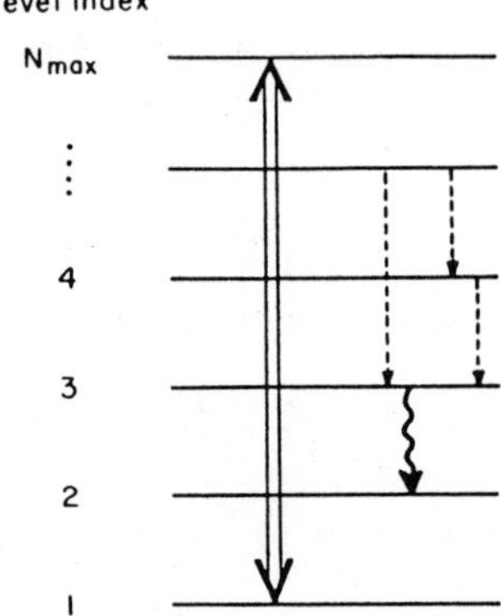

FIG. 2. Angular distribution of de-excitation gamma rays. The angular distribution of a γ quantum, e.g., $3 \rightarrow 2$, is determined by the various ways in which state 3 can be populated: (1) Coulomb excitation of state 3; (2) excitation of state 4 and subsequent unobserved γ decay or internal conversion de-excitation to 3; (3) excitation of still higher states with cascades of unobserved γ's or conversion electrons.

To compute the angular distributions under these circumstances one must know the parameters $\delta_{N\rightarrow M}(L)$ and $\epsilon^2_{N\rightarrow M}(L)$. The quantity δ is defined by

$$\underset{N\rightarrow M}{\delta(L)} = \left\{\frac{8\pi(L+1)}{L[(2L+1)!!]^2}\frac{1}{\hbar}\left(\frac{\omega}{c}\right)^{2L+1}\right\}^{1/2} (2I_M+1)^{-1/2}\langle I_M || i^{s(L)}\mathfrak{M}(L) || I_N\rangle \qquad (42)$$

with $s(L) = L$ for electric and $s(L) = L + 1$ for magnetic multipoles. The square of δ is the L-pole γ-transition rate (in $\sec^{-1}$), while $\epsilon^2_{N\rightarrow M}(L)$ denotes the corresponding L-pole internal conversion rate (in $\sec^{-1}$), summed over all electron shells. For $L \geq 1$, ϵ is given by

$$\epsilon^2_{N\rightarrow M}(L) = \delta^2_{N\rightarrow M}(L)\alpha_{N\rightarrow M}(L), \qquad (43)$$

where α is the usual total L-pole conversion coefficient.

In an arbitrary coordinate system, characterized by the superscript (i), the angular distribution for the gamma quanta from N to M may be written as

$$\frac{dW^{(i)}(\gamma_{N\to M})}{d\Omega_\gamma} = (4\pi)^{-1/2} \sum_{\substack{k=0,2,4 \\ -k\leq\kappa\leq k}} A_{k\kappa}^{(i)} \times F_k(I_M,I_N) \times Y_{k\kappa}[\vartheta_\gamma^{(i)}\varphi_\gamma^{(i)}], \tag{44}$$

where $\vartheta_\gamma^{(i)}$ and $\varphi_\gamma^{(i)}$ denote the polar coordinates of the gamma counter in coordinate system (i) and

$$\begin{aligned} A_{k\kappa}^{(i)} = {} & \alpha_{k\kappa}^{(i)}(N) + \alpha_{k\kappa}^{(i)}(N+1)G_k(I_N,I_{N+1}) \\ & + \alpha_{k\kappa}^{(i)}(N+2)[G_k(I_N,I_{N+2}) + G_k(I_N,I_{N+1})G_k(I_{N+1},I_{N+2})] \\ & + \alpha_{k\kappa}^{(i)}(N+3)[G_k(I_N,I_{N+3}) + G_k(I_N,I_{N+2})G_k(I_{N+2},I_{N+3}) \\ & + G_k(I_N,I_{N+1})G_k(I_{N+1},I_{N+3}) + G_k(I_N,I_{N+1})G_k(I_{N+1},I_{N+2})G_k(I_{N+2},I_{N+3})] \\ & + \cdots . \end{aligned} \tag{45}$$

The coefficients F and G are given by

$$F_k(I_M,I_N) = H(N) \sum_{L'L''} \delta_{N\to M}(L')\delta_{N\to M}(L'')F_k(L',L'',I_M,I_N), \tag{46}$$

$F_k(L',L'',I_M,I_N)$ being the usual gamma–gamma correlation coefficients,[3] and

$$G_k(I_M,I_N) = H(N)[(2I_M+1)(2I_N+1)]^{1/2} \sum_L (-1)^{I_M+I_N+L+k} \times \begin{Bmatrix} I_M & I_M & k \\ I_N & I_N & L \end{Bmatrix} [\epsilon_{N\to M}^2(L) + \delta_{N\to M}^2(L)]. \tag{47}$$

The quantity $H(N)$ is given by

$$H(N) = \left\{\sum_K \sum_L [\epsilon_{N\to K}^2(L) + \delta_{N\to K}^2(L)]\right\}^{-1}, \tag{48}$$

where the sum over K has to include all levels into which level N can decay and the sum over L includes all multipoles in the transition $N \to K$.

The quantities $\alpha_{k\kappa}^{(i)}(N)$ are the statistical tensors in the coordinate system (i). If the angular distribution (44) is calculated in the same coordinate system (Fig. 1) in which the excitation amplitudes $a_{I_f M_f}(M_i)$ were evaluated, the statistical tensors take the form

$$\alpha_{k\kappa}^{(0)}(N) = (2I_N+1)^{1/2}(2I_1+1)^{-1} \times \sum_{\substack{M_N M'_N \\ M_N = M'_N+\kappa}} (-1)^{I_N+M_N} \begin{pmatrix} I_N & I_N & k \\ -M_N & M'_N & \kappa \end{pmatrix} \sum_{M_1} a_{I_N M_N'}^*(M_1) \times a_{I_N M_N}(M_1). \tag{49}$$

For a new coordinate system (i) which is obtained from this coordinate system (0) by a rotation through the Eulerian angles (α, β, γ), the statistical tensors are related to $\alpha^{(0)}$ by the expression

$$\alpha_{k\kappa}^{(i)}(N) = \sum_{-k \leq \kappa' \leq k} D_{\kappa'\kappa}^{k}(\alpha,\beta,\gamma) \times \alpha_{k\kappa'}^{(0)}(N), \tag{50}$$

where the D function is defined by

$$D_{\kappa'\kappa}^{k}(\alpha,\beta,\gamma) = e^{i\kappa'\alpha} d_{\kappa'\kappa}^{k}(\beta) e^{i\kappa\gamma} \tag{51}$$

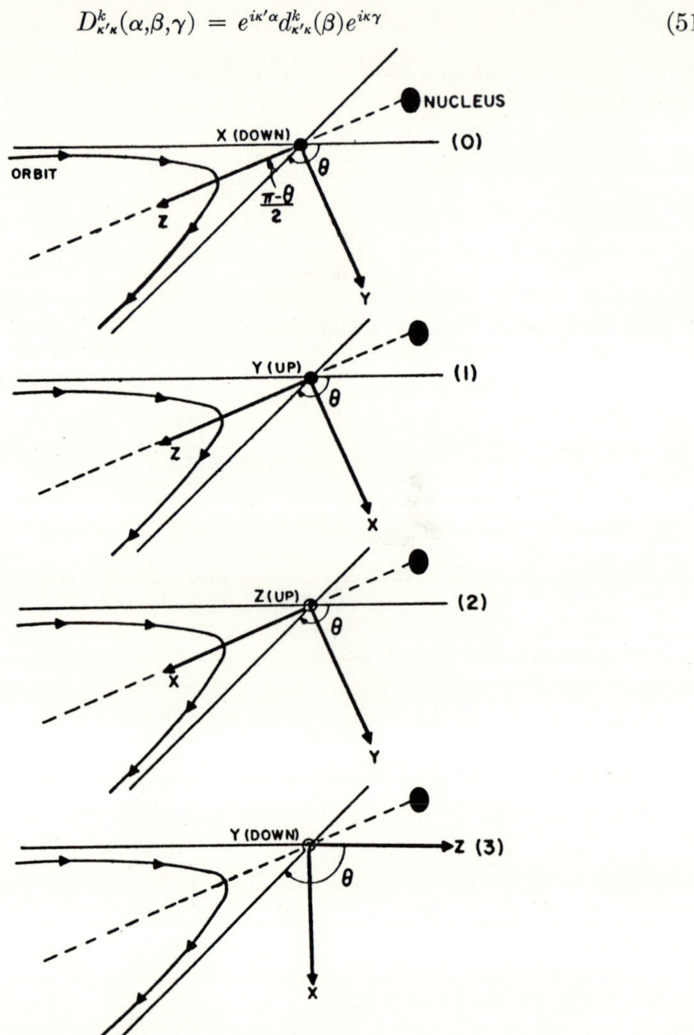

FIG. 3. The coordinate systems in which the angular distributions of the de-excitation gamma rays are given. Coordinate system (0): Used in Ref. 1 and in the present paper for the calculation of the excitation amplitudes. The quantities $\alpha^{(0)}$ are complex. Coordinate system (1): Used in Ref. 4. The quantities $\alpha^{(1)}$ are real. Coordinate system (2): Used in Ref. 2. The quantities $\alpha^{(2)}$ are complex. Coordinate system (3): Z axis in beam direction. The quantities $\alpha^{(3)}$ are real.

with

$$d^k_{\kappa'\kappa}(\beta) = [(k+\kappa')!(k-\kappa')!(k+\kappa)!(k-\kappa)!]^{1/2} \times \sum_{\sigma \geq 0} \frac{(-1)^{k-\kappa-\sigma}(\cos\beta/2)^{2\sigma+\kappa+\kappa'}(\sin\beta/2)^{2k-2\sigma-\kappa-\kappa'}}{(\kappa+\kappa'+\sigma)!\sigma!(k-\kappa'-\sigma)!(k-\kappa-\sigma)!}. \tag{52}$$

It is noted that the statistical tensors $\alpha^{(i)}_{k\kappa}(N)$ satisfy the symmetry relation

$$\alpha^{*(i)}_{k\kappa}(N) = (-1)^{\kappa}\alpha^{(i)}_{k-\kappa}(N). \tag{53}$$

In the computer program the quantities $\alpha^{(i)}_{k\kappa}(N)$ can be obtained in any of the four coordinate systems depicted in Fig. 3. According to Eqs. (50)–(52), one has the following explicit expressions for $\alpha^{(i)}_{k\kappa}(N)$ in these coordinate systems:

$$\alpha^{(1)}_{k\kappa}(N) = i^{\kappa}\alpha^{(0)}_{k\kappa}(N) \tag{54}$$

$$\alpha^{(2)}_{k\kappa}(N) = \sum_{-k\leq\kappa'\leq k} (-i)^{\kappa'} d^k_{\kappa'\kappa}\left(-\frac{\pi}{2}\right)\alpha^{(1)}_{k\kappa'}(N) \tag{55}$$

$$\alpha^{(3)}_{k\kappa}(N) = (-1)^{\kappa} \sum_{-k\leq\kappa'\leq k} d^k_{\kappa'\kappa}\left(\frac{\pi+\vartheta}{2}\right)\alpha^{(1)}_{k\kappa'}(N). \tag{56}$$

The coordinate system (2) was used in Ref. 2. In coordinate system (1) and (3) the statistical tensors are real quantities.

For practical purposes the coordinate system (3) is usually the most convenient one. Figure 4 gives an illustration of the angles ϑ_γ and φ_γ in this coordinate system. For the geometry in which the projectiles are detected in an annular counter symmetric around the beam axis, one finds the following angular distribution in system (3):

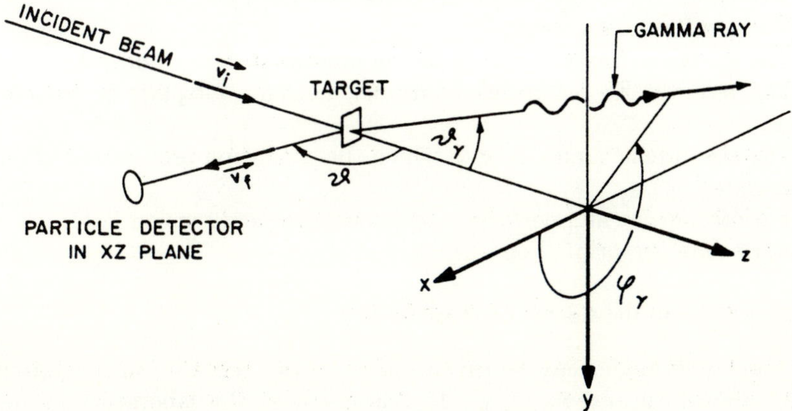

FIG. 4. Illustration of the angles ϑ_γ and φ_γ for coordinate system (3).

$$\frac{dW^{(3)}_{\text{ring}}(\gamma_{N\to M})}{d\Omega_\gamma} = \frac{1}{4\pi} \sum_{k=0,2,4} A^{(3)}_{k0} \times F_k(I_M I_N) \times (2k+1)^{1/2} P_k(\cos\vartheta_\gamma). \tag{57}$$

The quantities $W(\gamma_{M\to N})$ defined in Eqs. (44) and (57) are normalized in such a way that

$$d\sigma_{\text{Rutherford}}/d\Omega_p \times dW(\gamma_{N\to M})/d\Omega_\gamma = d^2\sigma/d\Omega_p\, d\Omega_\gamma, \tag{58}$$

where $d^2\sigma/(d\Omega_p\, d\Omega_\gamma)$ is the doubly-differential cross section for the scattering of an inelastic particle into $d\Omega_p$ and an emission of a gamma ray into $d\Omega_\gamma$.

In applying the formulas for the angular distributions, one should keep in mind that the coordinate system has the origin in the center of the target nucleus. Especially in heavy ion bombardment the target nuclei may receive an appreciable recoil and appropriate corrections should be applied.

V. Computer Program

The method discussed above has been utilized in computer programs for the California Institute of Technology IBM 7094 and the Rutgers University IBM 7040 computers. The program is written in Fortran IV.

The differential equations (16) for $\lambda = 2$ are solved for up to 10 nuclear states with a total of up to 90 magnetic substates. In the following sections, the details of the computation are described.

A. Input Data

The input consists of the following information:

(1) The masses A_1 and A_2 and the charge numbers Z_1 and Z_2 of projectile and target, respectively.

(2) The energy $\mathcal{E}_p$ of the projectile in the laboratory system.

(3) The scattering angle ϑ.

(4) The energies E_N and spins I_N of the nuclear states (10 or less).

(5) The reduced E2 matrix elements M_{rs} between all pairs of nuclear states (See V.O).

(6) The absolute accuracy a_c with which the excitation probabilities $P(N)$ have to be calculated.

(7) Various output and performance controls to be discussed later.

All input data are printed out.

B. Transformation to the Center-of-Mass System

The scattering angle may be given either in the center-of-mass system (ϑ_{CM}) or in the laboratory system (ϑ_{Lab}). If ϑ is given in the laboratory system, ϑ_{CM} depends somewhat on the energy of the final state, i.e., $\vartheta_{\text{CM}} = \vartheta_{\text{CM}}(N)$. The relation between ϑ_{Lab} and $\vartheta_{\text{CM}}(N)$ is given by:

$$\frac{\sin [\vartheta_{\text{CM}}(N) - \vartheta_{\text{Lab}}]}{\sin \vartheta_{\text{Lab}}} = \frac{A_1}{A_2}\left[1 + \frac{A_1 + A_2}{A_2}\frac{E(N)}{\mathcal{E}_p}\right]^{-1/2} \tag{59}$$

For most applications $E(N) \ll \mathcal{E}_p$ and the difference between the various values of $\vartheta_{\text{CM}}(N)$ is small. A special input quantity, called N_{CM}, is provided which specifies which one of the $\vartheta_{\text{CM}}(N)$ is used for the calculation of the orbit, i.e., $\vartheta = \vartheta_{\text{CM}}(N_{\text{CM}})$.

The angles, energies, and solid angle ratios for the projectile and for the recoiling target nucleus are calculated in a subroutine for all $\vartheta_{\text{CM}}(N)$. In addition, the recoil angle ζ_{Lab} is calculated in the laboratory system. All values are printed out. If ϑ is given in the center-of-mass system, the subroutine is bypassed.

C. The Matrices $\xi_{rs}^{(\text{symm})}$ and $\psi_{rs}^{(\text{symm})}$

From the energy spectrum, the matrix of the symmetrized ξ values is computed. Expressed in terms of energies rather than velocities, the numerical formula for Eq. (30) is:

$$\xi_{rs}^{(\text{symm})} = \frac{Z_1 Z_2 A_1^{1/2}}{6.325}\{[\mathcal{E}_p - (1 + A_1/A_2)E_r]^{-1/2} - [\mathcal{E}_p - (1 + A_1/A_2)E_s]^{-1/2}\}. \tag{60}$$

If the reduced matrix element M_{rs} is zero, the ξ for this transition is uninteresting, and the nominal value 100 is assigned to ξ. The ξ_{rs} matrix can be printed out.

The dimensionless symmetrized $\psi_{rs}^{(\text{symm})}$ matrix is calculated according to Eq. (37) for $\lambda = 2$ by the numerical expression

$$\psi_{rs}^{(\text{symm})} = \frac{14.36 A_1^{1/2}}{(1 + A_1/A_2)^2 Z_1 Z_2{}^2} \{[\mathcal{E}_p - (1 + A_1/A_2)E_r] \times [\mathcal{E}_p - (1 + A_1/A_2)E_s]\}^{3/4} M_{rs}^{(2)}. \tag{61}$$

The $\psi_{rs}^{(\text{symm})}$ matrix can be printed out.

D. Range and Step Width of the Integration

The integration of the differential equations (16) starts with the initial conditions (nucleus in ground state) at $w = -U_p$ and extends to $w = +U_p$. The range U_p and the initial step width $2\Delta w$ of the integration depend in the following way on the accuracy a_c with which the integration has to be performed:

$$U_p = \ln (2a_c^{-1/2}\epsilon^{-1}) \tag{62}$$

where $\epsilon = 1/\sin(\vartheta/2)$ and

$$\Delta w = 40 a_c^{1/5}(10 + 48\xi_m - 16\xi_m\epsilon) \tag{63}$$

with

$$\xi_m = \operatorname*{Max}_{r,s}\{\xi_{rs}\};\ \xi_m < \xi_{\max} \tag{64}$$

Since one level of high energy might lead to a very small step width although the excitation of this state through the channel of large ξ might be unimportant, we have introduced a special input $\xi_{\max}$ which determines the largest ξ_m to be considered in Eq. (63). To avoid taking the nominal value 100 for ξ seriously (see V.C), one must choose $\xi_{\max} < 100$. Due to the accuracy control (see V.K) the step width may be changed during the integration. The initial step width $2\Delta w$, and the estimated number of steps, $2U_p/2\Delta w$, are printed out.

E. Catalog of Magnetic Substates

A catalog is established which is a numbering (by positive integers s) of all nuclear magnetic substates to be considered. The catalog gives a connection between the index s and the numbers N_s, I_s, and M_s, where N defines the sequence of nuclear states. The levels appear in the same sequence as in the input, i.e., in order of increasing energy. The magnetic substates within each of the levels appear in the sequence of increasing magnetic quantum number. The computer can only accommodate 90 substates. If $s_{\max} = \text{Max}\ (s)$ exceeds 90 an error return is released. In any case $s_{\max}$ is printed on the output.

For excited states with spins much higher than the ground state spin, the numerically large magnetic quantum numbers are uninteresting especially for backward scattering angles (compare Ref. 1). To reduce the number of differential equations in such situations, we have introduced an input control $M_{\max}$ which is the largest (numerical) value of M which will be included in the computation. The catalog may therefore have the form

$$
\begin{array}{rcccccccccccl}
s = & 1 & 2 & 3 & 4 & 5 & 6 & 7 & 8 & 9 & 10 & 11 & \ldots \\
N = & 1 & 1 & 2 & 2 & 2 & 2 & 3 & 3 & 3 & 3 & 4 & \ldots \\
I_N = & \frac{1}{2} & \frac{1}{2} & \frac{3}{2} & \frac{3}{2} & \frac{3}{2} & \frac{3}{2} & \frac{5}{2} & \frac{5}{2} & \frac{5}{2} & \frac{5}{2} & \frac{7}{2} & \ldots \\
M_N = & -\frac{1}{2} & \frac{1}{2} & -\frac{3}{2} & -\frac{1}{2} & \frac{1}{2} & \frac{3}{2} & -\frac{3}{2} & -\frac{1}{2} & \frac{1}{2} & \frac{3}{2} & -\frac{3}{2} & \ldots
\end{array}
$$

for $M_{\max} = \frac{3}{2}$. Note that $M_{\max}$ must always be chosen larger than or equal to the ground state spin I_1.

F. The Matrix $\zeta_{rs}^{(\text{symm})}$

The $\zeta_{rs}^{(\text{symm})}$ matrix is dimensionless, real, and independent of w. Its maximum dimension is 90×90. It is stored in the memory during the integration and it is not printed out. A subroutine for the three-j symbol is used to compute $\zeta_{rs}^{(\text{symm})}$ according to Eq. (17) for $\lambda = 2$ as

$$
\zeta_{rs}^{(\text{symm})} = (5)^{1/2} \cdot (-1)^{I_s - M_s} \begin{pmatrix} I_s & 2 & I_r \\ -M_s & \mu & M_r \end{pmatrix} \psi_{rs}^{(\text{symm})}. \qquad (65)
$$

G. Initial Conditions for Integration

The preparations for the solution of the differential equations are now finished. The general scheme for the integration is the following:

For $w = -U_p$ the values of all amplitudes are known to be:

$$a_r = \delta_{r1} \text{ if initial } M_1 = -I_1$$
$$a_r = \delta_{r2} \text{ if initial } M_1 = -I_1 + 1$$
$$\cdot$$
$$\cdot$$
$$\cdot$$
$$a_r = \delta_{r2I_1+1} \text{ if initial } M_1 = +I_1.$$

However, one need not consider cases with $M_1 > 0$. The symmetry relation (25) can be applied to give the amplitudes for positive M_1 from the knowledge of the amplitudes where the initial M_1 is negative.

We thus have to consider only the following initial conditions:

$$a_{rl}(-U_p) = \delta_{rl} \qquad \text{where} \qquad l = 1, 2, \ldots l_{\max} \tag{66}$$

with

$$l_{\max} = \begin{cases} I_1 + 1 \text{ for integer } I_1 \\ I_1 + \frac{1}{2} \text{ for half integer } I_1. \end{cases} \tag{67}$$

The differential equations are solved in parallel, i.e., simultaneously for all $l_{\max}$ initial polarizations. The computer can accommodate up to 4 such initial conditions, i.e., $l_{\max} \leq 4$ or $I_1 \leq \frac{7}{2}$.

H. Integration Routine

A combination of a Runge–Kutta–Gill[5] and an Adams–Moulton[6] routine is used. First one generates, by means of three cycles of the Runge–Kutta–Gill routine, the values of the amplitudes and their derivatives in four consecutive points $-U_p$, $-U_p + 2\Delta w$, $-U_p + 4\Delta w$, and $-U_p + 6\Delta w$. These values then form the basis for an Adams–Moulton routine which generates $a(w + 8\Delta w)$ from the knowledge of the four preceding derivatives $\dot{a}(w)$, $\dot{a}(w + 2\Delta w)$, $\dot{a}(w + 4\Delta w)$, and $\dot{a}(w + 6\Delta w)$, and the amplitude $a(w + 6\Delta w)$. After each step of the integration, the computer estimates the truncation error (see Sec. V.K) and adjusts the step width to maintain a specified accuracy. The computation of $\dot{a}_r$ by Eq. (16) from a knowledge of w and $a_s(w)$ is incorporated in a subroutine within which another subroutine is used to compute the quantities $Q_{2\mu}(w, \xi_{rs})$ according to Eq. (20).

I. The Runge–Kutta–Gill Integration Procedure

The initial conditions (66) for a_{rl} are stored and the quantities $\dot{a}_{rl}(-U_p, a_{rl})$ are computed. For the use of the Adams–Moulton routine these numbers are also stored under the name $f_{rl}(1)$.

From the knowledge of $a_{rl}(-U_p)$, the values of $a_{rl}(-U_p + 2\Delta w)$ are computed

[5] S. Gill, *Proc. Cambridge Phil. Soc.* **46,** 96 (1950).

[6] See e.g., L. Collaz, "The Numerical Treatment of Differential Equations," 3rd ed., p. 82. Springer-Verlag, Berlin, 1960.

by means of the Runge–Kutta–Gill integration procedure making continuous use of the $Q_{2\mu}$ and $\dot{a}_{rl}$ subroutines. The Runge–Kutta–Gill equations[5] are

$$\left\{\begin{aligned} Q_{1r} &= \Delta w \times \dot{a}_r(w, a_s(w)) \\ a_r^{(1)} &= a_r(w) + Q_{1r} \end{aligned}\right. \tag{68}$$

$$\left\{\begin{aligned} K_{1r} &= \Delta w \times \dot{a}_r[w + \Delta w, a_s^{(1)}] \\ a_r^{(2)} &= a_r^{(1)} + b_1(K_{1r} - Q_{1r}) \\ Q_{2r} &= b_1 K_{1r} + c_1 Q_{1r} \end{aligned}\right.$$

$$\left\{\begin{aligned} K_{2r} &= \Delta w \times \dot{a}_r[w + \Delta w, a_s^{(2)}] \\ a_r^{(3)} &= a_r^{(2)} + b_2(K_{2r} - Q_{2r}) \\ Q_{3r} &= b_2 K_{2r} + c_2 Q_{2r} \end{aligned}\right.$$

$$\left\{\begin{aligned} K_{3r} &= \Delta w \times \dot{a}_r[w + 2\Delta w, a_s^{(3)}] \\ a_r(w + 2\Delta w) &= a_r^{(3)} + \tfrac{1}{3}K_{3r} - \tfrac{2}{3}Q_{3r} \end{aligned}\right.$$

with

$$\begin{aligned} b_1 &= 2 - \sqrt{2} \qquad & c_1 &= -2 + \tfrac{3}{2}\sqrt{2} \\ b_2 &= 2 + \sqrt{2} \qquad & c_2 &= -2 - \tfrac{3}{2}\sqrt{2}. \end{aligned} \tag{69}$$

The derivatives $\dot{a}_{rl}(-U_p + 2\Delta w)$ are stored under the name $f_{rl}(2)$, and the routine is repeated three times making available also

$$\begin{aligned} f_{rl}(3) &= \dot{a}_{rl}(-U_p + 4\Delta w) \\ f_{rl}(4) &= \dot{a}_{rl}(-U_p + 6\Delta w). \end{aligned} \tag{70}$$

J. The Adams–Moulton Routine

The computation now proceeds by means of the faster Adams–Moulton predictor-corrector formulas[6] in which, by a two-stage iteration, the value of $a_{rl}(w + 8\Delta w)$ is obtained from the amplitude $\dot{a}_{rl}(w + 6\Delta w)$, and the derivatives $\dot{a}_{rl}(w + 6\Delta w) = f_{rl}(4)$, $\dot{a}_{rl}(w + 4\Delta w) = f_{rl}(3)$, $\dot{a}_{rl}(w + 2\Delta w) = f_{rl}(2)$, and $\dot{a}_{rl}(w) = f_{rl}(1)$ by the equations:

$$a_{rl}^{(p)} = a_{rl}(w + 6\Delta w) + (\Delta w/12)[55f(4) - 59f(3) + 37f(2) - 9f(1)] \tag{71}$$

$$\left.\begin{aligned} a_{rl}(w + 8\Delta w) = a_{rl}(w + 6\Delta w) + (\Delta w/12)[9a_{rl}(w + 8\Delta w)a_{rl}^{(p)} \\ + 19f(4) - 5f(3) + f(2)] \end{aligned}\right\} \tag{72}$$

K. Accuracy Control

At the end of each step of the Adams–Moulton routine the values of $a_{rl}^{(p)}$ and a_{rl} are compared. As an estimate of the truncation error one uses the quantity

$$F = \tfrac{1}{14} \operatorname*{Max}_{rl} |a_{rl}^{(p)} - a_{rl}|. \tag{73}$$

If this number exceeds the required accuracy a_c, the step width is reduced by a factor of two, and if F is less than $a_c/50$, the step width is increased by a factor of 2. In either case a return is made to the Runge–Kutta–Gill routine for computing new initial values for the Adams–Moulton routine. If F lies between a_c and $a_c/50$, the integration proceeds without changing the step width. The integration routines are repeated until w exceeds U_p.

L. Excitation Probabilities

The Adams–Moulton procedure is repeated i times, i being an integer specified in the input. At this stage one may obtain a print out of the current value of the excitation probabilities. These are evaluated according to expression (28), making use of the symmetry relation (25), and one obtains

$$P(N) = \frac{2}{2I_1+1} \sum_{M_1<0<M_N} |a_{I_N M_N}(M_1)|^2 + \frac{1}{2I_1+1} \sum_{M_N} |a_{I_N M_N}(0)|^2. \qquad (74)$$

The second of the two terms in Eq. (74) occurs only if I_1 is integer.

As a check of the accuracy of the integration, the total probability, $P_{\text{tot}} = \Sigma P(N)$, is compared to unity. If $|P_{\text{tot}} - 1| > 20a_c$, an error message is given and the integration is discontinued.

Since it may be quite difficult to estimate the time it takes to perform the complete integration, the computer makes a time estimate based on the time required to make the first i steps of the integration. The time estimate is then compared to an upper time limit which is given in the input. If it exceeds the limit the problem is discontinued.

At the end of the integration, the final amplitudes for each initial polarization may be printed out.

M. Cross Sections

For a ready comparison with experiments, the cross sections are computed from the excitation probabilities according to Eq. (41). The numerical expression for the cross sections in the center-of-mass system for $\lambda = 2$ is:

$$\frac{d\sigma(N)}{d\Omega_{\text{CM}}} = 0.001296 \left[\frac{\mathcal{E}_p}{\mathcal{E}_p - (1 + A_1/A_2)E_N}\right]^{\frac{1}{2}} \times \left\{(1 + A_1/A_2)\frac{Z_1 Z_2}{\mathcal{E}_p}\right\}^2 P(N) \sin^{-4}(\vartheta/2). \qquad (75)$$

If, at the input, the scattering angle is given in the laboratory system, the laboratory cross sections are calculated by the formula

$$\frac{d\sigma_{\text{Lab}}(N)}{d\Omega_{\text{Lab}}} = \frac{d\sigma(N)}{d\Omega_{\text{CM}}} \cdot \frac{d\Omega_{\text{CM}}(N)}{d\Omega_{\text{Lab}}} \qquad (76)$$

The excitation probabilities, the cross sections, and the solid angle ratios are printed out.

N. Angular Distribution Tensors

At the end of the main program, the quantities $\alpha_{k\kappa}^{(i)}(N)$ are calculated in any of the four coordinate systems shown in Fig. 2. The coordinate system may be chosen by means of special output controls. They are calculated for $\kappa \geq 0$ according to relations (49) to (56). The summation over the initial magnetic substates M_1 in (49) is split up into three parts

$$\sum_{-I_1 \leq M_1 \leq I_1} = \sum_{-I_1 \leq M_1 \leq 0} + \sum_{0 > M_1 \geq I_1} + \sum_{M_1 = 0} . \tag{77}$$

The first part can be calculated directly, whereas the computation of the second part is accomplished by the use of the symmetry relation (25). The last term occurs only for integer values of the ground state spin.

The summation over κ' in Eqs. (55) and (56) for negative values of κ' is accomplished by the use of the symmetry relation (53).

The calculation of the function $d_{\kappa\kappa'}^{k}(\beta)$ is incorporated in a subroutine.

O. Subroutines and Functions

In the following a list is given of the subroutines and the function routines used in the program.

(1) *Electric quadrupole matrix elements.* In order to make the program more easily adaptable to nuclear model calculations, the read-in of the E2 matrix M_{rs} from data cards or its computation from nuclear model parameters is incorporated in a subroutine rather than in the main program. This makes it possible to use the same compilation of the main program with different ways of generating the E2 matrix.

(2) *Factorial-function.* Factorials are used in the subroutines for the 3-j symbol and for the d function. The calculation of these functions may require the values of factorials of large numbers in cases where the spins of the excited states are large. Since, for $n \geq 34$, $n!$ is larger than the largest floating-point number that can be accommodated in the computer (10^{38}), the quantity $f_{10}(n) = n!/10^n$ is computed in this factorial-function rather than the factorial itself. The value of $f_{10}(n)$ is smaller than 10^{38} for $n \leq 79$, and an error message is given if the argument of $f_{10}(n)$ exceeds 79. The powers of 10 that arise in the calculations of the d-function and of the 3-j symbol due to the replacement of $n!$ by $f_{10}(n)$, cancel (except for a factor of 10 in the 3-j symbol).

(3) *Three-j symbol.* The formula quoted in Ref. 7 is used in conjunction with the $f_{10}(n)$ function. Error messages are given for inconsistent arguments.

(4) *Laboratory and center-of-mass system.* For the transition from the laboratory to the center-of-mass system, the unrelativistic formulas compiled in Ref. 8 are

[7] M. Rotenberg, R. Bivins, N. Metropolis, and J. K. Wooten, "The 3-j and 6-j Symbols." The Technology Press, Mass. Inst. Technol., Cambridge, Massachusetts, 1959.

[8] H. E. Gove, "Nuclear Reactions" (P. M. Endt and M. Demeur, eds.). North-Holland Publ. Co., Amsterdam, and Wiley (Interscience), New York, 1959.

used in this subroutine. Error messages are given if the scattering angle or the excitation energy are kinematically impossible.

(5) *Subroutines for* $Q_{2\mu}(w,\xi)$. The real and imaginary parts of Q_{20}, Q_{21}, and Q_{22} are calculated according to Eq. (20) as a function of w and ξ. The output of the subroutine is a $6 \times N_{\max} \times N_{\max}$ matrix, where $N_{\max}$ is the number of nuclear levels (≤ 10). The quantity $Q(w,\xi)$ is set equal to zero for the uninteresting transitions in which ξ exceeds $\xi_{\max}$. It thus always disregards the cases where ξ has the nominal value 100 (see V.C).

(6) *Subroutine for* $\dot{a}_r(w)$. The real and imaginary parts of $\dot{a}_r(w)$ are computed by Eq. (16) from w and $a_s(w)$. The subroutine uses the ζ_{rs} matrix, the $Q_{2\mu}$ subroutine and the values of all amplitudes $a_{s,l}$. The complex amplitudes $a_{s,l}$ are stored as a two-dimensional array of size $2s_{\max} \times l_{\max}$ in which the real parts of $a_{s,l}$ are found among the first $s_{\max}$ numbers for each l while the imaginary parts are found in the same sequence at the locations $s_{\max} + 1$ to $2s_{\max}$. The derivatives $\dot{a}_{rl}$ are computed in parallel for all values of l, and the results are stored in a $(2s_{\max} \times l_{\max})$ two-dimensional array. The routine neglects terms in the sum (16) for which ζ_{rs} is smaller than the prescribed accuracy a_c, specified at the input.

(7) *The d function.* The value of $d^j_{mm'}(\beta)$ is computed as a function of the arguments (β, j, m, m') according to Eq. (52). Error messages are given for inconsistent arguments.

VI. User's Manual

This section describes the details of the administration of the program. This includes a description of the output and performance controls and of the form in which the input data should be entered into the computer. The input data are discussed in the sequence in which they appear on the data cards of the program. A FORTRAN listing of the complete program as well as an output example, are given in Chapt. VII.

A. Performance Controls

Data Card #1

INTERV = i (c.f. Sec. V.L) is an integer specifying the number of integration steps taken between print-outs of the probabilities $P(N)$ as a function of w. It also gives the number of integration steps between the P_{tot} controls.

NTIME is a fixed-point number indicating the maximum time, in minutes, allowed for the execution of the computation. During the beginning of the calculation, the computer estimates the total time needed for the integration. If this estimate exceeds NTIME, the execution is terminated. If a real-time routine is not available on the computer, the cards listed under "TIME ESTIMATE" should be removed.

NCM = N_{CM} (c.f. Sec. V.B) is an integer specifying the index of the level, the energy of which determines the kinematics of the orbit used in the integration. This input is only required in the case where the scattering angle is given in the laboratory system. (c.f. input data card #5).

ACCUR = a_c (c.f. Sec. V.K) is a floating-point number specifying the absolute accuracy to which the final probabilities should be computed.

Data Card #2

XIMAX = $\xi_{\max}$ (c.f. Sec. V.D) is a floating-point number which specifies the largest value of ξ which is considered in the computation. A value of $\xi_{\max} < 100$ must be entered.
EMMAX = $M_{\max}$ (c.f. Sec. V.E) is a floating-point number indicating the numerical value of the largest magnetic quantum number considred and has to be $\leq I_1$.
NMAX is an integer specifying the number of nuclear states. Its maximum value is 10.

B. Output Controls

All output controls are fixed-point numbers taking the values 1 or 0 according to whether the output in question is desired or not.

Data Card #3

OUXI controls the print-out of the ξ-matrix.
OUPSI controls the print-out of the ψ-matrix.
OUAMP controls the print-out of all excitation amplitudes for each initial polarization of the ground state.
OUPROW controls the print-out of the excitation probabilities during the integration after every *i*th step.

Data Card #4

OUANG 0
OUANG 1
OUANG 2
OUANG 3

control the print-out of the angular distribution coefficients $\alpha^{(0)}$, $\alpha^{(1)}$, $\alpha^{(2)}$, $\alpha^{(3)}$, respectively (see Fig. 3).

C. Input Data

Data Card #5

Z1 = Z_1 is a floating-point number specifying the charge number of the projectile.
A1 = A_1 is a floating-point number specifying the mass of the projectile in a.m.u.
Z2 = Z_2 is a floating-point number specifying the charge number of the target.
A2 = A_2 is a floating-point number specifying the mass of the target in a.m.u.
EP = $\mathcal{E}_p$ is a floating-point number specifying the laboratory energy of the projectile in MeV.

TLBDG = ϑ_{Lab} (c.f. Sec. V.B) is a floating-point number specifying the deflection angle (in degrees) in the laboratory system. If ϑ is given in the center-of-mass system, TLBDG must be entered as 0.

THETA = ϑ (c.f. Sec. I, Eq. 7) is a floating-point number specifying the deflection angle (in degrees) in the center-of-mass system. If the deflection angle is given in the laboratory system, THETA must be entered as 0.

Data Cards #6 to 6 + NMAX

Each card contains the information about the energy and spin of the Nth nuclear state. The levels should follow each other in the order of increasing excitation energy. The number of cards must equal NMAX.

EN(N) = E_N is a floating-point number specifying the excitation energy of the Nth nuclear state in MeV.

SPIN(N) = I_N is a floating-point number specifying the spin quantum number of the Nth nuclear state.

Data Cards #7 + NMAX to 7 + 2 × NMAX

These data cards are read in a subroutine. They contain the E2 matrix elements M_{rs} (c.f. Eq. 13 and Sec. V.O), given in units of $e.10^{-24}$ cm^2. Each card contains one row of the matrix whose size is NMAX × NMAX.

VII. Fortran Listing and Output Example

On the following pages, the FORTRAN-listing of the entire program and of all subroutines and functions is reproduced. The statements are arranged in a sequence which corresponds closely to the one followed in the description of the program in Chapt. V. Flow diagrams for the more important parts of the program are given in Chapt. VIII.

In order for the user to be able to check the proper operation of his program, an example of a computer output is given at the end of this chapter. It deals with the Coulomb excitation of the lowest rotational levels in Sm^{152} by O^{16} ions, backscattered through 160°. The Coulomb excitation of this particular nucleus has recently been treated by the method of numerical integration by U. Smilansky and G. Goldring[9] and by I. Berson.[10]

[9] U. Smilansky, "Calculation of Coulomb Excitation Probabilities, Part I, Backscattering $\vartheta = 180°$," Internal Rept., Weizmann Inst. of Sci., Rehovoth, Israel (May 1965); G. Goldring and U. Smilansky, "The Quadrupole Moment of the 2^+ state at 122 keV in Sm^{152}," preprint.

[10] I. Berson, *Soviet J. Nucl. Phys.* **1,** 325 (1965).

```
                                   FORTRAN SOURCE LIST
      SOURCE STATEMENT

$IBFTC JDBCXP  NODECK
C
C     DEBOER-WINTHER MULTIPLE COULOMB EXCITATION PROGRAM
C
      COMMON COT1,COT2
      COMMON/Y/ACC10
      COMMON /XX/NMAX,EN(10),SPIN(10),MATRIX(10,10),FMT4(6),XNUM(10)
      REAL MATRIX
      INTEGER OUANGO,OUANG1,OUANG2,OUANG3
      INTEGER OUXI,OUPSI,OUAMP,OUPROW
      DIMENSION XI(10,10),PSI(10,10),QU(10,10,6),P(10)
      DIMENSION Q1(180,4),PROB(90),AMPP(180,4),F(180,4,4)
      DIMENSION CAT(90,3),ZETA(90,90),AMPDOT(180,4),AMP(180,4)
      DIMENSION ANGO(10,6,5), ANG1(10,6,5), ANG2(10,6,5), ANG3(10,6,5)
      DIMENSION DSIG(10),DSIGLB(10),TCMDG(10),ZLBDG(10)
      DIMENSION R3(10),R4(10),EPP(10)
      DIMENSION FMT(7),FMT1(9),FMT2(7),FMT3(6)
      DIMENSION FT(17),FT1(5),FT2(6),FT3(8),FT4(13),FT5(5)
      DATA XNUM/1H1,1H2,1H3,1H4,1H5,1H6,1H7,1H8,1H9,2H10/
      DATA FMT/41H(1H1,6X1HW2X       (6X2HP(I2,1H))6X4HPTOT)/
      DATA FMT1/54H(34HOENERGY SPECTRUM      LEVEL INDEX N,          (I 9)
     1)/
      DATA FMT2/37H(1HO,20X13HENERGY IN MEV      (F9.4))/
      DATA FMT3/32H(1HO,20X4HSPIN,9X       (F9.4)/)/
      DATA FMT4/33H(1HO,10X,2HM=           (I2,10X))/
      DATA FT/98H(28HOEXECUTION TERMINATED AT W= F10.6,6HPTOT= F14.
     1      ,33H ERROR IN PTOT EXCEEDS 20 * ACCUR)/
      DATA FT1/25H(1XF10.3,11F11.        )/
      DATA FT2/32H(1XF5.1,F12.1,1X2(1XF20.       ))/
      DATA FT3/43H(5XI3,11XF15.          ,8XE15.          )/
      DATA FT4/73H(2XI3,13XF7.2,5XF15.          ,8XE15.          , 8XF8
     1.3,12XF7.2,12XF9.5)/
      DATA FT5/25H(1H ,3I7,2F20.         )/
C
C     READ IN STARTS
C
  500 READ(5,106) INTERV,NTIME,NCM,ACCUR
  106 FORMAT(3I5,F10.7)
C     INTERV - PRINT OUT OF P(N) WILL OCCUR EVERY INTERV-TH STEP
C     NTIME IS THE TIME IN MIN ALLOWED FOR THE JOB
C     NCM PICKS LEVEL FOR WHOSE CM SCATTERING ANGLE OBBIT IS CALCULATED
C     ACCUR IS ACCURACY WITH WHICH INTEGRATION IS PERFORMED
C
      IF(INTERV) 46,46,47
   46 CALL EXIT
C     INTERV .LE. ZERO ON LAST DATA CARD INDICATES NO MORE DATA
   47 READ(5,102)XIMAX,EMMAX,NMAX
  102 FORMAT(2F10.0,I5 )
C     XIMAX IS LARGEST XI CONSIDERED
C     EMMAX IS LARGEST MAGNETIC QUANTUM NUMBER CONSIDERED
C     NMAX IS NUMBER OF NUCLEAR LEVELS CONSIDERED
      IF(NMAX.GT.10) GO TO 816
      READ(5,104) OUXI,OUPSI,OUAMP,OUPROW
      READ(5,104)OUANGO,OUANG1,OUANG2,OUANG3
  104 FORMAT(4I5)
```

```
C      IF THE OUTS ARE ENTERED AS ZERO NO OUTPUT WILL OCCUR
C
       READ(5,100) Z1,A1,Z2,A2,EP,TLBDG,THETA
C      ONLY CNE OF THE TWO ANGLES MAY BE .NE. 0
   100 FORMAT(7F10.2)
   101 READ(5,105)(EN(N),SPIN(N),N=1,NMAX)
   105 FORMAT(2F10.0)
C      PRINT OUT OF INPUT DATA
       WRITE(6,430)
   430 FORMAT(54H1DE BOER - WINTHER MULTIPLE COULOMB EXCITATION PROGRAM/)
       WRITE(6,431) Z1,A1,EP
   431 FORMAT(32H0PROJECTILE CHARGE NUMBER  Z1 = F6.2,14H,  MASS  A1 =
      1F8.3,23HAMU,  LAB ENERGY  EP = F8.3,3HMEV)
       IF(TLBDG)432,433,432
   433 WRITE(6,434) THETA
   434 FORMAT(52H0SCATTERING ANGLE IN CENTER OF MASS SYSTEM  THETA = F7.2
      1,8H DEGREES)
       GO TO 435
   432 WRITE(6,436) TLBDG
   436 FORMAT(48H0SCATTERING ANGLE IN LABORATORY SYSTEM  TLBDG = F7.2,
      1 8H DEGREES)
   435 WRITE(6,437) Z2,A2
   437 FORMAT(28H0TARGET CHARGE NUMBER  Z2 = F6.2,14H,  MASS  A2 = F8.3,
      13HAMU/)
       FMT1(8) = XNUM(NMAX)
       WRITE(6,FMT1)(N,N=1,NMAX)
       FMT2(5) = XNUM(NMAX)
       WRITE(6,FMT2)(EN(N),N=1,NMAX)
       FMT3(4) = XNUM(NMAX)
       WRITE(6,FMT3)(SPIN(N),N=1,NMAX )
C      THE QUADRUPOLE MATRIX ELEMENTS ARE PRODUCED IN THE MAT SUBROUTINE
       CALL MAT
C
C      DECIMAL PLACES FOR PRINTOUT
       IJ=0
       AC=ACCUR
    30 AC=AC*10.0
       IJ=IJ+1
       IF(1.C-AC) 31,30,30
    31 CONTINUE
C      IJ HOLDS NO. OF DECIMAL PLACES WANTED IN PRINT OUT OF RESULTS
C      ACCORCING TO ACCURACY OF INTEGRATION
C
C      TRANSITION FROM LAB TO CM COORDINATE SYSTEM
C
       IF(TLBDG.EQ.0.0.AND.THETA.EQ.0.0) GO TO 810
       IF(TLBDG.EQ.0.0.AND.THETA.NE.0.0) GO TO 451
       IF(TLBDG.NE.0.0.AND.THETA.NE.0.0) GO TO 103
       CALL CMLAB(A1,A2,A1,A2,EP,EN,NMAX,TLBDG,TCMDG,ZLBDG,R3,R4,EPP)
       THETA=TCMDG(NCM)
   451 CONTINUE
C      COMPUTATION OF XI MATRIX
C
       CXI = Z1*Z2*SQRT(A1)/6.325
       DO 3 M=1,NMAX
```

```
      DO 3 N=1,NMAX
      XI(N,M) = 100.0
      IF(MATRIX(N,M))4,3,4
    4 RX1 = 1.0/SQRT(EP-(1.+(A1/A2))*EN(N))
      RX2 = 1.0/SQRT(EP-(1.+(A1/A2))*EN(M))
      XI(N,M) = CXI*(RX1-RX2)
    3 CONTINUE
      IF(OUXI)5,452,5
    5 WRITE(6,203)
  203 FORMAT(10HOXI MATRIX)
      DO 7 N = 1,NMAX
    7 WRITE(6,207) N,(XI(N,M),M=1,NMAX)
  207 FORMAT(4HON =I2,10F12.4)
C
C     DETERMINATION OF THE LARGEST XI VALUE IN XI MATRIX
  452 XIM = 0.0
      DO 25 M = 1,NMAX
      DO 25 N = 1,NMAX
      IF(XI(N,M).GT.XIMAX)GO TO 25
      IF(XI(N,M).LE.XIM)GO TO 25
      XIM = XI(N,M)
   25 CONTINUE
      WRITE(6,300)XIM
  300 FORMAT(7HOXIM = F10.4)
C
C     COMPUTATION OF PSI MATRIX
C
      CPSI = 14.36*SQRT(A1)/((1.+(A1/A2))**2 * Z1*Z2*Z2)
      DO 8 M=1,NMAX
      DO 8 N = 1,NMAX
      PP1 = (EP- (1.+(A1/A2))*EN(N))**.75
      PP2 = (EP- (1.+(A1/A2))*EN(M))**.75
    8 PSI(N,M)=CPSI*PP1*PP2*MATRIX(N,M)
      IF(OUPSI)10,11,10
   10 WRITE(6,204)
  204 FORMAT(11HOPSI MATRIX)
      DO 12 N = 1,NMAX
   12 WRITE(6,207) N,(PSI(N,M),M=1,NMAX)
   11 CONTINUE
C
      WRITE(6,442)
  442 FORMAT(84HOPERFORMANCE CONTROLS  NMAX,    INTERV,    NCM,    EMMAX,
     1 XIMAX,      ACCUR,      NTIME)
      WRITE(6,443) NMAX,INTERV,NCM,EMMAX,XIMAX,ACCUR,NTIME
  443 FORMAT(1HO,19X,I6,2X,2I8,2F9.2,F13.7,I7)
      EMCCK=EMMAX-SPIN(1)+0.001
      IF(EMCCK.LT.0.0) GO TO 814
      WRITE(6,444)
  444 FORMAT(81HOOUTPUT CONTROLS         OUXI, OUPSI, OUAMP, OUPROW, OUANG
     10, OUANG1, OUANG2, OUANG3)
      WRITE(6,445) OUXI,OUPSI,OUAMP,OUPROW,OUANGO,OUANG1,OUANG2,OUANG3
  445 FORMAT(1HC,17X3I7,5I8)
C
C     RANGE AND STEP WIDTH OF THE INTEGRATION
C
```

```
                                        FORTRAN SOURCE LIST JDBCXP
       SOURCE STATEMENT

       WRITE(6,781) THETA
   781 FORMAT(55HOCM SCATTERING ANGLE USED FOR INTEGRATION IS TCMDG(NCM)
      13H = F7.2,8H DEGREES)
       TRAD=THETA/57.295779
       STR = SIN(TRAD/2.0)
       EPS = 1.0/STR
       UP=ALOG(1.0/(EPS*SQRT(ACCUR)))
       COT1 = COS(TRAD/2.0)*EPS
       COT2 = COT1*COT1
       ACC10=ACCUR/10.0
       DW=40.0*(ACCUR**0.2)/(10.0+48.0*XIM+16.0*XIM*EPS)
       IF((5.0*DW).GT.UP) DW=0.4*UP
       ISTEP=UP/DW
       WRITE(6,202)EPS,UP,ISTEP
   202 FORMAT(7HOEPS = F7.3,/
      128HORANGE OF INTEGRATION, UP = F6.2 /
      236HOESTIMATED NUMBER OF STEPS, ISTEP = I4)
       D2W = DW + DW
       WRITE(6,251)D2W
   251 FORMAT(27HOINITIAL STEP WIDTH, D2W = F8.5)
C
C      CATALOGUE OF MAGNETIC SUBSTATES
C
       IS = 1
       DO 18 N= 1,NMAX
       QUAN = SPIN(N)
       IF(QUAN.GT.EMMAX)QUAN = EMMAX
       MSTOP = 2.0*QUAN + 1.0
       QUAN = -QUAN
       DO 15 I = 1,MSTOP
       CAT(IS,1) = N
       CAT(IS,2)= SPIN(N)
       CAT(IS,3) = QUAN
       QUAN = QUAN +1.0
       IS = IS + 1
    15 CONTINUE
    18 CONTINUE
       ISMAX = IS -1
       WRITE(6,250)ISMAX
   250 FORMAT(45HOTOTAL NUMBER OF MAGNETIC SUBSTATES, ISMAX = I3)
       IF(ISMAX-90)254,254,252
   252 WRITE(6,253)
   253 FORMAT(24HO ERROR ISMAX EXCEEDS 90)
       GO TO 500
   254 CONTINUE
       IF(SPIN(1).GT.3.6) GO TO 812
C
C      COMPUTATION OF ZETA MATRIX
C
       AA2 =2.0
       DO 29 IS = 1,ISMAX
       IEX = CAT(IS,2)-CAT(IS,3) + 0.0001
       PHZ = (-1.0)**IEX
       JP = CAT(IS,1) + 0.001
       B1 = -CAT(IS,3)
```

```
                                         FORTRAN SOURCE LIST JDBCXP
      SOURCE STATEMENT

      AA1 = CAT(IS,2)
      DO 20 IR = 1,ISMAX
      B2 = CAT(IS,3)-CAT(IR,3)
      IF(ABS(B2).GT.2.001)GO TO 815
      A3 = CAT(IR,2)
      B3 = CAT(IR,3)
      IP = CAT(IR,1) + 0.001
      ZETA(IR,IS)=PHZ*PSI(IP,JP)*2.236068*THREEJ(AA1,B1,AA2,B2,A3,B3)
      GO TO 20
  815 ZETA(IR,IS)=0.0
   20 CONTINUE
   29 CONTINUE
C
C     INTEGRATION OF THE DIFFERENTIAL EQUATIONS STARTS HERE
C
C     INITIAL CONDITIONS FOR INTEGRATION
C
      ISTEPS=0
C     ISTEPS COUNTS ACTUAL NUMBER OF STEPS
      KAST=C
      LMAX =  SPIN(1) + 1.001
      IZMAX = ISMAX+ISMAX
      W = -UP
      DO 80 L = 1,LMAX
      DO 81 IZR = 1,IZMAX
      AMP(IZR,L)=1.0 E-30
      IF(L.EQ.IZR)AMP(IZR,L)=1.0
   81 CONTINUE
   80 CONTINUE
C
C     INTEGRATION ROUTINE
C
C     CONSTANTS USED IN RUNGE-KUTTA EQUATIONS
      RB1=0.5857864
      C1=0.1213204
      RB2=3.4142136
      C2=-4.1213204
C
C     INITIAL TIMING OF INTEGRATION
C     CALL ICLOCK(TIME1)
C     ASSIGN 255 TO ITI
C     FOR COMPUTERS WITH ICLOCK-ROUTINE, REPLACE ABOVE TWO CARDS
C     OMITTING C FOR COMMENT
C
C     HEADING FOR PROW (VARIABLE FORMAT)
      IF(OUPROW.EQ.0) GO TO 399
      FMT(3)=XNUM(NMAX)
      WRITE(6,FMT)(N,N=1,NMAX)
      P(1)=1.00
      PTOT=1.C0
      DO 70 N=2,NMAX
   70 P(N)=C.00
      FT1(4)=XNUM(IJ)
      WRITE(6,FT1) W,(P(N),N=1,NMAX),PTOT
C
```

b

```
          SOURCE STATEMENT

C         THE RUNGE-KUTTA-GILL INTEGRATION PROCEDURE
      399 CONTINUE
          CALL Q(W,EPS,XIMAX,XI,NMAX,QU)
          CALL AMPDER(CAT,ZETA,AMP,QU,LMAX,ISMAX,AMPDOT)
          DO 400 L = 1,LMAX
          DO 400 IZR = 1,IZMAX
      400 F(IZR,L,1) = AMPDOT(IZR,L)
          DO 401 NAM = 2,4
          DO 90 L = 1,LMAX
          DO 91 IZR = 1, IZMAX
          Q1(IZR,L) = DW*AMPDOT(IZR,L)
          AMP(IZR,L)=AMP(IZR,L)+Q1(IZR,L)
       91 CONTINUE
       90 CONTINUE
          W = W + DW
          CALL Q(W,EPS,XIMAX,XI,NMAX,QU)
          CALL AMPDER(CAT,ZETA,AMP,QU,LMAX,ISMAX,AMPDOT)
          DO 96 L = 1,LMAX
          DO 97 IZR = 1,IZMAX
          RK1 = DW*AMPDOT(IZR,L)
          AMP(IZR,L) = AMP(IZR,L) + RB1*(RK1 - Q1(IZR,L) )
          Q1(IZR,L) = RB1*RK1 + C1*Q1(IZR,L)
       97 CONTINUE
       96 CONTINUE
          CALL AMPDER(CAT,ZETA,AMP,QU,LMAX,ISMAX,AMPDOT)
          DO 98 L = 1,LMAX
          DO 99 IZR = 1,IZMAX
          RK2 = DW*AMPDOT (IZR,L)
          AMP(IZR,L) = AMP(IZR,L) + RB2*(RK2 - Q1(IZR,L))
          Q1(IZR,L) = RB2*RK2 + C2*Q1(IZR,L)
       99 CONTINUE
       98 CONTINUE
          W = W + DW
          CALL Q(W,EPS,XIMAX,XI,NMAX,QU)
          CALL AMPDER(CAT,ZETA,AMP,QU,LMAX,ISMAX,AMPDOT)
          DO 110 L = 1,LMAX
          DO 111 IZR = 1,IZMAX
          RK3 = DW*AMPDOT(IZR,L)
          AMP(IZR,L) = AMP(IZR,L) + RK3/3.0 - 2.0*Q1(IZR,L)/3.0
      111 CONTINUE
      110 CONTINUE
          CALL AMPDER(CAT,ZETA,AMP,QU,LMAX,ISMAX,AMPDOT)
          DO 402 L = 1,LMAX
          DO 402 IZR = 1,IZMAX
      402 F(IZR,L,NAM) = AMPDOT(IZR,L)
          ISTEPS=ISTEPS+1
          KAST=KAST+1
      401 CONTINUE
C         WE NOW HAVE THE 4 STARTING VALUES OF THE DERIVATIVES AND CAN
C         PROCEED BY THE ADAMS-MOULTON METHOD
C
C         THE ACAMS-MOULTON ROUTINE
C
       95 CONTINUE
          DO 403 L = 1,LMAX
```

SOURCE STATEMENT

```
      DO 403 IZR = 1,IZMAX
      AMPP(IZR,L) = AMP(IZR,L) + DW/12. *(55.*F(IZR,L,4)
     1-59.*F(IZR,L,3) + 37.*F(IZR,L,2) - 9.*F(IZR,L,1) )
  403 CONTINUE
      W = W + DW + DW
      KAST = KAST + 1
      ISTEPS=ISTEPS+1
      CALL C(W,EPS,XIMAX,XI,NMAX,QU)
      CALL AMPDER(CAT,ZETA,AMPP,QU,LMAX,ISMAX,AMPDOT)
      DO 404  L = 1,LMAX
      DO 404 IZR = 1,IZMAX
      AMP(IZR,L) = AMP(IZR,L) + DW/12. * (9.*AMPDOT(IZR,L)
     1+ 19.C*F(IZR,L,4) - 5.*F(IZR,L,3) + F(IZR,L,2))
  404 CONTINUE
      CALL AMPDER(CAT,ZETA,AMP,QU,LMAX,ISMAX,AMPDOT)
      DO 405 L = 1,LMAX
      DO 405 IZR = 1,IZMAX
      F(IZR,L,1) = F(IZR,L,2)
      F(IZR,L,2) = F(IZR,L,3)
      F(IZR,L,3) = F(IZR,L,4)
      F(IZR,L,4) = AMPDOT(IZR,L)
  405 CONTINUE
C
C     TIMING OF INTEGRATION AFTER FIRST TEN STEPS
C     IF(ISTEPS.LT.10) GO TO 579
C     GO TO ITI(255,257)
C 255 CALL ICLOCK(TIME2)
C     TSTEP=ISTEP
C     TLAPSE=(TIME2-TIME1)/60.0
C     ESTTIM=TLAPSE*TSTEP/10.0
C     WRITE(6,256) ESTTIM
C 256 FORMAT(19HOESTIMATED TIME =  F10.4,1X3HSEC)
C     ASSIGN 257 TO ITI
C     IF(ESTTIM/60.0.LT.FLOAT(NTIME)) GO TO 257
C     WRITE(6,258)
C 258 FORMAT(45HO EXCEEDS ALLOWED TIME - EXECUTION TERMINATED)
C     GO TO 500
C 257 CONTINUE
C     FOR COMPUTERS WITH ICLOCK-ROUTINE, REPLACE ABOVE 13 CARDS
C     OMITTING C FOR COMMENT
  579 IF(W.GT.UP) GO TO 571
C
C     ACCURACY CONTROL
C
      FF=0.0
C     FIND LARGEST AMPP - AMP
      DO 573 L=1,LMAX
      DO 574 IR=1,ISMAX
      IMAG = IR + ISMAX
      FZR = AMPP(IR,L) - AMP(IR,L) +1.0 E-30
      FZI=AMPP(IMAG,L)-AMP(IMAG,L) + 1.0 E-30
      FZ=(SCRT(FZR*FZR+FZI*FZI))/14.0
      IF(FZ-FF)574,574,21
   21 FF=FZ
  574 CONTINUE
```

SOURCE STATEMENT

```
  573 CONTINUE
      ACC05C = ACCUR/ 50.0
      IF(FF.LT.ACC050) GO TO 575
      IF(FF.GT.ACCUR) GO TO 577
      IF(KAST-INTERV)95,571,571
  575 DW=2.C*DW
      D2W=DW+DW
      WRITE(6,576) W,D2W
  576 FORMAT(8HCAT W = F7.3,36H, STEP WIDTH WAS DOUBLED TO BE D2W =F8.5)
      GO TO 399
  577 DW=DW/2.0
      D2W=DW+DW
      WRITE(6,578) W,D2W
  578 FORMAT(8HCAT W = F7.3,36H, STEP WIDTH WAS HALVED TO BE D2W = F8.5)
      GO TO 399
C
C     THE EXCITATION PROBABILITIES DURING INTEGRATION
C
  571 LLMAX = 2.0*(SPIN(1) + 1.001)
      DO 52 IR = 1,ISMAX
      IMAG = IR+ISMAX
      PROB(IR)=0.0
      L = 1
      DO 54 LL =  2,LLMAX,2
C     LL IS INDEX COUNTER ONLY
      IF(LL-LLMAX)50,51,52
   50 FAC = 2.0
      GO TO 53
   51 FAC = 1.0
   53 CONTINUE
      PROB(IR)=PROB(IR)+(FAC/(2.0*SPIN(1)+1.0))*(AMP(IR,L)*AMP(IR,L)+
     1AMP(IMAG,L)*AMP(IMAG,L))
      L=L+1
   54 CONTINUE
   52 CONTINUE
      IR=1
      DO 61 N = 1,NMAX
      P(N) = 0.0
   60 P(N) = P(N) + PROB(IR)
      IR = IR + 1
      IF(IR.GT.ISMAX) GO TO 62
      ICAT = CAT(IR,1) + .01
      IF(ICAT-N) 60,60,61
   61 CONTINUE
   62 CONTINUE
      PTOT=C.0
      DO 41C N = 1,NMAX
  410 PTOT = PTOT + P(N)
      IF(W.GT.UP) GO TO 93
C
C     PTOT CHECK
      ABW = ABS(PTOT - 1.0)/20.0
      IF(ABW.LT.ACCUR) GO TO 93
      FT(10)=XNUM(IJ)
      WRITE(6,FT) W,PTOT
```

SOURCE STATEMENT

```
      GO TO 500
   93 IF(OUPROW)801,802,801
  801 WRITE(6,FT1) W,(P(N),N=1,NMAX),PTOT
  802 KAST=C
      IF(W-UP)95,92,92
   92 CONTINUE
      WRITE(6,790) ISTEPS
  790 FORMAT(34HOACTUAL NUMBER OF STEPS, ISTEPS = I4)
C
C     INTEGRATION COMPLETED
C
C     PRINT-OUT OF THE FINAL AMPLITUDES AMP(W=+UP)
  572 IF(OUAMP)350,122,350
  350 DO 120 L=1,LMAX
      WRITE(6,313) CAT(L,3)
  313 FORMAT(13H1INITIAL M = F4.1 )
      WRITE(6,314)
  314 FORMAT(1H0,1X4HSPIN5X12HMAG.QUAN.NO.6X14HREAL AMPLITUDE7X
     114HIMAG AMPLITUDE/)
      VAL = CAT(1,1)
      DO 12C IZR = 1,ISMAX
      IMAG = IZR + ISMAX
      IF(CAT(IZR,1) - VAL) 842,842,840
  840 VAL = CAT(IZR,1)
      WRITE(6,841)
  841 FORMAT(1H0)
  842 FT2(5)=XNUM(IJ)
      WRITE(6,FT2)(CAT(IZR,LC),LC=2,3),AMP(IZR,L),AMP(IMAG,L)
  120 CONTINUE
  122 CONTINUE
C
C     COMPUTATION OF THE DIFFERENTIAL CROSS-SECTIONS
C
      IF(TLBDG)261,298,261
C     HEADING FOR CM CROSS SECTIONS
  298 WRITE(6,510)
  510 FORMAT(1H1,12H LEVEL INDEX10X10HEXCITATION11X16HCM CROSS SECTION/
     122X13HPROBABILITIES9X15HBARNS/STERADIAN//7X1HN18X4HP(N)18X
     27HDSIG(N))
      GO TO 297
C     HEADING FOR LAB CROSS SECTIONS
  261 WRITE(6,471)
  471 FORMAT(1H1,6H LEVEL9X10HSCATTERING9X10HEXCITATION7X10HLAB CROSS
     17HSECTION3X19HENERGY OF SCATTERED5X10HLAB RECOIL 8X11HSOLID ANGLE/
     27H  INDEX6X16HANGLE-CM DEGREES5X13HPROBABILITIES6X13HBARNS/STERADI
     32HAN6X15HPROJECTILE, MEV5X13HANGLE DEGREES5X16HRATIO FOR RECOIL//
     44X1HN13X8HTCMDG(N)13X4HP(N)13X9HDSIGLB(N)13X6HEPP(N)13X8HZLBDG(N)
     513X5HR4(N))
  297 CDSIG1 = (1.0+A1/A2)
      CDSIG=SQRT(EP)*(CDSIG1*Z1*Z2/EP)**2
      DO 47C N = 1,NMAX
      DSIG(N)=.001296*CDSIG*SQRT(1./(EP-CDSIG1*EN(N)))
     1*P(N)*(EPS**4)
      IF(TLBDG) 296,295,296
  295 FT3(4) = XNUM(IJ)
```

SOURCE STATEMENT

```
      FT3(7) = XNUM(IJ+1)
      WRITE(6,FT3) N,P(N),DSIG(N)
      GO TO 470
  296 DSIGLB(N)=DSIG(N)*R3(N)
      FT4(5)=XNUM(IJ)
      FT4(8) = XNUM(IJ+1)
      WRITE(6,FT4)N,TCMDG(N),P(N),DSIGLB(N),EPP(N),ZLBDG(N),R4(N)
  470 CONTINUE
C
C     THE ANGULAR DISTRIBUTION TENSORS
C
C     ANGULAR DISTRIBUTION TENSORS ANGO(N,KA,KAPPA)
      IF(OUANGO.EQ.0.AND.OUANG1.EQ.0.AND.OUANG2.EQ.0.AND.OUANG3.EQ.0)
     1 GO TC 500
      IF(OUANGO) 740,741,740
  740 WRITE(6,711)
  711 FORMAT(1H1,9X49HTHE ANGULAR DISTRIBUTION TENSORS ANGO(N,KA,KAPPA))
      WRITE(6,754)
  754 FORMAT(7X1HN 6X2HKA4X5HKAPPA9X9HREAL ANGO,11X9HIMAG ANGO/)
  741 CE3=1.0/(2.0*SPIN(1)+1.0)
      IR = 2.0*SPIN(1) + 2.01
C     INDEX OF FIRST MAGNETIC SUBLEVEL IN FIRST EXCITED STATE
C     CALCULATION OF ANGO FOR ALL N.GE.2
      DO 701 NI = 2,NMAX
      CE2 =    SQRT(2.0*SPIN(NI)+ 1.0)
      IRI = IR
      KAMAX = 2.02*CAT(IR,2)
      IF(KAMAX.GT.4)KAMAX=4
      KASTOP = KAMAX + 1
C     CALCULATION OF ANGO FOR KA=0,2,KAMAX
      KA = C
      DO 702 KAI = 1,KASTOP,2
C     KAI IS INCEX COUNTER ONLY
      KAPSTP = KA + 1
C     CALCULATION OF ANGO FOR KAPPA=K,K-1,K-2,-----,0
      KAPPA = KA
      DO 703 KAPPI = 1,KAPSTP
C     KAPPI IS INDEX COUNTER ONLY
      IR = IRI
      BERE = 0.C
      BEIM = 0.0
C     SUMMATION OVER M(N) AND MPRIME(N)
  704 IRP = IR + KAPPA
      IF(IRP.GT.ISMAX) GO TO 705
      JN = CAT(IRP,1)+0.01
      IF(JN.GT.NI) GO TO 705
C     DEFINITION OF THE ARGUMENTS OF THE THREE-J SYMBOL
      AA1 = CAT(IR,2)
      BB1 = -CAT(IRP,3)
      BB2 = CAT(IR,3)
      AA3 = KA
      BB3 = KAPPA
      IEXP = CAT(IR,2)+CAT(IRP,3)+0.01
      FAC1 = (-1.0)**IEXP * THREEJ(AA1,BB1,AA1,BB2,AA3,BB3)
C     SUMMATION OVER GROUND STATE POLARIZATIONS
```

```
      L = 1
  709 CONTINUE
      LL = L+L
      LLL = LLMAX-LL
      IF(LLL)708,707,706
C     706 STARTS SUMMATION FOR NEGATIVE QUANTUM NUMBERS
  706 JR=2.C02*CAT(IR,3)
      JRP=2.002*CAT(IRP,3)
      IRNEG = IR-JR
      IRPNEG = IRP-JRP
      IMAG = IRNEG+ISMAX
      IMAGP = IRPNEG+ISMAX
      BERE = BERE+FAC1*
     1(AMP(IRNEG,L)*AMP(IRPNEG,L)+AMP(IMAG,L)*AMP(IMAGP,L) )
      BEIM = BEIM + FAC1*
     1(AMP(IRNEG,L)*AMP(IMAGP,L)-AMP(IMAG,L)*AMP(IRPNEG,L))
  707 CONTINUE
C     707 CCMPLETES SUMMATION FOR MAGNETIC QUANTUM NUMBERS .GE. 0
      IMA = IR+ISMAX
      IMAP = IRP+ISMAX
      BERE = BERE + FAC1*(AMP(IR,L)*AMP(IRP,L)+AMP(IMA,L)*AMP(IMAP,L))
      BEIM = BEIM + FAC1*(AMP(IR,L)*AMP(IMAP,L)-AMP(IMA,L)*AMP(IRP,L))
C     THE AMPS USED IN THE PRECEDING SECTIONS ARE GIVEN IN
C     CCORDINATE SYSTEM 0
      L = L+1
      GO TO 709
  708 CONTINUE
      IR = IR + 1
      GO TO 704
  705 CONTINUE
      KARE = KA + 1
      KAPH = KAPPA + 1
      KAIM = KA + 2
      ANGO(NI,KARE,KAPH)=CE2*CE3*BERE
      ANGO(NI,KAIM,KAPH)=CE2*CE3*BEIM
      FT5(4)=XNUM(IJ)
      IF(OUANGO)751,758,751
  751 WRITE(6,FT5) NI,KA,KAPPA,ANGO(NI,KARE,KAPH),ANGO(NI,KAIM,KAPH)
C
C     ANGULAR DISTRIBUTION TENSORS ANG1(N,KA,KAPPA)
  758 IF(KAPPA.EQ.0.OR.KAPPA.EQ.4) GO TO 301
      IF(KAPPA.EQ.1) GO TO 302
      IF(KAPPA.EQ.2) GO TO 303
      IF(KAPPA.EQ.3) GO TO 304
  301 ANG1(NI,KARE,KAPH) = ANGO(NI,KARE,KAPH)
      ANG1(NI,KAIM,KAPH) = ANGO(NI,KAIM,KAPH)
      GO TO 305
  302 ANG1(NI,KARE,KAPH) = -ANGO(NI,KAIM,KAPH)
      ANG1(NI,KAIM,KAPH) = ANGO(NI,KARE,KAPH)
      GO TO 305
  303 ANG1(NI,KARE,KAPH) = -ANGO(NI,KARE,KAPH)
      ANG1(NI,KAIM,KAPH) = -ANGO(NI,KAIM,KAPH)
      GO TO 305
  304 ANG1(NI,KARE,KAPH) = ANGO(NI,KAIM,KAPH)
      ANG1(NI,KAIM,KAPH) =-ANGO(NI,KARE,KAPH)
```

```
                                     FORTRAN SOURCE LIST JOBCXP
          SOURCE STATEMENT

      305 CONTINUE
          KAPPA = KAPPA-1
      703 CONTINUE
          KA = KA + 2
      702 CONTINUE
    C     FOLLOWING INSTRUCTION SKIPS A LINE IN PRINTOUT WHEN NI CHANGES
          WRITE(6,800)
      800 FORMAT(1H0)
      701 CONTINUE
          IF(OUANG1) 770,780,770
      770 WRITE(6,771)
      771 FORMAT(1H1,9X49HTHE ANGULAR DISTRIBUTION TENSORS ANG1(N,KA,KAPPA))
          WRITE(6,772)
      772 FORMAT(7X1HN 6X2HKA4X5HKAPPA9X9HREAL ANG1/)
          DO 773 NI=2,NMAX
          KA=0
          KAMAX=2.02*SPIN(NI)
          IF(KAMAX.GT.4) KAMAX=4
          KASTOP=KAMAX+1
          DO 774 KAI=1,KASTOP,2
    C     KAI IS INDEX COUNTER ONLY
          KARE=KA+1
          KAPPA=KA
          KAPSTP=KA+1
          DO 775 KAPPI=1,KAPSTP
    C     KAPPI IS INDEX COUNTER ONLY
          KAPH=KAPPA+1
          WRITE(6,FT5)NI,KA,KAPPA,ANG1(NI,KARE,KAPH)
      775 KAPPA=KAPPA-1
      774 KA=KA+2
    C     FOLLOWING INSTRUCTION SKIPS A LINE IN PRINTOUT WHEN NI CHANGES
          WRITE(6,800)
      773 CONTINUE
    C
    C     ANGULAR DISTRIBUTION TENSORS ANG2(N,KA,KAPPA)
      780 IF(OUANG2) 960,959,960
      960 WRITE(6,917)
      917 FORMAT(1H1,9X49HTHE ANGULAR DISTRIBUTION TENSORS ANG2(N,KA,KAPPA))
          WRITE(6,930)
      930 FORMAT(7X1HN 6X2HKA4X5HKAPPA9X9HREAL ANG2,11X9HIMAG ANG2/)
          DO 982 NI=2,NMAX
          KA=0
          KAMAX=2.02*SPIN(NI)
          IF(KAMAX.GT.4) KAMAX=4
          KASTOP=KAMAX+1
          DO 983 KAI=1,KASTOP,2
    C     KAI IS INDEX COUNTER ONLY
          DJ=KA
          KARE=KA+1
          KAIM=KA+2
          KAPPA=KA
          KAPSTP=KA+1
          DO 984 KAPPI=1,KAPSTP,2
    C     KAPPI IS INDEX COUNTER ONLY
          DMP=KAPPA
```

SOURCE STATEMENT

```
      KAPH=KAPPA+1
      SUM2RE=0.0
      SUM2IM=0.0
C     SUMMATION OVER KAPPAPRIME
      IF(KA.LT.4) GO TO 992
      SUM2RE=ANG1(NI,KARE,5)*(DJMM(-90.0,DJ,4.0,DMP)+DJMM(-90.0,DJ,-4.0,
     1DMP))
      SUM2IM=ANG1(NI,KARE,4)*(DJMM(-90.0,DJ,3.0,DMP)+DJMM(-90.0,DJ,-3.0,
     1DMP))
  992 IF(KA.LT.2) GO TO 990
      SUM2RE=SUM2RE-ANG1(NI,KARE,3)*(DJMM(-90.0,DJ,2.0,DMP)+DJMM(-90.0,
     1DJ,-2.0,DMP))
      SUM2IM=SUM2IM-ANG1(NI,KARE,2)*(DJMM(-90.0,DJ,1.0,DMP)+DJMM(-90.0,
     1DJ,-1.0,DMP))
  990 SUM2RE=SUM2RE+ANG1(NI,KARE,1)*DJMM(-90.0,DJ,0.0,DMP)
      ANG2(NI,KARE,KAPH)=SUM2RE
      ANG2(NI,KAIM,KAPH)=SUM2IM
      WRITE(6,FT5)NI,KA,KAPPA,ANG2(NI,KARE,KAPH),ANG2(NI,KAIM,KAPH)
  984 KAPPA=KAPPA-2
  983 KA=KA+2
C     FOLLOWING INSTRUCTION SKIPS A LINE IN PRINTOUT WHEN NI CHANGES
  982 WRITE(6,800)
C
C     ANGULAR DISTRIBUTION TENSORS ANG3(N,KA,KAPPA)
  959 IF(OUANG3) 760,759,760
  760 WRITE(6,217)
  217 FORMAT(1H1,9X49HTHE ANGULAR DISTRIBUTION TENSORS ANG3(N,KA,KAPPA))
      WRITE(6,230)
  230 FORMAT(7X1HN 6X2HKA4X5HKAPPA9X9HREAL ANG3/)
  759 DFARG =(180.0+THETA)/2.0
      DO 82 NI = 2,NMAX
      KA= 0
      KAMAX=2.02*SPIN(NI)
      IF(KAMAX.GT.4) KAMAX=4
      KASTOP=KAMAX+1
      DO 83 KAI = 1,KASTOP,2
C     KAI IS INDEX COUNTER ONLY
      DJ = KA
      KARE = KA + 1
      KAPPA = KA
      KAPSTP = KA + 1
      DO 84 KAPPI = 1,KAPSTP
C     KAPPI IS INDEX COUNTER ONLY
      DMP = KAPPA
      KAPH = KAPPA + 1
      SUM = 0.0
C     SUMMATION OVER INDICES .GE. 0 IN EQUATION (50)
      KPR = 0
   85 DM=KPR
      KPRI = KPR + 1
      A1KK = ANG1(NI,KARE,KPRI)
      DKKK = DJMM(DFARG,DJ,DM,DMP)
      SUM=SUM+A1KK*DKKK
      KPR = KPR + 1
      IF(KPR-KA)85,85,86
```

SOURCE STATEMENT

```
 86 CONTINUE
    SUM1 = 0.C
C   SUMMATION OVER NEGATIVE INDICES IN EQUATION (50)
    IF(KA.EQ.0)GO TO 307
    KPR = 1
306 DM=(-KPR)
    KPRI = KPR + 1
    A1KK = (-1.0)**KPR * ANG1(NI,KARE,KPRI)
    DKKK = DJMM(DFARG,DJ,DM,DMP)
    SUM1= SUM1+ A1KK*DKKK
    KPR = KPR + 1
    IF(KPR-KA)306,306,307
307 ANG3(NI,KARE,KAPH) = (-1.0)**KAPPA*(SUM+SUM1)
    IF(OUANG3)761,84,761
761 WRITE(6,FT5) NI,KA,KAPPA,ANG3(NI,KARE,KAPH)
 84 KAPPA =KAPPA -1
 83 KA = KA + 2
C   FOLLOWING INSTRUCTION SKIPS A LINE IN PRINTOUT WHEN NI CHANGES
    WRITE(6,8C0)
 82 CONTINUE
    GO TO 500
103 WRITE(6,782)
782 FORMAT(35HOERROR- CM AND LAB ANGLE BOTH GIVEN)
    GO TO 500
810 WRITE(6,811)
811 FORMAT(42HOERROR- NEITHER CM NOR LA3 ANGLE ARE GIVEN)
    GO TO 500
812 WRITE(6,813)
813 FORMAT(37HOERROR- GROUND STATE SPIN EXCEEDS 7/2)
    GO TO 500
814 WRITE(6,818)
818 FORMAT(43HOERROR EMMAX SMALLER THAN GROUND STATE SPIN)
    GO TO 500
816 WRITE(6,817)
817 FORMAT(22HOERROR NMAX EXCEEDS 10)
    GO TO 500
    END
```

FORTRAN SOURCE LIST

SOURCE STATEMENT

```
$IBFTC MAT
      SUBROLTINE MAT
C
C     READ IN AND PRINT OUT OF E2-MATRIX ELEMENTS
C
      REAL MATRIX
      COMMON /XX/NMAX,EN(10),SPIN(10),MATRIX(10,10),FMT4(6),XNUM(10)
      DO 36 N=1,NMAX
   36 READ(5,1)(MATRIX(N,M),M=1,NMAX)
    1 FORMAT(6F12.4)
      WRITE(6,441)
  441 FORMAT(53HOQUADRUPOLE MATRIX ELEMENTS IN BARNS, READ FROM DATA
     15HCARCS)
      FMT4(4) = XNUM(NMAX)
      WRITE(6,FMT4)(M,M=1,NMAX)
      DO 2 N=1,NMAX
    2 WRITE(6,207) N,(MATRIX(N,M),M=1,NMAX)
  207 FORMAT(4HON =I2,10F12.4)
      RETURN
      END
```

```
                                              FORTRAN SOURCE LIST
      SOURCE STATEMENT

$IBFTC FAC1C
      FUNCTION FAC10(N)
C
C     FAC10 GENERATES FACTORIAL(N)/10**N FOR INTEGERS N.LT.79
C
      IF(N)1,2,3
    1 WRITE(6,6)
    6 FORMAT(36HOERROR- FACTORIAL OF NEGATIVE NUMBER)
      RETURN
    2 FAC10=1.0
      RETURN
    3 IF(N.GT.79) GO TO 5
      FAC10=1.0
      Q=1.0
      DO 4 K=1,N
C     K IS INDEX COUNTER ONLY
      FAC10=FAC10*Q/10.0
      Q=Q+1.0
    4 CONTINUE
      RETURN
    5 WRITE(6,7)
    7 FORMAT(33HOERROR- FACTORIAL OF NUMBER.GT.79)
      RETURN
      END
```

```
                                         FORTRAN SOURCE LIST
      SOURCE STATEMENT

$IBFTC THREEJ
      FUNCTION THREEJ(A1, B1, A2, B2, A3, B3)
C
C     A FUNCTION FOR THREE J SYMBOLS WITH ARBITRARY ARGUMENTS
C
C     ROTENBERG ET. AL. (1.5) PAGE 2 WITH (2.4) FROM PAGE 13
C     INPUT IS THREE-J(J1,M1,J2,M2,J3,M3) FLOATING POINT ARGUMENTS
C     ROUTINE REQUIRES FACTORIAL FUNCTION ROUTINE
C
      DIMENSION A(3), B(3)
      A(1) = A1
      A(2) = A2
      A(3) = A3
      DO 19 N=1,3
      IF(A(N)+0.001) 20,19,19
   20 WRITE(6,60)
   60 FORMAT(28H0ERROR- NEGATIVE J IN THREEJ)
      CALL EXIT
   19 CONTINUE
   21 LA1=A1+A2+A3+0.001
      LA2=A1+A2+A3+0.6
      IF(LA2-LA1) 22,23,22
   22 WRITE(6,61)
   61 FORMAT(39H0ERROR- HALF INTEGER SUM OF J IN THREEJ)
      CALL EXIT
   23 B(1) = B1
      B(2) = B2
      B(3) = B3
      LB1=ABS(B1+B2+B3)+0.001
      LB2=ABS(B1+B2+B3)+0.6
      IF(LB2-LB1) 24,16,24
   24 WRITE(6,62)
   62 FORMAT(39H0ERROR- HALF INTEGER SUM OF M IN THREEJ)
      CALL EXIT
   16 NBA1=ABS(B1)-A1-100.001
      NBA2=ABS(B1)-A1-100.6
      IF(NBA2-NBA1) 30,40,30
   40 IF(NBA1.GT.(-100)) GO TO 26
      NBA1=ABS(B2)-A2-100.001
      NBA2=ABS(B2)-A2-100.6
      IF(NBA2-NBA1) 30,41,30
   41 IF(NBA1.GT.(-100)) GO TO 26
      NBA1=ABS(B3)-A3-100.001
      NBA2=ABS(B3)-A3-100.6
      IF(NBA2-NBA1)30,42,30
   42 IF(NBA1.GT.(-100)) GO TO 26
      GO TO 52
   30 WRITE(6,63)
   63 FORMAT(35H0ERROR- J-M HALF INTEGER IN THREE J)
      CALL EXIT
   26 WRITE(6,64)
   64 FORMAT(33H0ERROR- M LARGER THAN J IN THREEJ)
      CALL EXIT
   52 IF(LB1) 1,2,1
    2 S1 = A1 + A2 - A3 +0.001
```

```
      IF(S1)1,12,12
   12 S2 = A1 -A2 + A3  +0.001
      IF(S2)1,14,14
   14 S3 = A2 + A3 - A1+0.001
      IF(S3)1,15,15
   15 N=ABS(A1-A2-B3)+0.001
      PHZ = (-1.0)**N
      M = S1
      FS1=FAC10(M)
      M = S2
      FS2=FAC10(M)
      M = S3
      FS3=FAC10(M)
      FD=FAC10(LA1+1)
      DELTA=SQRT((FS1*FS2*FS3)/(FD*10.0))
      X=1.0
      DO 3 J = 1,3
      M=A(J)+B(J)+0.001
      FS=FAC10(M)
      X=X*FS
      M= A(J)-B(J)+0.001
      FSM=FAC10(M)
    3 X = X*FSM
      ROOT = SQRT(X)
      SUM = 0.0
      AK = C.0
   11 DS1 = AK+A3-A1-B2 +0.001
      IF(DS1)5,4,4
    4 M = DS1
      FDS1=FAC1C(M)
      DS2 = AK + B1 + A3 - A2 +0.001
      IF(DS2)5,6,6
    6 M = DS2
      FDS2=FAC10(M)
      DS3 = A1 - B1 - AK +0.001
      IF(DS3)8,7,7
    7 M = DS3
      FDS3=FAC1C(M)
      DS4 = A1 + A2 - A3 - AK+0.001
      IF(DS4)8,9,9
    9 M = DS4
      FDS4=FAC1C(M)
      DS5 = A2 + B2 - AK +0.001
      IF(DS5)8,10,10
   10 M = DS5
      FDS5=FAC10(M)
      M = AK +0.001
      FAK=FAC10(M)
      TCP=(-1.0)**M
      DENOM = FAK*FDS1* FDS2* FDS3* FDS4* FDS5
      SUM = SUM + (TOP/DENOM)
    5 AK = AK + 1.0
      GO TO 11
    8 THREEJ=PHZ*DELTA*ROOT*SUM
      RETURN
```

SOURCE STATEMENT

```
    1 THREEJ = 0.0
      RETURN
      END
```

FORTRAN SOURCE LIST

SOURCE STATEMENT

```
$IBFTC CMLAB
      SUBROUTINE CMLAB(A1,A2,A3,A4,EP,EN,NMAX,TLBDG,TCMDG,ZLBDG,R3,R4,
     1EPP)
C
C     SEE E.G.-J.B. MARION, NUCLEAR DATA TABLES PART 3 PAGE 162
C
      DIMENSION EN(10),R3(10),R4(10),EPP(10)
      DIMENSION TCMDG(10),TCMRAD(10),ZLBDG(10),ZLBRAD(10)
      TLBRAC=TLBDG/57.2957 95
      DO 10 N=1,NMAX
      ET=EP-EN(N)
      A = A1*A4*(EP/ET)/((A1+A2)*(A3+A4))
      B =A1*A3*(EP/ET)/((A1+A2)*(A3+A4))
      C = A2*A3/((A1+A2)*(A3+A4))*(1.-A1*EN(N)/(A2*ET))
      D = A2*A4/((A1+A2)*(A3+A4))*(1.-A1*EN(N)/(A2*ET))
      IF(C)2,2,1
    2 EMAX=EP*(A2/A1+A2)
      WRITE(6,3) EMAX
    3 FORMAT(37HOERROR- MAXIMUM EXCITATION ENERGY IS F8.4,4H MEV)
      CALL EXIT
    1 IF(B/C.LE.1.0) GO TO 4
      TMXDG= ARSIN((D/B)**0.5)*57.2957795
      WRITE(6,7)TMXDG
      IF(TMXDG.GE.TLBDG) GO TO 6
    7 FORMAT(36HOERROR- MAXIMUM SCATTERING ANGLE IS F7.2,8H DEGREES)
      CALL EXIT
    6 E3=ET*B*(COS(TLBRAD)+(D/B-(SIN(TLBRAD))**2)**0.5)**2
      TCMRAC(N)=ARCOS((E3/ET-B-D)/(2.0*((A*C)**0.5)))
      TCMDG(N)=TCMRAD(N)*57.2957795
      WRITE(6,8) TCMDG(N),N
    8 FORMAT(34HOSECOND POSSIBLE CM SCATTERING IS F7.2
     117H DEGREES FOR N = I3)
      E3=ET*B*(COS(TLBRAD)-(D/B-(SIN(TLBRAD))**2)**0.5)**2
      TCMRAC(N)=ARCOS((E3/ET-B-D)/(2.0*((A*C)**0.5)))
      TCMDG(N)=TCMRAD(N)*57.2957795
      GO TO 5
    4 E3=ET*B*(COS(TLBRAD)+(D/B-(SIN(TLBRAD))**2)**0.5)**2
      TCMRAC(N)=ARCOS((E3/ET-B-D)/(2.0*((A*C)**0.5)))
      TCMDG(N)=TCMRAD(N)*57.2957795
    5 E4=ET-E3
      EPP(N)=E3
      ZLBRAC(N)=ARCOS((SQRT(A1*EP)-SQRT(A3*E3)*COS(TLBRAD))/SQRT(A4*E4))
      ZLBDG(N)=ZLBRAD(N)*57.2957795
      R3(N)=SIN(TCMRAD(N))**2/(SIN(TLBRAD)**2*COS(TCMRAD(N)-TLBRAD))
      RAD=(180.00/57.2957795)-TCMRAD(N)-ZLBRAD(N)
      R4(N)=SIN(TCMRAD(N))**2/(SIN(ZLBRAD(N))**2*COS(RAD))
   10 CONTINUE
      RETURN
      END
```

```
                                            FORTRAN SOURCE LIST
         SOURCE STATEMENT

 $IBFTC Q
         SUBROUTINE Q(W,EPS,XIMAX,XI,NMAX,QU)
 C
 C       GENERATES QU(N,M,NU) FOR A GIVEN W
 C
         COMMON COT1,COT2
         REAL NOMW
         DIMENSION XI(10,10),QU(10,10,6)
         EW = EXP(W)
         COSHY= 0.5 * (EW+ 1.0/EW)
         SINHY= 0.5*(EW-1.0/EW)
         DW1 = EPS*COSHY + 1.0
         DWSQ = DW1*DW1
         DENW = DWSQ*DWSQ
         NOMW = COSHY + EPS
         PE1 = 0.75*(2.0*NOMW*NOMW - COT2*SINHY*SINHY)/DENW
         PE2 = 1.83711730 * COT1 * SINHY* NOMW/DENW
         PE3 = 0.91855865 * COT2 * SINHY*SINHY/DENW
         ALFA = EPS*SINHY+ W
         DO 36 N = 1,NMAX
         DO 34 M = 1,NMAX
         IF(ABS(XI(N,M)).GE.XIMAX)GO TO 30
         ALF-ALFA*XI(N,M)
         S = SIN(ALF)
         C= COS(ALF)
         QU(N,M,1) = PE1*C
         QU(N,M,2) = PE1*S
         QU(N,M,3) = PE2*S
         QU(N,M,4) = -PE2*C
         QU(N,M,5) = -PE3*C
         QU(N,M,6) = -PE3*S
         GO TO 34
      30 DO 35 I = 1,6
      35 QU(N,M,I) = 0.0
      34 CONTINUE
      36 CONTINUE
         RETURN
         END
```

```
                                          FORTRAN SOURCE LIST
        SOURCE STATEMENT

$IBFTC AMPDER
         SUBROUTINE AMPDER(CAT,ZETA,AMP,QU,LMAX,ISMAX,AMPDOT)
C
C       A SUBROUTINE TO GENERATE THE AMPDOT ARRAY
C
        COMMON/Y/ACC10
        DIMENSION CAT(90,3),ZETA(90,90),AMPDOT(180,4),AMP(180,4)
        DIMENSION QU(10,10,6)
        DO 43 IR = 1,ISMAX
        N = CAT(IR,1) + 0.001
        IMAG=IR+ISMAX
        DO 45 L = 1,LMAX
        AMPDOT(IR,L)=0.0
        AMPDOT(IMAG,L)=0.0
     45 CONTINUE
        DO 42 IS = 1,ISMAX
        M = CAT(IS,1) + 0.001
        IARG = IS+ISMAX
        MU = ABS(CAT(IR,3) - CAT(IS,3)) + 0.001
        IF(MU.GT.2) GO TO 42
        Z = ZETA(IR,IS)
        IF(ABS(Z).LE.ACC10) GO TO 42
        NU = MU+MU+2
        R1 = Z*QU(N,M,NU)
        NU = MU+MU+1
        R2 = Z*QU(N,M,NU)
        DO 41 L = 1,LMAX
        AMPDOT(IR,L)= AMPDOT(IR,L)-R1*AMP(IS,L)-R2*AMP(IARG,L)
        AMPDOT(IMAG,L)= AMPDOT(IMAG,L)-R1*AMP(IARG,L)+R2*AMP(IS,L)
     41 CONTINUE
     42 CONTINUE
     43 CONTINUE
        RETURN
        END
```

```
                                   FORTRAN SOURCE LIST
      SOURCE STATEMENT

$IBFTC DJMM
C       A FUNCTION FOR LOWER CASE D OF BETA
C
C       DEFINITION SAME AS IN EDMONDS, EQUATION (4.1.15) BUT WITH
C       M AND MPRIME INTERCHANGED
C
        FUNCTION DJMM(BETA,J,M,MPRIME)
        REAL J,M,MPRIME
        IF(J)1,20,20
     20 RJM=J+MPRIME+0.01
        IF(RJM)1,2,2
      1 WRITE(6,50)
     50 FORMAT(23HOERROR IN DJMM ARGUMENT)
        CALL EXIT
      2 JM=RJM
        F1=FAC10(JM)
        RJM=J-MPRIME+0.01
        IF(RJM) 1,3,3
      3 JM=RJM
        F1=F1*FAC10(JM)
        RJM=J+M+0.01
        IF(RJM)1,4,4
      4 JM=RJM
        F1=F1*FAC10(JM)
        RJM=J-M+0.01
        IF(RJM)1,5,5
      5 JM=RJM
        F1=F1*FAC10(JM)
        ROOT = SQRT(F1)
        BRAD = BETA/57.295779
        CB = COS(BRAD/2.0)
        SB = SIN(BRAD/2.0)
        SUM=0.0
        SIGMA=0.0
     11 RMMS=M+MPRIME+SIGMA
        IF(RMMS) 30,31,32
     30 MMS=RMMS-0.01
        GO TO 33
     31 MMS=0
        GO TO 33
     32 MMS=RMMS+0.01
     33 IF(MMS)6,7,7
      7 F2=FAC10(MMS)
        RJMPS=J-MPRIME-SIGMA
        IF(RJMPS)34,35,36
     34 JMPS=RJMPS-0.01
        GO TO 37
     35 JMPS=0
        GO TO 37
     36 JMPS=RJMPS+0.01
     37 IF(JMPS)8,9,9
      9 F2=F2*FAC10(JMPS)
        RJMS=J-M-SIGMA
        IF(RJMS)38,39,40
     38 JMS=RJMS-0.01
```

FORTRAN SOURCE LIST DJMM

```
    SOURCE STATEMENT

      GO TO 41
   39 JMS=0
      GO TO 41
   40 JMS=RJMS+0.01
   41 IF(JMS)8,10,10
   10 F2=F2*FAC10(JMS)
      IS = SIGMA+0.01
      DENOM=F2*FAC10(IS)
      JJ=J+J+0.01
      EMMP=M+MPRIME
      IF(EMMP)12,13,14
   12 MMP=EMMP-0.01
      GO TO 15
   13 MMP=0
      GO TO 15
   14 MMP=EMMP+0.01
   15 PHZ=(-1.0)**JMPS
      TOP=PHZ*(CB**(2*IS+MMP))*(SB**(JJ-2*IS-MMP))
      SUM = SUM + TOP/DENOM
    6 SIGMA = SIGMA + 1.0
      GO TO 11
    8 DJMM = ROOT*SUM
      RETURN
      END
```

```
PROJECTILE CHARGE NUMBER  Z1 =   8.00,  MASS  A1 =   16.000AMU,  LAB ENERGY  EP =   40.000MEV

SCATTERING ANGLE IN LABORATORY SYSTEM  TLBDG =  160.00 DEGREES

TARGET CHARGE NUMBER  Z2 =  62.00,  MASS  A2 =  152.000AMU

ENERGY SPECTRUM     LEVEL INDEX N        1         2         3         4         5

                    ENERGY IN MEV   0.0000    0.1220    0.3670    0.7050    1.1220

                    SPIN            0.0000    2.0000    4.0000    6.0000    8.0000

QUADRUPOLE MATRIX ELEMENTS IN BARNS, READ FROM DATA CARDS

          M= 1           2           3           4           5

N = 1   -0.0000     -1.8450     -0.0000     -0.0000     -0.0000

N = 2   -1.8450      2.2052     -2.9586     -0.0000     -0.0000

N = 3   -0.0000     -2.9586      2.8210     -3.7318     -0.0000

N = 4   -0.0000     -0.0000     -3.7318      3.3563     -4.3662

N = 5   -0.0000     -0.0000     -0.0000     -4.3662      3.8236

XI MATRIX

N = 1  100.0000     -0.0838    100.0000    100.0000    100.0000

N = 2    0.0838     -0.0000     -0.1696    100.0000    100.0000

N = 3  100.0000      0.1696     -0.0000     -0.2368    100.0000

N = 4  100.0000    100.0000      0.2368     -0.0000     -0.2969

N = 5  100.0000    100.0000    100.0000      0.2969     -0.0000

XIM =    0.2969
```

```
PSI MATRIX

N = 1      -0.0000     -0.7119     -0.0000     -0.0000     -0.0000

N = 2      -0.7119      0.8487     -1.1328     -0.0000     -0.0000

N = 3      -0.0000     -1.1328      1.0746     -1.4115     -0.0000

N = 4      -0.0000     -0.0000     -1.4115      1.2605     -1.6253

N = 5      -0.0000     -0.0000     -0.0000     -1.6253      1.4108

PERFORMANCE CONTROLS NMAX,   INTERV,   NCM,   EMMAX,   XIMAX,    ACCUR,     NTIME

                      5        10       4      4.00     2.00   0.0000010     10

CUTPUT CONTROLS      OUXI, OUPSI, OUAMP, OUPROW, OUANGO, OUANG1, OUANG2, OUANG3

                      1      1      1      1       1       1       1       1

CM SCATTERING ANGLE USED FOR INTEGRATION IS TCMDG(NCM) =  162.08 DEGREES

EPS =   1.012

RANGE OF INTEGRATION, UP =   6.90

ESTIMATED NUMBER OF STEPS, ISTEP =   79

INITIAL STEP WIDTH, D2W =  0.17369

TOTAL NUMBER OF MAGNETIC SUBSTATES, ISMAX =  33
```

```
       W          P( 1)       P( 2)       P( 3)       P( 4)       P( 5)       PTOT
    -6.895   1.0000000   0.0000000   0.0000000   0.0000000   0.0000000   1.0000000
AT W =   -5.853, STEP WIDTH WAS HALVED TO BE D2W =   0.08685
    -5.506   0.9999997   0.0000000   0.0000000   0.0000000   0.0000000   0.9999997
    -4.637   0.9999996   0.0000000   0.0000000   0.0000000   0.0000000   0.9999996

AT W =   -4.551, STEP WIDTH WAS DOUBLED TO BE D2W = 0.17369
    -2.987   0.9999866   0.0000128   0.0000000   0.0000000   0.0000000   0.9999994

AT W =   -1.424, STEP WIDTH WAS HALVED TO BE D2W =   0.08685
    -1.077   0.9886196   0.0113534   0.0000225   0.0000000   0.0000000   0.9999955
    -0.208   0.9216666   0.0771991   0.0011242   0.0000066   0.0000000   0.9999965
     0.660   0.7832802   0.2072638   0.0092863   0.0001644   0.0000015   0.9999962
     1.529   0.6864438   0.2926710   0.0203403   0.0005336   0.0000068   0.9999955
AT W =    2.050, STEP WIDTH WAS DOUBLED TO BE D2W = 0.17369
     2.744   0.6538434   0.3213230   0.0241907   0.0006312   0.0000073   0.9999955
AT W =    4.308, STEP WIDTH WAS HALVED TO BE D2W =   0.08685
     4.655   0.6541571   0.3215278   0.0236970   0.0006066   0.0000070   0.9999955
     5.523   0.6541047   0.3215799   0.0236973   0.0006064   0.0000070   0.9999954
     6.392   0.6540992   0.3215853   0.0236973   0.0006065   0.0000070   0.9999953
     6.913   0.6540995   0.3215848   0.0236975   0.0006065   0.0000070   0.9999952

ACTUAL NUMBER OF STEPS, ISTEPS =  122
```

INITIAL M = -0.0

SPIN	MAG.QUAN.NO.	REAL AMPLITUDE	IMAG AMPLITUDE
0.0	-0.0	0.8045946	0.0820188
2.0	-2.0	-0.0008823	0.0018095
2.0	-1.0	-0.0005703	0.0108872
2.0	-0.0	-0.1048562	-0.5570854
2.0	1.0	-0.0005703	0.0108872
2.0	2.0	-0.0008823	0.0018095
4.0	-4.0	-0.0000013	-0.0000012
4.0	-3.0	-0.0000235	-0.0000130
4.0	-2.0	0.0004316	0.0000978
4.0	-1.0	0.0080600	-0.0000316
4.0	-0.0	-0.1517439	0.0232580
4.0	1.0	0.0080600	-0.0000316
4.0	2.0	0.0004316	0.0000978
4.0	3.0	-0.0000235	-0.0000130
4.0	4.0	-0.0000013	-0.0000012
6.0	-4.0	0.0000000	0.0000000
6.0	-3.0	-0.0000031	0.0000075
6.0	-2.0	-0.0000038	-0.0000192
6.0	-1.0	0.0002650	-0.0022970
6.0	-0.0	0.0004050	0.0244049
6.0	1.0	0.0002650	-0.0022970
6.0	2.0	-0.0000038	-0.0000192
6.0	3.0	-0.0000031	0.0000075
6.0	4.0	0.0000000	0.0000000
8.0	-4.0	-0.0000000	-0.0000000
8.0	-3.0	0.0000010	0.0000003
8.0	-2.0	0.0000093	0.0000071
8.0	-1.0	-0.0003603	-0.0001085
8.0	-0.0	0.0025602	0.0004306
8.0	1.0	-0.0003603	-0.0001085
8.0	2.0	0.0000093	0.0000071
8.0	3.0	0.0000010	0.0000003
8.0	4.0	-0.0000000	-0.0000000

LEVEL INDEX	SCATTERING ANGLE-CM DEGREES	EXCITATION PROBABILITIES	LAB CROSS SECTION BARNS/STERADIAN
N	TCMDG(N)	P(N)	DSIGLB(N)
1	162.06	0.6540995	0.13568408E 00
2	162.07	0.3215848	0.66796199E-01
3	162.07	0.0236975	0.49352465E-02
4	162.08	0.0006065	0.12676751E-03
5	162.10	0.0000070	0.14749129E-05

ENERGY OF SCATTERED PROJECTILE, MEV	LAB RECOIL ANGLE DEGREES	SOLID ANGLE RATIO FOR RECOIL
EPP(N)	ZLBDG(N)	R4(N)
26.548	8.97	3.95104
26.449	8.96	3.95782
26.249	8.94	3.97155
25.974	8.91	3.99083
25.634	8.88	4.01502

THE ANGULAR DISTRIBUTION TENSORS ANGO(N,KA,KAPPA)

N	KA	KAPPA	REAL ANGO	IMAG ANGO
2	0	0	0.3215848	-0.0000000
2	2	2	-0.0009009	-0.0000000
2	2	1	-0.0000000	0.0007688
2	2	0	-0.1718221	-0.0000000
2	4	4	0.0000030	-0.0000000
2	4	3	-0.0000000	-0.0000090
2	4	2	-0.0006988	-0.0000000
2	4	1	0.0000000	-0.0014276
2	4	0	0.1716789	-0.0000000
3	0	0	0.0236975	0.0000000
3	2	2	-0.0000345	0.0000000
3	2	1	0.0000000	0.0000504
3	2	0	-0.0120673	0.0000000
3	4	4	0.0000000	-0.0000000
3	4	3	-0.0000000	-0.0000027
3	4	2	0.0000035	-0.0000000
3	4	1	-0.0000000	-0.0000728
3	4	0	0.0095069	0.0000000
4	0	0	0.0006065	0.0000000
4	2	2	0.0000027	0.0000000
4	2	1	-0.0000000	0.0000014
4	2	0	-0.0003056	0.0000000
4	4	4	-0.0000000	-0.0000000
4	4	3	0.0000000	-0.0000000
4	4	2	-0.0000019	-0.0000000
4	4	1	0.0000000	-0.0000020
4	4	0	0.0002337	-0.0000000
5	0	0	0.0000070	0.0000000
5	2	2	0.0000001	0.0000000
5	2	1	0.0000000	0.0000000
5	2	0	-0.0000035	-0.0000000
5	4	4	-0.0000000	-0.0000000
5	4	3	0.0000000	0.0000000
5	4	2	-0.0000001	-0.0000000
5	4	1	0.0000000	-0.0000000
5	4	0	0.0000027	-0.0000000

THE ANGULAR DISTRIBUTION TENSORS ANG1(N,KA,KAPPA)

N	KA	KAPPA	REAL ANG1
2	0	0	0.3215848
2	2	2	0.0009009
2	2	1	-0.0007688
2	2	0	-0.1718221
2	4	4	0.0000030
2	4	3	-0.0000090
2	4	2	0.0006988
2	4	1	0.0014276
2	4	0	0.1716789
3	0	0	0.0236975
3	2	2	0.0000345
3	2	1	-0.0000504
3	2	0	-0.0120673
3	4	4	0.0000000
3	4	3	-0.0000027
3	4	2	-0.0000035
3	4	1	0.0000728
3	4	0	0.0095069
4	0	0	0.0006065
4	2	2	-0.0000027
4	2	1	-0.0000014
4	2	0	-0.0003056
4	4	4	-0.0000000
4	4	3	-0.0000000
4	4	2	0.0000019
4	4	1	0.0000020
4	4	0	0.0002337
5	0	0	0.0000070
5	2	2	-0.0000001
5	2	1	-0.0000000
5	2	0	-0.0000035
5	4	4	-0.0000000
5	4	3	0.0000000
5	4	2	0.0000001
5	4	1	0.0000000
5	4	0	0.0000027

THE ANGULAR DISTRIBUTION TENSORS ANG2(N,KA,KAPPA)

N	KA	KAPPA	REAL ANG2	IMAG ANG2
2	0	0	0.3215848	0.0000000
2	2	2	-0.1056696	-0.0007688
2	2	0	0.0848076	0.0000000
2	4	4	0.0893113	0.0013386
2	4	2	-0.0682094	-0.0004963
2	4	0	0.0649352	0.0000000
3	0	0	0.0236975	0.0000000
3	2	2	-0.0074069	-0.0000504
3	2	0	0.0059914	0.0000000
3	4	4	0.0049736	0.0000690
3	4	2	-0.0037562	-0.0000233
3	4	0	0.0035624	0.0000000
4	0	0	0.0006065	0.0000000
4	2	2	-0.0001858	-0.0000014
4	2	0	0.0001561	0.0000000
4	4	4	0.0001210	0.0000019
4	4	2	-0.0000933	-0.0000007
4	4	0	0.0000891	0.0000000
5	0	0	0.0000070	0.0000000
5	2	2	-0.0000021	-0.0000000
5	2	0	0.0000019	0.0000000
5	4	4	0.0000013	0.0000000
5	4	2	-0.0000011	-0.0000000
5	4	0	0.0000011	0.0000000

THE ANGULAR DISTRIBUTION TENSORS ANG3(N,KA,KAPPA)

N	KA	KAPPA	REAL ANG3
2	0	0	C.3215848
2	2	2	-0.0015429
2	2	1	-0.0317755
2	2	0	-0.1658358
2	4	4	0.0000748
2	4	3	0.0010845
2	4	2	0.0097839
2	4	1	C.0551445
2	4	0	C.1523255
3	0	0	0.0236975
3	2	2	-0.00C1374
3	2	1	-0.0022307
3	2	0	-0.0116463
3	4	4	0.0000032
3	4	3	0.0000506
3	4	2	0.0005050
3	4	1	0.0030719
3	4	0	0.0084263
4	0	0	0.0006065
4	2	2	-0.0000070
4	2	1	-0.0000558
4	2	0	-0.0002951
4	4	4	0.0000001
4	4	3	0.0000018
4	4	2	0.0000141
4	4	1	0.0000747
4	4	0	C.0002075
5	0	0	0.0000070
5	2	2	-0.0000002
5	2	1	-0.0000006
5	2	0	-0.0000034
5	4	4	C.0000000
5	4	3	0.0000000
5	4	2	0.0000002
5	4	1	0.0000008
5	4	0	0.0000024

Data Cards for Example

```
#1          10    10     4 0.000001


#2       2.00      4.0           5


#3          1     1     1     1
#4          1     1     1     1


#5          8.00      16.00       62.0       152.00       40.00      160.00


#6       0.0       0.0
#7       0.122     2.0
#8       0.367     4.0
#9       0.705     6.0
#10      1.122     8.0


#11            -0.0000     -1.8450     -0.0000     -0.0000     -0.0000
#12            -1.8450      2.2052     -2.9586     -0.0000     -0.0000
#13            -0.0000     -2.9586      2.8210     -3.7318     -0.0000
#14            -0.0000     -0.0000     -3.7318      3.3563     -4.3662
#15            -0.0000     -0.0000     -0.0000     -4.3662      3.8236


#16      Last data card should always be blank card.
```

VIII. Flow Diagrams

In order to facilitate changes or additions to the program, the flow diagrams of the more important parts of the main program and of the functions and subroutines are given in this chapter.

A. Overall organization of the program

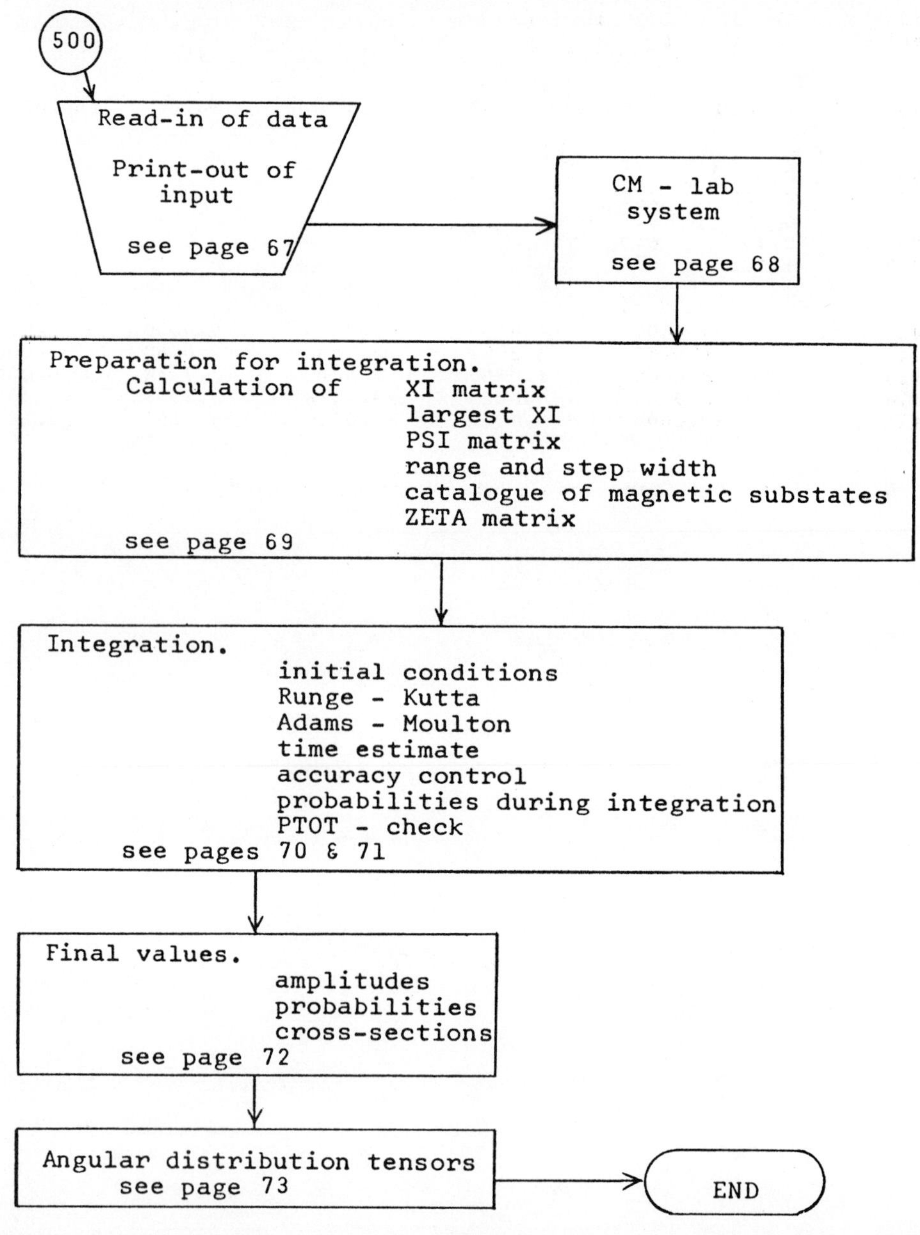

66

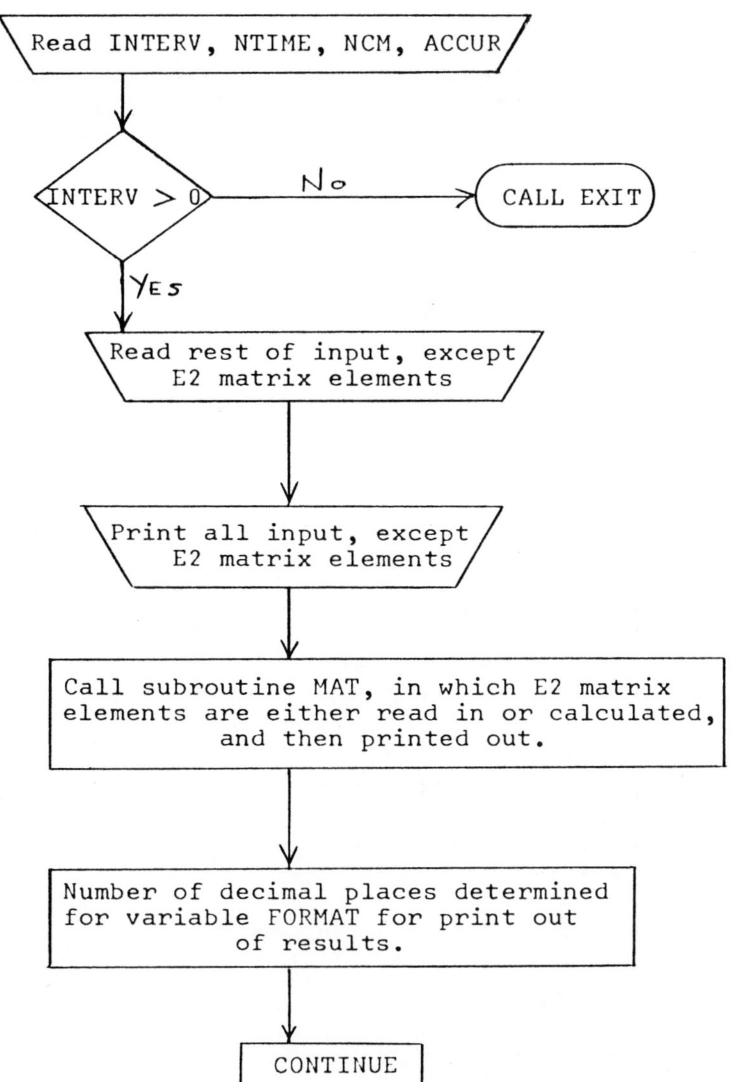
Read INTERV, NTIME, NCM, ACCUR
INTERV > 0
No
CALL EXIT
YES
Read rest of input, except E2 matrix elements
Print all input, except E2 matrix elements
Call subroutine MAT, in which E2 matrix elements are either read in or calculated, and then printed out.
Number of decimal places determined for variable FORMAT for print out of results.
CONTINUE

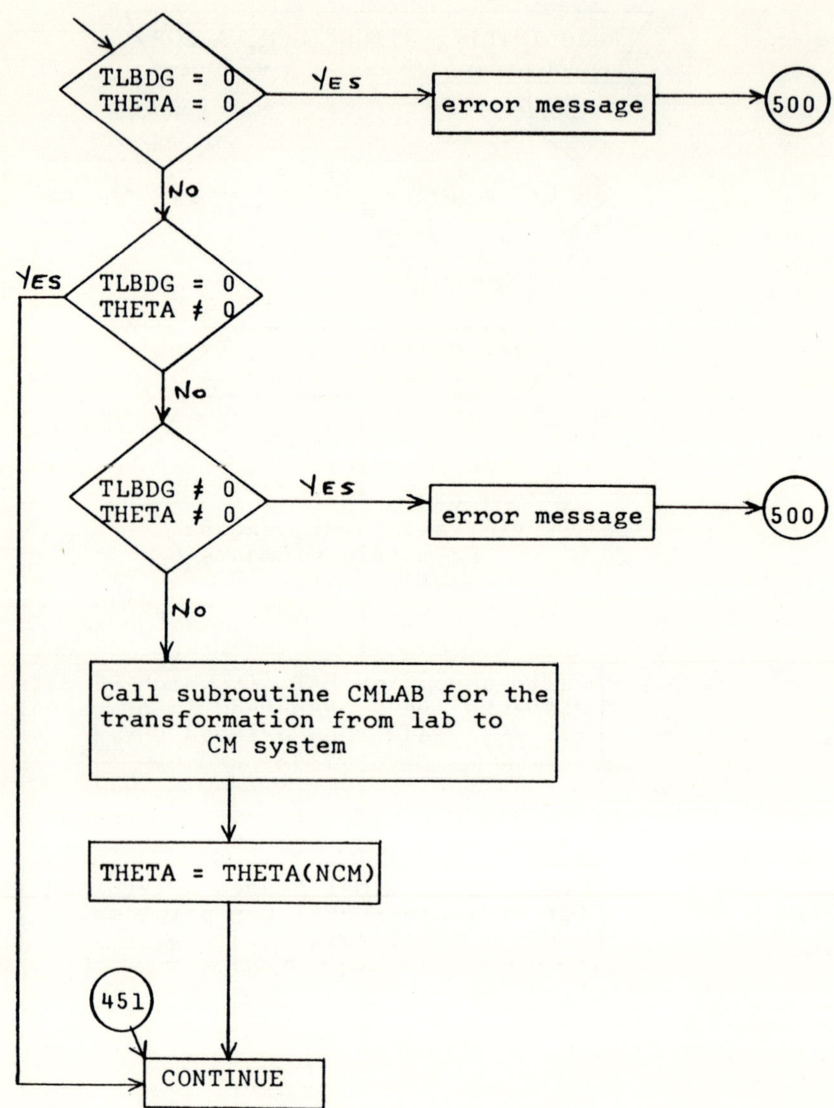
TLBDG = 0
THETA = 0
YES
error message
500
NO
YES
TLBDG = 0
THETA ≠ 0
NO
TLBDG ≠ 0
THETA ≠ 0
YES
error message
500
NO
Call subroutine CMLAB for the transformation from lab to CM system
THETA = THETA(NCM)
451
CONTINUE

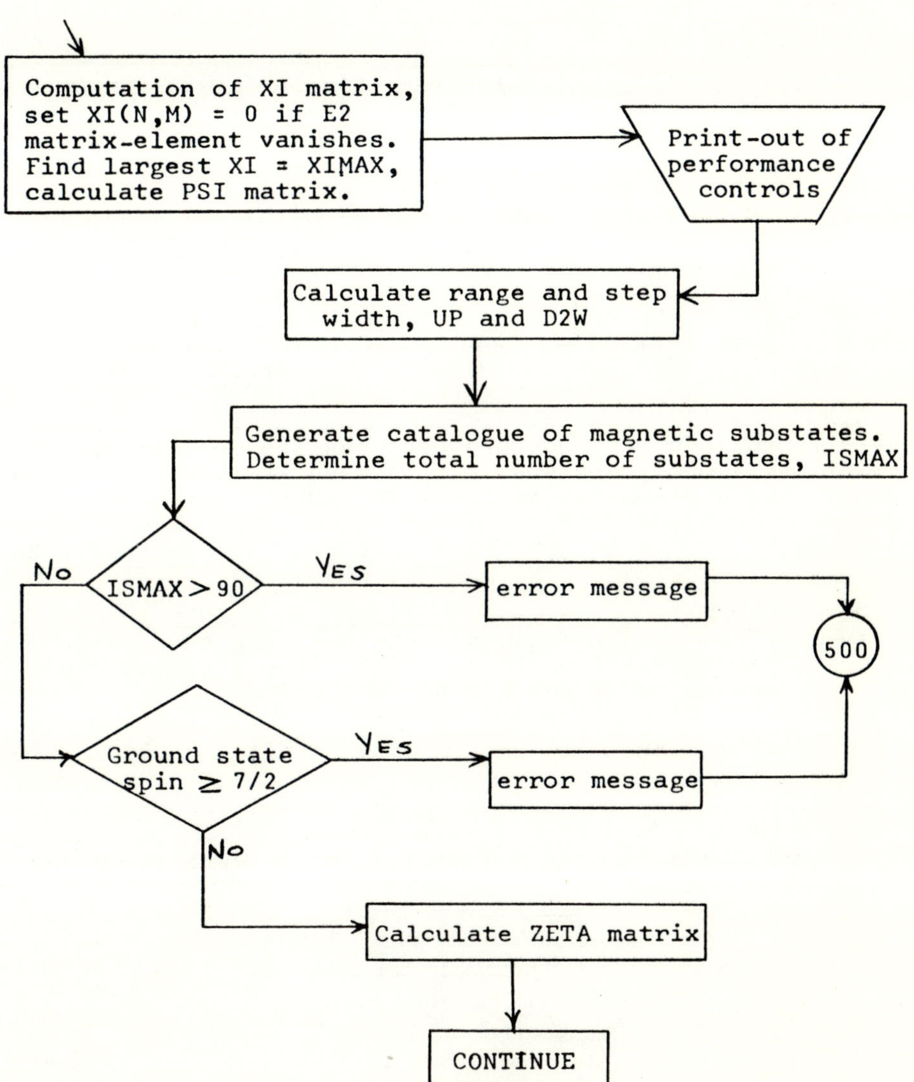
Computation of XI matrix, set XI(N,M) = 0 if E2 matrix-element vanishes. Find largest XI = XIMAX, calculate PSI matrix.
Print-out of performance controls
Calculate range and step width, UP and D2W
Generate catalogue of magnetic substates. Determine total number of substates, ISMAX
No
ISMAX > 90
Yes
error message
500
Ground state spin ≥ 7/2
Yes
error message
No
Calculate ZETA matrix
CONTINUE

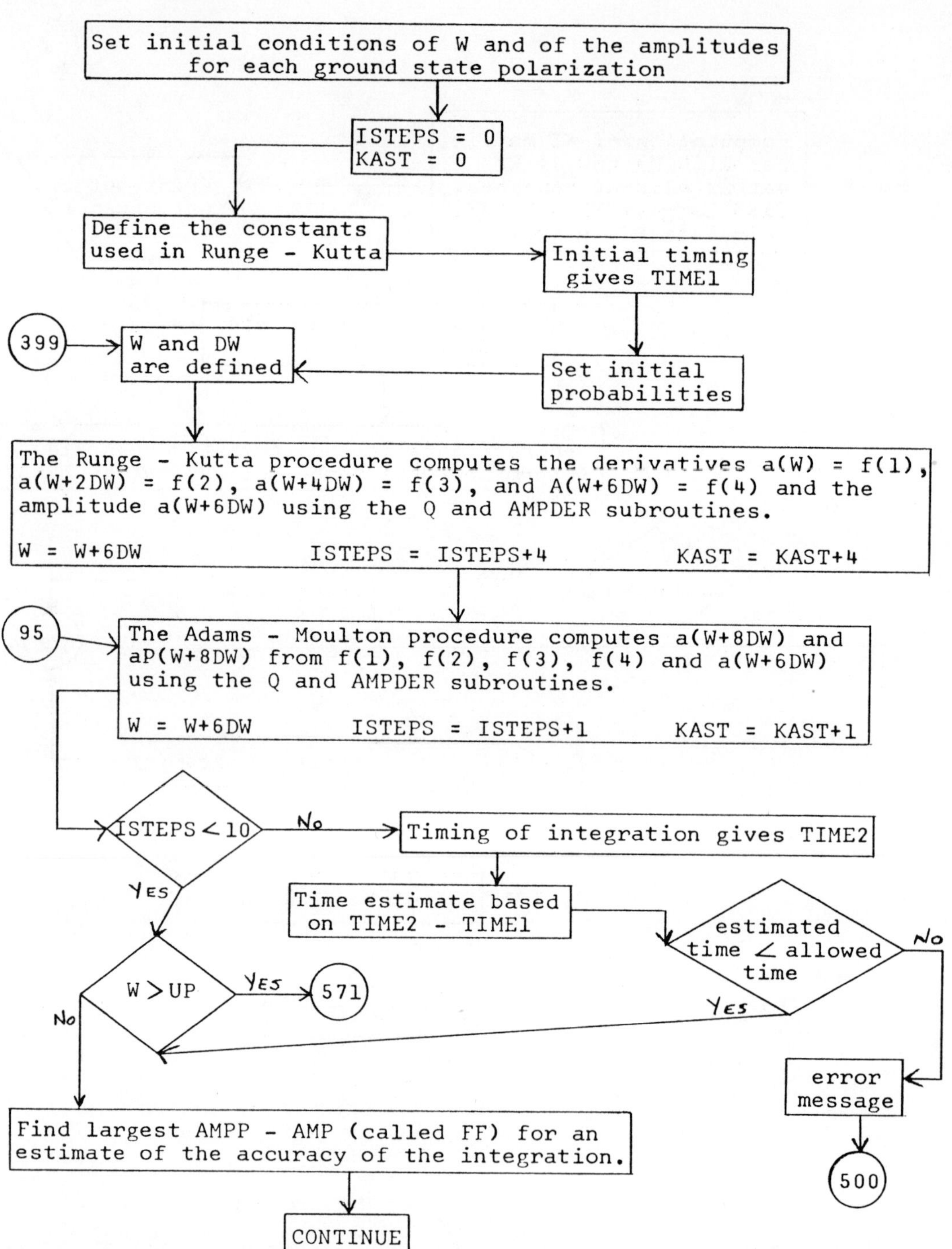
Set initial conditions of W and of the amplitudes for each ground state polarization
ISTEPS = 0
KAST = 0
Define the constants used in Runge - Kutta
Initial timing gives TIME1
399
W and DW are defined
Set initial probabilities
The Runge - Kutta procedure computes the derivatives a(W) = f(1), a(W+2DW) = f(2), a(W+4DW) = f(3), and A(W+6DW) = f(4) and the amplitude a(W+6DW) using the Q and AMPDER subroutines.
W = W+6DW ISTEPS = ISTEPS+4 KAST = KAST+4
95
The Adams - Moulton procedure computes a(W+8DW) and aP(W+8DW) from f(1), f(2), f(3), f(4) and a(W+6DW) using the Q and AMPDER subroutines.
W = W+6DW ISTEPS = ISTEPS+1 KAST = KAST+1
ISTEPS < 10
No
Timing of integration gives TIME2
Yes
Time estimate based on TIME2 - TIME1
estimated time < allowed time
No
W > UP
Yes
571
Yes
No
error message
Find largest AMPP - AMP (called FF) for an estimate of the accuracy of the integration.
500
CONTINUE

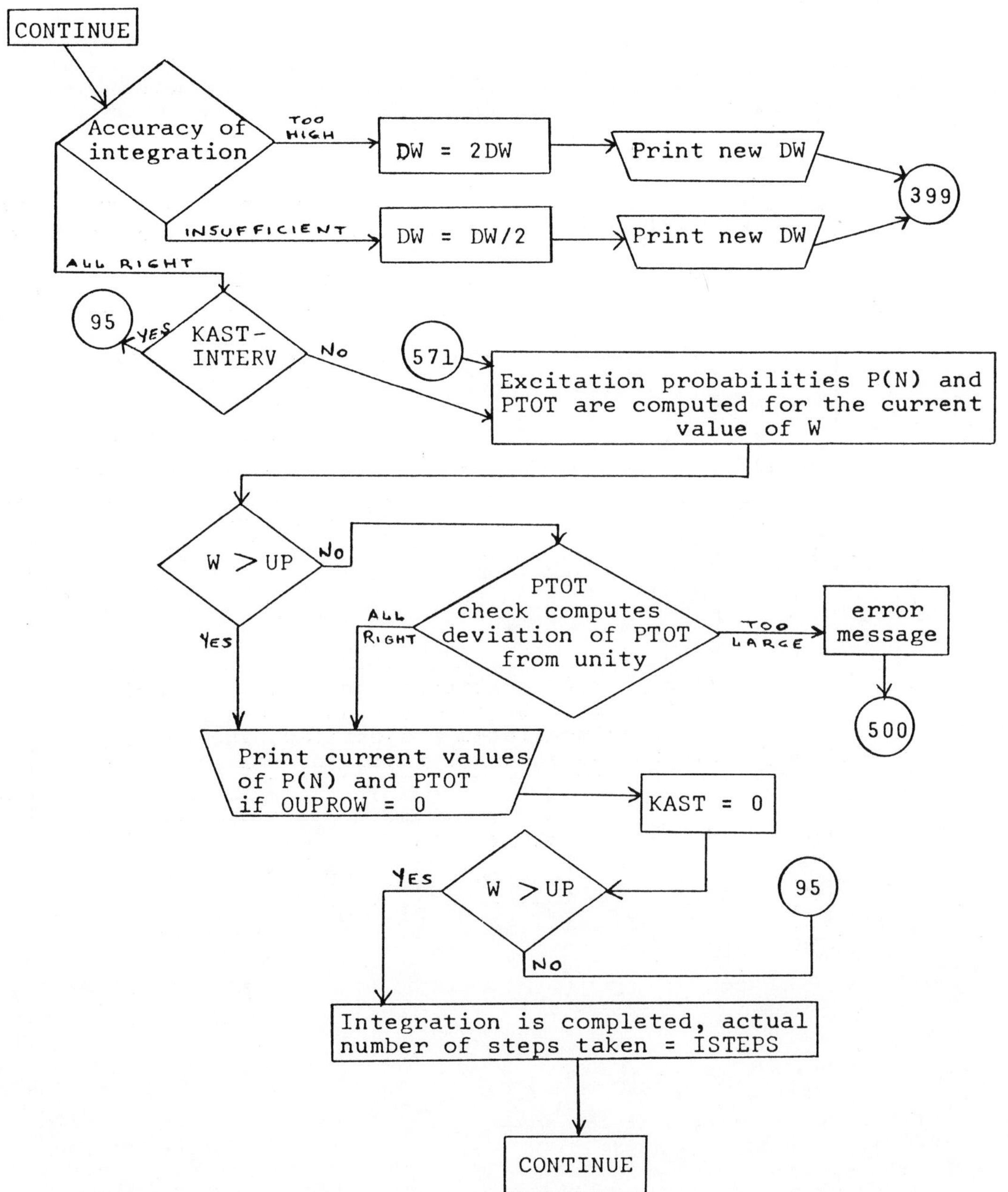
CONTINUE
Accuracy of integration
TOO HIGH
DW = 2DW
Print new DW
INSUFFICIENT
DW = DW/2
Print new DW
399
ALL RIGHT
95
YES
KAST-INTERV
NO
571
Excitation probabilities P(N) and PTOT are computed for the current value of W
W > UP
NO
PTOT check computes deviation of PTOT from unity
ALL RIGHT
TOO LARGE
error message
500
YES
Print current values of P(N) and PTOT if OUPROW = 0
KAST = 0
YES
W > UP
95
NO
Integration is completed, actual number of steps taken = ISTEPS
CONTINUE

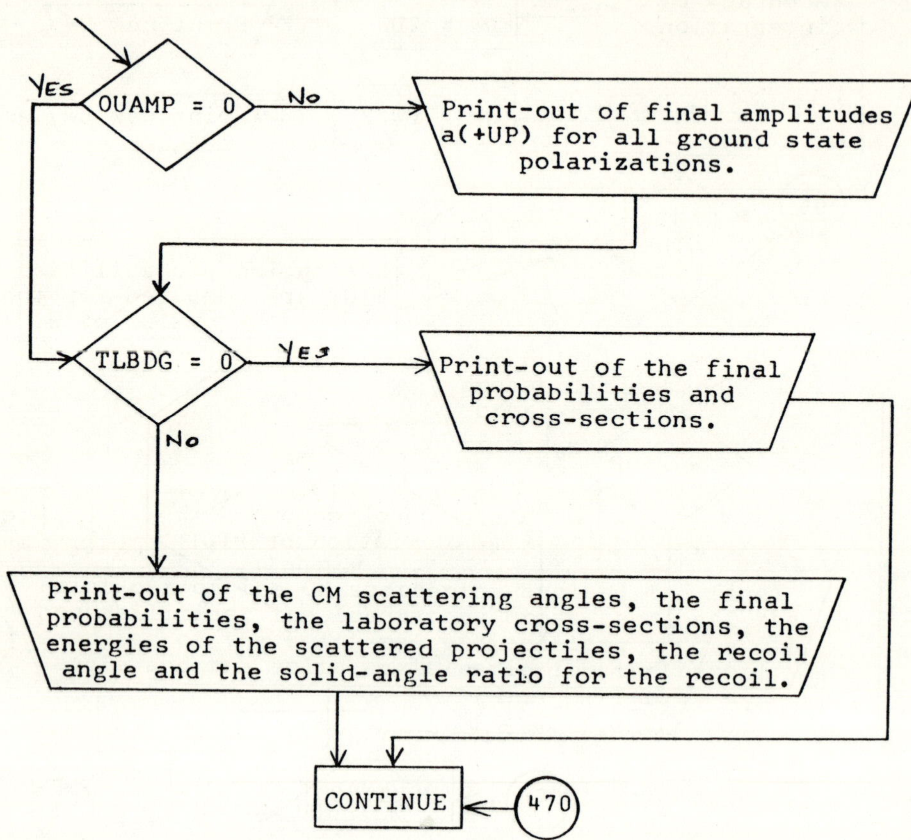

YES
OUAMP = 0
NO
Print-out of final amplitudes a(+UP) for all ground state polarizations.
TLBDG = 0
YES
Print-out of the final probabilities and cross-sections.
NO
Print-out of the CM scattering angles, the final probabilities, the laboratory cross-sections, the energies of the scattered projectiles, the recoil angle and the solid-angle ratio for the recoil.
CONTINUE
470

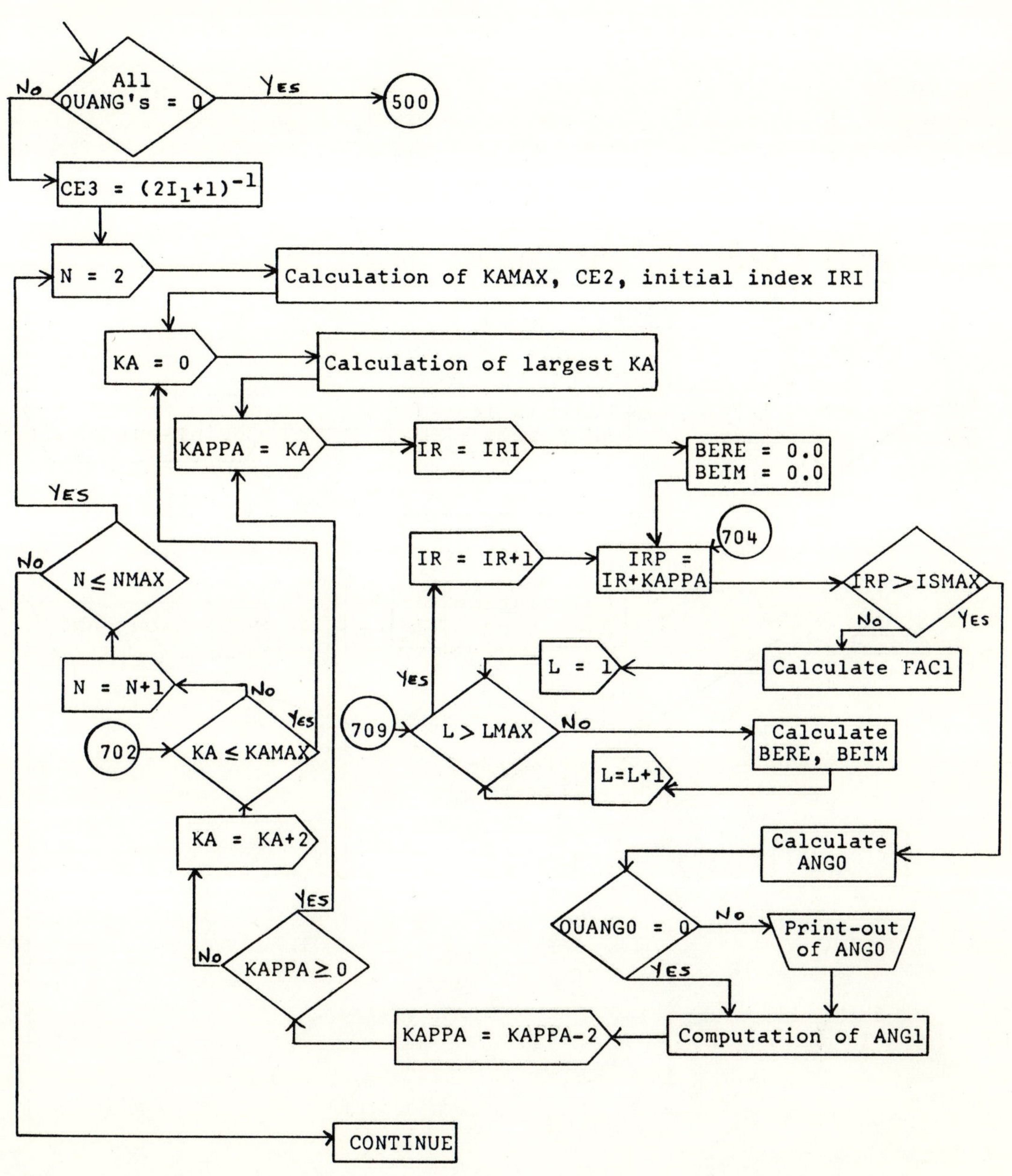
All QUANG's = 0
Yes
No
500
CE3 = (2I1+1)-1
N = 2
Calculation of KAMAX, CE2, initial index IRI
KA = 0
Calculation of largest KA
KAPPA = KA
IR = IRI
BERE = 0.0
BEIM = 0.0
Yes
N ≤ NMAX
No
704
IR = IR+1
IRP = IR+KAPPA
IRP > ISMAX
No
Yes
Calculate FAC1
L = 1
N = N+1
No
Yes
Yes
702
KA ≤ KAMAX
709
L > LMAX
No
Calculate BERE, BEIM
L=L+1
KA = KA+2
Calculate ANG0
Yes
QUANG0 = 0
No
Print-out of ANG0
No
KAPPA ≥ 0
Yes
KAPPA = KAPPA-2
Computation of ANG1
CONTINUE

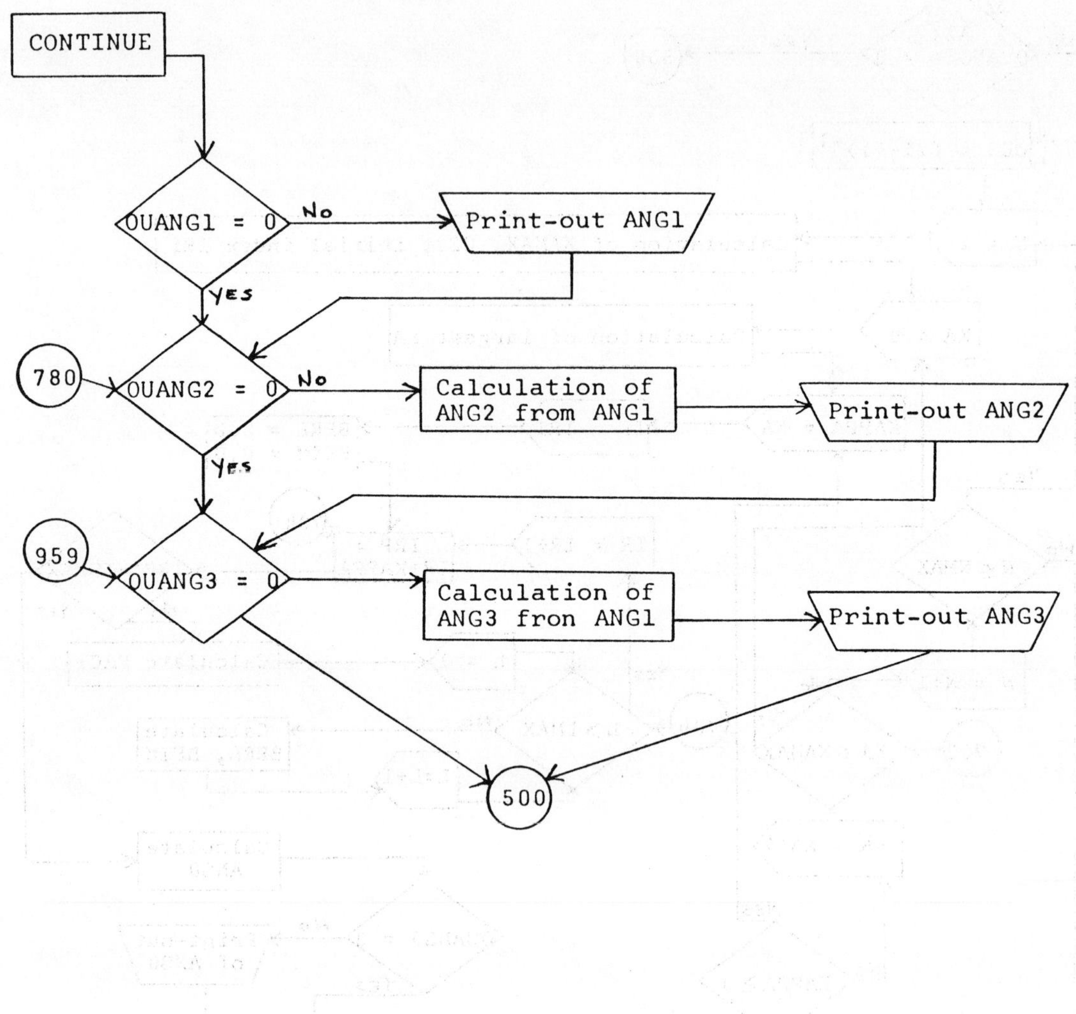
CONTINUE
OUANG1 = 0
No
Print-out ANG1
YES
780
OUANG2 = 0
No
Calculation of ANG2 from ANG1
Print-out ANG2
YES
959
OUANG3 = 0
Calculation of ANG3 fron ANG1
Print-out ANG3
500

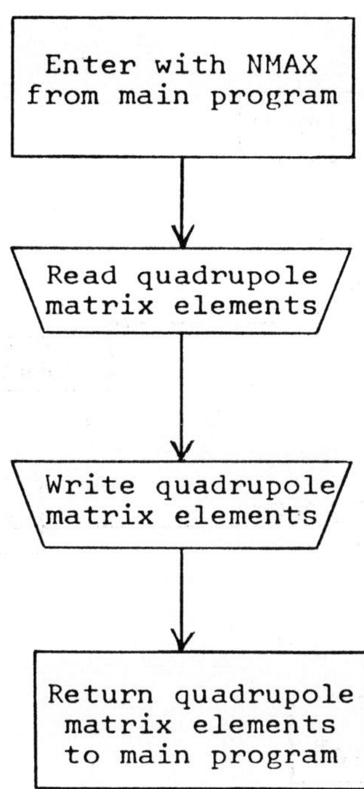
Enter with NMAX
from main program
Read quadrupole
matrix elements
Write quadrupole
matrix elements
Return quadrupole
matrix elements
to main program

I. Function FAC10

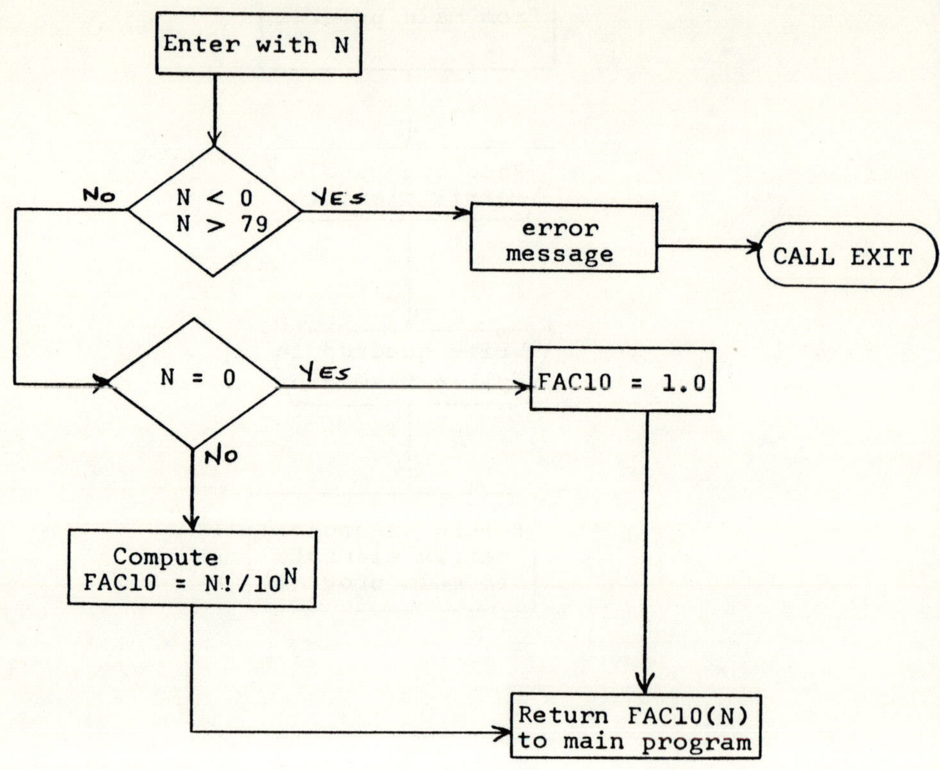

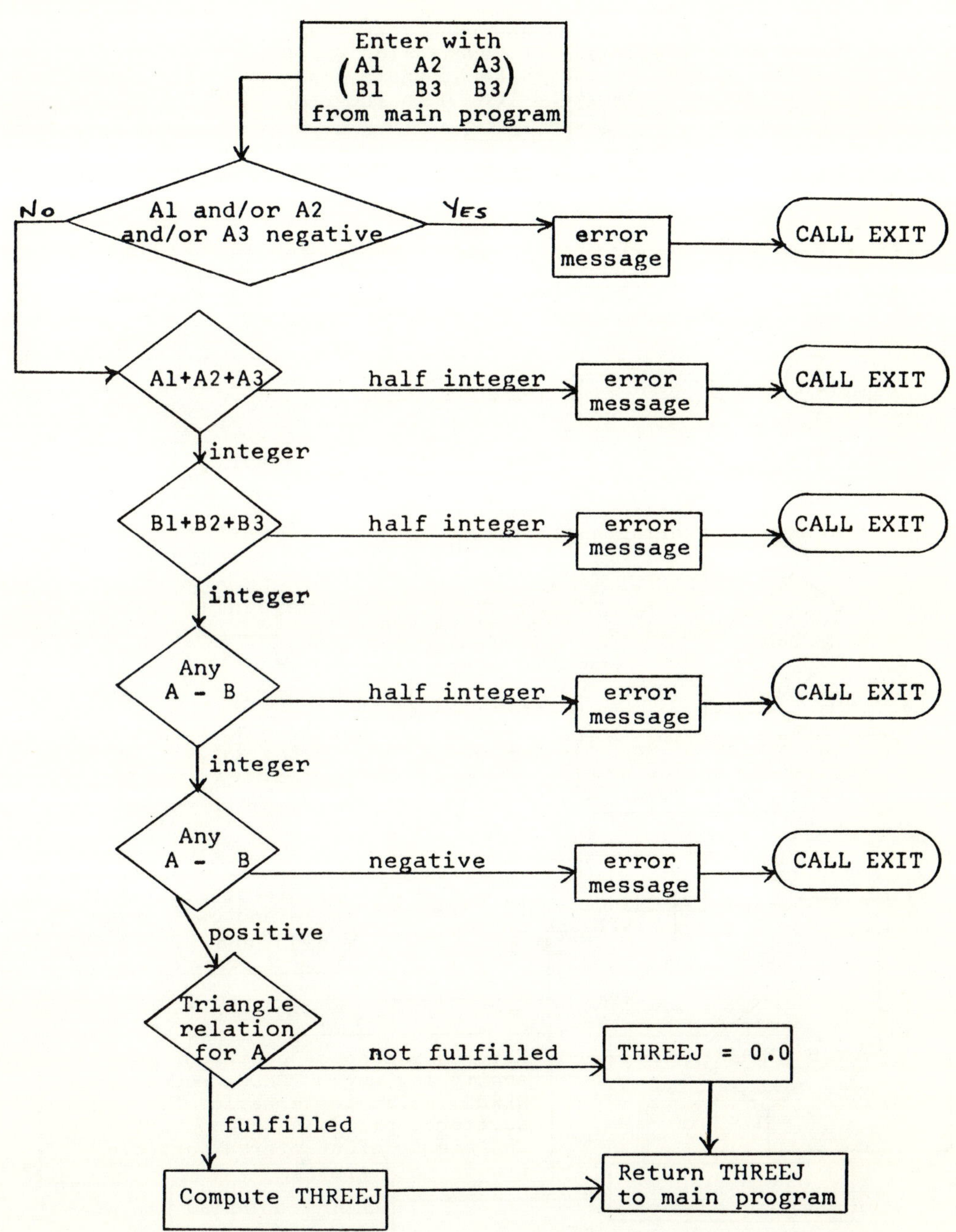
Enter with
(A1 A2 A3
B1 B3 B3)
from main program
A1 and/or A2
and/or A3 negative
No
Yes
error message
CALL EXIT
A1+A2+A3
half integer
error message
CALL EXIT
integer
B1+B2+B3
half integer
error message
CALL EXIT
integer
Any
A - B
half integer
error message
CALL EXIT
integer
Any
A - B
negative
error message
CALL EXIT
positive
Triangle
relation
for A
not fulfilled
THREEJ = 0.0
fulfilled
Compute THREEJ
Return THREEJ
to main program

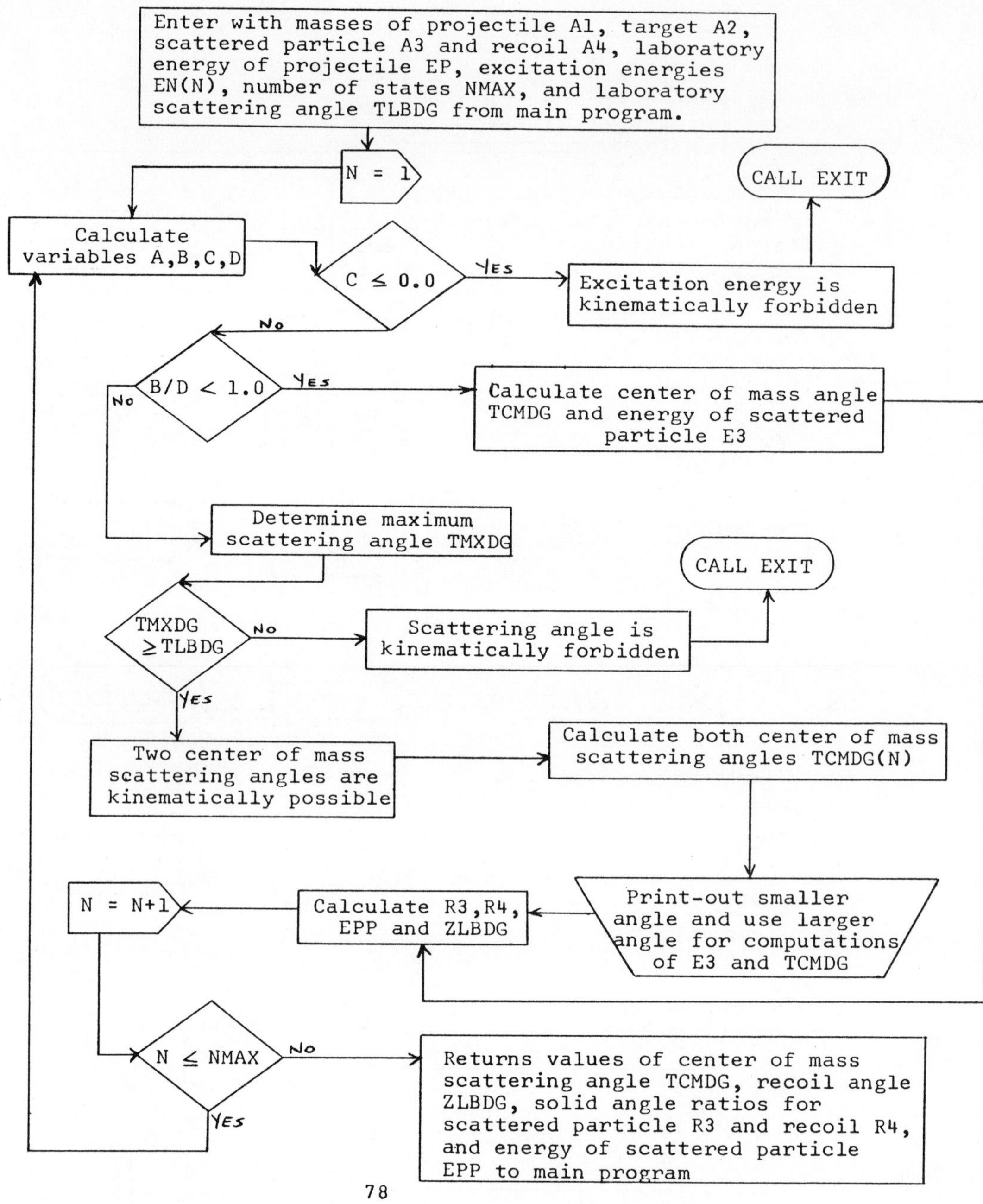
Enter with masses of projectile A1, target A2, scattered particle A3 and recoil A4, laboratory energy of projectile EP, excitation energies EN(N), number of states NMAX, and laboratory scattering angle TLBDG from main program.
N = 1
CALL EXIT
Calculate variables A,B,C,D
C ≤ 0.0
YES
Excitation energy is kinematically forbidden
NO
B/D < 1.0
YES
NO
Calculate center of mass angle TCMDG and energy of scattered particle E3
Determine maximum scattering angle TMXDG
CALL EXIT
TMXDG ≥ TLBDG
NO
Scattering angle is kinematically forbidden
YES
Two center of mass scattering angles are kinematically possible
Calculate both center of mass scattering angles TCMDG(N)
N = N+1
Calculate R3, R4, EPP and ZLBDG
Print-out smaller angle and use larger angle for computations of E3 and TCMDG
N ≤ NMAX
NO
Returns values of center of mass scattering angle TCMDG, recoil angle ZLBDG, solid angle ratios for scattered particle R3 and recoil R4, and energy of scattered particle EPP to main program
YES

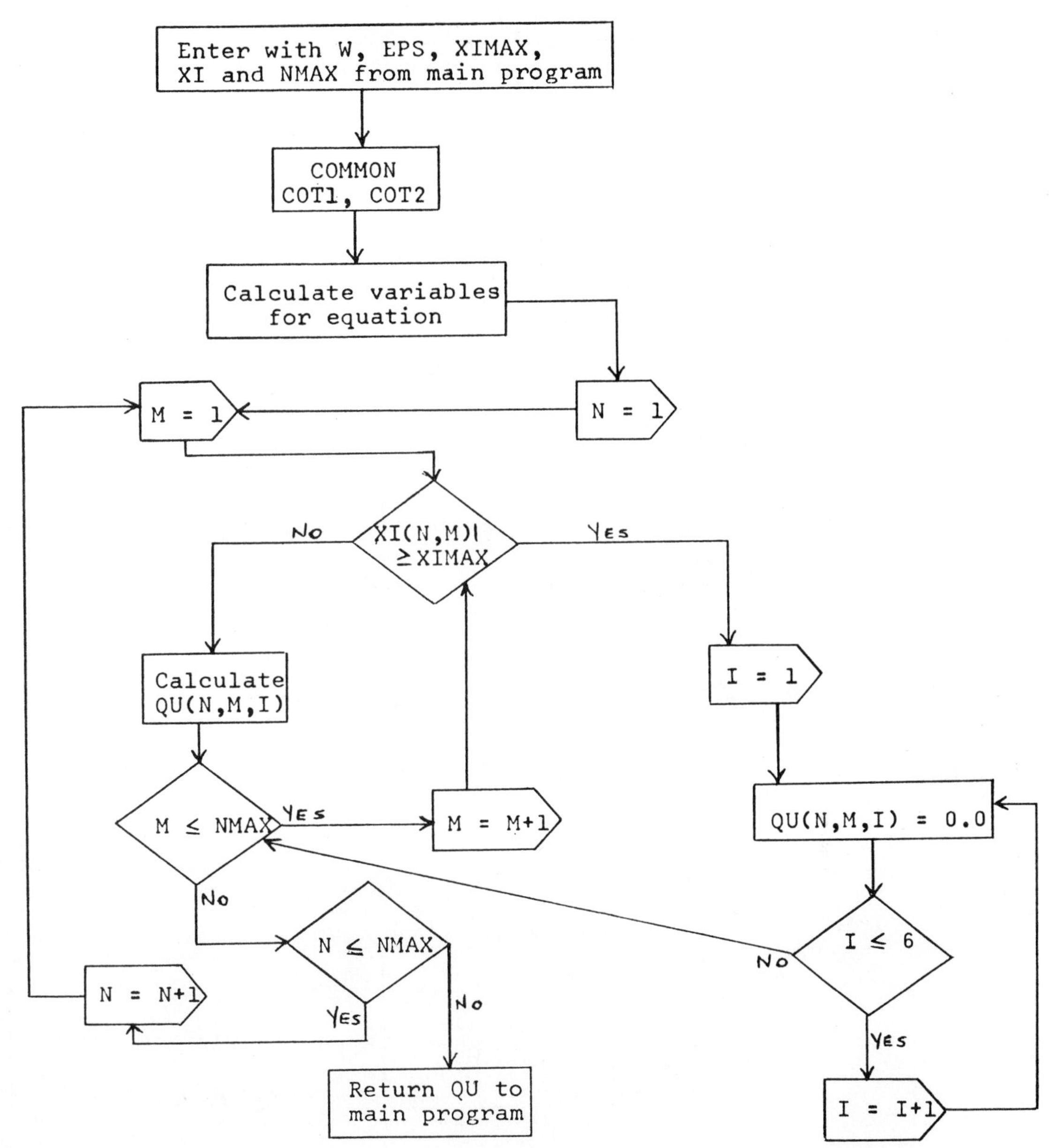
Enter with W, EPS, XIMAX, XI and NMAX from main program
COMMON COT1, COT2
Calculate variables for equation
N = 1
M = 1
|XI(N,M)| ≥ XIMAX
No
Yes
Calculate QU(N,M,I)
I = 1
M ≤ NMAX
Yes
M = M+1
QU(N,M,I) = 0.0
No
N ≤ NMAX
I ≤ 6
No
N = N+1
Yes
No
Yes
Return QU to main program
I = I+1

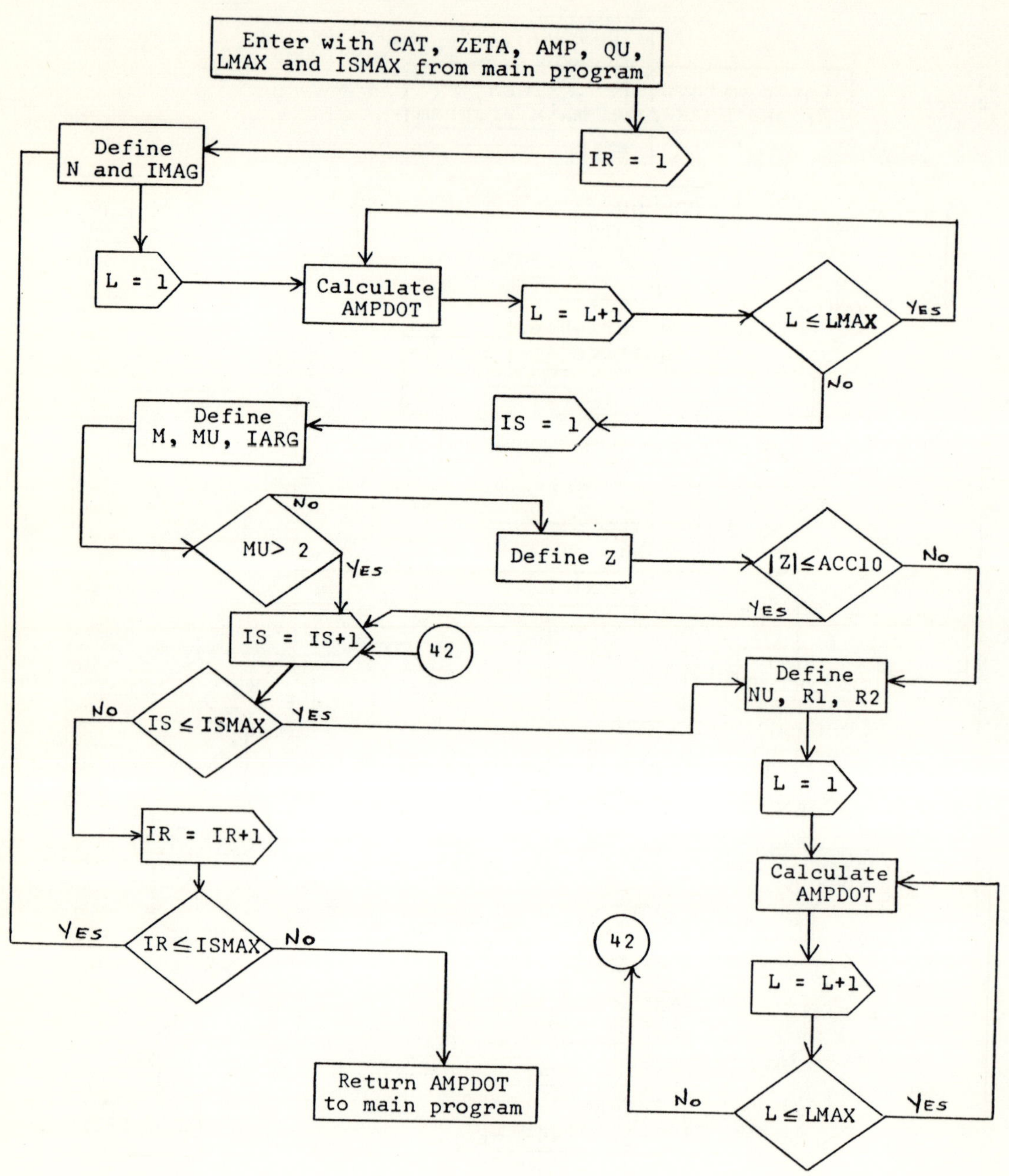
Enter with CAT, ZETA, AMP, QU, LMAX and ISMAX from main program
IR = 1
Define N and IMAG
L = 1
Calculate AMPDOT
L = L+1
L ≤ LMAX
Yes
No
IS = 1
Define M, MU, IARG
MU> 2
No
Yes
Define Z
|Z|≤ACC10
No
Yes
IS = IS+1
42
IS ≤ ISMAX
No
Yes
Define NU, R1, R2
L = 1
Calculate AMPDOT
L = L+1
L ≤ LMAX
No
Yes
42
IR = IR+1
IR≤ISMAX
Yes
No
Return AMPDOT to main program

N. Function DJMM

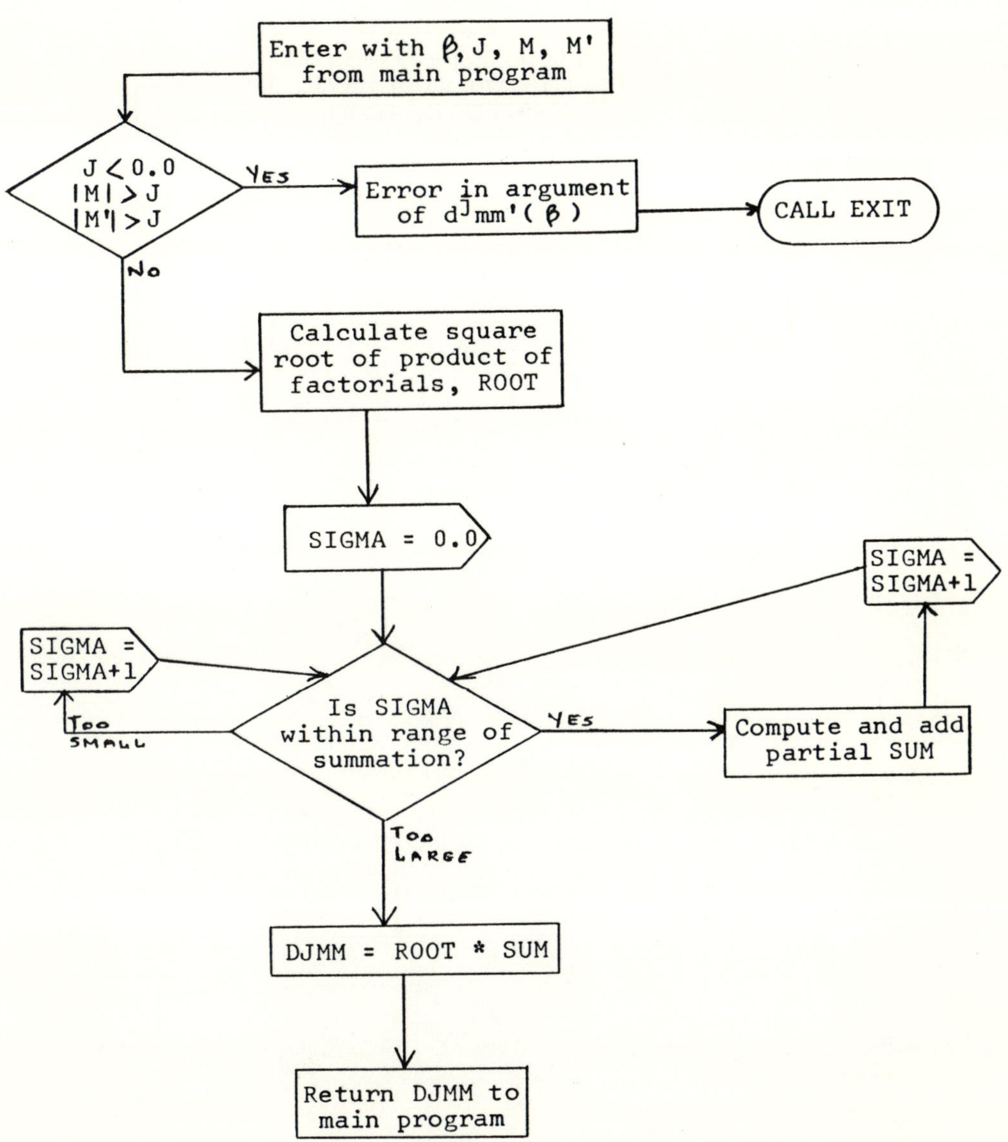

Acknowledgments

The authors are very grateful for the invaluable cooperation received from both Barbara Zimmerman at the California Institute of Technology who initially coded the entire program, and Shirley Feldman at Rutgers, The State University, who added many details. They also want to thank Janette Rasmussen for the typing of this paper. Aage Winther acknowledges fellowships sponsored at the California Institute of Technology by the Office of Naval Research and the Atomic Energy Commission, and at Rutgers, The State University, by the National Science Foundation.